Y0-BQE-172

MATHEMATICS

Applications and Connections

COURSE 1

GLENCOE

Macmillan/McGraw–Hill

Lake Forest. Illinois   Columbus. Ohio   Mission Hills. California   Peoria. Illinois

Copyright © 1993 by the Glencoe Division of Macmillan/McGraw-Hill Publishing Company.
All rights reserved. Printed in the United States of America. Except as permitted under the
United States Copyrights Act of 1976, no part of this publication may be reproduced or
distributed in any form or by any means, or stored in a database or retrieval system, without
prior written permission of the publisher.

Send all inquiries to:
Glencoe Division, Macmillan/McGraw-Hill
936 Eastwind Drive
Westerville, Ohio 43081

ISBN: 0-02-800582-1 (Student Edition)
ISBN: 0-02-824041-3 (Teacher's Wraparound Edition)

3 4 5 6 7 8 9 10 RRD-W 01 00 99 98 97 96 95 94 93

Dear Students, Teachers, and Parents,

Welcome to middle school! You're special, and we want your middle school mathematics experience to be special, too. That's why we've written the first and only mathematics program in the United States designed specifically for you. *Mathematics: Applications and Connections* picks up on your special interests with exciting content and a layout that was prepared with your age group in mind. We feel sure the text will show you why it's important to study mathematics every day.

One of the first things you'll notice as you page through the text is the variety of ways in which mathematics is presented. You'll also see many connections made among mathematical topics. And you'll note how easily and naturally mathematics fits into other subject areas and with technology.

Please take a look at how the content for each lesson is clearly labeled up front. We think you'll also appreciate the easy-to-follow lesson format. New concepts in all lessons are introduced with an interesting application followed by clear examples.

It won't take you long to realize the practical value of mathematics as you read *Mathematics: Applications and Connections* and complete the activities. You will also see how often mathematics is used in real-world situations that relate directly to your life. If you don't already understand the importance of mathematics in your life, you soon will!

Sincerely, The Authors

Kay Balch

Linda Dritsas

David D'Palma

Patricia Frey-Mason

Arthur C. Howard

Ron Pelfrey

Beatrice Moore-Harris

Jack M. Ott

Patricia S. Wilson

Barbara L. Smith

Kay Balch teaches mathematics at Mountain Brook Junior High in Birmingham, Alabama. She is also the Mathematics Department Chairperson. Ms. Balch received her B.A. and M.A. from Auburn University in Alabama. She also has an Educational Specialist degree from the University of Montevallo. Ms. Balch is a member of the National Council of Teachers of Mathematics and is active in several other mathematics organizations at the national, state, and local levels.

Linda Dritsas is the Mathematics Coordinator for the Fresno Unified School District in Fresno, California. She also taught at California State University at Fresno for two years. Ms. Dritsas received her B.A. and M.A.(Education) from California State University at Fresno. Ms. Dritsas has published numerous mathematics workbooks and other supplementary materials. She has been the Central Section President of the California Mathematics Council and is a member of the National Council of Teachers of Mathematics and the Association for Supervision and Curriculum Development.

Arthur C. Howard is Consultant for Secondary Mathematics at the Aldine School District in Houston, Texas. He received his B.S. and M.Ed. from the University of Houston. Mr. Howard has taught in grades 7-12 and in college. He is Master Teacher in the Rice University School Mathematics Project in Houston. Mr. Howard is also active in numerous professional organizations at the national and state levels, including the National Council of Teachers of Mathematics. His publications include curriculum materials and articles for newspapers, books, and *The Mathematics Teacher.*

William Collins teaches mathematics at James Lick High School in San Jose, California. He has served as the Mathematics Department Chairperson at James Lick and Andrew Hill High Schools. He received his B.A. from Herbert H. Lehman College and is a Masters candidate at California State University, Hayward. Mr. Collins has been a consultant for the National Assessment Governing Board. He is a member of the National Council of Teachers of Mathematics and is active in several professional mathematics organizations at the state level. Mr. Collins is currently a mentor teacher for the College Board's EQUITY 2000 Consortium in San Jose, California.

Patricia Frey-Mason is the Mathematics Department Chairperson at the Buffalo Academy for Visual and Performing Arts in Buffalo, New York. She received her B.A. from D'Youville College in Buffalo, New York, and her M.Ed. from the State University of New York at Buffalo. Ms. Frey- Mason has published several articles in mathematics journals. She is a member of the National Council of Teachers of Mathematics and is active in other professional mathematics organizations at the state, national, and international levels. Ms. Frey-Mason was named a 1991 Woodrow Wilson Middle School Mathematics Master Teacher.

David D. Molina is a professor at Trinity University in San Antonio, Texas. He received his M.A. and Ph.D. in Mathematics Education from the University of Texas at Austin. Dr. Molina has been a speaker both at national and international mathematics conferences. He has been a presenter for the National Council of Teachers of Mathematics, as well as a conductor of workshops and in services for other professional mathematics organizations and school systems.

Beatrice Moore-Harris is the EQUITY 2000 Project Administrator and former Mathematics Curriculum Specialist for K-8 in the Fort Worth Independent School District in Fort Worth, Texas. She is also a consultant for the National Council of Teachers of Mathematics. Ms. Moore-Harris received her B.A. from Prairie View A & M University in Prairie View, Texas. She has also done graduate work there and at Texas Southern University in Houston, Texas, and Tarleton State University in Stephenville, Texas. Ms. Moore-Harris is active in many state and national mathematics organizations. She also serves on the editorial Board of NCTM's *Mathematics and the Middle Grades* journal.

Ronald S. Pelfrey is the Mathematics Coordinator for the Fayette County Public Schools in Lexington, Kentucky. He has taught mathematics in Fayette County Public Schools, with the Peace Corps in Ethiopia, and at the University of Kentucky in Lexington, Kentucky. Dr. Pelfrey received his B.S., M.A., and Ed.D. from the University of Kentucky. He is also the author of several publications about mathematics curriculum. He is an active speaker with the National Council of Teachers of Mathematics and is involved with other local, state, and national mathematics organizations.

Barbara Smith is the Mathematics Supervisor for Grades K-12 at the Unionville-Chadds Ford School District in Unionville, Pennsylvania. Prior to being a supervisor, she taught mathematics for thirteen years at the middle school level and three years at the high school level. Ms. Smith received her B.S. from Grove City College in Grove City, Pennsylvania and her M.Ed. from the University of Pittsburgh in Pittsburgh, Pennsylvania. Ms. Smith has held offices in several state and local organizations, has been a speaker at national and state conferences, and is a member of the National Council of Teachers of Mathematics.

Jack Ott is a Professor of Mathematics Education at the University of South Carolina in Columbia, South Carolina. He has also been a consultant for numerous schools in South Carolina as well as the South Carolina State Department of Education and the National Science Foundation. Dr. Ott received his A.B. from Indiana Wesleyan University, his M.A. from Ball State University, and his Ph.D. from The Ohio State University. Dr. Ott has written articles for *The Mathematics Teacher* and *The Arithmetic Teacher* and has been a speaker at national and state mathematics conferences.

Jack Price has been active in mathematics education for over 40 years, 38 of those in grades K-12. He is currently the Co-Director of the Center for Science and Mathematics Education at California State Polytechnic University at Pomona, California, where he teaches mathematics and methods courses for preservice teachers and consults with school districts on curriculum change. Dr. Price received his B.A. from Eastern Michigan University, and has a Doctorate in Mathematics Education from Wayne State University. He is active in state and national mathematics organizations and is a past director of the National Council of Teachers of Mathematics.

Patricia S. Wilson is an Associate Professor of Mathematics Education at the University of Georgia in Athens, Georgia. Dr. Wilson received her B.S. from Ohio University and her M.A. and Ph.D. from The Ohio State University. She has received the Excellence in Teaching Award from the College of Education at the University of Georgia and is a published author in several mathematics education journals. Dr. Wilson has taught middle school mathematics and is currently teaching middle school mathematics methods courses. She is on the Editorial Board of the *Journal for Research in Mathematics Education*, published by the National Council of Teachers of Mathematics.

Editorial Advisor

Dr. Piyush C. Agrawal
Supervisor for Mathematics Programs
Dade County Public Schools
Miami, Florida

Consultants

Winifred G. Deavens
Mathematics Supervisor
St. Louis Public Schools
St. Louis, Missouri

Leroy Dupee
Mathematics Supervisor
Bridgeport Public Schools
Bridgeport, Connecticut

Marieta W. Harris
Curriculum Coordinator
Memphis City Schools
Memphis, Tennessee

Deborah Haver
Mathematics Supervisor
Chesapeake Public Schools
Chesapeake, Virginia

Dr. Alice Morgan-Brown
Assistant Superintendent for
  Curriculum Development
Baltimore City Public Schools
Baltimore, Maryland

Dr. Nicholas J. Rubino, Jr.
Program Director/Citywide
  Mathematics
Boston Public Schools
Boston, Massachusetts

Telkia Rutherford
Mathematics Coordinator
Chicago Public Schools
Chicago, Illinois

Phyllis Simon
Supervisor, Computer Education
Conway Public Schools
Conway, Arkansas

Beverly A. Thompson
Mathematics Supervisor
Baltimore City Public Schools
Baltimore, Maryland

Jo Helen Williams
Director, Curriculum and
  Instruction
Dayton Public Schools
Dayton, Ohio

Reviewers

David Bradley
Mathematics Teacher
Thomas Jefferson Junior High
  School
Kearns, Utah

Joyce Broadwell
Mathematics Teacher
Dunedin Middle School
Dunedin, Florida

Charlotte Brummer
K-12 Mathematics Supervisor
School District Eleven
Colorado Springs, Colorado

Richard Buckler
Mathematics Coordinator K-12
Decatur School District
Decatur, Illinois

Josie A. Bullard
MSEN PCP Lead Teacher/
  Math Specialist
Lowe's Grove Middle School
Durham, North Carolina

Joseph A. Crupie, Jr.
Mathematics Department
  Chairperson
North Hills High School
Pittsburgh, Pennsylvania

Kathryn L. Dillard
Mathematics Teacher
Apollo Middle School
Antioch, Tennessee

Rebecca Efurd
Mathematics Teacher
Elmwood Junior High School
Rogers, Arkansas

Nevin Engle
Supervisor of Mathematics
Cumberland Valley School District
Mechanicsburg, Pennsylvania

Rita Fielder
Mathematics Teacher
Oak Grove High School
North Little Rock, Arkansas

Henry Hull
Adjunct Assistant Professor
Suffolk County Community
  College
Selden, New York

Elaine Ivey
Mathematics Teacher
Adams Junior High School
Tampa, Florida

Donna Jamell
Mathematics Teacher
Ramsey Junior High School
Fort Smith, Arkansas

Augustus M. Jones
Mathematics Teacher
Tuckahoe Middle School
Richmond, Virginia

Marie Kasperson
Mathematics Teacher
Grafton Middle School
Grafton, Massachusetts

Larry Kennedy
Mathematics Teacher
Kimmons Junior High School
Fort Smith, Arkansas

Patricia Killingsworth
Math Specialist
Carver Math/Science Magnet
  School
Little Rock, Arkansas

Al Lachat
Mathematics Department
  Chairperson
Neshaminy School District
Feasterville, Pennsylvania

Kent Luetke-Stahlman
Resource Scholar Mathematics
J. A. Rogers Academy of Liberal
  Arts & Sciences
Kansas City, Missouri

Dr. Gerald E. Martau
Deputy Superintendent
Lakewood City Schools
Lakewood, Ohio

Nelson J. Maylone
Assistant Principal
Maltby Middle School
Brighton, Michigan

Irma A. Mayo
Mathematics Department
  Chairperson
Mosby Middle School
Richmond, Virginia

Daniel Meadows
Mathematics Consultant
Stark County Local School
  System
Canton, Ohio

Dianne E. Meier
Mathematics Supervisor
Bradford Area School District
Bradford, Pennsylvania

Rosemary Mosier
Mathematics Teacher
Brick Church Middle School
Nashville, Tennessee

Judith Narvesen
Mathematics Resource Teacher
Irving A. Robbins Middle School
Farmington, Connecticut

Raymond A. Nichols
Mathematics Teacher
Ormond Beach Middle School
Ormond Beach, Florida

William J. Padamonsky
Director of Education
Hollidaysburg Area School
  District
Hollidaysburg, Pennsylvania

Delores Pickett
Instructional Supervisor
Vera Kilpatrick Elementary
  School
Texarkana, Arkansas

Thomas W. Ridings
Team Leader
Gilbert Junior High School
Gilbert, Arizona

Sally W. Roth
Mathematics Teacher
Francis Scott Key Intermediate
  School
Springfield, Virginia

Dr. Alice W. Ryan
Assistant Professor of Education
Dowling College
Oakdale, New York

Fred R. Stewart
Supervisor of
  Mathematics/Science
Neshaminy School District
Langhorne, Pennsylvania

Terri J. Stillman
Mathematics Department
  Chairperson
Boca Raton Middle School
Boca Raton, Florida

Marty Terzieff
Secondary Math Curriculum
  Chairperson
Mead Junior High School
Mead, Washington

Tom Vogel
Mathematics Teacher
Capital High School
Charleston, West Virginia

Joanne Wilkie
Mathematics Teacher
Hosford Middle School
Portland, Oregon

Larry Williams
Mathematics Teacher
Eastwood 8th Grade School
Tuscaloosa, Alabama

Deborah Wilson
Mathematics Teacher
Rawlinson Road Middle School
Rock Hill, South Carolina

Francine Yallof
Mathematics Teacher
East Middle School
Brentwood, New York

Table of Contents

Chapter 2

Graphs and Statistics

High Interest Features

Did You Know?
4, 9, 12, 61

Teen Scene
5, 64

**When Am I Ever Going
To Use This?**
10, 68

Save Planet Earth
37

Journal Entry
7, 17, 19, 24, 46, 63, 71

Mini-Lab
54

Applications and Connections

Have you ever asked yourself this question?

"When am I ever going to use this stuff?"

It may be sooner than you think! Here are two of the many ways this textbook will help you answer that question.

Applications

You can find mathematics in all of the subjects you study in school and in your life outside of school. Did you know that the smallest egg laid by any bird weighed only 0.0128 ounces? You can find out what bird laid it by reading Lesson 3-2 on page 82. In Lesson 10-5 on page 368, you can go to the movies while learning about percents and fractions.

These and other applications provide you with fascinating information that connects math to the real world and other school subjects and gives you a reason to learn math. Here are some more application topics.

entertainment	science
sports	social studies
smart shopping	music
hobbies	health
ecology	art

Five **DECISION MAKING** features further enable you to connect math to your real-life experiences as a consumer.

Connections

You'll discover that various areas of mathematics are very much interrelated. Lesson 5-7 on page 179 shows how you can use algebra to help solve a division problem. Lesson 6-7 on page 210 connects measurement with fractions.

Connections to algebra, geometry, statistics, measurement, and probability help show the power of mathematics.

The **Mathematics Labs** and **Mini-Labs** also help you connect what you've learned before to new concepts. You'll use counters, measuring tapes, geoboards, and many other objects to help you discover these concepts.

Chapter 5

Decimals: Division

Chapter 6

Patterns and Number Sense

High Interest Features

When Am I Ever Going To Use This?
158, 171, 221

Teen Scene
160, 219

Cultural Kaleidoscope
162

Did You Know?
179, 207, 216

Save Planet Earth
181, 215

Journal Entry
173, 181, 206, 218, 220

Mini-Labs
174, 207, 210, 213

Chapter 7

Fractions: Addition and Subtraction

Chapter 8

Fractions: Multiplication and Division

High Interest Features

When Am I Ever Going to Use This?
232, 289

Did You Know?
236, 257, 261, 280, 300

Cultural Kaleidoscope
248

Teen Scene
253, 285

Journal Entry
235, 264, 283, 291

Mini-Labs
240, 253

Chapter 9

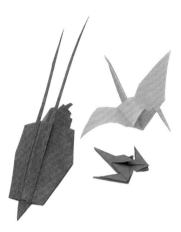

Investigations in Geometry

Chapter 10

Ratio, Proportion, and Percent

High Interest Features

Did You Know?
321, 334, 368, 371

Teen Scene
374

Cultural Kaleidoscope
376

When Am I Ever Going to Use This?
318, 361

Journal Entry
317, 320, 339, 354, 373, 379

Mini-Labs
311, 315, 318, 319, 321, 322, 334, 337, 369

Technology

Labs, examples, computer-connection problems, and other features help you become an expert in using computers and calculators as problem-solving tools. You'll also learn how to read data bases, use spreadsheets, and use BASIC and LOGO programs. On many pages, **Calculator Hints** and printed keystrokes illustrate how to use a calculator.

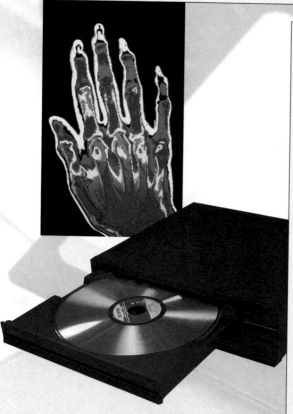

Here are some highlights.

Chapter 11

Area and Volume

Chapter 12

Investigations with Integers

High Interest Features

Did You Know?
388, 391, 427, 448

**When Am I Ever Going
To Use This?**
404, 435

Teen Scene
398, 443

Save Planet Earth
394

Journal Entry
394, 407, 422, 434

Mini-Labs
389, 391, 395, 400, 401,
404, 408, 449, 450

Chapter 13

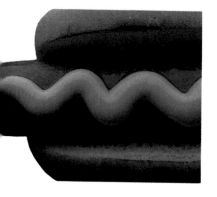

An Introduction to Algebra

Chapter 14

Probability

High Interest Features

Teen Scene
467, 503

Did You Know?
480, 487, 498, 507

When Am I Ever Going To Use This?
483, 506

Cultural Kaleidoscope
486

Save Planet Earth
513

Journal Entry
465, 486, 501, 513

Mini-Labs
487, 499, 507, 510

End of Text Materials

How many people are aware of the hazards to their health of second-hand smoke? You'll have the opportunity to work in a cooperative group to create and use a questionnaire to determine the answer.

The **Extended Projects Handbook** consists of interesting long-term projects that involve serious issues that impact all citizens in the United States and the world.

Previewing Your Text

Have you ever taken a long car trip? Chances are the driver consulted a map beforehand to chart the course of your journey. Just as a map directs travelers, the next four pages show you what mathematics holds in store for you this year. Please read them carefully so you can make the most of the text's many interesting features!

These features will help to guide you through each lesson.

Applications opening nearly every lesson provide you with fascinating information that connects math to the real world and give you a reason to learn math.

Objectives tell you exactly what you'll learn in each lesson.

Words to Learn lists the new words you'll encounter.

When Am I Ever Going to Use This? connects mathematics to real-life careers. Other interesting margin features are **Did You Know?** and **Teen Scene**.

8-5 Sequences

Objective
Recognize and extend sequences.

Words to Learn
sequence

When am I ever going to use this?

Record Industry
"Single specialists" make sure single records are in the stores once they are added to the play lists at radio stations. They also work with entertainers when they go on tour and get involved with merchandising.

These positions require organizational and problem-solving skills.

For more information contact:
Sales and Marketing Executives
Statler Office Tower #458
Cleveland, OH 44115

What type of person is a music composer? The image that probably comes to mind is a white-haired gentleman. But not anymore! The Chicago Symphony Orchestra and its *Young Composers Project* have given teenagers and children from the Chicago area the opportunity to write and compose music. In 1991, the Chicago Symphony presented the works of Rich Carte, who was 7 years old, John Orfe, 14, and Jeff Letterly, 18. These three composers based their works on melodies written by students from Howland School of the Arts.

When composers write music, they need to indicate how long each note should be held. Look at the notes below.

Notes

whole (1) $\frac{1}{2}$ $\frac{1}{4}$ $\frac{1}{8}$ $\frac{1}{16}$ $\frac{1}{32}$ $\frac{1}{64}$

The numbers $1, \frac{1}{2}, \frac{1}{4}, \frac{1}{8}, \frac{1}{16}, \frac{1}{32},$ and $\frac{1}{64}$ form a **sequence**. A sequence of numbers is a list in a specific order.

Notice that each number in the sequence, or pattern, is multiplied by $\frac{1}{2}$ to get the next number.

$1, \frac{1}{2}, \frac{1}{4}, \frac{1}{8}, \frac{1}{16}, \frac{1}{32}, \frac{1}{64}, \cdots$

Examples

Find the next number in each sequence.

1 2, 6, 18, 54

Each number in the sequence is multiplied by 3. The next number is 54×3, or 162.

2 125, 25, 5, 1

Each number in the sequence is multiplied by $\frac{1}{5}$. The next number is $1 \times \frac{1}{5}$, or $\frac{1}{5}$.

Lesson 8-5 Sequences **289**

Problem Solving and **Applications** in each lesson directly link math to real-world fields like engineering, and to science, history, art, and other subjects.

Problem-Solving Hints provide clues about which strategy to use to solve problems. Some lessons also include **Mental Math Hints**, **Estimation Hints**, and **Calculator Hints**.

Mixed Reviews present problems that help you remember what you've learned. Lesson references tell you exactly where to look in previous lessons to restudy important concepts.

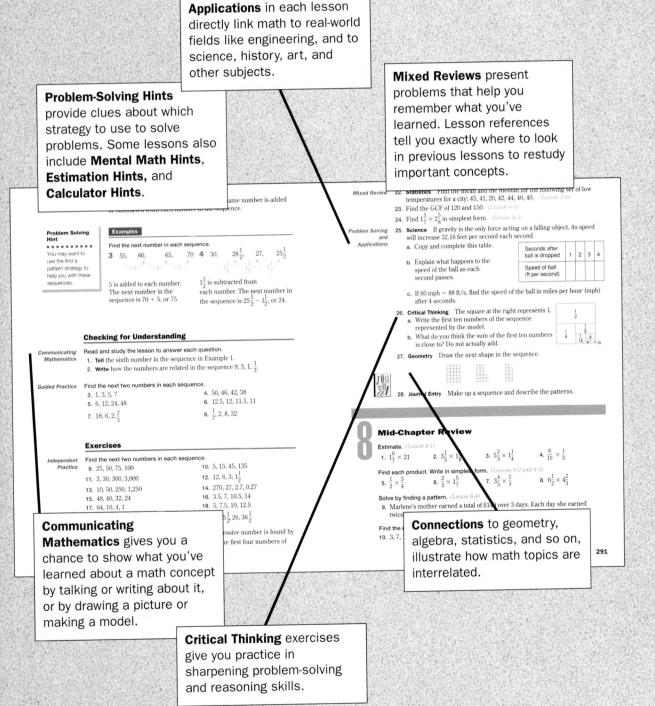

Problem Solving Hint

You may want to use the find a pattern strategy to help you with these sequences.

Examples

Find the next number in each sequence.

3 55, 60, 65, 70 **4** 30, $28\frac{1}{2}$, 27, $25\frac{1}{2}$

5 is added to each number. The next number in the sequence is 70 + 5, or 75.

$1\frac{1}{2}$ is subtracted from each number. The next number in the sequence is $25\frac{1}{2} - 1\frac{1}{2}$, or 24.

Checking for Understanding

Communicating Mathematics Read and study the lesson to answer each question.

1. **Tell** the sixth number in the sequence in Example 1.
2. **Write** how the numbers are related in the sequence 9, 3, 1, $\frac{1}{3}$.

Guided Practice Find the next two numbers in each sequence.

3. 1, 3, 5, 7
4. 50, 46, 42, 38
5. 6, 12, 24, 48
6. 12.5, 12, 11.5, 11
7. 18, 6, 2, $\frac{2}{3}$
8. $\frac{1}{2}$, 2, 8, 32

Exercises

Independent Practice Find the next two numbers in each sequence.

9. 5, 25, 50, 75, 100
10. 5, 15, 45, 135
11. 3, 30, 300, 3,000
12. 12, 6, 3, $1\frac{1}{2}$
13. 10, 50, 250, 1,250
14. 270, 27, 2.7, 0.27
15. 48, 40, 32, 24
16. 3.5, 7, 10.5, 14
17. 64, 16, 4, 1
18. 5, 7.5, 10, 12.5

Mixed Review 22. **Statistics** Find the mean and the median for the following set of low temperatures for a city: 45, 41, 20, 42, 44, 40, 40. *(Lesson 2-8)*
23. Find the GCF of 120 and 150. *(Lesson 6-4)*
24. Find $1\frac{5}{7} \times 2\frac{5}{8}$ in simplest form. *(Lesson 8-3)*

Problem Solving and Applications 25. **Science** If gravity is the only force acting on a falling object, its speed will increase 32.16 feet per second each second.

a. Copy and complete this table.

b. Explain what happens to the speed of the ball as each second passes.

Seconds after ball is dropped	1	2	3	4
Speed of ball (ft per second)				

c. If 60 mph = 88 ft/s, find the speed of the ball in miles per hour (mph) after 4 seconds.

26. **Critical Thinking** The square at the right represents 1.
a. Write the first ten numbers of the sequence represented by the model.
b. What do you think the sum of the first ten numbers is close to? Do not actually add.

27. **Geometry** Draw the next shape in the sequence.

28. **Journal Entry** Make up a sequence and describe the patterns.

Mid-Chapter Review

Estimate. *(Lesson 8-1)*

1. $1\frac{6}{7} \times 21$
2. $3\frac{1}{3} \times 1\frac{3}{4}$
3. $5\frac{2}{3} \times 1\frac{1}{4}$
4. $\frac{6}{10} \times \frac{1}{3}$

Find each product. Write in simplest form. *(Lessons 8-2 and 8-3)*

5. $\frac{1}{2} \times \frac{3}{4}$
6. $\frac{2}{3} \times 1\frac{5}{7}$
7. $3\frac{4}{5} \times \frac{2}{3}$
8. $6\frac{1}{2} \times 4\frac{2}{3}$

Solve by finding a pattern. *(Lesson 8-4)*

9. Marlene's mother earned a total of $14 over 3 days. Each day she earned twice

Find the
10. 3, 7,

291

Communicating Mathematics gives you a chance to show what you've learned about a math concept by talking or writing about it, or by drawing a picture or making a model.

Connections to geometry, algebra, statistics, and so on, illustrate how math topics are interrelated.

Critical Thinking exercises give you practice in sharpening problem-solving and reasoning skills.

Getting Into Each Chapter

Chapter 3

Decimals: Addition and Subtraction

Spotlight on Movie Videos and Technology

Have You Ever Wondered...

- Which movies are the most popular as video rentals?
- How common are different forms of entertainment technology?

Movie Rentals - 1980-1989

Title	Year	Rental (Millions)
E.T.-The Extra Terrestrial	1982	228.40
Return of the Jedi	1983	168.00
The Empire Strikes Back	1980	141.60
Ghostbusters	1984	128.30
Raiders of the Lost Ark	1981	115.60
Indiana Jones and the Last Crusade	1989	115.50
Indiana Jones and the Temple of Doom	1984	109.00
Beverly Hills Cop	1984	108.00
Back to the Future	1985	104.20
Tootsie	1982	95.30
Rain Man	1989	86.00

Chapter Project

Movie Videos and Technology

Work in a group.

1. Poll your classmates, family, and friends to find their favorite movies.
2. Make a chart or graph presenting your findings.
3. Pretend that you are opening a video rental store. List the first movies that you would purchase and any other important details. Show on a poster.

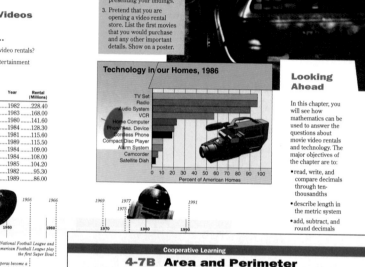

Technology in our Homes, 1986

TV Set, Radio, Audio System, VCR, Home Computer, Phone Ans. Device, Cordless Phone, Compact Disc Player, Alarm System, Camcorder, Satellite Dish

Percent of American Homes: 0 10 20 30 40 50 60 70 80 90 100

Looking Ahead

In this chapter, you will see how mathematics can be used to answer the questions about movie video rentals and technology. The major objectives of the chapter are to:

- read, write, and compare decimals through ten-thousandths
- describe length in the metric system
- add, subtract, and round decimals

1956   1966   1969   1977   1975   1991
1950   1960   1970   1980   1990

National Football League and American Football League play the first Super Bowl

Soap Operas become a television kit

A **Spotlight** feature opens each chapter in *Mathematics: Applications and Connections*. **Spotlight** focuses on topics you and your friends will find interesting. Each **Spotlight** includes:

- "Have You Ever Wondered.". . .questions to engage your interest
- interesting **data** related to the feature
- a **timeline** or **comic** connected to relevant events
- a **Chapter Project** related to the **Spotlight** feature
- a preview of what's coming up called **Looking Ahead**

Cooperative Learning

4-7B Area and Perimeter

A Follow-Up of Lesson 4-7

Objective
Find relationships between perimeter and area.

Materials
centimeter grid paper
scissors

In this lab, you will use grid paper to experiment with area and perimeter and to see the relationship between them.

Try this!

Work in groups of three.

- Each member of the group should use a sheet of centimeter grid paper. Draw a rectangular shape with a perimeter of 48, staying on the lines. Examples of rectangular shapes with a perimeter of 14 centimeters are presented below.

1 / 6   2 / 5

- Cut out your rectangular shape. Find the area by counting the number of squares. Compare with other members of the group.

What do you think?

1. Describe the perimeter of each rectangular shape.
2. Describe the area of each rectangular shape.
3. What can you conclude about the relationship between area and perimeter?

Extension

4. Suppose these shapes represent a dog pen you want to build with 48 feet of fence. Which rectangular shape would you choose? Why?
5. On centimeter grid paper, draw a rectangular shape with the

---

Mathematics Labs prior to or after some lessons give you hands-on experience, with a partner or group, in discovering a math concept on your own. You may also participate in shorter **Mini-Labs** in which you will investigate math concepts within a lesson.

Previewing Your Text

Wrapping Up Each Chapter

A **Study Guide and Review** at the end of each chapter helps you connect the whys and hows of reviewing what you've learned. How? By presenting objectives and examples in the left column and review exercises in the right. This unique format makes it easier for you to remember and apply what you've learned.

Chapter 4 Study Guide and Review

Communicating Mathematics

Choose the letter that best matches each phrase.

1. distance from the center of a circle to any point on the circle
2. distance around any figure
3. twice the radius
4. $a(b+c) = ab+ac$
5. $l \cdot w$
6. $2\pi r$
7. one of the numbers being multiplied
8. number of times a base is used as a factor

a. factor
b. area
c. circumference
d. radius
e. exponent
f. perimeter
g. diameter
h. distributive property

9. Tell why you can multiply two decimal numbers and get an answer that is less than the factors you used.
10. Explain the two methods used when multiplying decimals for placing the decimal point correctly. Why is it a good idea to do both?

Skills and Concepts

Objectives and Examples

Upon completing this chapter, you should be able to:

Review Exercises

Use these exercises to review and prepare for the chapter test.

- estimate products using rounding and compatible numbers *(Lesson 4-1)*

  a. $31.78 \rightarrow 30$
     $\times 4 \rightarrow \times 4$
     120
     31.78 × 4 is about 120.

  b. $398.1 \rightarrow 400$
     $\times 236 \rightarrow \times 250$
     $100,000$
     398.1 × 236 is about 100,000.

Estimate each product.
11. 6.9×88
12. 4.2×39.6
13. 38.5×791

Estimate each product.
14. 4.2×240.7
15. $51.2 \times 1,891$
16. 121×7.9

- use powers and exponents in

Evaluate each expression.

22. 0.15
    $\times 227$

ation $m = 201 \times 3.94$.

distributive property *(Lesson 4-2)*

$90 \times 3.8 = 90(3 + 0.8)$
$ = 90 \times 3 + 90 \times 0.8$
$ = 270 + 72$
$ = 342$

mentally. Use the distributive property.
24. 40×8.9
25. 30×10.76
26. 9×16
27. 5×6.8

- multiply decimals *(Lesson 4-5)*

  38.76 ← *two decimal places*
  $\times 4.2$ ← *one decimal place*
  7752
  15504
  162.792 ← *three decimal places*

Multiply.
28. $8.74 \cdot 2.23$
29. $0.04 \cdot 5.1$
30. $0.04 \cdot 0.0063$
31. $11.089 \cdot 5.6$
32. Evaluate $ca(b-a)$ if $a=2$, $b=5$, and $c=3$.

- find perimeter *(Lesson 4-6)*

  $P = 2l + 2w$
  $ = 2 \times 6.15 + 2 \times 3.21$
  $ = 12.30 + 6.42$
  $ = 18.72 \text{ m}$

Find the perimeter of each figure.
33.
34.
35. A square has a perimeter of 100 meters. What is the length of each side?

- find the area of rectangles and squares *(Lesson 4-7)*

  $A = l \cdot w$
  $ = 8.3 \times 4$
  $ = 33.2 \text{ m}^2$

  $A = s^2$
  $ = 6^2$
  $ = 36 \text{ in}^2$

Find the area of each figure.
36. the rectangle in Exercise 34.
37. the square in Exercise 35.
38. a rectangle with $l = 11.2$ mi and $w = 9.3$ mi
39. square 7.6 m

Chapter 4 Study Guide and Review **153**

Objectives and Examples

Review Exercises

- find the circumference of circles *(Lesson 4-9)*

  $c = \pi d = 2\pi r$
  $ = 2 \cdot 3.14 \cdot 2.4$
  $ = 15.072$

Find the circumference. Round to the nearest
40.

Applications and Problem Solving

42. Find the area of the figure at the right. *(Lesson 4-8)*

43. **Elevators** The express elevator to the 60th floor of the "Sunshine 60" building in Tokyo, Japan, operates at the incredible speed of 22.72 miles per hour. If one mile per hour is the same as 88 feet per minute, what is the speed of this elevator in feet per minute? *(Lesson 4-3)*

44. **Fitness** Janice exercises every morning by running once around a circular path surrounding a park in her neighborhood. The diameter of the circle of the path she follows is 1.7 kilometers. How far does she run? *(Lesson 4-9)*

Curriculum Connection Projects

- **Science** Find the distance of three of our satellites from Earth. Then find the circumference of the orbit of each. Remember, the surface of Earth is about 6,400 kilometers from its center.
- **Language Arts** Use a copy of today's newspaper to find, measure, and list the areas of various rectangular and square ads.

Read More About It

Korman, Gordon. *Son of Interflux.*
Rockwell, Thomas. *How To Get Fabulously Rich.*
Wilkinson, Elizabeth. *Making Cents: Every Kid's Guide to Making Money.*

Curriculum Connection Projects at the end of each chapter suggest short research projects and field trips. They link math to science, language arts, physical education, and other subjects, and to real-world situations.

Now that you know what's ahead, you're ready to begin using the text. You'll soon see for yourself that you can do math, and you'll realize why math is useful every day. Textbooks, and your teachers and parents who help you interpret them, are great resources. But, remember, attitude counts. The most important factor for your success in mathematics is you. Have a wonderful year!

Read More About It lists fiction and nonfiction books you may want to read.

Tools for Problem Solving

Spotlight on Food and Fitness

Have You Ever Wondered...

- How long you would have to run to burn off the number of calories in a slice of apple pie?

- How people's eating habits have changed over the years?

Calories and Energy Equivalents in Food

This table shows the calories in some foods, and the minutes it would take a 150-pound person to use them up in various ways.

Food	Calories	Minutes to Burn Off Food				
		lying down	walking (4 mph)	bicycle riding (9 mph)	swimming	running (7 mph)
Apple, large	125	83	23	20	14	14
Bread and butter	100	67	18	16	11	11
Hamburger sandwich	245	163	45	39	27	27
Ice cream, vanilla, $\frac{1}{2}$ cup	135	90	25	22	15	15
Malted milk shake, 10 ounces	335	223	61	54	37	37
Pancake with syrup	121	81	22	19	13	13
Pie, apple, $\frac{1}{6}$ of pie	405	270	74	65	44	44
Pork chop, pan fried	335	223	61	54	37	37

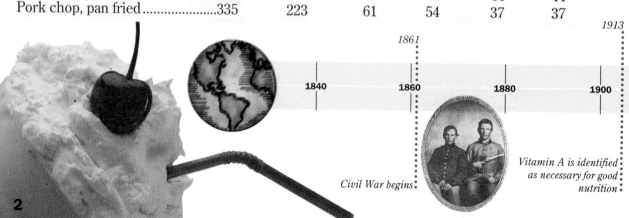

1840   *1861*   1860   1880   1900   *1913*

Civil War begins

Vitamin A is identified as necessary for good nutrition

Eating Trends

Average yearly consumption, in pounds, of various foods.

Red Meat

Pounds

1970	1975	1980	1985	1989
132.4	124.2	123.4	120.9	111.3

Years

Fish/Shellfish

Pounds

1970	1975	1980	1985	1989
11.8	12.2	12.8	14.4	15.7

Years

Boneless Chicken

Pounds

1970	1975	1980	1985	1989
27.7	27.5	34.3	39.7	47

Years

Sugar

Pounds

1970	1975	1980	1985	1989
101.8	89.2	83.8	63.4	62.2

Years

Broccoli

Pounds

1970	1975	1980	1985	1989
0.5	1	1.6	2.9	4.5

Years

Rice

Pounds

1970	1975	1980	1985	1989
6.7	7.6	9.5	9.1	16.6

Years

Chapter Project

Food and Fitness

Work in a group.

1. Throughout the course of one week, make a list of the foods you eat often and the number of calories each food provides.

2. List the activities you do and the number of calories each activity burns.

3. Do you think you take in more calories than you burn? Add food and activities that may balance out the list. Show your data in a chart.

Looking Ahead

In this chapter, you will see how mathematics can be used to answer questions about food and nutrition. The major objectives of the chapter are to:

- solve problems using the four-step plan

- estimate sums, differences, products, and quotients

- evaluate expressions

- solve equations using mental math and the guess-and-check strategy

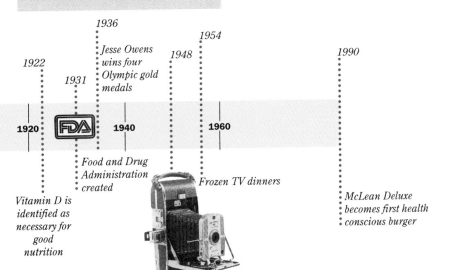

1922

Vitamin D is identified as necessary for good nutrition

1931

Food and Drug Administration created

1920

1940

1936

Jesse Owens wins four Olympic gold medals

1948

1954

Frozen TV dinners

1960

1990

McLean Deluxe becomes first health conscious burger

3

1-1 A Plan for Problem Solving

Objective

Solve problems using the four-step plan.

DID YOU KNOW

The fifth graders from Deep River Elementary School in Sanford, North Carolina, not only calculated the number of pennies in a mile, they collected them and donated the money to charity. Their school received a $5,000 Good Neighbor Award.

In 1991, Mrs. Nall, a teacher from North Carolina, challenged her class to determine how many pennies are in a mile. You can use a four-step plan to help solve this problem and other problems.

1 Explore
- Read the problem carefully.
- Ask yourself questions like, "What facts do I know?" and "What do I need to find out?"

2 Plan
- See how the facts relate to each other.
- Make a plan for solving the problem.
- Make an estimate of the answer.

3 Solve
- Use your plan to solve the problem.
- If your plan does not work, revise it or make a new plan.

4 Examine
- Reread the problem.
- Ask, "Is my answer close to my estimate?"
- Ask, "Does my answer make sense for the problem?"
- If not, solve the problem another way.

Example 1

Let's use this plan to help solve the problem above.

Explore You need to find out how many pennies are in one mile.

Plan One way to solve the problem would be to find out how many pennies are in one foot. You can then multiply that number by the number of feet in one mile. Before you begin, guess the number of pennies you think are in one mile.

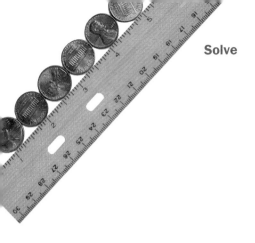

Solve Lay as many pennies as you can end-to-end next to a foot ruler. It takes 16 pennies.

number of pennies   *number of feet*
in one foot   *in one mile*

$$16 \quad \times \quad 5{,}280 \quad = \quad ?$$

16 ⊠ 5280 ⊟ 84480

There are 84,480 pennies in one mile.

Examine To see if your answer makes sense, think 5,280 is about 5,000. $16 \times 5{,}000 = 80{,}000$. So 84,480 is reasonable.

Teenagers between the ages of 12 and 17 watch about 3 hours of TV a day, or an average of about 21 hours a week.

Example 2

On the average, Americans buy 52,000 new televisions per day. If 31,200 of these are color televisions, how many black and white televisions are sold each day?

Explore You need to find the average number of black-and-white televisions sold each day. You know the total number of televisions sold each day. You also know how many of the total are color televisions.

Plan Since you know the total and one of the parts, you will need to subtract to find the other part.

You can estimate the answer by rounding.

$$50{,}000 - 30{,}000 = 20{,}000.$$

Your answer should be about 20,000.

Solve *number of*   *minus*   *number of*
televisions sold   *color televisions*
each day   *sold each day*

$$52{,}000 \qquad - \qquad 31{,}200$$

$$52{,}000 - 31{,}200 = 20{,}800$$

The average number of black-and-white televisions sold each day is 20,800.

Examine   The number of black-and-white televisions sold each day added to the number of color televisions sold each day should equal 52,000.

$$20,800 + 31,200 \stackrel{?}{=} 52,000$$
$$52,000 = 52,000 \quad ✔$$

Checking for Understanding

Communicating Mathematics

Read and study the lesson to answer each question.

1. **Write** a paragraph explaining the four-step plan for problem solving.

2. **Tell** why you should check to see whether your answer is reasonable.

3. **Write** one or two sentences explaining why you think there would be more or fewer nickels than pennies in one mile.

Guided Practice

Explain how you would solve each problem.

4. Four hundred fifty teenagers registered for the all-city table tennis tournament. If 307 participants are boys, how many participants are girls?

5. Jill delivers newspapers each Saturday and Sunday. If Jill works for one year, how many days does she work?

6. Luis is going on a school trip to Washington, D.C., this spring. He is taking $75 for food, $30 for entertainment, and $25 to buy gifts for his family and friends. How much money is Luis taking on his trip?

7. Bryan is on the basketball team. Last season, his height was 6 feet 2 inches. He has grown $1\frac{1}{2}$ inches since last season. How tall is Bryan now?

Use the four-step plan to solve each problem. Use a calculator if necessary.

8. The student council at Belmont Middle School is hosting their annual fall dance. Tickets to the dance cost $3 each. There are 345 students in the school. If each student buys one ticket, how much money would the student council take in?

9. Sarah sold 322 boxes of candy in two weeks for her drill team. If Sarah sold the same number of boxes of candy each day, how many boxes of candy did she sell a day?

10. Last weekend, Jessie spent $3 for lunch, $6 for dinner, and $7 at the movies. How much money did he spend?

11. Jamal needs to hand out an equal number of chips to 23 classmates. There are 522 chips. How many chips should Jamal give each classmate?

12. Movie tickets cost $6.50 for movies shown after 3:00 P.M. and $4.50 for movies shown before 3:00 P.M. How much will three tickets cost for a movie starting at 1:20 P.M.?

Exercises

Use the four-step plan to solve each problem. Use a calculator if necessary.

13. Freida and Janie are shopping for a gift for their mother. The girls have decided to buy a purse for $47. If Freida and Janie share the cost equally, how much will each girl spend?

14. Oko is in charge of refreshments for her class party. She cannot spend more than $2 per person. If there are 28 students in her class, what is the most Oko can spend on refreshments?

15. Rob bought a cassette tape for $9. It is now on sale for half price. How much does the cassette tape cost now?

16. Brett's baseball team won twice as many games as they lost. They lost 11 games. How many games did they win?

17. **Critical Thinking** In high school football, a team can score five different ways. A touchdown is 6 points, a field goal is 3 points, a safety is 2 points, a two-point conversion is 2 points, and a one-point conversion is 1 point. To score in either of the last two ways, a team must have scored a touchdown on the preceding play. Can a team score 23 points? Explain your answer.

18. **Science** You can estimate the temperature in degrees Fahrenheit by counting the number of times a cricket chirps in 15 seconds. Then add that number to 40. What is the temperature if a cricket chirps 20 times in 15 seconds?

19. **Geography** On Haley's map, each inch represents 8 miles. Atlanta is about 12 inches from Waycross. About how many miles is this?

20. **Journal Entry** Write a problem in which the solution involves the use of multiplication and the answer is 400. Explain how to solve the problem using the four-step plan.

1-2A Estimation

A Preview of Lesson 1-2

Objective

Use estimation techniques in a problem-solving situation.

Materials

beans
scale
large bowl
measuring cup

A new CD store advertises this promotion. "You can win two free tickets to the concert of your choice by guessing the number of jelly beans in this jar." In order to make a good guess, you must know how to estimate.

Try this!

Work in groups of four.

- Fill a bowl with beans until you think you have one pound of beans.
- Weigh the beans.
- Continue to add or take away beans until you reach one pound.
- Estimate the number of beans in your bowl. Record and explain your estimate.
- Use a measuring cup to measure one cup of beans from your pound. Count the beans in the cup. Use this number to adjust your estimate for the number of beans in a pound.
- Decide on your best estimate. Now find the exact number by counting the beans in one pound.

What do you think?

1. Was it difficult to estimate the number of beans in a pound in the beginning? Explain.

2. Explain why using the measuring cup could help you make a better estimate.

3. Which estimate was better, the initial estimate or the adjusted one? Explain.

Extension

4. Compare your exact answer to the exact answer of another group. Were the answers the same? If not, why might the answers be different?

5. Estimate how many beans it would take to make 3 pounds.

1-2 Using Rounding

Objective

Estimate sums and differences of whole numbers using rounding.

DID YOU KNOW

The electricity saved from one recycled aluminum can will operate a TV for three hours.

The average person in the United States uses about 586 pounds of paper, 55 pounds of aluminum cans, 190 pounds of plastic, and 325 pounds of glass bottles and jars in one year. If all these materials were recyclable, *about* how many pounds could be recycled by the average person in one year?

You can estimate the number of pounds by using rounding. Round the numbers to the nearest hundred. Then add mentally.

586	600
55	100
190	200
+ 325	+ 300
	1,200

The average person could recycle about 1,200 pounds in one year.

Let's review the rules for rounding.

Examples

1   Round 37 to the nearest ten.   *37 is closer to 40 than to 30.*

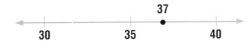

Look at the digit to the right of the tens place. Since 7 > 5, round the digit in the tens place up. 37 rounded to the nearest ten is 40.

2   Round 126 to the nearest hundred.

Look at the digit to the right of the hundreds place. Since 2 < 5, the digit in the hundreds place stays the same. 126 rounded to the nearest hundred is 100.

3   Round 1,564 to the nearest thousand.

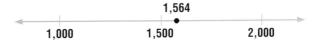

Look at the digit to the right of the thousands place. Since the digit is 5, round the digit in the thousands place up. 1,564 rounded to the nearest thousand is 2,000.

Estimation is a useful skill that provides a quick and easy answer when an exact answer is not necessary. Estimation also allows you to check the reasonableness of an answer.

When am I ever going to use this?

Suppose you were on a date and planned to spend $35 on dinner. You choose a restaurant where the average price of an entree is $12. Estimate the cost of appetizers, desserts, beverages, tax, and tip. Can you afford it?

Example 4 *Problem Solving*

Consumer Math Suppose you are eating out. You order a dinner for $6.89, a soft drink for $1.25, and a piece of pie for $2.25. The waiter gives you a bill for $12.39. Is this reasonable?

Round each item to the nearest dollar amount.

$6.89 rounds to $7.
$1.25 rounds to $1.
$2.25 rounds to $2.

The total bill should be about $7 + $1 + $2 or $10. Therefore, a total of $12.39 is *not* reasonable.

Checking for Understanding

Communicating Mathematics

Read and study the lesson to answer each question.

1. **Write** two reasons for using estimation.

2. **Tell** how you would round 734 to the nearest ten by using the number line below.

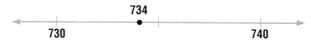

3. **Tell** about a situation where you would use estimation.

Guided Practice

Round to the nearest ten.

4. 32   5. 456   6. 86   7. 297

Round to the nearest hundred.

8. 244   9. 398   10. 2,345   11. 459

Round to the nearest thousand.

12. 2,345   13. 5,983   14. 7,356   15. 1,099

Estimate. State whether the answer shown is reasonable.

16. $4.56 + $7.99 = $10.95

17. 598 + 345 + 198 = 1,141

18. 54 + 67 + 22 + 177 = 371

19. 765 − 632 = 233

Exercises

Round to the nearest hundred.

20. 555   21. 934   22. 834   23. 175

24. 460   25. 202   26. 2,895   27. 1,344

Round to the nearest thousand.

28. 3,339   29. 4,578   30. 9,387   31. 1,010

32. 5,712   33. 8,050   34. 2,444   35. 6,686

Estimate. Use the data at the right.

36. What is the distance from Atlanta to Boston?

37. What is the distance from Chicago to Atlanta?

38. If you leave Columbus and travel to Boston, how far have you traveled?

39. Which trip is shorter, Boston to Denver or Chicago to Atlanta?

Mileage Chart

Approximate Mileages	Atlanta, GA	Baltimore, MD	Boston, MA	Chigago, IL	Columbus, OH	Denver, CO	Houston, TX
Atlanta, GA		673	1,065	675	576	1,395	814
Baltimore, MD	673		395	675	385	1,650	1,475
Boston, MA	1,065	395		965	725	1,960	1,845
Chicago, IL	675	675	965		310	1,036	1,080
Columbus, OH	576	385	725	310		1,230	1,125
Denver, CO	1,395	1,650	1,960	1,036	1,230		1,025
Houston, TX	814	1,475	1,845	1,080	1,125	1,025	

40. **Strange Facts**   In 1948, Jack O'Leary began a case of the hiccups which did not end until 1956. If he tried 60,000 remedies in those eight years, what is the average number of remedies he tried each year? *(Lesson 1-1)*

41. **Business**   In one month, Pam collected $51 from the customers on her paper route. If she owed the newspaper company $32, how much was her profit? *(Lesson 1-1)*

42. **Critical Thinking**   It has been said that each day, the population of Detroit, Michigan, drops by 79 people. Suppose the population is 1,403,339 today. Estimate what the population will be in one month.

43. **History**   The best-known Chinese mathematician of the third century was Liu Hui. In A.D. 263, he wrote *Sea Island Arithmetic Classic.* About how long ago was this?

44. **History**   Sir Isaac Newton was a famous British mathematician. He was born December 25, 1642, and died March 20, 1727. About how old was he when he died?

1-3 Front-End Estimation

Objective

Estimate sums and differences using front-end estimation.

The top three running backs in the National Football League in 1990 carried the ball 1,304 yards, 1,297 yards, and 1,225 yards. Was their combined yardage greater than 4,000 yards?

You can use **front-end estimation** to answer this question.

Words to Learn

front-end estimation

Front-End Estimation	1. Add or subtract the front-end digits. 2. Adjust by estimating the sum or difference of the remaining digits. 3. Add the two values.

DID YOU KNOW

Barry Sanders of the Detroit Lions led the NFL in rushing in 1990. He was followed by Thurman Thomas of the Buffalo Bills, and Marion Butts of the San Diego Chargers.

Step 1

$$1,304$$
$$1,297$$
$$+ 1,225$$
$$\overline{3,000}$$

Step 2

$$1,304$$
$$1,297$$
$$+ 1,225$$
$$\overline{700}$$

Step 3 $3,000 + 700 = 3,700$

Since 3,700 is less than 4,000, the combined yardage for the three running backs was not greater than 4,000 yards.

Example 1 *Problem Solving*

School The Miller Middle School Drama Club is presenting a play next week. The play will be held in the auditorium, which seats 650 people. There are 179 sixth graders, 188 seventh graders, and 213 eighth graders enrolled in the school. Is the auditorium large enough to hold all of the students?

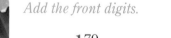

Add the front digits.

$$179$$
$$188$$
$$+ 213$$
$$\overline{400}$$

Adjust your estimate.

$$179$$
$$188$$
$$+ 213$$
$$\overline{160}$$

$$400 + 160 = 560$$

Since 560 is less than 650, the auditorium is large enough to hold the students.

Example 2 *Problem Solving*

Transportation   Valleyview Junior High has an enrollment of 442 students. One hundred twenty-six students ride the bus. Estimate how many students do not ride the bus.

Subtract the front digits.   *Adjust your estimate.*

$$
\begin{array}{r}
442 \\
-\ 126 \\
\hline
300
\end{array}
\qquad
\begin{array}{r}
442 \\
-\ 126 \\
\hline
20
\end{array}
$$

$$300 + 20 = 320$$

About 320 students do not ride the bus.

Checking for Understanding

Communicating Mathematics

Read and study the lesson to answer each question.

1. **Tell** the difference between front-end estimation and estimation using rounding.

2. **Write** in your own words how adjusting made the estimate closer in Example 1.

3. **Tell** whether more or less than one half of the students ride the bus in Example 2. Explain.

Guided Practice   Estimate using front-end estimation.

4. $\begin{array}{r} 423 \\ +\ 134 \\ \hline \end{array}$
   5. $\begin{array}{r} 205 \\ +\ 215 \\ \hline \end{array}$
   6. $\begin{array}{r} 334 \\ -\ 137 \\ \hline \end{array}$
   7. $\begin{array}{r} 642 \\ +\ 348 \\ \hline \end{array}$

8. $544 + 486$
   9. $272 - 242$
   10. $675 - 352$

Exercises

Independent Practice   Estimate using front-end estimation.

11. $\begin{array}{r} 115 \\ 329 \\ +\ 165 \\ \hline \end{array}$
    12. $\begin{array}{r} 4{,}599 \\ -\ 3{,}278 \\ \hline \end{array}$
    13. $\begin{array}{r} 239 \\ 455 \\ +\ 611 \\ \hline \end{array}$
    14. $\begin{array}{r} 1{,}245 \\ +\ 4{,}453 \\ \hline \end{array}$

15. $\begin{array}{r} 2{,}345 \\ +\ 1{,}239 \\ \hline \end{array}$
    16. $\begin{array}{r} 1{,}333 \\ +\ 2{,}199 \\ \hline \end{array}$
    17. $\begin{array}{r} 4{,}798 \\ -\ 2{,}578 \\ \hline \end{array}$
    18. $\begin{array}{r} 567 \\ -\ 348 \\ \hline \end{array}$

19. $342 + 262$
    20. $867 - 464$
    21. $3{,}484 - 1{,}255$

22. $8{,}321 + 854$
    23. $568 - 421$
    24. $888 - 570$

25. Estimate five hundred forty-three plus seven hundred sixty-four.

26. Estimate eight hundred eighty-eight minus five hundred thirty.

27. Estimate the sum of nine thousand nine hundred sixty-one and four hundred seventy-six.

28. Estimate the difference of seven thousand nine hundred thirty-four and six hundred thirty-one.

Mixed Review 29. **Computers** The Changs' dot matrix printer prints four pages each minute. Hiro Chang has a history report that is 12 pages long. How long will it take the printer to print the report? *(Lesson 1-1)*

30. Round 989 to the nearest hundred. *(Lesson 1-2)*

31. Round 9,046 to the nearest thousand. *(Lesson 1-2)*

Problem Solving and Applications 32. **Critical Thinking** Make up a problem using front-end estimation where the estimate is 9,000.

33. **Sports** The Toronto Blue Jays play baseball in a 54,000-seat stadium that has a moveable roof. If there are 4,873 no-shows for a game, about how many seats are filled?

34. **History** During the 1991 Persian Gulf War, the United States government bought 338,768 bottles of sunscreen and 760,000 tubes of Chap Stick®. About how many more tubes of Chap Stick® than bottles of sunscreen did the government buy?

35. **Research** Bring in an article from a newspaper or magazine that uses an estimation, not an exact answer. Explain why an exact answer was not used.

36. **Mathematics and Science** Read the following paragraphs.

The world is round for the same reason that a raindrop is round. A liquid naturally shapes itself into a ball. When Earth was formed, it was hot and liquid. Because it was floating in space, it became round and stayed round after it cooled and hardened.

Our Earth spins on its axis and revolves around the sun as it spins. It has been noted that Earth spins a little faster in September than it does in March.

As Earth spins, a point on the equator moves around at about 1,037 miles per hour. About how many miles does the point travel each day?

1-4 Using Patterns

Objective

Estimate products and quotients using patterns.

You've probably heard of the three R's, but have you heard of the "three D's?" They stand for dedication, determination, and desire. They were the keys to winning the 1991 World Double Dutch Invitational Tournament for four Columbus, Ohio, students. The YWCA team, "Double Forces," competed all over the nation, defeating about 150 teams along the way.

The World Double Dutch jump rope competition was held at the University of Maryland. Suppose the tournament was a three-day, sold-out event. If the event was held in a gymnasium that seated 1,375 people, about how many people attended the competition during the three days?

We need to estimate the product of 1,375 and 3. You can use patterns to estimate products. Round factors to their greatest place-value position. Do not change 1-digit factors. Multiply by multiples of 10, 100, or 1,000.

Examples

1 Estimate $3 \times 1,375$.

Step 1 Round 1,375 to 1,000.

Step 2 Look for a pattern. Solve mentally.

$3 \times 1 = 3$
$3 \times 10 = 30$
$3 \times 100 = 300$
$3 \times 1,000 = 3,000$

About 3,000 people attended the competition.

2 Estimate 42×234.

Step 1 Round 42 to 40.
Round 234 to 200.

Step 2 Look for a pattern. Solve mentally.

$40 \times 2 = 80$
$40 \times 20 = 800$
$40 \times 200 = 8,000$

42×234 is about 8,000.

3 Estimate $6 \times 3,988$.

Round 3,988 to 4,000.

Look for a pattern. Solve mentally.

$6 \times 4 = 24$
$6 \times 40 = 240$
$6 \times 400 = 2,400$
$6 \times 4,000 = 24,000$
$6 \times 3,988$ is about 24,000.

You can also use patterns to estimate quotients. Round the dividend to its greatest place-value position.

Examples

4 Estimate $345 \div 5$.

   Step 1 Round 345 to 300.

   Step 2 Look for a pattern. Solve mentally.
   $$30 \div 5 = 6$$
   $$300 \div 5 = 60$$

   $345 \div 5$ is about 60.

5 Estimate $8{,}763 \div 3$.

   Step 1 Round 8,763 to 9,000.

   Step 2 Look for a pattern. Solve mentally.
   $$9 \div 3 = 3$$
   $$90 \div 3 = 30$$
   $$900 \div 3 = 300$$
   $$9{,}000 \div 3 = 3{,}000$$

   $8{,}763 \div 3$ is about 3,000.

Example 6 *Problem Solving*

Sports   Suppose 780 rope jumpers competed for the World Double Dutch Invitational Tournament championship. If there are four members on each team, about how many teams competed?

Estimate $780 \div 4$.

Step 1 Round 780 to 800.

Step 2 Look for a pattern. Solve mentally.
$$8 \div 4 = 2$$
$$80 \div 4 = 20$$
$$800 \div 4 = 200$$

About 200 teams competed.

Checking for Understanding

Communicating Mathematics   Read and study the lesson to answer each question.

1. **Tell** about and explain the patterns in the examples. Compare the zeros in the factors to the zeros in the products.

2. **Tell** how you would use patterns to estimate 31×485.

Estimate using patterns.

3. 3×45 4. $2 \times 7,899$ 5. 5×31 6. 8×7

7. 84×6 8. 521×9 9. 38×49 10. 17×21

11. 78×34 12. 610×6 13. 277×8 14. 4×532

15. $618 \div 5$ 16. $379 \div 2$ 17. $108 \div 4$ 18. $492 \div 5$

19. $207 \div 4$ 20. $566 \div 3$ 21. $721 \div 7$ 22. $975 \div 2$

23. $3,216 \div 3$ 24. $6,044 \div 3$ 25. $7,855 \div 4$

Exercises

Estimate using patterns.

26. 128×4 27. 685×5 28. 21×404 29. $109 \div 4$

30. $78 \div 4$ 31. 76×51 32. $567 \div 6$ 33. $6,189 \times 6$

34. $8,157 \div 2$ 35. 43×19 36. $1,125 \div 4$

37. Estimate 256 divided by 5.

38. Estimate 436 times 8.

39. **History**   In July 1776, General George Washington had nearly 30,000 troops guarding New York City. After retreating through New Jersey to the Delaware River, only about 3,000 troops remained with him. How many troops did he lose? *(Lesson 1-1)*

40. Round 6,817 to the nearest hundred. *(Lesson 1-2)*

41. Estimate $835 + 917$ using front-end estimation. *(Lesson 1-3)*

42. Estimate $6,926 - 5,050$ using front-end estimation. *(Lesson 1-3)*

43. **Smart Shopping**   Keith's mother needs to buy one new tire for her car. Tires are on sale for 4 for $189. About how much will one tire cost?

44. **Recreation**   The Elliot's are planning a vacation to Myrtle Beach, South Carolina. The hotel they have chosen will cost $76 per night. They need a room for 5 nights. About how much will they spend for the hotel room?

45. **Critical Thinking**   Janna is shopping for a stereo. She is trying to decide between two different stereos sold at Hyperstore®. One stereo is on sale for $429 with 9 months to pay. The other one costs $339 with 6 months to pay. Janna has budgeted $50 a month to pay for the stereo. Which stereo can she afford?

46. **Journal Entry**   Write a paragraph explaining how patterns can help you estimate the product of 29 and 9,134.

1-5 Using Compatible Numbers

Objective
Estimate quotients using compatible numbers.

Words to Learn
compatible numbers

In 1987, Richard Branson and Per Lindstrand completed the first transatlantic hot-air balloon flight. They flew 2,790 miles from Maine to Ireland in 3 days. About how many miles did they travel a day? You need to estimate the quotient 2,790 ÷ 3.

You can estimate the quotient by using **compatible numbers.** Compatible numbers are two numbers that are easy to divide mentally. They are often members of fact families.

To estimate 2,790 ÷ 3, round 2,790 to 2,700.

$$2{,}700 \div 3 = 900 \qquad \textit{27 and 3 are compatible numbers.}$$

Branson and Lindstrand traveled about 900 miles a day.

Examples

1 Estimate 57 ÷ 7.

$57 \div 7 \rightarrow 56 \div 7$ *56 and 7 are compatible numbers.*

$56 \div 7 = 8$

$57 \div 7$ is *about* 8.

2 Estimate 778 ÷ 8.

$778 \div 8 \rightarrow 800 \div 8$ *800 and 8 are compatible numbers.*

$800 \div 8 = 100$

$778 \div 8$ is *about* 100.

Checking for Understanding

Communicating Mathematics

Read and study the lesson to answer each question.

1. **Write** an example of a compatible number for 12. Explain your answer.

2. **Tell** of a situation where using compatible numbers would be useful.

Use compatible numbers to choose the better estimate.

3. $532 \div 9$ a. $540 \div 9$ b. $450 \div 9$

4. $273 \div 7$ a. $350 \div 7$ b. $280 \div 7$

5. $4,135 \div 6$ a. $4,200 \div 6$ b. $3,600 \div 6$

Estimate using compatible numbers.

6. $36 \div 7$ 7. $31 \div 5$ 8. $46 \div 7$ 9. $75 \div 9$

10. $147 \div 5$ 11. $75 \div 6$ 12. $832 \div 4$ 13. $187 \div 3$

14. $6,789 \div 9$ 15. $4,398 \div 5$ 16. $5,432 \div 7$

Exercises

Estimate using compatible numbers.

17. $24 \div 5$ 18. $213 \div 8$ 19. $529 \div 6$ 20. $219 \div 7$

21. $734 \div 9$ 22. $876 \div 3$ 23. $3,758 \div 6$ 24. $459 \div 5$

25. $6,345 \div 8$ 26. $2,240 \div 7$ 27. $530 \div 9$ 28. $686 \div 7$

29. $1,546 \div 4$ 30. $413 \div 6$ 31. $8,287 \div 9$ 32. $123 \div 5$

33. Estimate 324 divided by 11.

34. Estimate 6,076 divided by 3.

35. **Transportation** Michelle's father rides the bus to work and back home every day. Each round trip ride costs 85 cents. If Michelle's father worked 20 days last month, how much did he spend for bus fare? *(Lesson 1-1)*

36. Estimate $9,180 - 3,333$ using rounding. *(Lesson 1-2)*

37. Estimate $7,961 + 3,815$ using front-end estimation. *(Lesson 1-3)*

38. Estimate $6,257 \times 22$ using patterns. *(Lesson 1-4)*

39. **School** Kelly Oak Middle School prepared 1,987 lunches last week. About how many lunches were prepared each day?

40. **Travel** Greg is driving with his uncle from Toronto, Canada, to Dallas, Texas. The trip is 2,323 kilometers long. About how far would they need to drive each day to complete the trip in 9 days? Is 9 days a reasonable amount of time to allot for this trip? Explain.

41. **Critical Thinking** Which is the better estimate for $233 \div 13$?

   a. more than 20   b. less than 20

42. **Journal Entry** The following problem appeared on the math test in Mr. Purdue's class.

Estimate the quotient $2,643 \div 6$.

Arlene's answer was $3,000 \div 6 = 500$.

Manuel's answer was $2,400 \div 6 = 400$.

What method did each student use? Explain.

DECISION MAKING

Choosing a Keyboard

Situation
You have decided to spend up to $450 of the money you earned this summer to buy a keyboard. Which keyboard best fits your needs and your budget?

Hidden Data
Cost of a stand: Do you need to buy a keyboard whose price includes a stand or can you afford to spend extra money for one?

Cost of an AC adapter: Does the model you want come with an adapter or do you need to buy one?

Skills: Do you already know how to play a keyboard or will you need lessons? How much will lessons cost?

State tax: This may apply to your situation.

SOME SOUND IDEAS FOR SAVING MONEY

ROCK TO NEW SOUNDS

A.

A. The Multi-Accompaniment Keyboard with stand. Features 61 full-size keys. With the touch of a few buttons, you can arrange high-quality professional type backgrounds. Features 2 built-in speakers, and you can play up to 12 keys at the same time! Sounds great for rock, reggae, or cha cha. Complete with memory, full MIDI-capability, stand, and AC adapter. **$399⁰⁰**
Warranted..................

B. The Rapper Keyboard features 32 mid-size keys and all the latest rap effects. Play any one of 25 pre-set tones with the push of a button. Three-effects pad give you those drum or orchestral hits. Microphone included with voice modification capabilities. Uses 5 AA batteries, not included, or AC adapter, not included. **$199⁰⁰**
Warranted..................

C. The Mighty Miracle Keyboard is the piano teaching system just for you. With the help of your PC or game system, you can learn how to play this keyboard with customized instruction. 32 mid-size keys, monitors your progress and features a built-in orchestra. AC adapter included. **$399⁰⁰**
Warranted..................

B.

C.

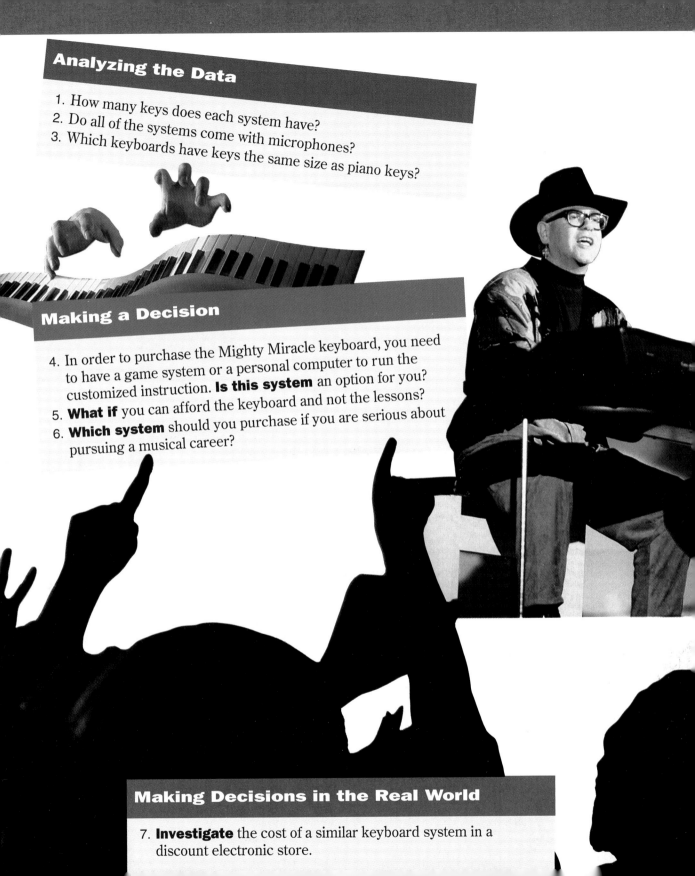

Analyzing the Data

1. How many keys does each system have?
2. Do all of the systems come with microphones?
3. Which keyboards have keys the same size as piano keys?

Making a Decision

4. In order to purchase the Mighty Miracle keyboard, you need to have a game system or a personal computer to run the customized instruction. **Is this system** an option for you?
5. **What if** you can afford the keyboard and not the lessons?
6. **Which system** should you purchase if you are serious about pursuing a musical career?

Making Decisions in the Real World

7. **Investigate** the cost of a similar keyboard system in a discount electronic store.

1-6 Order of Operations

Objective

Evaluate expressions using the order of operations.

Words to Learn

order of operations

Pat and Phil are helping plan a class trip to the Buhl Science Center Planetarium. They want to find the total cost of twenty-eight $4-tickets and eleven $6-tickets.

First, they make an estimate.

$$28 \times 4 \quad \rightarrow \quad 30 \times 4 \text{ or } 120$$
$$11 \times 6 \quad \rightarrow \quad 10 \times 6 \text{ or } 60$$

The total cost will be about $120 + 60$ or $180.

Pat and Phil then enter the following numbers into their calculators to find the exact answer.

$$28 \enspace \boxed{\times} \enspace 4 \enspace \boxed{+} \enspace 11 \enspace \boxed{\times} \enspace 6 \enspace \boxed{=}$$

Pat's calculator display shows an answer of 178. Phil's display shows 738. Which answer is correct? They reason that Pat's answer is correct because it is close to the estimate.

When more than one operation is used, we need to know which one to perform first so that everyone gets the same result. Mathematicians have come up with rules called the **order of operations.**

Calculator Hint

• • • • • • • • • •

Calculators that follow the order of operations have algebraic logic. To find out if your calculator has algebraic logic, enter $7 + 3 \times 2$. If your calculator displays 13, it has algebraic logic.

Order of Operations	1. Multiply and divide in order from left to right.
	2. Add and subtract in order from left to right.

Some calculators follow the order of operations. In Pat and Phil's case, Pat's calculator follows the order of operations. The correct answer is indeed $178.

Examples

Find the value of each expression.

1 $17 - 5 + 3 \times 8$

$$\begin{aligned} 17 - 5 + 3 \times 8 &= 17 - 5 + 24 \\ &= 12 + 24 \\ &= 36 \end{aligned}$$

    Multiply 3 and 8.
    Subtract 5 from 17.
    Add 12 and 24.

2 $12 \div 4 + 5 \times 3 - 16$

$$12 \div 4 + 5 \times 3 - 16 = 3 + 5 \times 3 - 16 \qquad \textit{Divide 12 by 4.}$$
$$= 3 + 15 - 16 \qquad \textit{Multiply 5 and 3.}$$
$$= 18 - 16 \qquad \textit{Add 3 and 15.}$$
$$= 2 \qquad \textit{Subtract 16 from 18.}$$

In mathematics, there are other ways, besides using $\times$, to show multiplication. One way is to use a raised dot.

$$2 \cdot 3 \quad \boxed{\text{means}} \!\!\longrightarrow\ 2 \times 3$$

Example

3 Evaluate $4 \cdot 6 - 2$.

Think: $4 \cdot 6 - 2$ is the same as $4 \times 6 - 2$.

$$4 \cdot 6 - 2 = 24 - 2 \qquad \textit{Multiply 4 and 6.}$$
$$= 22 \qquad \textit{Subtract 2 from 24.}$$

Checking for Understanding

Communicating Mathematics

Read and study the lesson to answer each question.

1. **Write** a sentence explaining why it is important to have order of operations rules.
2. **Tell** how estimation helped determine whether Pat or Phil had the correct answer.
3. **Show** the value of $28 \times 4 + 11 \times 6$ on your calculator to see if it follows the order of operations.

Guided Practice

Name the operation that should be done first. Then find the value.

4. $4 + 6 \times 2$
5. $3 \times 5 + 2$
6. $2 \times 7 - 3$
7. $24 \div 4 + 5$
8. $16 + 4 - 12$
9. $24 - 3 + 8$
10. $16 \div 4 + 3$
11. $5 \cdot 7 \cdot 2$
12. $18 \div 9 + 16$

Exercises

Independent Practice

Find the value of each expression.

13. $17 - 8 + 2$
14. $15 - 11 + 4 - 3$
15. $48 - 6 + 3 \times 4$
16. $13 \times 9 - 5$
17. $18 - 2 \times 5 - 1$
18. $34 + 16 - 5 \times 2$
19. $76 \div 19 \cdot 4$
20. $121 \div 11 + 4 \cdot 8$
21. $72 \div 12 - 4$
22. $8 + 56 \div 8$
23. $14 + 16 \div 2 - 3$
24. $18 + 12 \cdot 3 \div 4$

Mixed Review

25. **Smart Shopping** Juanita called several computer dealers to compare prices before buying a computer. The highest price given was $2,349 and the lowest was $1,788. Round to the nearest hundred to estimate the amount she could save by buying the less expensive computer. *(Lesson 1-2)*

26. Estimate 4,078 ÷ 71 using compatible numbers. *(Lesson 1-5)*

27. Estimate 93 ÷ 9 using patterns. *(Lesson 1-4)*

Problem Solving and Applications

28. **Entertainment** Write an expression that would help you find the total cost of twenty adult zoo tickets at $6 each and ten children's zoo tickets at $2 each.

29. **Journal Entry** Write down any three numbers. Then write as many expressions as possible using those three numbers and any or all of the operations +, −, ×, ÷. Evaluate each expression.

Mid-Chapter Review

Use the four-step plan to solve each problem. *(Lesson 1-1)*

1. If a car travels 65 miles each hour, how long will it take it to travel 455 miles?

2. Lara baby-sat six hours last Saturday and earned $15. How much did Lara earn per hour?

Round to the nearest hundred. *(Lesson 1-2)*

3. 608
4. 479
5. 123
6. 943
7. 1,247
8. 3,173
9. 855
10. 8,474

Estimate using front-end estimation. *(Lesson 1-3)*

11. 285
    $+ 313$
12. 475
    $- 140$
13. 312
    $- 211$
14. 3,433
    $+ 5,028$

Estimate using patterns. *(Lesson 1-4)*

15. 102×5
16. $68 \div 2$
17. 48×50
18. $9,038 \div 3$
19. $388 \div 8$
20. 623×6

Estimate using compatible numbers. *(Lesson 1-5)*

21. $35 \div 8$
22. $1,243 \div 3$
23. $536 \div 4$
24. $723 \div 6$
25. $6,493 \div 7$
26. $567 \div 8$

Find the value of each expression. *(Lesson 1-6)*

27. $2 + 3 - 4$
28. $5 \times 3 + 3$
29. $18 \div 2 \cdot 3$
30. $14 - 3 \times 2$

1-7A Variables and Expressions

A Preview of Lesson 1-7

Objective

Model algebraic expressions.

Materials

paper bags
popcorn kernels

Have you ever seen a magician pull rabbits out of a hat and wonder how many were in there? The number of rabbits in the hat is unknown to us.

In algebra, we often use unknown values. For example, in the phrase *the sum of three and some number,* the "some number" is unknown to us. When you find out the value of "some number", then you can find the sum of three and that number.

Activity One

Work with a partner.

- Each pair will need a paper bag and about 20 popcorn kernels.
- Place 3 kernels and the empty bag in front of you. This represents the phrase *the sum of 3 and some number.* The 3 kernels represent the known value, 3. The bag can contain any number of kernels. Right now, it is empty.

- Have one partner place any number of the remaining kernels in the bag.

- Empty the bag and count all the kernels.

What do you think?

1. What is the "some number"?
2. Complete the phrase *the sum of three and some number* using your known number.
3. How many kernels are there in all?
4. Change the number of kernels in the bag. Complete the phrase using your new number.

Suppose you want to model the phrase *two times some number.* You do not know the value of the number, but you need to show it two times. Therefore, you need two bags to model the phrase.

Activity Two

- Place two paper bags in front of you.
- Have one partner place the same number of kernels in each bag.

- Empty the bags and count all the kernels.

What do you think?

5. What is the "some number"?
6. Complete the phrase *two times some number* using your known number.
7. How many kernels are there in all?
8. How many bags would you need to model the phrase *four times a number?*

Extension

9. How could you model the phrase *4 more than 2 times a number?* How many bags and kernels do you need?
10. Show two ways to model the phrase *the sum of 6 and a number.*

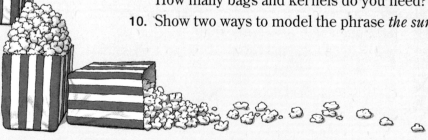

1-7 **Variables and Expressions**

Objective

Evaluate numerical and simple algebraic expressions.

Words to Learn

algebra
variable
algebraic expression
evaluate

Did you know you can estimate how far away a lightning bolt is by counting the number of seconds there are between the lightning and when you hear the thunder? Just count the seconds and multiply by 1,000 feet.

In mathematics, **algebra** is a language that uses symbols. Algebra uses the symbols for addition, subtraction, multiplication, and division that you already know. To figure out the lightning problem above, you can write the relationship as $1,000 \times$ seconds.

You can make a table to show the pattern between the number of seconds after you see the lightning and the number of feet the sound travels.

Feet	Seconds	Distance (Feet)
1,000	× 1	1,000
1,000	× 2	2,000
1,000	× 3	3,000
1,000	× 4	4,000
1,000	× 5	5,000
1,000	× 6	6,000

As lightning gets closer or farther away, the number of seconds you count will vary. A **variable** is a symbol, usually a letter, used to represent a number that changes or varies. Any letter can be used as a variable. In this case, we can use the letter s to stand for seconds. Now we can write the relationship as $1,000 \times s$.

$1,000 \times s$ is called an **algebraic expression.** Algebraic expressions are a combination of variables, numbers, and at least one operation.

The variable s can be replaced with a number so that the value of the expression can be found. If you counted 6 seconds between the lightning and the sound of thunder, you could replace the variable s with 6. The relationship would now be written as $1,000 \times 6$.

Once the variable has been replaced with a number, you can **evaluate,** or find the value of, the expression. The lightning bolt is about $1,000 \times 6$, or 6,000 feet away.

1 Evaluate $17 + x$ if $x = 24$.

$$17 + x = 17 + 24 \qquad \textit{Replace x with 24.}$$
$$= 41 \qquad\qquad \textit{Add 17 and 24.}$$

2 Evaluate $r - s$ if $r = 32$ and $s = 20$.

$$r - s = 32 - 20 \qquad \textit{Replace r with 32 and s with 20.}$$
$$= 12 \qquad\qquad \textit{Subtract 20 from 32.}$$

There is another notation, besides the $\times$ or raised dot, that mathematicians use for multiplication with variables.

$$2a \quad \text{means} \quad 2 \times a \text{ or } \textit{two times a number}$$
$$ab \quad \text{means} \quad a \times b$$

Mental Math Hint

• • • • • • • • • •

Use the order of operations to evaluate.

3 Evaluate $3 + 2a$ if $a = 6$.

$$3 + 2a = 3 + 2 \times 6 \qquad \textit{Replace a with 6.}$$
$$= 3 + 12 \qquad\quad \textit{Multiply 2 and 6.}$$
$$= 15 \qquad\qquad \textit{Add 3 and 12.}$$

4 Evaluate ab if $a = 436$ and $b = 72$.

$$ab = 436 \times 72 \qquad \textit{Replace a with 436 and b with 72.}$$
$$= 31{,}392 \qquad \textit{Multiply 436 and 72.}$$

Checking for Understanding

Communicating Mathematics

Read and study the lesson to answer each question.

1. **Write** an algebraic expression that means the same as $4r$.
2. **Tell** the variables, numbers, and operations in the algebraic expression $3c + 4 - p$.

Guided Practice

Evaluate each expression if $x = 12$.

3. $x - 5$ 4. $x + 7$ 5. $3x$ 6. $x \div 4$

7. $6 + x$ 8. $20 - x$ 9. $48 \div x$ 10. $2x$

Evaluate each expression if $g = 4$ and $h = 10$.

11. $7 - g$ 12. $h + 6$ 13. $g + h$

14. $h \times 2$ 15. $h - 2$ 16. $4g$

Exercises

Evaluate each expression if $b = 3$ and $c = 9$.

17. $b \times 5$ **18.** $c + b$ **19.** $c \div b$ **20.** $c + 3$

21. $b \times 0$ **22.** $15 \div b$ **23.** $2b$ **24.** $c \div 1$

25. $c - 6$ **26.** $9 + b$ **27.** $c - b$ **28.** bc

Evaluate each expression if $m = 2$, $n = 5$, and $p = 10$.

29. $n + 6$ **30.** mn **31.** $40 \div p$ **32.** $p \div m$

33. $4 + 2m$ **34.** $20 \div p$ **35.** $6 + m + n$ **36.** $n \times 0$

37. $2n - 3m$ **38.** $m + n + p$ **39.** $p \div m - 4$ **40.** $3p - 2$

41. Evaluate the expression $p \times 4$ if $p = 9$.

42. Money Juan collected \$38 for the holiday children's fund. Phyllis collected half as much. Use the four-step plan to determine how much money Phyllis collected. *(Lesson 1-1)*

43. Estimate the sum of 61,832 and 75,901 using front-end estimation. *(Lesson 1-3)*

44. Estimate the quotient of 289 and five using patterns. *(Lesson 1-4)*

45. Estimate the quotient of 333 and 5 using compatible numbers. *(Lesson 1-5)*

46. Find the value of ten plus five divided by five. *(Lesson 1-6)*

47. Plumbing While installing water pipes, Mike used pieces of pipe that measured 3 feet, 4 feet, 1 foot, and 2 feet. If Mike cut the pipe from a 14-foot pipe, how much pipe is left?

48. Critical Thinking Philipe and Liliana have calculators. Philipe starts at zero and adds three each time. Liliana starts at 100 and subtracts seven each time. If they press their keys at the same time, will their displays ever show the same number at the same time? If so, what is the number?

49. Make Up a Problem Use at least two variables to make up a problem with an answer of 15.

50. Computer Connection In the BASIC language, LET statements are used to assign values to variables. In the program at the right, line 10 is the LET statement. The computer will replace X with 8, compute $8 + 5$, and print the output, 13. Find the output for each program. The symbol * means multiplication and / means division.

```
10 LET X = 8
20 PRINT X + 5
```

a.
```
10 LET Y = 5
20 PRINT 3*Y
```

b.
```
10 LET B = 10
20 LET A = 4
30 PRINT 2*A + 20/B
```

Lesson 1-7 Algebra Connection : Variables and Expressions   **29**

1-8 Guess and Check

Objective

Solve problems using the guess-and-check strategy.

Each hand in the human body has 27 bones. There are 6 more bones in your fingers than in your wrist. There are 3 less bones in your palm than your wrist. How many bones are in each part of your hand?

Explore   What do you know?

- There are 27 bones in each hand.
- The wrist, palm, and fingers contain bones.
- There are 6 more bones in the fingers than the wrist.
- There are 3 less bones in the palm than the wrist.

What do you need to find?

How many bones are in each part of your hand?

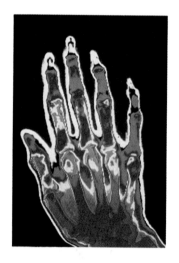

Plan   Begin by guessing the number of bones in the wrist. Then check your guess against the total number of bones in the hand. If the guess doesn't check, use what you know to make another guess.

Solve   Guess the number of bones in the wrist.
Try 5.

	Wrist	Fingers	Palm	Total	
	?	(+6)	(−3)	(27)	*The guess is too low because 18 is less than 27.*
Guess	5 +	11 +	2 =	18	

					The guess is too high because 33 is greater than 27.
Try 10.					
Guess	10 +	16 +	7 =	33	

Pick a number between 5 and 10. Try 8.

Guess	8 +	14 +	5 =	27	*The guess checks.*

Examine   There are 8 bones in the wrist, 14 in the fingers, and 5 in the palm. The answer makes sense because the answer shows that there are 6 more bones in the fingers than the wrist and 3 less bones in the palm than the wrist.

Checking for Understanding

Communicating Mathematics

Read and study the lesson to answer each question.

1. **Tell** why it is important to use the information you learn from each guess when using the guess-and-check strategy.

2. **Write** a brief explanation of how the guess-and-check strategy can be used to predict the most likely outcome of this tic-tac-toe game in progress.

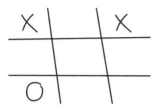

Guided Practice

Use the guess-and-check strategy to find each number.

3. The product of a number and itself is 121.

4. The sum of a number and its double is 27.

5. Ten more than a number is twice the number.

6. The product of a number and 23 is 138.

Problem Solving

Practice

Solve using any strategy.

7. Each hair on the scalp grows about 5 inches each year. About how many inches will your hair grow in 3 years?

Strategies

Look for a pattern.
Solve a simpler problem.
Act it out.
Guess and check.
Draw a diagram.
Make a chart.
Work backward.

8. Rita went to Kenneth's Hair Salon. She spent $48 for a perm, $16 for a manicure, and $9 for a bottle of spray gel. Estimate the amount of money Rita spent at Kenneth's.

9. Admission to St. Ann's health fair was $6 for adults, $4 for children under age 12, and $3 for senior citizens. Twelve people paid a total of $50 for admission. If eight children attended, how many adults and senior citizens were in the group?

10. Allison pays $124 each month on a bank loan. How much money does Allison pay in one year?

11. Jason earns $3 per hour babysitting. Last month, he worked 20 hours. How much money did Jason make?

12. Last week, the Delaware Animal Shelter sent a total of 21 cats and dogs to new homes. There were 5 more cats than dogs. How many of each were adopted?

DATA SEARCH

13. **Data Search** Refer to pages 2 and 3. Which foods have had the greatest changes in consumption over the years?

1-9 Solving Equations

Objective

Solve equations by using mental math and the guess-and-check strategy.

Words to Learn

equation
solve
solution

___?___ is the capital of Kentucky.

This sentence is neither true nor false until you fill in the blank. If you say Frankfort, the sentence is true. If you say Louisville, or any other city, the sentence is false.

In mathematics, a sentence that contains an equals sign, =, is called an **equation.**

$$22 + 10 = 32 \qquad \textit{This equation is true.}$$
$$10 - 2 = 6 \qquad \textit{This equation is false.}$$

An equation can contain a variable.

$$x - 5 = 7$$

This equation is neither true nor false until x is replaced with a number. When you replace the variable with a number that makes the equation true, you **solve** the equation. Any number that makes the equation true is called a **solution.** The solution of $x - 5 = 7$ is 12 since $12 - 5 = 7$.

Example 1

Which of the numbers 4, 5, or 6 is the solution of $x + 5 = 11$?

Replace x with 4.

$$x + 5 = 11$$
$$4 + 5 \stackrel{?}{=} 11$$
$$9 = 11$$

This sentence is false.

Replace x with 5.

$$x + 5 = 11$$
$$5 + 5 \stackrel{?}{=} 11$$
$$10 = 11$$

This sentence is false.

Replace x with 6.

$$x + 5 = 11$$
$$6 + 5 \stackrel{?}{=} 11$$
$$11 = 11$$

This sentence is true.

The solution is 6.

The equation in the example above was solved by replacing the variable, x, with each number until a true sentence was found.

Sometimes, you can use the strategy you learned in the previous lesson, guess-and-check, to solve equations.

Example 2 *Problem Solving*

Estimation Hint

● ● ● ● ● ● ● ● ● ●

To solve this equation, start with a guess. Since 78 is about 80 and 104 is about 100, think: 80 + 20 = 100 so the answer should be close to 20.

Consumer Math   Tomás purchased a pair of rollerblades for $78. He also bought a helmet. He spent a total of $104. If h represents the cost of the helmet, the equation $78 + h = 104$ represents the total cost of the two items. What is the cost of the helmet?

$78 + h = 104$	$78 + h = 104$	$78 + h = 104$
$78 + 20 \stackrel{?}{=} 104$	$78 + 23 \stackrel{?}{=} 104$	$78 + 26 \stackrel{?}{=} 104$
$98 = 104$	$101 = 104$	$104 = 104$

This sentence is false.   This sentence is false.   This sentence is true.

The solution is 26. The helmet costs $26.

Some equations also can be solved mentally by using basic facts or arithmetic skills you already know well.

Example 3

Mental Math Hint

● ● ● ● ● ● ● ● ● ●

You can use a multiplication table to help you solve mental math problems.

Solve $3x = 12$ mentally.

$3 \cdot 4 = 12$ *You know $3 \cdot 4 = 12$.*
   $12 = 12$

The solution is 4, so the value of x is 4.

Example 4

Solve $t = 9 + 7$ mentally.

$16 = 9 + 7$ *You know that $9 + 7 = 16$.*
$16 = 16$

The solution is 16, so the value of t is 16.

Checking for Understanding

Communicating Mathematics   Read and study the lesson to answer each question.

1. **Tell** the solution of $x + 3 = 8$.

2. **Write** a definition of *equation*.

Guided Practice   Tell whether the equation is *true* or *false* by replacing the variable with the given value.

3. $r - 12 = 40; r = 52$

4. $m + 7 = 84; m = 77$

5. $n \div 6 = 42; n = 252$

6. $14 + p = 39; p = 53$

Name the number that is a solution of the given equation.

7. $4 - t = 1; 2, 3, 5$ **8.** $y + 37 = 64; 25, 26, 27$

9. $14p = 56; 3, 4, 5$ **10.** $a = 9 + 42; 41, 51, 61$

Solve each equation using basic facts.

11. $2 + x = 7$ **12.** $7x = 14$

13. $b - 3 = 6$ **14.** $6 + 6 = s$

15. $r = 5 + 8$ **16.** $8 \div y = 2$

Exercises

Independent Practice

Name the number that is a solution of the given equation.

17. $22 + c = 30; 6, 8, 9$ **18.** $e - 26 = 14; 12, 40, 364$

19. $t \div 3 = 18; 54, 55, 56$ **20.** $9x = 108; 12, 13, 14$

21. $11 = d \div 4; 41, 43, 44$ **22.** $64 = 16a; 3, 4, 5$

23. $45 - 9 = w; 36, 37, 38$ **24.** $91 = 7x; 11, 12, 13$

Solve each equation mentally.

25. $a = 3 + 11$ **26.** $y = 12 - 6$ **27.** $x - 6 = 13$

28. $m + 4 = 11$ **29.** $4 + n = 12$ **30.** $13 = p + 8$

31. $3n = 21$ **32.** $18 = 2x$ **33.** $6q = 42$

Mixed Review

34. Distance The steamer *Alaska* traveled 182 miles in 5 days. Use rounding to estimate how many miles it averaged each day. *(Lesson 1-2)*

35. Use patterns to estimate the product of 792 and 11. *(Lesson 1-4)*

36. Use compatible numbers to estimate 137 divided by 5. *(Lesson 1-5)*

37. Evaluate $3x - y$, if $x = 4$ and $y = 1$. *(Lesson 1-7)*

Problem Solving and Applications

38. Banking Currency rates are calculated daily when exchanging money between countries. If $200d$ indicates the number of Greek drachmas for each United States dollar represented by d, how many drachmas will be received in exchange for $50?

39. Critical Thinking Translate each sentence into an equation and solve.

     a. *Nine less than w is equal to 6.*

     b. *The sum of r and 7 is equal to 22.*

     c. *The product of 3 and 13 is equal to s.*

     d. *The difference of t and 16 is equal to 11.*

40. Smart Shopping Rosilyn is saving for a pair of jeans that costs $42. She has saved $16. How much more money does Rosilyn need to save?

1-10 Classifying Information

Objective

Solve problems by classifying information.

In 1990 and 1991, 541,000 American soldiers were sent to the Persian Gulf in the Middle East. Never before had so many women gone to war. They served in jobs such as ferry pilots, shipboard navigators, and truck drivers. If 509,000 men served in the Persian Gulf, how many women served?

Explore

What do you know?

- 541,000 Americans were sent to the Persian Gulf.
- 509,000 American men served in the Persian Gulf.
- Women served as ferry pilots, shipboard navigators, and truck drivers.

What do you need to find?

You need to find out how many women served in the Persian Gulf.

Plan

Classify the information into what you need to know to solve the problem and what you do not need to know.

What you need to know:

- 541,000 Americans were sent to the Persian Gulf.
- 509,000 Americans were men.

What you do not need to know:

- Women served in jobs such as ferry pilots, shipboard navigators, and truck drivers.

Lesson 1-10 Problem-Solving Strategy: Classifying Information **35**

Solve 541,000 − 509,000 = 32,000

32,000 women served in the Persian Gulf.

Examine The number of men plus the number of women equals the total number of Americans who served in the Persian Gulf. The answer makes sense. The extra information was not needed to solve the problem.

Checking for Understanding

Communicating Mathematics

Read and study the lesson to answer each question.

1. **Tell** how you know if a problem contains extra information.
2. **Write** what you might do to solve a problem that does not have enough information.

Guided Practice

Solve by classifying information.

3. A 5-CD package is on sale for $100 at SchoolKids Music. The sale lasts until Friday. How much is each CD?

4. Isabella spent $125 last month on piano lessons. How many lessons did she have?

5. A single-sided, single density $5\frac{1}{4}$-inch computer disk has 180K of storage space. A double-sided, double density $5\frac{1}{4}$-inch disk holds 360K of information. A double-sided, double density $3\frac{1}{2}$-inch disk holds 720K of information. How much more information does the $3\frac{1}{2}$-inch disk hold than the double density $5\frac{1}{4}$-inch disk?

Problem Solving

Practice

Solve using any strategy.

6. The fastest jet aircraft in the world is believed to be the Lockheed SR-71A, a military airplane formerly flown by the United States Air Force. It holds the air speed record of 2,190 miles per hour. How many miles will the aircraft travel in five hours?

Strategies

• • • • • • •

Look for a pattern.
Solve a simpler problem.
Act it out.
Guess and check.
Draw a diagram.
Make a chart.
Work backward.

7. The smallest animals are protozoans. They are so tiny that they can only be seen with a powerful microscope. If 5,000 protozoans measure about one-half inch, how many inches would 100,000 protozoans measure?

8. Matthew listens to the radio for four hours every day after school. How many hours does he listen to the radio each week?

9. Carla spent $58 on a sweater and $34 on shoes. How much did she spend in all?

10. Blue whales are the largest animals that ever lived. Blue whales have the biggest babies. A baby blue whale is about 25 feet long when it is born. It grows to more than 100 feet in length as an adult and weighs about 150 tons. How much more does a grown blue whale weigh than a baby blue whale?

11. Mical spent $1.29 for apples, $4.59 for chicken breasts, and $3.67 for potatoes. About how much change should he receive from a $20-dollar bill?

Save Planet Earth

Radon in and Around the Classroom In 1989, the Environmental Protection Agency (EPA) urged all school buildings to be tested for radon contamination. Radon is a radioactive gas. Basements and rooms on the first floor have a greater risk of radon contamination. Poor ventilation in schools may mean that the contaminated air enters the rooms from the foundation of the building and builds up instead of leaving.

The EPA expects that at least 10 percent of schools in this country have a radon problem.

How You Can Help

- Encourage school officials to test frequently-used rooms such as the basement, ground-level floors, gymnasiums, cafeterias, libraries, and offices on a regular basis.

- Conduct tests in the winter months when doors and windows are likely to be closed.

1 Study Guide and Review

Communicating Mathematics

Fill in the blanks with the appropriate term(s).

1. In mathematics, ___?___ is a language using symbols.
2. The expression $2a$ means 2 ___?___ a.
3. An equation is a sentence in mathematics that contains an ___?___ sign.
4. Any number that makes the equation true is a ___?___ .
5. In the equation $x + 6 = 11$, x is called the ___?___ .

State whether each statement is true or false.

6. The solution of the equation $x + 6 = 11$ is 11.
7. $14 - 2 \times 3$ is 8.
8. A reasonable answer for $\$579 + \$712$ is $\$1,491$.
9. Explain the fourth step of the four-step plan and why it is important.
10. Amy and Sachiko both worked the problem $809 + 468$ using estimation. Explain why Amy's answer, 1,260, differs from Sachiko's, which is 1,300.

Skills and Concepts

Objectives and Examples	Review Exercises
Upon completing this chapter, you should be able to:	*Use these exercises to review and prepare for the chapter test.*
• solve problems using the four-step plan *(Lesson 1-1)* The four steps are *Explore, Plan, Solve,* and *Examine.*	Use the four-step plan to solve. 11. **History** In the presidential election of 1932, Roosevelt won with 472 electoral votes, while Hoover had 59. How many electoral votes were cast?
• round numbers *(Lesson 1-2)* To the nearest hundred, 5,769 is 5,800 because 6 is greater than 5. To the nearest thousand, 7,218 is 7,000 because 2 is less than 5.	Round as indicated. 12. 3,257 to the nearest hundred 13. 5,287 to the nearest thousand
• estimate sums and differences using rounding *(Lesson 1-2)* $\$429 + \$789$ rounded to the nearest hundred is $\$400 + \$800$ or $\$1,200$.	14. **Smart Shopping** On a shopping trip, Emil buys a coat for $\$91$, a shirt for $\$17$, and jeans for $\$24$. The cashier asks for $\$163$. Is this reasonable?

Objectives and Examples

- estimate sums and differences using front-end estimation *(Lesson 1-3)*

$$\begin{array}{r} 523 \\ + 379 \end{array} \qquad \begin{array}{r} 523 \\ + 379 \\ \hline 800 \end{array} \qquad \begin{array}{r} 523 \\ + 379 \\ \hline 90 \end{array}$$

$800 + 90 = 890$

- estimate products and quotients using patterns *(Lesson 1-4)*

239×3
$\qquad 2 \times 3 = 6$
$\qquad 20 \times 3 = 60$
$\qquad 200 \times 3 = 600$

- estimate quotients using compatible numbers *(Lesson 1-5)*

$440 \div 9 \;\rightarrow\; 450 \div 9$

450 and 9 are compatible.
$450 \div 9 = 50$

- evaluate expressions using the order of operations *(Lesson 1-6)*

Multiply and divide left to right, then add and subtract left to right.
$$6 + 12 \div 3 = 6 + 4 = 10$$

A raised dot means multiply.
$$5 \cdot 7 = 5 \times 7 = 35$$

- evaluate numerical and simple algebraic expressions *(Lesson 1-7)*

If $a = 7$ and $b = 14$, then
$a + b = 7 + 14 = 21$.

- solve equations using mental math and the guess-and-check strategy *(Lesson 1-9)*

The solution of $m - 2 = 7$ is 9, because $9 - 2 = 7$ is a true statement.

Review Exercises

Estimate using front-end estimation.

15. $\begin{array}{r} 498 \\ + 356 \end{array}$ **16.** $\begin{array}{r} 3{,}479 \\ - 2{,}105 \end{array}$ **17.** $\begin{array}{r} 1{,}347 \\ + 7{,}792 \end{array}$

Estimate using patterns.

18. $317 \div 6$

19. 42×707

Estimate using compatible numbers.

20. $271 \div 4$

21. $6{,}231 \div 7$

22. $711 \div 8$

Find the value of each expression.

23. $9 - 3 \cdot 2 - 1$

24. $3 \cdot 7 + 5 \times 9 \div 3$

25. Consumer Math   Write an expression that would indicate the cost of 3 T-shirts at \$7 each and four sweatshirts at \$11 each.

Evaluate each expression if $a = 4$, $b = 3$, and $c = 12$.

26. $c + a$

27. $21 \div b$

Name the number that is a solution of the equation.

28. $14a = 42$; 3, 4, 5

29. $b + 23 = 31$; 7, 8, 9

Solve each equation mentally.

30. $72 \div x = 8$ **31.** $m - 13 = 11$

Applications and Problem Solving

32. **Health** Together, two brothers weigh 300 pounds. One weighs 50 pounds more than the other. How much does each brother weigh? *(Lesson 1-8)*

33. **Transportation** The fastest railway trains can reach a top speed of 170 miles per hour. How many hours would it take to travel from Newark to Cleveland on one of these trains? *(Lesson 1-10)*

34. **History** John F. Kennedy was first elected to the House of Representatives in 1946. How long had he been in the legislature when he was elected president in 1961? *(Lesson 1-1)*

35. **Travel** Kanya and her family went on a vacation with a tour group to Hawaii. A total of 63 people went on the trip at a cost of $964 each. About how much money was collected in all? *(Lesson 1-4)*

36. **Business** For the week of October 21, 1991, the New York Stock Exchange traded 882,460,000 shares of stock. For the same week in 1990, 719,890,000 shares were traded. About what was the difference? *(Lesson 1-3)*

37. **Jobs** A survey of over 6,000 U.S. doctors showed that their average income increased $15,000 in 1989 to $133,000 per year. Use front-end estimation to estimate what their average income was in 1988. *(Lesson 1-3)*

38. Paul's father is 4 times as old as Paul. His grandfather is twice as old as Paul's father. If the sum of their ages is 104, how old is Paul, his father, and his grandfather? *(Lesson 1-8)*

Curriculum Connection Projects

- **Consumer Awareness** From ads in tonight's newspaper, make a list of the prices of cars, groceries, clothes, and so on. Round off each price in a second column.

- **Physical Education** Identify and write down one or more orders of operation for each of three sports or athletic activities.

Read More About It

Burns, Marilyn. *The Book of Think (or How to Solve a Problem Twice Your Size).*
Konisburg, E.L. *From the Mixed-Up Files of Mrs. Basil E. Frankweiler.*
Montgomery, R.A. *The Mystery of the Maya.*

Study Guide and Review

1 Test

1. **Geography**   Rick and Jason went hiking and canoeing down the Danube River, Europe's second longest river. They traveled from its source in the Black Forest 1,760 miles to its end at the Black Sea. If they averaged 11 miles each day, how long was the trip?

Round to the nearest hundred.

2. 754

3. 4,839

Round to the nearest thousand.

4. 4,839

5. 7,070

Estimate using any strategy.

6. $\begin{array}{r} 1,492 \\ + 1,941 \\ \hline \end{array}$

7. $\begin{array}{r} 7,356 \\ - 4,384 \\ \hline \end{array}$

8. $198 \div 5$

9. 28×41

10. **Consumer Math**   Hector and Carlos have decided to put their money together to buy a new video game. Hector has $19 and Carlos has $12. About how much can they spend?

11. **Weather**   Throughout the world, 16,000,000 thunderstorms occur each year. About how many thunderstorms occur each week?

Find the value of each expression.

12. $48 - 24 \div 3 \times 6$

13. $63 \div 7 - 2 \times 3$

Evaluate each expression if $m = 2$, $n = 7$, and $p = 21$.

14. $p - m$

15. $m + p \div n$

Name the number that is a solution of the given equation.

16. $32 \div x = 2$; 16, 30, 64

17. $y + 17 = 23$; 5, 6, 7

Solve each equation mentally.

18. $5z = 55$

19. $15 - w = 9$

Solve.

20. Music To Go sold 75 rock tapes, 64 rap tapes, and 43 jazz tapes on Wednesday. How many jazz and rock tapes were sold?

Bonus   **Geology**   Around the world, erosion by rivers averages 12 inches every nine thousand years. Charles studied two rivers. One had eroded 36 inches, and the other had eroded 72 inches. About what is the difference in the ages of the rivers?

Graphs and Statistics

Spotlight on Population

Have You Ever Wondered...

- What it means when people say that the world population is growing at a faster and faster rate?

- What it means when newspapers report that the median age of the United States population is increasing?

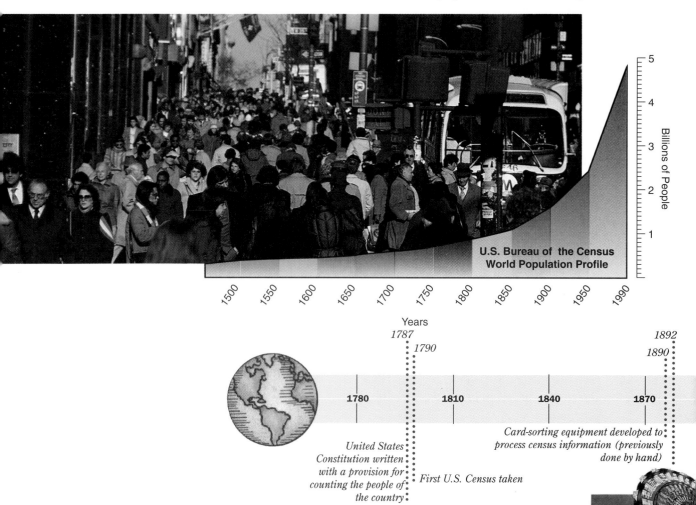

U.S. Bureau of the Census
World Population Profile

Billions of People

1500 1550 1600 1650 1700 1750 1800 1850 1900 1950 1990

Years
1787
1790
1892
1890
1780   1810   1840   1870

United States Constitution written with a provision for counting the people of the country

First U.S. Census taken

Card-sorting equipment developed to process census information (previously done by hand)

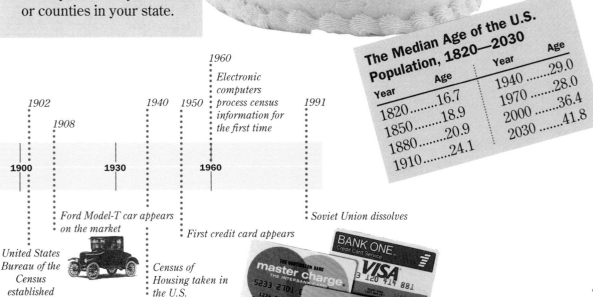

Looking Ahead

In this chapter, you will see how mathematics can be used to answer questions about the population.

The major objectives of the chapter are to:

- read and write numbers through trillions
- solve problems by interpreting bar graphs and pictographs
- organize data into tables
- construct line graphs and bar graphs
- find the mean, median, and mode of a set of data

Chapter Project

Population
Work in a group.

1. Make a map or chart showing the populations of the major cities or counties in your state.

2. Find out the area of one city or county.

3. Divide the population by the area to find the population density, or the number of people per square mile.

4. Make a class graph showing the population density of the major cities or counties in your state.

The Median Age of the U.S. Population, 1820—2030			
Year	Age	Year	Age
1820	16.7	1940	29.0
1850	18.9	1970	28.0
1880	20.9	2000	36.4
1910	24.1	2030	41.8

1902

1908

1940 *1950*

1960
Electronic computers process census information for the first time

1991

1900 **1930** **1960**

Ford Model-T car appears on the market

First credit card appears

Soviet Union dissolves

United States Bureau of the Census established

Census of Housing taken in the U.S.

2-1 **Reading Greater Numbers**

Objective

Read and write numbers through trillions.

Words to Learn

place value

The English language uses letters and their position to name or convey meaning. The same letters in different positions have different meanings.

For example:

Cats means domesticated mammals often kept as pets.

Cast means something that is formed by casting in a mold or form.

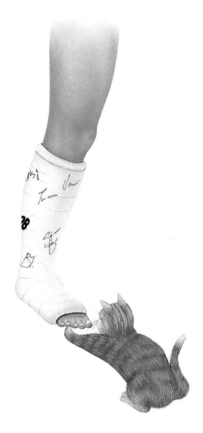

The language of mathematics uses digits and **place-value** positions to name numbers. Just like the English language, the same digits in different place-value positions have different meanings.

The digit 4 and its place-value position in 254,380 name the number four thousand or 4,000.

The digit 4 and its place-value position in 4,572,001 name the number four million or 4,000,000.

Trillions			Billions			Millions			Thousands			Ones		
Hundred Trillions	Ten Trillions	Trillions	Hundred Billions	Ten Billions	Billions	Hundred Millions	Ten Millions	Millions	Hundred Thousands	Ten Thousands	Thousands	Hundreds	Tens	Ones
									2	5	4,	3	8	0
								4,	5	7	2,	0	0	1

254,380 and 4,572,001 are written in standard form. Both of these numbers can also be written in words.

Standard Form		**Words**
254,380	→	two hundred fifty-four thousand, three hundred eighty
4,572,001	→	four million, five hundred seventy-two thousand, one

1 Write the place-value position of the digits 1 and 8 in 35,716,904,820,420.

1 is in the ten billions place-value position.
8 is in the hundred thousands place-value position.

2 Write each number in words.

Standard Form	Words
a. 400,010,200	four hundred million, ten thousand, two hundred
b. 403,000,060,000,000	four hundred three trillion, sixty million

Numbers in standard form can be renamed by using place-value names.

$$20,000 = 20 \text{ thousand} \qquad 50,000,000 = 50 \text{ million}$$

Checking for Understanding

Communicating Mathematics

Read and study the lesson to answer each question.

1. **Tell** the place-value position of 4 in 3,647,270,012.
2. **Write** a number with a 6 in the hundred thousands place-value position.
3. **Write** the number 73,405,125,630,089 in words.

Guided Practice

Write the place-value position for each digit in 572,049,736,168,776.

4. 9   5. 5   6. 8   7. 0
8. 3   9. 2   10. 1   11. 4

Write each number or amount in words.

12. 74,000,000
13. 30,004
14. 6,860,030
15. 2,031,000,500,000
16. $5,000
17. $76,908,000

Replace each ■ with a number to make a true sentence.

18. 50,000 = ■ hundreds
19. 370 = ■ tens
20. 46,000,000 = ■ hundred thousands
21. 89,000,000,000,000 = ■ billions

Exercises

Write the place-value position for each digit in 620,390,071,504,850.

22. 4   23. 8   24. 9   25. 1

26. 6   27. 7   28. 3   29. 2

Write each number in words.

30. 23,600   31. 5,082

32. 370,008,000   33. 9,031,000,000

34. 321   35. 72,000,800,000,000

Write each number in standard form.

36. eighty-two million   37. seven trillion

38. four hundred thousand   39. sixty-eight

40. twenty-one trillion   41. seven billion

Replace each ■ with a number to make a true sentence.

42. 39,000 = ■ hundreds

43. 76,000,000 = ■ hundred thousands

44. 2,800,000,000,000 = ■ millions

45. ■ = 14 millions

46. **School**   The number of students at Aurora Junior High is half the number it was twenty years ago. If there were 758 students twenty years ago, how many students are there now? *(Lesson 1-1)*

47. Using rounding, estimate $398 + 688 + 241$. Is 1,337 a reasonable answer? *(Lesson 1-2)*

48. Estimate 632 divided by 76 using compatible numbers. *(Lesson 1-5)*

49. Find the value of $23 - 6 + 2 \times 5$ *(Lesson 1-6)*

50. Evaluate $m - 3p$ if $m = 22$ and $p = 7$. *(Lesson 1-7)*

51. Solve $17 = 9 + z$ mentally. *(Lesson 1-9)*

52. **Banking**   As a safeguard against error, the dollar amount on a check is written in both standard form and in words. Write a dollar amount of $5,732.35 in words.

53. **Science**   Astronomers use the light-year, the distance light travels in one year, as a unit to measure distance to the stars. A light-year is about 6 trillion miles. Write 6 trillion in standard form.

54. **Critical Thinking**   Which number is greater, 9,120,000 or 9,210,000? Why?

55. **Journal Entry**   Find an example in a newspaper of a number written partially in words, such as 50 thousand. Write your own newspaper headline and article using this number.

2-2 Use a Graph

Objective

Solve problems by interpreting bar graphs and pictographs.

Words to Learn

statistics
bar graph
pictograph

In 1989, 11,556 consumer magazines were published. Organizations like the Audit Bureau of Circulations (A.B.C.) collect, organize, and summarize **statistics** about the magazine industry. These statistics help publishers figure out which magazines are successful. Statistics involves compiling, analyzing, and presenting numerical data.

Bar graphs are used to compare numbers. The bar graph below shows magazine revenues for several popular magazines.

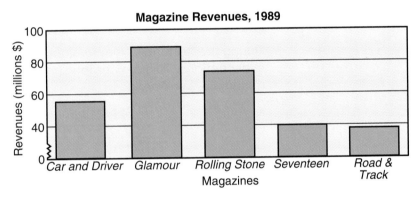

The broken line at the bottom of the vertical scale means that part of the scale has been left out.

About how much higher are the revenues of *Car and Driver* than *Road & Track?*

Explore   Before you try to solve the problem, look at the graph closely.

What do the numbers on the left side of the graph mean?

What numbers are missing along this scale?

How do you know that *Car and Driver* has higher revenues than *Road & Track?*

Plan   You can use estimation to solve this problem.

First, estimate the revenues for *Car and Driver.* What amount is represented by the bar above this magazine title?

Then estimate the revenues for *Road & Track.*

Find the difference between these two amounts.

Solve Car and Driver - about $55 million
Road & Track - about $40 million

$55 million − $40 million = $15 million

Car and Driver revenues are about $15 million more than *Road & Track*.

Examine The scale jumps from $0 to $30 million to make the graph easier to read. The bar for *Car and Driver* shows about $15 million more than the bar for *Road & Track*, so *Car and Driver's* revenues are higher than the revenues of *Road & Track*.

A **pictograph** is another type of graph used to compare numbers.

Example

During the fall fund-raising project, Robert's class sold magazine subscriptions. This pictograph shows how many magazines they sold during a 6-week period.

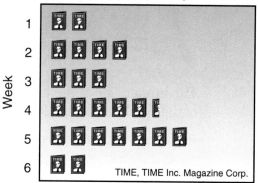

Subscriptions sold per Week

TIME, TIME Inc. Magazine Corp.

= 4 subscriptions

a. **How many subscriptions were sold during week three?**

Locate week three on the left side of the pictograph. There are 3 . Each means 4 subscriptions, so they sold 12 subscriptions during week three.

b. **What does the half-magazine shown during week four mean?**

Since each means 4 subscriptions, a half-magazine means 2 subscriptions.

c. **During what week did the class sell 16 subscriptions?**

Each means 4 subscriptions, so 16 subscriptions means 4 . They sold 16 subscriptions during week two.

Checking for Understanding

Communicating Mathematics

Read and study the lesson to answer each question.

1. **Tell** why some graphs have a broken line on the scale between 0 and the first interval.
2. **Tell** why the publishers of *Road & Track* magazine would be more concerned about the gap in revenues between their publication and *Car and Driver* than with *Rolling Stone.*

Guided Practice

Solve by using the bar graph on page 47.

3. About how much higher were *Glamour* magazine's revenues than *Seventeen's?*
4. Which magazine has the highest revenue? About how much more revenue did this magazine have than *Rolling Stone?*

Problem Solving

Practice

Solve using any strategy.

5. During lunch, the cafeteria sold 123 cartons of 2% milk, 50 cartons of orange juice, and 25 cartons of grape juice. About how many cartons of drinks did they sell?

Strategies

● ● ● ● ● ● ●

Look for a pattern.
Solve a simpler problem.
Act it out.
Guess and check.
Draw a diagram.
Make a chart.
Work backward.

6. Jared bought a backpack for $24.99. How much change did he receive?
7. Find the number if the product of the number and 28 is 252.
8. Regina drives 345 miles per week. How far does she drive in four weeks?

9. Mrs. Jiminez bought 10 packages of batteries for $6.00 each. How much did she spend on batteries?
10. Elaine's brother Chuck is 6 years older than Elaine. Her mother is 3 times as old as Chuck. The sum of their ages is 69. How old is each family member?
11. In the Hornets' first basketball game of the season, they scored 37 points in the first half and 19 points in the second half. How many points did they score in all?
12. Laura collected $2.00 from each student to buy a gift for their teacher. If 29 people contributed, how much money was collected?
13. Rivera High School will graduate 789 seniors on June 8. The ceremony will be held in the school gymnasium. The gymnasium holds 2,400 people in addition to the graduates. Is it reasonable to offer each graduate three tickets for family and friends?

2-3 Make a Table

Objective

Solve problems by organizing data into a table.

Words to Learn

frequency table

Marta wants to work as a literacy volunteer at her local library. She helps adults learn how to read. She gathered data about volunteer workers. During 1989, the total number of volunteers 16 years old and older was 38,142. Of this number, 16,680 were men and 21,462 were women.

In order to see the relationship between age and sex of volunteer workers, Marta decided to organize the information in a table. She wanted to know about how many more women than men in the 25-34 age range were volunteers.

Volunteer Workers, 1989

Age	Men	Women
16-19	879	1,023
20-24	935	1,129
25-34	3,678	5,002
35-44	4,683	5,665
45-54	2,601	3,069
55-64	1,987	2,468
65+	1,917	3,106
Totals	16,680	21,462

Explore   How is the table organized?
Is the information easy to locate?

What are you asked to find?
About how many more women than men in the 25-34 age range worked as volunteers?

Plan   You can estimate the difference. Round each number to the nearest thousand.

Solve   Find the 25-34 age range in the table.

Women: 5,002 rounds to 5,000
Men: 3,678 rounds up to 4,000.

5,000 − 4,000 = 1,000

Examine   In this age range, about 1,000 more women than men worked as volunteers.

Sometimes you need to organize data in a way that makes it easy to study the results. You can use a **frequency table.**

Coach Franklin makes up the starting lineup before the start of each baseball game. He looks at the number of hits each player has made the previous game before deciding in which order they will bat.

Number of Hits

Player	Tally	Frequency
Burns	I	1
Parker	III	3
Martinez	II	2
Cruz	IIII	4
Plesich	II	2
Higgins	III	3
Reid	I	1
Hartley		0
Wilson	II	2

To make a frequency table:
- Draw a table with three columns.
- In the first column, list the items in the set of data.
- In the second column, mark the tallies.
- In the third column, write the frequency or number of tallies.

Checking for Understanding

Communicating Mathematics

Read and study the lesson to answer each question.

1. **Tell** why it is important to organize data.

2. **Write** a sentence telling what each column represents in a frequency table.

Guided Practice

3. Copy the table and complete the frequency column.

Favorite TV Show

TV Show	Tally	Frequency
The Simpsons	꧖꧖꧖꧖	⸻
Wonder Years	꧖꧖꧖꧖ III	⸻
Full House	꧖꧖꧖꧖ ꧖꧖꧖꧖ I	⸻
Life Goes On	꧖꧖꧖꧖ III	⸻

Problem Solving

Practice

Solve using any strategy.

4. Kendra spent $5.69 on makeup. About how much change could she expect to receive if she paid with a $20 bill; $4 or $14?

5. David works five hours per week after school. How many hours does he work in six weeks?

6. Survey your class to find their favorite radio station. Organize the data in a frequency table.

2-3B Data Bases

A Follow-Up of Lesson 2-3

Objective

Explore the uses of a computer data base.

Words to Learn

data base
field
record

A **data base** is a collection of data that is organized and stored. The data can be added to, updated, changed, and retrieved easily. The data can be listed from greatest to least, least to greatest, as well as put in alphabetical order.

The data base below contains employee information. Each column heading in a data base is called a **field**. Each row of information is called a **record.**

NUMBER	NAME	RATE	HOURS	EARNINGS
155	C. SMITH	$15.50	35	$542.50
290	M. JONES	$18.25	30	$547.50
265	S. GARCIA	$17.00	40	$680.00
311	G. AVILLO	$22.50	32.5	$731.25
189	R. GREEN	$12.50	45.5	$568.75

What do you think?

1. How many fields are there in the data base above?

2. What are the fields?

3. How many records does the data base contain?

4. As the new employee manager, suppose you wanted to know which employee has the highest rate. Arrange the data so that the rate appears in order from greatest to least.

5. Suppose you needed to update your employee data base by adding the three new employees below. Write a new data base, arranging the entries from greatest employee number to the least.

NUMBER	NAME	RATE	HOURS	EARNINGS
142	R. HINES	$15.50	33	$511.50
255	P. SEUSS	$16.75	30	$502.50
301	B. MCKAY	$14.25	40	$570.00

Extension

6. Create your own data base about your family. You could include names, years of birth, addresses, telephone numbers, or important dates such as birthdays.

2-4 Range and Scales

Objective

Choose appropriate scales and intervals for frequency tables.

Words to Learn

scales
range
interval

One of Carl's favorite pastimes is playing the video game Dr. Mario™. His last sixteen scores were: 10,200; 3,500; 65,300; 56,000; 28,200; 17,000; 43,200; 23,300; 41,900; 32,000; 29,100; 46,400; 25,000; 37,800; 49,700; and 34,300.

Carl has decided to organize this data into a frequency table so it is easier to study his scores.

Carl needs to determine the **scale** for his frequency table. To do this, Carl first needs to find the **range** of his scores.

Range	The range of a set of numbers is the difference between the least number and the greatest number in the set.

Problem Solving Hint

● ● ● ● ● ● ● ● ● ●

Carl is using the problem-solving strategy make a table.

Carl's greatest score is 65,300. His least score is 3,500.

range   → $65,300 - 3,500 = 61,800$

The range of Carl's scores is 61,800.

Mental Math Hint

● ● ● ● ● ● ● ● ● ●

Scale intervals are usually one, ten, one hundred, one thousand, and so on.

The next step is to choose an **interval** for the data. The intervals separate the scale into equal parts. Since the data includes scores from 3,500 to 65,300, Carl decides to use intervals of 10,000. Carl begins his scale with zero and ends with 70,000. Using this scale, all of his scores will be included in his table.

Carl's Dr. Mario™ Scores		
Score	**Tally**	**Frequency**
60,100 - 70,000	I	1
50,100 - 60,000	I	1
40,100 - 50,000	IIII	4
30,100 - 40,000	III	3
20,100 - 30,000	IIII	4
10,100 - 20,000	II	2
0 - 10,000	I	1

The scale must contain the range of the scores. The scale is appropriate since the least score is 3,500 and the greatest is 65,300.

Now Carl can easily see that most of his scores were between 20,100 and 50,000.

1 Find the range for this set of mathematics test scores.
69, 95, 15, 36, 75, 81, 28, 7, 54, 23, 39, 43, 79, 60, 58, 72, 88, 62, 48, 67, 37, 68, 55, 49, 58, 70, 83, 79, 65, 45

The greatest score is 95. The least score is 7.
The range is 95 − 7 or 88.

2 Use the scores in Example 1 to make a frequency table.

Choose a scale from 0 to 100 since the least score is 7 and the greatest is 95. To keep the number of intervals small, an interval of 10 is used.

Mathematics Scores		
Score	**Tally**	**Frequency**
91-100	I	1
81-90	III	3
71-80	IIII	4
61-70	JHT I	6
51-60	JHT	5
41-50	IIII	4
31-40	III	3
21-30	II	2
11-20	I	1
0-10	I	1

 Mini-Lab

Work with a partner.
Materials: paper, pencil

- Survey ten classmates. Find the number of hours a week each person spends on the telephone.
- Find the range.
- Use the range to choose the scale and intervals.
- Make a frequency table.

Talk About It

a. What does your frequency table tell you?

b. Compare your frequency table with a frequency table from another pair of classmates. Are the scale and intervals the same? Explain why or why not.

Checking for Understanding

Communicating Mathematics

Read and study the lesson, and use the frequency table at right to answer each question.

Team 6-2		
Number of Absences	Tally	Students
5	I	1
4		0
3	III	3
2	II	2
1	⊬⊬⊤	5
0	⊬⊬⊤ II	7

1. **Tell** the greatest number of absences that a student has had this year.
2. **Tell** the interval.
3. **Write** the number of students that have not been absent this year.

Guided Practice

Use this set of data to answer Exercises 4-5.
138, 152, 112, 127, 173, 136, 98, 125, 145, 115, 159, 131, 123, 149, 135.

4. Find the range for the set of data.
5. Would an interval of 10 or 100 be better to use? Why?

Exercises

Independent Practice

Find the range for each set of data.

6. 4, 7, 6, 9, 2, 3, 6
7. 13, 38, 9, 23, 17, 43, 23
8. 58, 92, 23, 14, 62, 79, 43, 85, 91, 18, 50, 47, 25, 88
9. 2,450; 144; 789; 3,488; 360; 2,840; 1,500; 936; 221; 1,958
10. $12, $37, $59, $23, $8, $43, $15, $25, $53

Choose the better scale for a frequency table for each set of data.

11. the data in Exercise 6: 0 to 10 or 0 to 100
12. the data in Exercise 7: 0 to 50 or 0 to 100
13. the data in Exercise 8: 0 to 100 or 0 to 1,000
14. the data in Exercise 9: 0 to 1,000 or 0 to 4,000

Choose the best interval for a frequency table for each set of data.

15. the data in Exercise 6
    a. 1 b. 10 c. 100
16. the data in Exercise 7
    a. 1 b. 5 c. 100
17. the data in Exercise 8
    a. 1 b. 5 c. 10
18. the data in Exercise 9
    a. 10 b. 100 c. 1,000

Mixed Review

19. Write the number sixty-two million, thirty-seven in standard form. *(Lesson 2-1)*
20. Evaluate $3x - 2 \cdot 4 + y$ if $x = 7$ and $y = 1$. *(Lesson 1-7)*
21. Name the number that is the solution of the equation $42 \div h = 14$; 3, 4, or 5. *(Lesson 1-9)*

22. **Sports**   In a recent professional golf tournament the top twenty scores were 62, 62, 66, 67, 64, 65, 67, 68, 68, 64, 65, 68, 69, 66, 66, 66, 67, 67, 68, 69.

    a. What is the range of the scores?

    b. To form a scale for a frequency table, which interval would be best to use?

       a. 1   b. 5   c. 10

23. **Reading**   Trek took a survey of ten of his classmates to find the number of hours each person reads for pleasure each week. The results were 3, 0, 6, 10, 2, 20, 19, 8, 2, and 4.

    a. What is the range of the data?

    b. To form a scale for a frequency table, which interval would be best to use?

       a. 1   b. 5   c. 100

24. **Critical Thinking**   Write a set of at least 15 pieces of data that have a range of ten.

25. **Research**   Find an example of a set of data in a newspaper or magazine. Write one or more sentences to describe what scale you would use if you were to represent the set in a graph.

26. **Statistics**   Twenty people from Kaitlin's home town made care packages to send to people in Moscow. The ages of the people that made the packages were 13, 29, 32, 9, 12, 15, 6, 45, 69, 34, 25, 16, 19, 13, 14, 44, 17, 77, 34, and 50. Find the range of the ages.

2 Mid-Chapter Review

Write each number in words. *(Lesson 2-1)*

1. 43,800   2. 7,305   3. 8,764,483,333

4. 503   5. 47,858,488,002,000

6. Refer to the bar graph at the right. About how many more students studied Spanish than French? *(Lesson 2-2)*

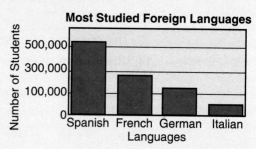

7. Survey your class to find their favorite book. Show the data in a frequency table. *(Lesson 2-3)*

Find the range of each set of data. *(Lesson 2-4)*

8. 48, 37, 4, 17, 55, 53, 28, 38

9. 389, 283, 299, 517, 273, 188, 407

2-5 Bar Graphs and Line Graphs

Objective

Construct bar graphs and line graphs.

Words to Learn

line graph

For a class assignment, Niki took a poll in her class to see which comic strips her classmates read. She recorded the results in a frequency table as shown below.

Comic Strip	Frequency
Peanuts	28
THE FAR SIDE	13
Calvin and Hobbes	9
Garfield	26
Funky Winkerbean	19

If Niki wanted to show the class which comic strips are read most, she would probably want to draw a bar graph. Remember, a bar graph is used to compare quantities. The length of each bar represents the number of classmates that read each comic strip.

Example 1

Use the data from Niki's poll to draw a vertical bar graph.

LOOKBACK

You can review bar graphs on pages 47 and 48.

The range of the frequencies is 28 − 9 or 19. Begin the scale at zero. Use an interval of 5. Then draw and label the bar graph.

If Niki would collect data from her classmates each week, she could show them how the number of classmates reading a certain comic strip changes from week to week. To do that, she would use a **line graph.** A line graph is usually used to show the change and direction of change over time.

Example 2 *Problem Solving*

Television   Anton takes a class poll each morning for a week to see how many class members watch two or more hours of TV each evening. Draw a line graph using the data collected.

Day	Frequency
Monday	26
Tuesday	24
Wednesday	18
Thursday	11
Friday	6

Step 1   Label the graph with a title.

Step 2   Draw and label the vertical scale and the horizontal scale.

Step 3   Find the range of the data. Choose an interval that best represents the data and mark off equal spaces on the vertical scale. The scale should begin with zero.

Step 4   Mark off equal spaces on the horizontal scale and label it with the appropriate day.

Step 5   Draw a dot to show the frequency for each day. Draw line segments to connect the points.

The range is 26 − 6 or 20. The vertical scale goes from 0 to 30. Choose an interval of five.

Anton notes that the number of students that watched two or more hours of TV each evening decreased toward the end of the week.

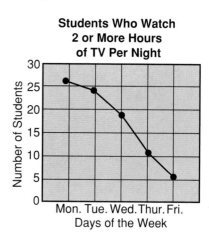

Checking for Understanding

Communicating Mathematics

Read and study the lesson to answer each question.

1. **Tell** which comic strip was read the most by using the bar graph in Example 1.

2. **Tell** which day students watched the least amount of TV by using the line graph in Example 2.

3. **Write** a sentence explaining how to find the range for a set of data.

Guided Practice Use the following tables to answer Exercises 4-10.

TABLE A
World's Most Populous Countries (in thousands)

Country	Population
China	1,119,900
India	863,400
Russia	291,000
United States	251,400

TABLE B
Average Normal Temperature for Helena, Montana

Month	Temperature
Jan.	18
Mar.	32
May	52
July	68
Sept.	56
Nov.	31

4. What is the range of population in Table A?
5. What is the range of temperatures in Table B?
6. Make a horizontal bar graph for the data in Table A.
7. Make a line graph for the data in Table B.
8. Which graph shows the change and direction of change over time?
9. Which country is the most populous?
10. Does the average monthly temperature in Helena increase or decrease from July to September? By how much?

Exercises

Independent Practice Make a bar graph for each set of data.

11. a vertical bar graph

Tasha's Semester Grades

Subject	Grade
English	89
Math	94
Science	83
Social Studies	78

12. a horizontal bar graph

Favorite Color

Color	Frequency
red	23
blue	16
yellow	2
green	5
pink	13

Make a line graph for each set of data.

13.

Grading Period	Score
1st	84
2nd	75
3rd	82
4th	89
5th	79
6th	92

14.

Month	Bill
Jan.	$149.25
Feb.	$132.75
March	$119.20
April	$ 83.50
May	$ 95.38
June	$115.65

Solve. Use the bar graph.

National Average Income

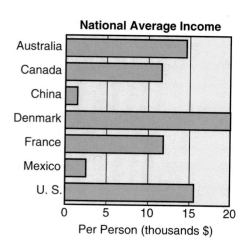

15. Which country has the highest average income per person?

16. What is the approximate average income per person in the country with the lowest average income?

17. What is the difference between the average incomes per person in Canada and Denmark?

Mixed Review

18. Write 207,657,109 in words. *(Lesson 2-1)*

19. Find the range for the following set of data: 21, 79, 11, 9, 55, 38, 111, 92. *(Lesson 2-4)*

20. **Travel** Setsu went on a trip with her family. The first day they traveled from Chicago to Toronto, 829 kilometers. The second day they traveled 745 kilometers to Quebec. About how far did they travel in all? Use front-end estimation. *(Lesson 1-3)*

21. Survey your class to find their favorite radio station. Make a bar graph to show the data. *(Lesson 2-2)*

Problem Solving and Applications

22. **Critical Thinking** Would you use a bar graph or a line graph to show how popular rap music was between 1990 and 1992? Explain.

Use the line graph to solve Exercises 23-24.

23. **Economics** Estimate the number of years it will take for the savings account to double in value.

24. **Consumer Math** If the money is withdrawn after 20 years, how much should the customer receive?

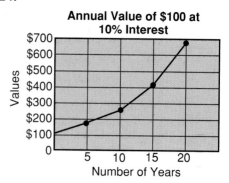

25. **Draw a Model** Find a set of data in a newspaper or a magazine and use a bar graph or a line graph to represent that data.

26. **Data Search** Refer to pages 42 and 43. According to the graph, what impact do you think the changes in the world population has had on the environment?

Making Predictions

Objective

Make predictions from line graphs.

Jansen dropped a rubber ball from different distances and recorded the highest point of each bounce on the line graph below. If Jansen drops the ball from 120 centimeters, how high do you think the ball will bounce?

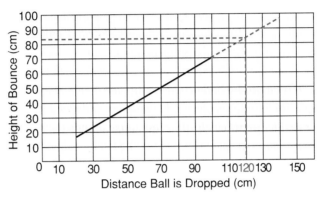

Line graphs can be used to assist in making predictions. If you extend the graph and draw a vertical line above 120 centimeters, you will see that the vertical line intersects the graph line at about 84 centimeters. So, the ball will bounce about 84 centimeters.

DID YOU KNOW

The life expectancy of people living in Eastern and Western Africa is 51 years, the lowest in the world.

Example

The line graph below shows life expectancy by age and sex. How many more years can a 30-year-old woman expect to live?

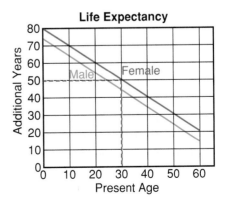

A vertical line from 30 on the horizontal scale intersects the female life-expectancy line at about 50 on the vertical scale. So a 30-year-old woman can expect to live 50 more years.

Checking for Understanding

Communicating Mathematics

Read and study the lesson to answer each question.
Refer to the graphs on page 61.

1. **Write** how high you think the ball will bounce if dropped from 135 centimeters.

2. **Tell** how many more years a 25-year-old man can expect to live.

3. **Tell** how many more years you can expect to live.

4. To what age can a 30-year-old woman expect to live, according to the graph in the example?

Guided Practice

Use the line graph in Example 1 to find the additional years each of the following can expect to live.

5. newborn baby girl
6. newborn baby boy
7. 55-year-old man
8. 5-year-old girl

Find the total number of years each of the following can expect to live.

9. 50-year-old man
10. 15-year-old girl
11. 35-year-old woman
12. 65-year-old man

13. How much longer can a 10-year-old girl expect to live than a 10-year-old boy?

Exercises

Independent Practice

Use the temperature graph to solve each problem.

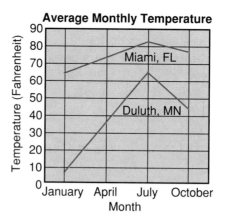

Average Monthly Temperature

14. Give the expected temperature for the given location and month.
    a. Duluth, July
    b. Miami, April

15. How much warmer is it in Miami than in Duluth in July?

16. How much colder is it in Duluth than in Miami in January?

17. What would you expect the temperature to be in Duluth in November?

Mixed Review   **18. Animals**   Make a vertical bar graph for the following set of data.
(Lesson 2-5)

Top Animal Speeds (mph)	
Antelope	61
Cheetah	70
Coyote	43
Elk	45
Gazelle	50
Gray Fox	42
Hyena	40
Lion	50
Quarterhorse	47.5
Wildebeest	50
Zebra	40

19. Estimate the quotient 72 divided by 7 using patterns. *(Lesson 1-4)*

20. Choose an appropriate interval to form a scale for a frequency table for the following set of data: 21, 79, 11, 9, 55, 38, 111, 92. *(Lesson 2-4)*

Problem Solving and Applications   **21. Travel**   Refer to the graph on page 62. Should a tourist take a swim suit on a visit to Miami in January? Explain your answer.

Use the growth and weight graphs to complete Exercises 22-26.

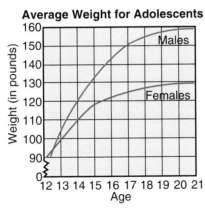

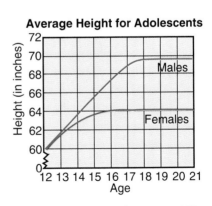

22. About how much should a woman expect to weigh at age 18?

23. About how much should a man expect to weigh at age 21?

24. About how much taller is a boy than a girl at age 15?

25. About how tall should a man expect to be when he is 25 years old?

26. Critical Thinking   When do the curves on the graphs cross? What does that mean?

27. Journal Entry   Write a paragraph explaining how information in a line graph can help you make predictions.

2-7 Mean, Median, Mode

Objective

Find the mean, median, and mode of a set of data.

Words to Learn

mean
average
mode
median

TEEN SCENE

The first music video aired on MTV was "Video Killed the Radio Star," by The Buggles on August 1, 1981.

Do you have someone in your household that constantly uses the remote control to change channels? If your household has cable television, you most likely have many channels from which to choose. The chart at the right lists the number of channels available from different cable companies. What is the average number of channels available through these cable companies?

Coaxial	23
American	27
Variety	33
Commax	31
Toppers	30
Million	27
Truax	28

You can analyze this set of data by using three measures of central tendency: mean, mode, and median.

Mean	The mean of a set of data is the sum of the data divided by the number of addends.

The mean is often called the **average** and is the arithmetic average of a set of data. The mean of the number of channels available is

$$\frac{23 + 27 + 33 + 31 + 30 + 27 + 28}{7} \approx 28.43$$

The symbol $\approx$ means is about equal to.

You can also use your calculator to find the mean.

$(23 + 27 + 33 + 31 + 30 + 27 + 28) \div 7 = 28.428571$

Mode	The mode of a set of data is the number or item that appears most often.

The mode is the number of channels that appears the most times. The number 27 appears twice and is the mode. If another company had 23 channels, then 23 would have been listed twice also, and the data would have two modes.

Median	The median is the middle number when the data are arranged in order.

To find the median of the data in the chart, first order the data. Then find the middle number.

23   27   27   28   30   31   33

3 numbers   *3 numbers*

The middle number is 28. The median is 28.

When a set of data has an even number of data, the set has two middle numbers. In such cases, the median is the mean of the two numbers.

Example *Problem Solving*

Smart Shopping   Gail checked the prices of 10 of her favorite CD's at the Record Shack. The prices are listed below. Find the mean, mode, and median of the prices.

$12 $17 $16 $16 $15 $14 $16 $15 $17 $13

The mean is

$$\frac{12 + 17 + 16 + 16 + 15 + 14 + 16 + 15 + 17 + 13}{10} = 15.1 \text{ or } \$15.10.$$

The price of $16 appears most often, so it is the mode.

Mental Math Hint

● ● ● ● ● ● ● ● ● ●

You can find the mean of 15 and 16 by thinking of the number halfway between, 15.5

The set of data in order is

12   13   14   15   15   16   16   16   17   17

4 numbers   *4 numbers*

There are two middle numbers, 15 and 16. To find the median, you need to find the mean of these two numbers.

$\boxed{(}$ 15 $\boxed{+}$ 16 $\boxed{)}$ $\boxed{\div}$ 2 $\boxed{=}$ 15.5

The median is 15.5, so the median price is $15.50.

Checking for Understanding

Communicating Mathematics

Read and study the lesson to answer each question.

1. **Tell** the median and mode for the scores 5, 9, 13, 14, and 14.
2. **Tell** how to find the mean of the set of data given in Exercise 1.
3. **Show** how to find the median if a score of 8 is added to the set of data given in Exercise 1.

List the data in each set from least to greatest. Then find the median, mode, and mean.

4. 8, 5, 2, 9, 3, 6, 9

5. 126, 136, 110

6. 15, 20, 16, 20, 15, 22

7. 536, 684, 536, 536

8. 7, 3, 8, 7, 9, 3, 6, 7, 4

9. 52, 74, 83, 59, 68, 72

Exercises

Find the mean, median, and mode for each set of data.

10. 3, 9, 1, 4, 7, 6

11. 8, 5, 9, 3, 5, 4, 8

12. 4, 15, 9, 7, 12, 9, 15, 9

13. 18, 5, 7, 8, 5, 6, 7, 12

14. 26, 18, 31, 40, 18, 25, 31

15. 38, 54, 40, 32, 48, 52

Use the data at right to answer Exercises 16-19.

16. What is the mode?

17. What is the median?

18. What is the mean?

19. If the high temperature on June 5 was 92 instead of 86, would it change
    a. the mode? Why?
    b. the median? Why?
    c. the mean? Why?

Daily High Temperature for June	
Temperature (Fahrenheit)	Frequency
92	1
90	4
89	10
87	9
86	5
85	1

20. Choose the best number to begin a scale for a frequency table for the following set of data: 24, 67, 11, 9, 52, 38, 114, 98. *(Lesson 2-4)*

21. Evaluate $2 + 3m$ if $m = 6$. *(Lesson 1-7)*

22. **Science** To obtain accurate measurements, a scientist will discard measurements that are greatly different from the rest and find the mean of the remaining measures. Use this technique to find an accurate measurement from the following: 26 mm, 17 mm, 28 mm, 28 mm, 27 mm, 26 mm, 35 mm, 27 mm.

23. **Sports** Ramsey scored 13, 21, 18, 23, and 20 points for each of the first five basketball games. Find the mean and range of the points he scored per game.

24. **Critical Thinking** Add a number to the following set of data such that the mode, median, and mean of the new set of data are the same as those for the original set: 91, 93, 93, 95, 95, 98, 100.

25. **Collect Data** Find the prices of 10 food items in your school cafeteria. Find the mean, median, and mode.

2-7B Measures of Central Tendency

A Follow-Up of Lesson 2-7

Objective

Explore the effect of extreme values on measures of central tendency.

Materials

shoes
beans
calculator

Measures of central tendency are numbers that represent a set of data. Researchers use these numbers when analyzing data.

Try this!

Work in groups of five.

- Have one person in your group take off one shoe.
- Fill the shoe with beans.
- Count the number of beans in the shoe and record the number.
- Record the data for each group.
- Use the data and your calculator to find the mean, mode, and median of the set of data.

What do you think?

1. If you add the number 5 to your set of data, which measure of central tendency is affected the most?
2. Find the mean, mode, and median of the new set of data.
3. Which measure of central tendency will be most representative of the data before adding 5? after adding 5?
4. If you add the number 400 to your set of data, which measure of central tendency is affected the most?
5. Find the mean, mode, and median of the new set of data.
6. Which measure of central tendency will be most representative of the data after adding 400?

Extension

7. Suppose you use a shoe from a first-grader and you fill it with beans. If you add this number to your set, will it affect the mean? the mode? the median?
8. Suppose you use a shoe from your teacher. What will happen to the mean? to the mode? to the median?
9. Choose your own numbers to add to your set of data. Study the effects on the mean, median, and mode.

Mathematics Lab 2-7B Measures of Central Tendency **67**

2-8 Misleading Statistics

Objective

Recognize when statistics and graphs are misleading.

Baseball Statistician

A statistician collects, organizes, and interprets numerical facts called data. For example, during a baseball game, a statistician keeps track of the number of at bats, hits, and runs each player gets. A knowledge of math is needed to calculate batting averages, and earned-run averages.

For more information, contact: National and American League Office 350 Park Avenue New York, NY 10022

Donna is writing a book report about Henry "Hank" Aaron. To show that Aaron is the lifetime home run leader, Donna draws the graphs shown below. Which one should she use for her presentation?

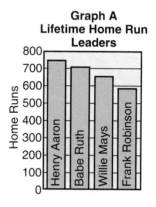

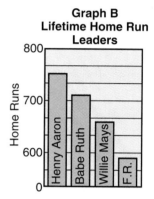

Both graphs present the same information, but on Graph B it looks like Aaron hit over twice as many home runs as Robinson.

Donna decided that Graph B was misleading because the distance from 0 to 600 on the vertical scale is less than the distance from 600 to 700. Donna chose Graph A.

Graph B can be corrected by showing that part of the scale between 0 and 600 has been left out as shown at right.

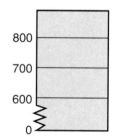

Example 1 *Problem Solving*

Stock Market The graphs below display the same information. Did the price drop drastically in March and then triple during April and May?

Graph A is misleading because the distance between 0 and 25 is the same as the distance between 25 and 26 on the vertical scale. Graph B shows that the price of the stock remained relatively stable.

Measures of central tendency can be misleading when the wrong measure is used. The following are some guidelines for the best use of measures of central tendency.

- The mode should be used when it is desired to know the most frequent item or number.

- The mean should be used when there are no numbers that differ greatly from the rest.

- The median should be used when there are numbers that are much greater or less than the majority of the numbers.

Example 2 *Problem Solving*

Travel The miles per gallon of different cars are shown in the frequency table at the right. The mode is 34, the median is 32, and the mean is 30. Which measure best represents the data? Explain.

Miles Per Gallon	Frequency
34	4
32	3
31	3
5	1

The mode is not the best representation because we do not want to know the most frequent mileage. The mean is not the best representation to use because the number 5 differs greatly from the rest of the numbers.

The median is the best measure of central tendency because the number 5 is much less than the other numbers.

Checking for Understanding

Communicating Mathematics

Read and study the lesson to answer each question.

1. **Tell** when the mode should be used.

2. **Tell** when the mean and median should be used.

3. **Draw** a vertical axis for a graph that is misleading.

Guided Practice A local telephone company employee made two graphs to show the cost of calling Portland. Both graphs show the same information. Use the graphs to answer Exercises 4-7.

Graph A
Cost of Portland Call
(weekend rate)

Graph B
Cost of Portland Call

4. Tell how the title of graph B may be misleading.
5. Which graph makes a call to Portland look the most economical? Why?
6. Which graph would you show to a telephone customer? Why?
7. Which graph would you show to a company stockholder? Why?

Tell whether the mean, median, or mode would be best to describe the following. Explain each answer.

8. 2, 300, 290, 305

9. 23, 23, 22, 21, 20

10. classmates' favorite subjects

11. populations of Texas, Utah, Idaho, Nevada, Wyoming

Exercises

Independent Practice A used car lot has the following vehicles for sale. Use the chart and a measure of central tendency to show the following.

Vehicle	Camaro	Civic	Escort	Ranger	Storm	Daytona
Doors	2	4	2	4	4	4
Miles	23,000	31,000	120,000	33,000	20,000	35,000
Color	black	blue	red	white	blue	blue
Price	$8,995	$7,999	$5,995	$7,995	$8,995	$6,995

12. number of doors

13. number of miles

14. color

15. price

16. Name other information that could be given in a table like this.

Solve. Use the line graph.

17. Make a misleading vertical scale for the graph.

18. Show how the graph title may be changed to be misleading.

19. Show how the scale labeling may be changed to be misleading.

**Cable TV
Average Monthly Rates**

Mixed Review 20. **School** Ten students took Mrs. Bonilla's geography test. The scores were 93, 70, 68, 84, 89, 78, 77, 84, 76, 91. Find the mean, median, and mode. *(Lesson 2-7)*

21. **Population** The population of China was estimated to be 1,045,537,000 in 1986. Write this population number in words. *(Lesson 2-1)*

Problem Solving and Applications

22. **Sales** An employer tells a prospective sales representative that the average income of their sales representatives is $35,000. How may the employer be misleading the prospective sales representative?

23. **Journalism** An entertainment writer has titled a line graph showing ticket sales this year at the Wimberly Playhouse as "Ticket Sales." Write a more descriptive title.

24. **Critical Thinking** Find a graph in a magazine or newspaper. Redraw the graph so that the data will appear to show different results.

25. **Journal Entry** Describe how the mean of a set of data can be used to mislead a consumer.

26. **Mathematics and History** Read the following paragraph.

The beginning of the Industrial Revolution was marked by a population explosion that had a major impact on Europe. Between 1750 and 1914, the population of Europe more than tripled. Because of the number of medical discoveries and the fact that there were no major wars, the death rate decreased and the population increased.

a. Between 1750 and 1914, the population of Europe grew from 140 million to 463 million. How many more people lived in Europe in 1914 than in 1750?

b. In 1750, the population of what is now Italy was approximately 13,150,000. By 1910, the population had increased to approximately 34,377,000. By about how much did the population of Italy grow during this time?

Lesson 2-8 Misleading Statistics   **71**

2 Study Guide and Review

Communicating Mathematics

State whether each statement is *true* or *false*.

1. The mode of a set of data is the number or item that appears most often.
2. The mean of 83, 84, 87, 90, and 97 is 87.
3. Graphs can be misleading if the intervals on the vertical scale are unequal.
4. The number 3,000 an be renamed as 300 tens.

Use a sentence or a diagram to answer the following questions.

5. Which type of graph (a line graph or a bar graph) would be best to use to predict temperature changes over a period of time?
6. When would the median be a better measure of central tendency to use than the mean?
7. How do you select a scale interval?

Skills and Concepts

Objectives and Examples	Review Exercises
Upon completing this chapter, you should be able to:	*Use these exercises to review and prepare for the chapter test.*
• read and write numbers through trillions *(Lesson 2-1)* The number 78,965,210,347,506 is seventy-eight trillion, nine hundred sixty-five billion, two hundred ten million, three hundred forty-seven thousand, five hundred six.	8. Write the place-value position of the digits 2, 4, 6, and 8 in the number 126,895,041. 9. Write 12,984,050,765 in words. 10. Write two million, three hundred four in standard form. 11. Write 16 tens in standard form.
• choose appropriate graph scales and intervals by using the range of the data *(Lesson 2-4)* Find the range of the following data: 7, 11, 24, 9, 16, 19, 18. The range is 24 − 7 = 17.	**Vacation** Four families went skiing together. The ages of the children were 16, 8, 9, 15, 10, 11, 17, 4, 7 and 15. 12. What is the range of the ages? 13. What scale would you use to make a frequency table? 14. What interval would you use?

Study Guide and Review

- construct line and bar graphs
  (Lesson 2-5)

Number of Pets in Animal Shelter

Season	Pets
Winter	65
Spring	75
Summer	85
Autumn	70

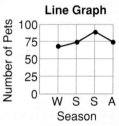

15. Make a line graph from the following data.

Allen's grades during Junior High

Grade	Frequency
A	3
B	9
C	5
D	1
F	0

16. Make a horizontal bar graph from the data in Exercise 15.

17. Make a vertical bar graph from the data in Exercise 15.

- make predictions from line graphs
  (Lesson 2-6)

Numbers of Pets in Animal Shelter

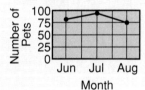

Estimate how many animals are at the shelter on July 15.

There are about 90 animals at the shelter on July 15.

18. Use the graph below to estimate sales on Friday at First Motors.

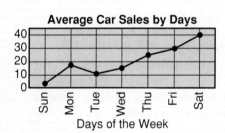

- find the mean, median, and mode of a set of data *(Lesson 2-7)*

Hashim's bowling scores for April were 117, 98, 104, 121, 105, 104, 120, and 111.

The mode is 104 because it is the only score that occurs twice. The mean is 110, the sum of the eight scores divided by 8. When the eight scores are in order, 105 and 111 are the two middle numbers. Therefore, the median is $\frac{105 + 111}{2}$ or 108.

Use the following set of data to complete Exercises 19, 20, and 21.
   22, 16, 31, 30, 34, 26, 28, 22, 15

19. Find the mode.

20. Find the median.

21. Find the mean.

Study Guide and Review

Objectives and Examples

- recognize when statistics are misleading *(Lesson 2-8)*

  Choose the best measure of central tendency for the set 4, 4, 20, 23, 24, 27, 30. The median would be best. 4 is much less than the other numbers.

- recognize when graphs are misleading *(Lesson 2-8)*

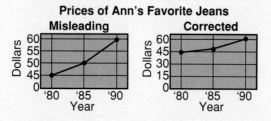

Prices of Ann's Favorite Jeans

Misleading · Corrected

Review Exercises

Tell whether the mean, median, or mode would be best to describe each set of data.

22. 10, 55, 60, 77, 79, 85, 85, 89

23. choices for Homecoming Queen

24. 81, 84, 85, 87, 88, 90, 91, 98

Use the line graph to complete.

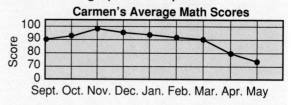

Carmen's Average Math Scores

25. Correct this misleading graph.
26. Which looks better for Carmen?

Applications and Problem Solving

Use the bar graph to answer each question. *(Lesson 2-2)*

27. **Marriage**   Was the marriage rate in 1980 higher or lower than in 1988?

28. **Marriage**   What is the difference between the marriage rate in 1970 and the divorce rate in the same year?

Use the frequency table to answer each question. *(Lesson 2-3)*

29. **Television**   In what interval did most of the responses fall?

30. **Television**   How many people watch TV less than 5 hours per week?

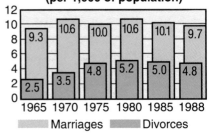

U.S. Marriage and Divorce Rates (per 1,000 of population)

Hours of TV per Week	
Interval	Frequency
0–2	10
3–4	11
5–6	5
7–8	3

Curriculum Connection Projects

- **Astronomy**   Find the average distances in miles from the sun of each of the nine planets in our solar system. Record each in numbers and in words.

Read More About It

Sarnoff, Jane and Reynold Ruffins. *The Code and Cipher Book.*
Stwertha, Eve and Albert. *Make It Graphic! Drawing Graphs for Science and Social Studies Projects.*

2 Test

Write each number in standard form.

1. thirty-one trillion, sixteen thousand

2. 37 hundreds

Write each number in words.

3. 452,000,007,000

4. 973,675,430

5. Below are the scores for 27 students on a recent history test. Organize the data in a frequency table.

72, 95, 87, 77, 88, 79, 92, 97, 100, 82, 75, 93, 92, 75, 77, 71, 98, 99, 85, 84, 77, 80, 91, 91, 85, 83, 89

Geography The lengths, in miles, of the Great Lakes are Lake Superior, 350; Lake Michigan, 307; Lake Huron, 206; Lake Erie, 241; and Lake Ontario, 193.

6. What is the range of this data?

7. What scale would you use for the vertical axis of a graph?

8. Make a bar graph for this data.

9. Find the mean length of the lakes.

10. Find the median length.

Jobs Willard mows lawns, rakes leaves, shovels snow, and does other outdoor jobs. Last year, his earnings were: January-$55, February-$70, March-$35, April-$23, May-$38, June-$55, July-$74, August-$78, September-$69, October-$55, November-$38, December-$58.

11. Make a line graph for this data.

12. Change your graph to make it misleading.

13. Find Willard's mean monthly earnings.

14. Find the mode.

15. Find the median.

16. Willard wants to buy a new bike for $179 in June. Will he earn enough in February, March, April, and May to buy the bicycle?

Use the graph at the right to answer the following questions.

17. What source generates the most U.S. electricity?

18. How much more electricity does nuclear power generate than natural gas?

19. Is this graph misleading? Explain your answer.

20. What is the range of billions of kilowatt-hours?

Bonus Find or make up a set of data and make two graphs from it that seem to show different results.

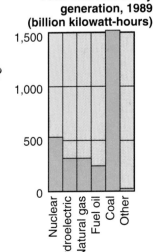

Total U.S. Electricity generation, 1989 (billion kilowatt-hours)

3

Decimals: Addition and Subtraction

Spotlight on Movie Videos and Technology

Have You Ever Wondered...

- Which movies are the most popular as video rentals?
- How common are different forms of entertainment technology?

Movie Rentals - 1980-1989

Title	Year	Rental (Millions)
E.T.-The Extra Terrestrial	1982	228.40
Return of the Jedi	1983	168.00
The Empire Strikes Back	1980	141.60
Ghostbusters	1984	128.30
Raiders of the Lost Ark	1981	115.60
Indiana Jones and the Last Crusade	1989	115.50
Indiana Jones and the Temple of Doom	1984	109.00
Beverly Hills Cop	1984	108.00
Back to the Future	1985	104.20
Tootsie	1982	95.30
Rain Man	1989	86.00

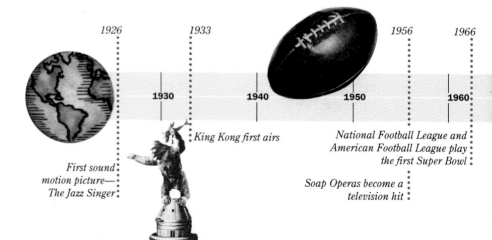

1926

1933

1956

1966

1930

1940

1950

1960

King Kong first airs

National Football League and American Football League play the first Super Bowl

First sound motion picture— The Jazz Singer

Soap Operas become a television hit

Chapter Project

Movie Videos and Technology

Work in a group.

1. Poll your classmates, family, and friends to find their favorite movies.

2. Make a chart or graph presenting your findings.

3. Pretend that you are opening a video rental store. List the first movies that you would purchase and any other important details. Show on a poster.

Technology in our Homes, 1986

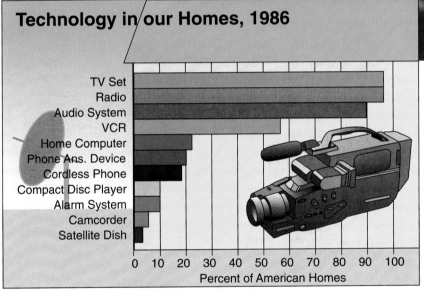

TV Set
Radio
Audio System
VCR
Home Computer
Phone Ans. Device
Cordless Phone
Compact Disc Player
Alarm System
Camcorder
Satellite Dish

0 10 20 30 40 50 60 70 80 90 100
Percent of American Homes

Looking Ahead

In this chapter, you will see how mathematics can be used to answer the questions about movie video rentals and technology. The major objectives of the chapter are to:

- read, write, and compare decimals through ten-thousandths

- describe length in the metric system

- add, subtract, and round decimals

- solve problems by using formulas

- read and interpret circle graphs

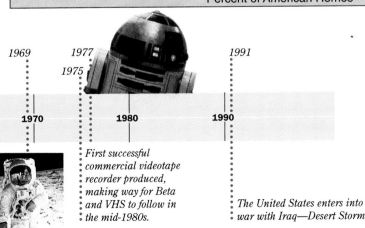

1969

1975   *1977*   *1991*

1970   **1980**   **1990**

First successful commercial videotape recorder produced, making way for Beta and VHS to follow in the mid-1980s.

The United States enters into war with Iraq—Desert Storm

77

3-1A Decimals Through Hundredths

A Preview of Lesson 3-1

Objective

Model decimals through hundredths.

Materials

base-ten blocks

Base-ten blocks can be used to show whole numbers. But did you know you can use base-ten blocks to show decimals too? When working with decimals, the blocks have different meanings, as shown below.

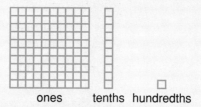

ones   tenths hundredths

Try this!

Work in groups of four.

- Place two tenths and three hundredths in front of your group.

- Trade the tenths for hundredths. Count the hundredths.

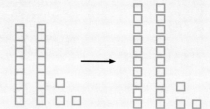

What do you think?

1. Two tenths and three hundredths is the same as how many hundredths?
2. Show five tenths and six hundredths with base-ten blocks. Trade the tenths for hundredths. How many hundredths do you have now?
3. Use blocks to show how many tenths are the same as seventy hundredths.
4. If you trade the tenths for hundredths, how many hundredths are in the blocks to the right?

3-1 Decimals Through Hundredths

Objective

Model, read, and write decimals through hundredths.

Did you know that the fingernails on the hand you use more grow faster than the nails on your other hand? Fingernails grow at an average rate of $\frac{4}{100}$ inch per week.

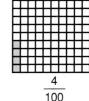

$\frac{4}{100}$

The fraction $\frac{4}{100}$ can be written as a decimal. Decimals are another way to write fractions when the denominators are 10, 100, and so on.

Write: 0.04   **Say:** four hundredths

The place-value chart used for whole numbers can be extended to the right of the ones place to name decimals. The decimal point is used to separate the ones and the tenths places.

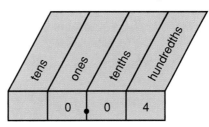

Example 1

One hundred sixth-grade students are asked their favorite ice cream flavor. Fifty-three students say chocolate.

a. Write the decimal for the model shown at the right.

The decimal is 0.53.

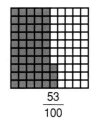

$\frac{53}{100}$

b. Write the decimal in a place-value chart.

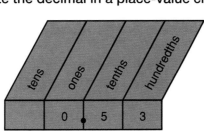

5 is in the tenths place.
3 is in the hundredths place.

c. Write the decimal in words.

Notice that the last digit, 3, is in the hundredths place. The decimal in words is *fifty-three hundredths.*

You can also write mixed numbers, such as $1\frac{74}{100}$, as decimals.

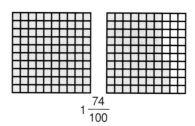

$1\frac{74}{100}$

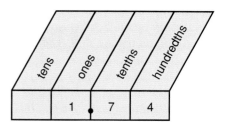

Write: 1.74   **Say:** one and seventy-four hundredths

↑   ↑

Read the decimal point as "and."

Example 2

Alicia lives $2\frac{3}{10}$ miles from her school.

a. Write the mixed number $2\frac{3}{10}$ as a decimal.

   The decimal is 2.3.

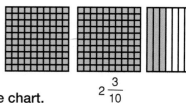

$2\frac{3}{10}$

b. Write the decimal in a place-value chart.

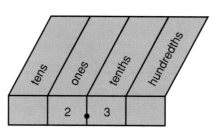

2 is in the ones place.
3 is in the tenths place.

c. Write the decimal in words.

   The decimal in words is *two and three tenths.*

Checking for Understanding

Communicating Mathematics

Read and study the lesson to answer each question.

1. **Tell** if the model below shows 0.6, 0.06, or 6.0.

2. **Show** the decimal 2.35 using base-ten blocks.

Write each fraction as a decimal.

3. $\frac{7}{10}$ 4. $\frac{6}{100}$ 5. $\frac{34}{100}$ 6. $\frac{3}{10}$

Write each expression as a decimal.

7. sixty-three hundredths   8. five hundredths

9. four and three tenths   10. two tenths

11. thirty-eight hundredths   12. four hundredths

Exercises

Independent Practice Write each fraction as a decimal.

13. $\frac{2}{10}$ 14. $\frac{66}{100}$ 15. $\frac{73}{100}$ 16. $\frac{20}{100}$

17. $\frac{5}{10}$ 18. $\frac{77}{100}$ 19. $\frac{92}{100}$ 20. $\frac{7}{100}$

Write each expression as a decimal.

21. three hundredths   22. twenty-one hundredths

23. one tenth   24. one and twelve hundredths

25. forty-seven and two tenths   26. six hundredths

Mixed Review 27. Solve the equation $3a = 30$ mentally. *(Lesson 1-9)*

28. Write the number 123,456,789 in words. *(Lesson 2-1)*

29. **Statistics**   Find the mean for the set of Tai's history test scores: 88, 90, 87, 91, 47. Why is this a misleading statistic? *(Lesson 2-8)*

Problem Solving and Applications 30. **Sports**   Carl Lewis broke the world record in the 100-meter dash in 1991 with a time of 9.98 seconds. Write this time in words.

31. **Critical Thinking**   Copy and complete each diagram to show 0.7.

a.

b.

c.

d.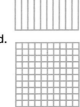

3-2 Decimals Through Ten-Thousandths

Objective

Read and write decimals through ten-thousandths.

The smallest egg laid by any bird was laid by the Vervain hummingbird of Jamaica. The egg weighed $\frac{128}{10,000}$ of an ounce.

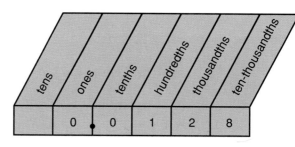

The fraction $\frac{128}{10,000}$ can be written as a decimal. You can extend the place-value chart and write the fraction as a decimal.

DID YOU KNOW

The Vervain hummingbird is about the size of a bumblebee.

tens	ones	tenths	hundredths	thousandths	ten-thousandths
	0	0	1	2	8

Write: 0.0128

Say: one hundred twenty-eight ten-thousandths

Example 1

A 100-yard football field is about $\frac{568}{10,000}$ of a mile long.

a. Write this fraction as a decimal in a place-value chart.

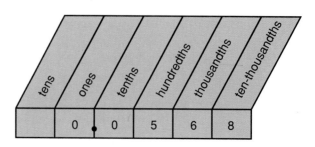

tens	ones	tenths	hundredths	thousandths	ten-thousandths
	0	0	5	6	8

b. Write the decimal in words.

Notice that the last digit, 8, is in the ten-thousandths place. The decimal in words is *five hundred sixty-eight ten-thousandths*.

 Example 2 *Problem Solving*

Auto Racing   Emerson Fittipaldi of Brazil set a one-lap record of two hundred twenty-five and five hundred seventy-five thousandths miles per hour at the Indianapolis 500-mile time trial on May 13, 1990.

a. Write the expression as a mixed number.

The mixed number is $225\frac{575}{1,000}$.

b. Write the mixed number as a decimal in a place-value chart.

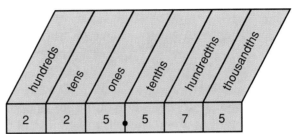

The last digit, 5, should be in the thousandths place.

The decimal is 225.575.

Checking for Understanding

Communicating Mathematics

Read and study the lesson to answer each question.

1. **Tell** if 0.0128 is less than or greater than 1.

2. **Show** 47.938 in a place-value chart.

Guided Practice

Write each fraction as a decimal.

3. $\frac{34}{1,000}$ 4. $\frac{24}{1,000}$ 5. $\frac{13}{10,000}$ 6. $\frac{564}{10,000}$

Write each expression as a decimal.

7. fifty-three thousandths

8. thirty-eight thousandths

9. nine hundred ten-thousandths

10. three and twenty-seven ten-thousandths

Exercises

Independent Practice

Write each fraction as a decimal.

11. $\frac{77}{1,000}$ 12. $\frac{201}{10,000}$ 13. $\frac{2,384}{1,000}$ 14. $\frac{92}{10,000}$

15. $\frac{837}{10,000}$ 16. $\frac{3}{1,000}$ 17. $\frac{64}{1,000}$ 18. $\frac{3,647}{10,000}$

Write each expression as a decimal.

19. twenty-one thousandths
20. two and two thousandths
21. nineteen and three hundred six ten-thousandths
22. forty-seven and fifteen thousandths
23. six hundred one and one thousandth

Mixed Review

24. Write twenty-one and sixteen thousandths as a decimal. *(Lesson 3-1)*
25. Find the value of $3x - 2y$ if $x = 6$ and $y = 4$. *(Lesson 1-7)*
26. **Statistics**   Find Allen's mean golf score for the season if he has played 9 times and his scores have been 84, 88, 78, 79, 84, 84, 86, 83, 81. *(Lesson 2-7)*

Problem Solving and Applications

27. **Biology**   An amoeba is one of the larger creatures of the microscopic world. The length of a typical amoeba is 0.0008 meters. Write 0.0008 in words.

28. **Biology**   The Chlamydomonas is much smaller than an amoeba. It has a length of about three thousandths of a centimeter. Write three thousandths as a decimal.

29. **Measurement**   The micron is a measure that is 0.0001 of a centimeter. Write 0.0001 in words.

30. **Critical Thinking**   Fill in each blank with a digit from 0 to 9 to make the greatest decimal possible. No two digits may be alike.

$$\underline{\hspace{1cm}} \; 3 . \; \underline{\hspace{1cm}} \; 8 \; \underline{\hspace{1cm}}$$

31. **Mathematics and Distance**   Read the following paragraphs.

> The term mile is used for various units of distance such as the statute mile of 5,280 feet. The term originated from the Roman *mille passus,* or "thousand paces," which measured 5,000 Roman feet.
>
> In about 1500, the "old London" mile was defined as eight furlongs. At that time, the furlong, measured by a larger northern German foot, was 625 feet, so the mile equaled 5,000 feet. During the reign of Queen Elizabeth I, the mile gained an additional 280 feet, to what it is today, 5,280 feet.
>
> A nautical mile is the length of Earth's surface of one minute of arc, or by international definition, 1,852 meters or 1.1508 statute miles.

a. One nautical mile is equal to 1.852 kilometers. Write the decimal in words.
b. The knot is one nautical mile per hour. How many kilometers does a ship travel in 2 hours at 2 knots?

3-3 Length in the Metric System

Objective

Show relationships between metric units of length and measure line segments.

Words to Learn

metric system
meter
millimeter
centimeter
kilometer

Tracy wants to use a box to build a holder for his cassette tapes. To measure its length, width, and height, he uses the **metric system** and decimals.

The basic unit of length in the metric system is the **meter.** All other metric units of length are defined in terms of the meter.

Decimals are used in the metric system. The chart below summarizes the most commonly used metric units of length.

unit	symbol	meaning	size	model
millimeter	mm	thousandth	0.001 m	thickness of spaghetti
centimeter	cm	hundredth	0.01 m	width of fingernail
meter	m	one	1.0 m	width of doorway
kilometer	km	thousand	1,000 m	six city blocks

Tracy measured and found that the length of one cassette tape was about 10 centimeters.

When am I ever going to use this?

In 1988, President Reagan signed a bill that required all government agencies to use the metric system in part by 1992. The metric system may become commonly used in the United States during your lifetime.

Examples

1 How long is a small paper clip in centimeters? in millimeters?

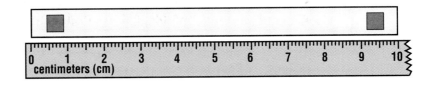

The paper clip is about 3.2 centimeters, or 32 millimeters long.

2 How long is the pencil in centimeters?

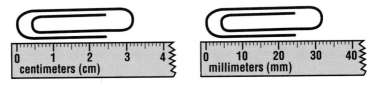

The pencil is about 12 centimeters long.

Example 3 *Connection*

Geometry   Ashley has a scale drawing of the upstairs of her home. Use your centimeter ruler to measure the dimensions of the scale drawing of Ashley's bedroom.

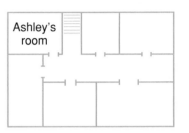

Ashley's room

In the drawing, the length of Ashley's bedroom is 1.5 centimeters and the width is 1.1 centimeters.

Checking for Understanding

Communicating Mathematics

Read and study the lesson to answer each question.

1. **Tell** which is longer, 1 centimeter or 1 millimeter.
2. **Draw** a line that measures 14.8 centimeters long.

Guided Practice

Use a centimeter ruler to measure each line segment.

3. ⎯⎯⎯⎯⎯⎯   4. ⎯⎯⎯⎯

5. ⎯⎯⎯⎯⎯⎯⎯⎯⎯   6. ⎯⎯⎯⎯⎯⎯

Exercises

Independent Practice

Use a centimeter ruler to measure each line segment.

7. ⎯   8. ⎯⎯⎯⎯⎯⎯⎯⎯⎯⎯

9. ⎯⎯   10. ⎯⎯⎯⎯⎯⎯

Measure a side of each square.

11. ☐   12. ☐   13. ☐

Mixed Review

14. Name the operation that should be done first. Then find the value of $12 - 15 \div 3$. *(Lesson 1-6)*

15. **Statistics**   Find the range for the following set of data: 101, 140, 211, 89, 167, 188, 203, 135. *(Lesson 2-4)*

Problem Solving and Applications

16. **Critical Thinking**   If the prefix *cent* means hundredth, what do you think centimeter means? What does century mean?

17. **Journal Entry**   Write one or two sentences explaining how decimals and the metric system are related.

86   **Chapter 3** Decimals: Addition and Subtraction

3-4 Comparing and Ordering Decimals

Objective
Compare two decimals.
Order a set of decimals.

Mike is doing a science project about bees and beekeeping. He went to the public library to get some information about bees. He is looking for the books with the call numbers 638.186 and 638.178. Which number comes first?

To compare decimals, line up the decimal points of the two numbers. Then start at the left, comparing the digits in the greatest place-value positions. Find the digits in the same place-value position that are not equal. The decimal with the greater digit is the greater decimal.

638.186
638.178

8 and 7 are not equal. 8 hundredths > 7 hundredths. So, the book with the call number 638.178 would come first.

You can also use a number line to compare decimals.

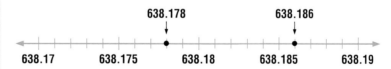

Mental Math Hint

You can also think of base-ten models when comparing decimals. For example, 1.63 would be:

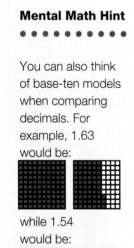

while 1.54 would be:

Example 1

Which is greater, 1.63 or 1.54?

1.63   *Line up the decimal points.*

1.54   *Start at the left and compare.*

6 and 5 are not equal. 6 tenths > 5 tenths, so 1.63 is greater than 1.54. You can check the answer by comparing these decimals on a number line.

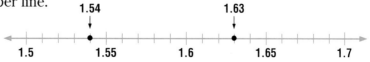

Mini-Lab

Work with a partner.

Materials: hundreds grids, pencil or crayon

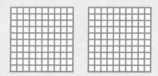

- Shade a grid to show 0.2, or two tenths.
- Shade another grid to show 0.20, or twenty hundredths.

Talk About It

a. Compare the two grids. What can you say about them?
b. Shade two more grids to show 0.4 and 0.40. How do they compare?

Sometimes it is easier to compare decimals when they have the same number of place-value positions or decimal places. You can annex, or place zeros to the right of a decimal so that each decimal has the same number of decimal places. As shown in the Mini-Lab, this will not change the value.

Example 2

Which is greater, 1.018 or 1.01?

| 1.018 | *Line up the decimal points.* |
| 1.010 | *Annex a zero so that each has the same number of decimal places. Compare.* |

Since 8 thousandths is greater than 0 thousandths, 1.018 is greater than 1.01.

Check the answer using a number line.

You can also use comparing to order decimals.

Many libraries have replaced the card catalog file with computer systems such as E-Z-Cat. The data bases help you find the information you need for your reports with a thorough and quick search of your topic.

Example 3 *Problem Solving*

Library Science Shannon Gray works at the library. She has a stack of books to return to the shelves. Help her order the call numbers on the books from least to greatest.

Compare two decimals at a time. Annex zeros when necessary.

745.231	745.231	745.200
745.23	745.230	745.230
745.2	745.200	745.231
745.412	745.412	745.412

The call numbers in order from least to greatest are 745.2, 745.23, 745.231, and 745.412.

Checking for Understanding

Communicating Mathematics

Read and study the lesson to answer each question.

1. **Tell** how the decimals 0.35 and 0.53, graphed on the number line, compare.

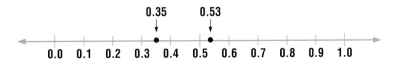

2. **Write** a response letter to a student who says 6.59 is greater than 6.8 because 59 is greater than 8. Pretend that you are the teacher.

3. **Draw** a number line to compare 4.52 and 4.49.

Guided Practice

State the greatest number in each group.

4. 2.05 or 2.50
5. 0.035 or 0.023
6. 38.95 or 38.83
7. 0.031 or 0.03
8. 0.2 or 0.19
9. 13.507, 13.05, or 13.84

Order each set of decimals from least to greatest.

10.	0.31	11.	12	12.	215.347
	0.09		12.3		215.374
	0.30		11.95		213.849
	0.10		12.01		216.003

Exercises

Independent Practice

State the greatest number in each group.

13. 0.118 or 0.109

14. 6.082 or 6.087

15. 97.408 or 9.4709

16. 0.062, 0.603, or 0.06

17. 0.0215 or 0.022

18. 574.848 or 574.8482

19. 16.098 or 160.99

20. 0.059, 0.06, or 0.6

21. Which is greater, 0.95 or 0.949?

22. Which is the least, 47.223, 47.2, or 47.023?

Order each set of decimals from least to greatest.

23.	379.87	24.	0.025	25.	43.8
	378.87		0.0316		42.998
	397.87		0.0306		43.16
	379.88		0.0249		44.022

Order each set of decimals from greatest to least.

26.	108.41	27.	27.31	28.	52.003
	106.77		26.95		52.3
	115.9		27.029		52.03
	109.04		26.88		53.2
	108.75		27.12		

Mixed Review

29. **Land Measure**   One acre is about 0.0016 square miles. Write this decimal in words. *(Lesson 3-2)*

30. Measure this line segment to the nearest whole centimeter:

   ━━━━━━━━━━━━━━━━━━━━━ .   *(Lesson 3-3)*

31. Find the value of $36 \div 9 + 7 \times 2$. *(Lesson 1-6)*

Problem Solving and Applications

32. **Statistics**   You learned in Chapter 2 that the median is the middle number when numbers are in order. Find the median of 23.7, 24.2, 23.69, 24.08, 24.13.

33. **Sports**   The chart at the right shows the times for five runners in a race. Who was the winner?

Runner	Time
Sarah	12.31 s
Camille	11.84 s
Fay	11.97 s
Debby	11.79 s

34. **Geometry**   In a right triangle, the longest side is called the *hypotenuse*. If 12.5513 meters, 0.2505 meters, and 12.5488 meters are the lengths of the sides of a right triangle, which is the hypotenuse?

35. **Critical Thinking**   Anita has more money than Corinna. Anita has less money than Megan. Justin has $0.10 more than Eric. Use the decimals at the right to find out how much money each person has.

   $0.89
   $1.18
   $1.07
   $0.79
   $1.70

Rounding Decimals

Objective
Round decimals.

LOOK BACK

You can review rounding whole numbers on page 9.

A radio station WKTT-FM can be found by tuning to 93.7 on the radio dial. Radio station DJ's often round their call numbers so it is easier for listeners to remember and find.

To round WKTT-FM's station number, look at the number line below.

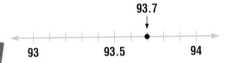

93.7

93.7 rounded to the nearest whole number is 94.

The DJ would probably refer to WKTT as FM 94 because, on the number line, 93.7 is closer to 94 than 93.

Although decimals are often rounded to the nearest whole number, they can be rounded to any place-value position.

Examples

1 Round 56.4 to the nearest whole number.

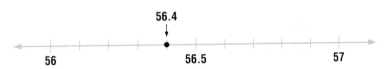

56.4

The number line shows that 56.4 is closer to 56 than 57.
56.4 rounded to the nearest whole number is 56.

You can also round decimals without using a number line.

Rounding Decimals	• Look at the digit to the right of the place being rounded. • The digit remains the same if the digit to the right is 0, 1, 2, 3, or 4. • Round up if the digit to the right is 5, 6, 7, 8, or 9.

2 Round 0.123 to the nearest hundredth.

Look at the digit to the right of the hundredths place.

$$0.12\underset{\uparrow}{3}$$

hundredths place

Since 3 < 5, the digit in the hundredths place stays the same.

0.123 rounded to the nearest hundredth is 0.12.

3 Round 2.983 to the nearest tenth.

Look at the digit to the right of the tenths place.

$$2.\underset{\uparrow}{9}83$$

tenths place

Round up since 8 > 5.

2.983 rounds to 3.0.

Example 4 *Problem Solving*

Money Sheri Williams works in the deli department at Swiftway Food Mart. She has calculated the cost per ounce of their store-made pasta to be $0.134. To the nearest cent, how much is this?

To round to the nearest cent, round to the nearest hundredth.

$$0.13\underset{\uparrow}{4}$$

hundredths place

Since 4 < 5, the digit in the hundredths place stays the same.

To the nearest hundredth, 0.134 rounds to 0.13. The price per ounce for the pasta is about $0.13 or 13 cents.

Checking for Understanding

Communicating Mathematics

Read and study the lesson to answer each question.

1. **Draw** a number line to show why 0.48 rounded to the nearest tenth is 0.5.
2. **Tell** how to round 3.4567 to the nearest hundredth.

Guided Practice

Draw a number line to show how to round each decimal to the nearest tenth.

3. 2.34   4. 6.487   5. 0.32   6. 34.06

Round to the underlined place-value position.

7. 2.1<u>8</u>5   8. 20.<u>1</u>5   0.3<u>4</u>9   10. 1<u>6</u>.44

Exercises

Independent Practice

Draw a number line to show how to round each decimal to the nearest tenth.

11. 4.36   **12.** 0.28   **13.** 7.202   **14.** 0.1487

15. 6.01   **16.** 23.04   **17.** 3.79   **18.** 8.99

Round to the underlined place-value position.

19. 12.2$\underline{8}$5   **20.** 49.$\underline{8}$02   **21.** 9.$\underline{7}$75   **22.** 58.$\underline{8}$6

23. 9.7$\underline{7}$5   **24.** 5.99$\underline{9}$8   **25.** 1.0$\underline{0}$4   **26.** 79.$\underline{9}$8

Mixed Review

27. Order the following set of decimals from least to greatest. 27.025, 26.98, 27.13, 27.9, 27.131 *(Lesson 3-4)*

28. Write *nine and sixteen thousandths* as a decimal. *(Lesson 3-2)*

29. **Statistics** Make up a set of data that has no mode and state why it has no mode. *(Lesson 2-7)*

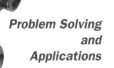

30. **Family Finances** Mr. Maldonado kept track of the phone calls each of his four children made during the month of March. Luis made 45 calls, Lorena made 40, Diana made 25, and Mirna made 33. Make a bar graph for this data. *(Lesson 2-5)*

Problem Solving and Applications

31. **Retail** Cynthia uses her calculator to find the sales tax on a purchase. If her calculator displays 2.3471, how much is the tax?

32. **Critical Thinking** Form a decimal out of the digits 0, 1, 3, 5, and 7 so that the decimal will round up if rounded to the nearest tenth, and it will stay the same if rounded to the nearest hundredth.

Mid-Chapter Review

Express each fraction as a decimal. *(Lessons 3-1 and 3-2)*

1. $\frac{3}{10}$ **2.** $\frac{51}{100}$ **3.** $\frac{101}{10,000}$ **4.** $\frac{584}{1,000}$

Measure each line segment to the nearest centimeter. *(Lesson 3-3)*

5. ————————————   **6.** ————————

Order each set of decimals from greatest to least. *(Lesson 3-4)*

7. 109.33, 190.32, 110.23, 100.34   **8.** 27.34, 27.35, 273.6, 27.03

Round to the underlined place-value position. *(Lesson 3-5)*

9. $\underline{5}$.6   **10.** 2$\underline{4}$.63   **11.** 8.7$\underline{0}$4   **12.** 1.32$\underline{4}$

3-6 Estimating Sums and Differences

Objective

Estimate decimal sums and differences using rounding, front-end estimation, and clustering.

Word to Learn

clustering

◀ **LOOK BACK**

You can review estimating whole numbers using rounding on page 9.

Susan had a sleep-over to celebrate her birthday. During the party Susan and her friends ordered two pizzas. One pizza cost $7.75 and the other cost $8.95. About how much did the two pizzas cost altogether?

Since you only need to know about how much the pizzas cost, you can estimate. One way to estimate is to round the numbers to the same place-value position.

Round each number to the nearest dollar amount.

$$
\begin{array}{rcr}
\$7.75 & \rightarrow & \$8.00 \\
+\ 8.95 & \rightarrow & +\ 9.00 \\
\hline
 & & \$17.00
\end{array}
$$

The pizzas cost about $17.00.

Example 1 *Connection*

Measurement   Emil has two samples of the same chemical he wants to store in a 0.5-liter container. One sample is 0.38 liter. The other sample is 0.21 liter. Can he store both samples in the container?

To estimate, round each number to the nearest tenth.

$$
\begin{array}{rcr}
0.38 & \rightarrow & 0.4 \\
+\ 0.21 & \rightarrow & +\ 0.2 \\
\hline
 & & 0.6
\end{array}
$$
8 > 4 so round up.
1 < 5 so digit stays the same.

Since 0.6 > 0.5, both samples cannot be stored in the 0.5-liter container.

Example 2 *Problem Solving*

Smart Shopping   Sonya pays for a $1.82 purchase with a twenty-dollar bill. About how much change should she receive?

$$
\begin{array}{rcr}
\$20.00 & & \$20.00 \\
-\ 1.82 & \rightarrow & -\ 2.00 \\
\hline
 & & \$18.00
\end{array}
$$
Round to the nearest dollar. Then subtract.

Sonya should receive about $18 in change.

LOOKBACK

You can review front-end estimation with whole numbers on page 12.

Another way to estimate decimal sums and differences is by using front-end estimation.

Example 3 *Problem Solving*

Smart Shopping   An action figure sells at Neumann's Department Store for $5.83. The same figure sells for $3.68 at Joe's Discount House. About how much less is the figure at Joe's?

Subtract the front digits.

Adjust your estimate by subtracting the next digits.

$$
\begin{array}{r}
\$5.83 \\
-\ 3.68 \\
\hline
\$2
\end{array}
\qquad \rightarrow \qquad
\begin{array}{r}
\$5.83 \\
-\ 3.68 \\
\hline
\$2.20
\end{array}
$$

The action figure is about $2.20 less at Joe's.

Example 4 *Problem Solving*

Recycling   Mr. Tate's sixth grade class has been recycling paper for two months. The first month the class received $18.52 for their recycled paper. The second month they received $31.30. About how much money did they receive in two months?

Add the front digits.

Adjust your estimate by adding the next digits.

$$
\begin{array}{r}
\$18.52 \\
+\ 31.30 \\
\hline
\$4
\end{array}
\qquad
\begin{array}{r}
\$18.52 \\
+31.30 \\
\hline
\$49.00
\end{array}
$$

Mr. Tate's class received about $49 in two months.

Clustering is another way you can estimate sums and differences. Clustering is used when numbers are close to the same number.

Example 5 *Problem Solving*

Recycling   Mr. Tate's class began turning in newspapers to the local recycling center in October. The table shows the amount collected each week. About how much money did Mr. Tate's class collect in October?

Week	Money
1	$6.88
2	$6.75
3	$7.02
4	$6.97

Since each amount is about $7, add this amount 4 times. Mr. Tate's class collected about $28.

Checking for Understanding

Communicating
Mathematics

Read and study the lesson to answer each question.

1. **Tell** a classmate how to use rounding to estimate the sum of 16.34 and 13.51.

2. **Write** an example of three decimals whose estimated sum would be 10 using rounding but 11 using front-end estimation.

Guided Practice

Estimate using rounding.

3. 3.75	4. 11.46	5. 7.43	6. 11.451
$+\,2.31$	$+\,10.57$	$-\,3.68$	$-\,10.077$

Estimate using front-end estimation.

7. 0.398	8. 0.94	9. 1.45	10. 0.490
-0.274	$+0.43$	$+0.36$	-0.082

11. Travis wants to buy a pack of baseball cards that cost $1.05. If he pays for the cards with a ten-dollar bill, about how much change should he receive?

Exercises

*Independent
Practice*

Estimate using rounding.

12. 0.203	13. 31.556	14. 0.612	15. 0.54
$+0.581$	-17.405	-0.185	$+0.47$
16. 0.506	17. 4.027	18. 5.677	19. 0.3
$+0.391$	-0.022	-2.34	-0.183

20. $5.55 - 5.42$

21. $0.454 + 1.685$

Estimate using front-end estimation.

22. 63.54	23. 29.09	24. 5.45	25. 0.998
-2.43	-18.05	$+0.57$	-0.5454
26. $\$45.98$	27. 9.88	28. 6.16	29. $\$39.95$
-30.15	$+6.42$	-3.07	$+64.50$

Estimate using any strategy.

30. $2.3465 + 434.49$

31. $\$57.98 - \$26.95$

32. $0.880 - 0.734$

33. $5.34 + 6.33 + 1.90$

34. $8.454 + 7.999 + 8.378 + 8.238 + 7.866$

35. Literature Lewis Carroll, author of *Alice's Adventures in Wonderland,* kept a list of all of the letters he received and sent during the last 37 years of his life. There were 98,721 letters. If about the same number were sent and received each year, about how many did he save each year? Estimate using compatible numbers. *(Lesson 1-5)*

36. Recruiting Jay Edwards was trying to recruit sales people to sell a brand of cleaning supplies. He told his prospects that they would be likely to earn about $475 a week, because that was the average amount his current employees were earning. In the past, his 6 salespeople have earned weekly amounts of $200, $400, $410, $260, $320 and $1,260. Was he misleading his prospective employees? Explain. *(Lesson 2-8)*

37. State the greater of the numbers 1.079 and 1.08. *(Lesson 3-2)*

38. Estimate 39.7 − 28.561 using rounding. *(Lesson 3-5)*

39. Weather Washington, D.C. has an average annual precipitation of 35.86 inches. Round this amount to the nearest tenth. *(Lesson 3-5)*

40. Express $\frac{8}{1,000}$ as a decimal. *(Lesson 3-2)*

41. History Use the table and rounding to solve.

    a. About how many cemetery acres are in Virginia?

    b. About how many cemetery acres are in Tennessee?

    c. About how many more cemetery acres are there in Fredericksburg than in Antietam?

    d. About how many more cemetery acres are there in Gettysburg than in Stones River?

National Cemeteries		
Name	State	Acres
Antietam	MD	11.36
Ft. Donelson	TN	15.34
Fredericksburg	VA	12.00
Gettysburg	PA	20.58
Poplar Grove	VA	8.72
Shiloh	TN	10.05
Stones River	TN	20.09
Vicksburg	MS	116.28
Yorktown	VA	2.91

42. Critical Thinking Three same-priced items are purchased. Based on rounding, the estimate of the total was $12. What is the maximum price each item could have cost? What is the minimum price?

43. Science The table at the right shows the five most common elements in Earth's crust along with the percent of the crust each element represents. About what percent of Earth's crust is either oxygen or silicon?

Earth's Crust	
Element	Percent
Oxygen	46.60
Silicon	27.72
Aluminum	8.13
Iron	5.00
Calcium	3.63

44. Journal Entry Write a paragraph explaining how front-end estimation would help you buy items at a store.

3-7 Determine Reasonable Answers

Objective

Determine whether answers are reasonable.

Marty needs to buy beach supplies to take on his vacation to Hawaii. He has found two pairs of sunglasses for $13.99 each, a swimsuit for $29.99, and a bottle of sunscreen for $6.49. Is $50.00 or $70.00 needed to pay for these supplies? Which amount is reasonable?

Explore What do you know?
You know the items Marty wants to buy and how much each item costs.

What do you need to know?
You need to know how much money Marty needs in order to buy the items.

Plan To solve the problem, you need to add the cost of each of the items to find the total cost. Then see if $50.00 or $70.00 is needed to pay for the items.

Estimate first.

$14 + $14 + $30 + $7 = $65

Solve

$13.99	*sunglasses*
13.99	*sunglasses*
29.99	*swimsuit*
6.49	*sunscreen*
$64.46	

The total cost of the items is $64.46. $70.00 is needed to buy the items.

Examine $50.00 would not be enough money since the items cost $64.46. The answer of $70.00 is the reasonable answer.

Checking for Understanding

Communicating Mathematics

Read and study the lesson to answer each question.

1. **Write** a reasonable total amount for the beach supplies if Marty also bought two beach towels for $24.99 each in the lesson introduction.

2. **Tell** how you know if an answer is reasonable.

Guided Practice   Solve.

3. When Sandy added 0.0597 and 0.00383, the calculator showed 0.6353. Is this answer reasonable?

4. There are 3,556 seats in the Blendon High School stadium. What is a reasonable number of rows in the stadium if each row holds about 53 people?

5. Max wants to spend his allowance on Nintendo games. He has saved $55.00. About how many games can Max buy if each game costs $19.99?

Problem Solving

Practice   Solve using any strategy.

6. The sixth grade is planning a field trip. There are 589 students in the sixth grade. Each school bus holds 48 people. How many buses will they need, 12 or 13?

Strategies
• • • • • • •
Look for a pattern.
Solve a simpler problem.
Act it out.
Guess and check.
Draw a diagram.
Make a chart.
Work backward.

7. Priority mail costs $2.90 for the first two pounds and $2.50 for each additional pound. Does a 3-pound package cost more or less than $5.50 to send?

8. If 223,532 people attended a 4-game home stand to see the Minnesota Twins during the 1991 season, which is a reasonable estimate for the number of people that attended each game: 55,000 or 5,500?

9. The sum of the ages of the Brimage twin brothers and the Webber twin sisters is 50 years. The Webbers are 7 years older than the Brimages. How old is each set of siblings?

10. A rectangular swimming pool is 20 feet wide and 38 feet long. Its depth goes from 2 feet to 6 feet. The price of a pool cover is $5 per square foot. How much will a pool cover cost?

3-8 Adding and Subtracting Decimals

Objective

Add and subtract decimals.

DID YOU KNOW

There are IronKids triathlons held each summer for kids ages 7 to 14. Senior division athletes, (ages 11 to 14) swim 200 meters, bike 6.2 miles, and run 1.2 miles.

If you like to swim, ride a bicycle, and run, then the Ironman triathlon is for you! You may want to start practicing now because in the Ironman, athletes swim 2.4 miles, bike 112 miles, and run a 26.2-mile marathon all in the same day! How many miles is that?

In order to find the total distance, you need to add the three numbers.

Adding decimals is like adding whole numbers. Make sure that you line up the decimal points before you add or subtract.

$$\begin{array}{r} 2.4 \\ 112.0 \\ +26.2 \\ \hline 140.6 \end{array}$$ *Estimate: 2 + 110 + 30 = 142.*

The athletes cover 140.6 miles during the triathlon. The estimate shows that the answer is reasonable.

Example 1

Find the difference of 154.2 and 23.4.

$$\begin{array}{r} \overset{3\,12}{15\cancel{4}.2} \\ -\;23.4 \\ \hline 130.8 \end{array}$$ *Estimate: 150 − 20 = 130.*

The estimate shows that the answer is reasonable.

Example 2   *Problem Solving*

Health   A man with a mass of 72.9 kilograms steps onto a balance scale while wearing clothes that have a mass of 1.36 kilograms. What is his total mass?

$$\begin{array}{r} 72.90 \\ +\;1.36 \\ \hline 74.26 \end{array}$$ *Estimate: 73 + 1 = 74.*
Line up the decimal points.
Annex a zero. Add.

His total mass is 74.26 kilograms. The estimate shows that the answer is reasonable.

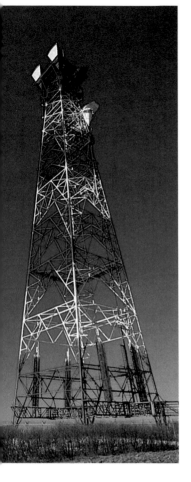

Example 3 *Problem Solving*

Communication Radio signals at a sporting event can arrive at your home quicker than the TV signal of the same event. This is because radio signals travel by telephone cable and TV signals travel by satellite. A radio signal sent from New York City to Houston via cable takes 0.009 seconds to reach the home listeners. A TV signal takes about 0.27 seconds to reach the viewers' eyes. How much of a delay is there between the two signals?

First, line up the decimal points. Then subtract.

$$\begin{array}{r} 0.27 \\ -\ 0.009 \\ \hline \end{array}$$ *Annex a zero.* *Rename and subtract.* $$\begin{array}{r} {\scriptstyle 6\ 10} \\ 0.270 \\ -0.009 \\ \hline 0.261 \end{array}$$

You can also add and subtract decimals using a calculator.

0.27 $\boxed{-}$ 0.009 $\boxed{=}$ 0.261

There is a delay of 0.261 seconds between the two signals that travel from New York City to Houston.

Example 4 *Connection*

Algebra Evaluate $x + y$ if $x = 0.342$ and $y = 4.907$.

$$\begin{array}{r} 0.342 \\ +\ 4.907 \\ \hline 5.249 \end{array}$$ *Estimate: 5 + 0 = 5.* *Line up the decimal points.* *Add.*

The value is 5.249.

Checking for Understanding

Communicating Mathematics Read and study the lesson to answer each question.

1. **Tell** how you would add 3.25 and 2.2 using pencil and paper.

2. **Write** a paragraph explaining whether the sum of 36.02 and 1.39 is 3.732.

Guided Practice Add or subtract.

3. $\begin{array}{r} 2.31 \\ +1.77 \\ \hline \end{array}$
4. $\begin{array}{r} 41.39 \\ -23.17 \\ \hline \end{array}$
5. $\begin{array}{r} 107.3 \\ -98.1 \\ \hline \end{array}$
6. $\begin{array}{r} 6.4 \\ +\ 3.3 \\ \hline \end{array}$

7. $\begin{array}{r} 86.405 \\ -37.527 \\ \hline \end{array}$
8. $\begin{array}{r} 0.34 \\ +0.69 \\ \hline \end{array}$
9. $\begin{array}{r} 3.7 \\ -2.95 \\ \hline \end{array}$
10. $\begin{array}{r} 1.04 \\ +0.98 \\ \hline \end{array}$

Exercises

Add or subtract.

11. 56.2
    −11.1

12. 42.3
    +33.5

13. 62.39
    +77.74

14. 71.34
    −43.78

15. 379.512 + 486.088

16. 4 − 1.9

17. 115 − 102.9

18. 3.702 + 0.49

19. 3.702 + 0.49 + 2.4

20. 84.34 − 67.235

21. 46 − 23.78

22. 2.438 + 5.0076 + 3.23

Solve each equation.

23. $x = 4.3 − 2.1$

24. $m = 89.57 + 13.5$

25. $292 − 278 = p$

26. $51.62 + 6.099 = t$

27. $c = 0.085 + 2.487$

28. $r = 46 + 478.98$

29. Find the difference of 67.38 and 37.46.

30. Evaluate the expression $r − s$ if $r = 34.2$ and $s = 22.4$.

31. Evaluate the expression $a + b$ if $a = 5.7$ and $b = 3.4$.

32. **Science**   Janet has a leaf collection that includes 15 birch, 8 willow, 5 oak, 10 maple, and 8 miscellaneous leaves. Make a bar graph showing this data. *(Lesson 2-5)*

33. Write thirteen hundredths as a decimal. *(Lesson 3-1)*

34. Estimate $6.97 − $2.49 using rounding. *(Lesson 3-7)*

35. **Statistics**   Find the mean of the numbers 120, 112, 88, 100, 141, and 147. *(Lesson 2-7)*

36. **Animals**   A giant tortoise can travel about 0.076 meters per second. A spider can travel about 0.189 meters per second. How much faster is the spider?

37. **Critical Thinking**   Work with a partner. Arrange the digits 1, 2, 3, 4, 5, 6, 7, and 8 into two decimals so that their sum is as close to 1 as possible. Use each digit only once. The sum cannot be equal to or greater than 1.

38. **Journal Entry**   Write a paragraph explaining how adding and subtracting decimals compares to adding and subtracting whole numbers.

3-9 **Reading Circle Graphs**

Objective

Interpret circle graphs.

Words to Learn

circle graph

Does the thought of having a big plate of spaghetti make your mouth water? The circle graph below shows that spaghetti is a favorite of pasta lovers.

A **circle graph** is used to compare parts of a whole. The circle represents the whole. The circle is separated into parts of the whole. In the circle graph below, the whole is the total pasta sales.

By looking at the pieces of the circle graph you can tell that long pasta is the most popular in retail sales because it is the largest part of the graph.

Retail Sales by Pasta Shape

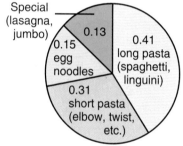

You can also tell that long pasta is the most popular because it has the greatest number. Long pasta accounts for 0.41 or $\frac{41}{100}$ of the sales.

Example *Problem Solving*

Environment   Plastic is the worst form of pollution found on beaches. The next worse is metal. What part of beach pollution comes from plastic and metal?

**1990 Cleanup
Trash on U.S. Beaches**

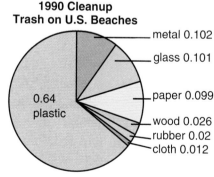

To solve, add the parts for plastic and metal.

$$\begin{array}{r} 0.640 \\ +0.102 \\ \hline 0.742 \end{array}$$

0.742 or $\frac{742}{1,000}$ of beach pollution comes from plastic and metal.

Checking for Understanding

Communicating Mathematics

Read and study the lesson to answer each question.

1. **Tell** how much more the sales of short pasta are than the sales of egg-noodle pasta by using the graph in the lesson introduction.
2. **Write** a paragraph explaining how a circle graph is used.

Guided Practice

Solve. Use the circle graph.

3. Which group has the largest enrollment?
4. Which group has the smallest enrollment?
5. Which two groups are about the same size?
6. What is the enrollment for grades 1 through 12?

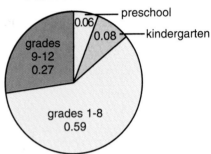

1989 School Enrollment
Public and Private

preschool 0.06
kindergarten 0.08
grades 9-12 0.27
grades 1-8 0.59

Exercises

Independent Practice

Solve. Use the circle graph.

7. Which sport is the most popular?
8. Which sport is the least popular?
9. Which two sports received about the same number of votes?
10. Which two sports together have about the same number of votes as football?
11. Which two sports together received about half of the votes?

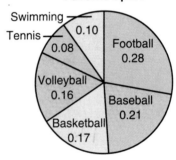

Student Survey
Favorite Sport

Swimming 0.10
Tennis 0.08
Football 0.28
Volleyball 0.16
Baseball 0.21
Basketball 0.17

Mixed Review

12. Solve the equation $y = 43.1 + 7.256$. *(Lesson 3-7)*
13. Solve the equation $33 - a = 30$ mentally. *(Lesson 1-9)*
14. **Fitness** The line graph shows Andy's progress doing sit-ups. When he started in January, he was doing 40 per day. Extend the graph line and predict how many he will be doing in October. *(Lesson 2-6)*

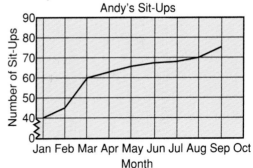

Andy's Sit-Ups

Number of Sit-Ups

90
80
70
60
50
40
0

Jan Feb Mar Apr May Jun Jul Aug Sep Oct
Month

Vacations The circle graph at the right shows how a sample of people spend their vacations.

15. How do most people spend their vacations?

16. What part of vacationers either visit family or friends or stay at home?

17. How much greater is the part of vacationers that travel than the part that goes to a resort?

no vacation/ don't know 0.06

summer/winter resort 0.17

visit family/ friends 0.30

stay at home 0.21

travel/ sightsee 0.26

How People Spend Vacations

18. **Critical Thinking** Which would be the most appropriate data to show in a circle graph? Explain.

 a. data showing the distance classmates walk to school

 b. data showing the time spent on homework for a week

 c. data showing the population change of North Carolina from the year 1900 to the year 2000

Sales of Favorite Cookies

19. **Cookies** The circle graph at the right shows sales of favorite cookies at the cookie store. Which type is most popular?

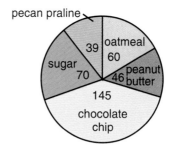

pecan praline

sugar 70

39

oatmeal 60

peanut 46 butter

145 chocolate chip

DATA SEARCH

20. **Data Search** Refer to pages 76 and 77. How many more rentals were there for E.T. than for Ghostbusters?

CULTURAL KALEIDOSCOPE

Booker T. Washington

Booker Taliaferro Washington was an educator, the founder and principal of Tuskegee Institute, and one of the most influential spokespersons for African-Americans between 1895 and 1915.

With Washington's guidance, Tuskegee Institute became a model of industrial education. The school specialized in such trades as carpentry, farming, and mechanics. Washington believed that hard work and learning work skills would help African-Americans economically. Once African-Americans had a strong economic background, Washington believed they would be given civil and political rights.

Washington served as an advisor on civil rights for both President Theodore Roosevelt and William Howard Taft.

3-10 Use a Formula

Objective

Solve problems by using a formula.

The Pentagon is a large, five-sided building in Arlington County, Virginia. It was built between 1941 and 1943 and serves as the headquarters for the United States Department of Defense. With five floors, a mezzanine, and a basement, the height of the Pentagon is far less than its horizontal dimensions. Each of the five sides are the same length, 276.3 meters. What is the distance around, or the perimeter of, the Pentagon?

Explore   What do you know?
The Pentagon has five equal sides.
Each side measures 276.3 meters.

Plan   The perimeter is the distance around a figure. You can use addition and a formula to find the perimeter.

Solve   Use the letter P to stand for the perimeter and the letter s to stand for each of the sides.

$P = s + s + s + s + s$

Since you know each side equals 276.3 meters, replace each s with 276.3.

$P = 276.3 + 276.3 + 276.3 + 276.3 + 276.3$
$\quad = 1{,}381.5$

The perimeter of the Pentagon is 1,381.5 meters.

Examine   To see if the answer makes sense, you can estimate the perimeter.

Think: $300 \times 5 = 1{,}500$

Since 1,500 is close to 1,381.5, the answer makes sense.

Checking for Understanding

Read and study the lesson to answer each question.

1. **Write** a formula for finding the perimeter of a square.
2. **Tell** how using a formula can help you solve a problem.

Guided Practice

Solve using a formula.

3. The length *(l)* of a flower garden is 8 feet and its width *(w)* is 6 feet. How much fencing will Raymond need to put around the garden? Use $P = l + l + w + w$ where *P* is the perimeter.

4. Travis is carpeting the family room. The length *(l)* of the room is 12 feet and the width *(w)* is 7 feet. What is the area *(A)* of the room he wants to carpet? Use $A = l \times w$.

5. In bowling, teams are given handicaps based on the averages of individual team members. This allows each team to have an equal chance of winning. The final handicap score *(s)* equals the game score *(g)* plus the handicap *(h)*. Quentin had a handicap bowling score of 178. His handicap was 12. What was his game score?

Problem Solving

Practice

Solve using any strategy.

6. The two long sides of a kite measure 56 centimeters each. The two short sides measure 34 centimeters each. What is the perimeter of the kite?

7. Organize this information into a frequency table.

Strategies

● ● ● ● ● ● ●

Look for a pattern.
Solve a simpler problem.
Act it out.
Guess and check.
Draw a diagram.
Make a chart.
Work backward.

Temperatures during September (in degrees F)									
77	55	52	67	45	82	60	61	50	70
70	72	59	68	66	79	77	60	63	58
53	55	59	62	63	58	51	49	54	63

8. The sale price *(s)* of an item is equal to the list price *(l)* less the discount *(d)*. Find the sale price of a sweater that costs $49.99 with a discount of $12.49.

9. Marlin bought three pairs of athletic shoes at Just For Feet. The shoes cost $39.99, $75.50, and $89.90. Estimate the amount of money Marlin spent at Just For Feet.

10. Find the number if the sum of the number and 6.2 is 14.6.

11. Lance wants to buy a heavy bag that costs $64.95, boxing gloves that cost $29.99, and a helmet that costs $34.50. Estimate how much Lance will spend.

3 Study Guide and Review

Communicating Mathematics

Choose the best answer.

1. Six hundred and twelve thousandths written as a decimal is
   a. 600.012   b. 0.612   c. 612.00
2. Which number is the greatest?
   a. 49.248   b. 49.3   c. 48.998
3. Three ways to estimate decimal sums and differences are rounding, clustering, and
   a. back-end estimation   b. finding the perimeter   c. front-end estimation
4. The best graph to compare parts of a whole is a
   a. line graph   b. bar graph   c. circle graph
5. The basic unit of length in the metric system is the
   a. foot   b. centimeter   c. meter
6. Draw a number line and show how 1.17 would be rounded to the nearest tenth.

Skills and Concepts

Objectives and Examples	Review Exercises
Upon completing this chapter, you should be able to:	Use these exercises to review and prepare for the chapter test.
• model, read, and write decimals through hundredths *(Lesson 3-1)*	7. Write $\frac{8}{100}$ as a decimal.
$0.24 = \frac{24}{100} =$ twenty-four hundredths	Write each expression as a decimal. 8. two tenths 9. thirty-four hundredths 10. six and five tenths
• model, read, and write decimals through ten-thousandths *(Lesson 3-2)*	Write each expression as a decimal. 11. fifty-two thousandths
$0.313 = \frac{313}{1,000} =$ three hundred thirteen thousandths	12. one hundred seven and sixty-three hundredths

Objectives and Examples

- show relationships between metric units of length and measure line segments *(Lesson 3-3)*

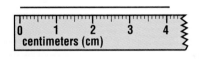

The line segment at the left measures 4 cm or 41 mm.

- compare two decimals *(Lesson 3-4)*

Why is 3.43 greater than 3.427?

3.4<u>3</u>0   Because 3 is
3.4<u>2</u>7   greater than 2.

- order a set of decimals *(Lesson 3-4)*

Write 45.93, 46.4, 45.89, and 45.311 in order from least to greatest.

The decimals in order from least to greatest are 45.311, 45.89, 45.93, and 46.4.

- round decimals *(Lesson 3-5)*

4.<u>7</u>39 is 4.7 to the nearest tenth, because 3 < 5.
4.7<u>3</u>9 is 4.74 to the nearest hundredth, because 9 > 5.

- estimate decimal sums and differences using rounding, front-end estimation, and clustering *(Lesson 3-6)*

  a. rounding
  $7.79 - 2.32 \rightarrow 7.8 - 2.3 = 5.5$

  b. front-end estimation

  $$\begin{array}{ccc} 7.79 & 7.79 & 7.79 \\ \underline{-\ 2.32} & \underline{-\ 2.32} & \underline{-\ 2.32} \\ 5 & & 5.40 \end{array}$$

  c. clustering
  $7.96 + 8.1 + 8.23 + 7.7$
  $\rightarrow 8 \times 4 = 32$

Review Exercises

Measure each line segment to the nearest whole centimeter.

13. ━━━━━━━━━━━━━━━

14. ━━━━━━━━━━━━━━━━━━━━

15. ━━━━━

16. Measure the line segment in Exercise 13 to the nearest millimeter.

State the greater number in each pair.

17. 5.218 or 5.207

18. 11.6 or 11.13

Order each set of decimals from least to greatest.

19. 0.0319, 0.31, 0.032, 0.0289

20. 75.3, 7.598, 7.8, 75.6, 75.09

Round to the underlined place-value position.

21. <u>7</u>.29   22. 76.8<u>0</u>2

23. 13.<u>5</u>81   24. 69.<u>9</u>99

Estimate.

25. $4.86 - 1.131$. Use rounding.

26. $34.29 + $17.58. Use front-end estimation.

27. $6.19 + 5.98 + 5.7 + 6 + 6.3$

28. **Money** Iola worked three days last week, earning $17.75, $21.30, and $19.85. About how much did she earn in all?

Study Guide and Review

Objectives and Examples

- add and subtract decimals
  (Lesson 3-8)

Subtract 2.89	7.30
from 7.3.	− 2.89
	4.41

- interpret circle graphs *(Lesson 3-9)*

  Magazine Sales by Grade

  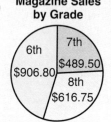

  Sixth graders sold the most subscriptions, earning $906.80.

Review Exercises

Add or subtract.

29. 25.6 + 47.92

30. 12 − 3.45

31. Solve $x = 3.1 + 0.48$.

Use the graph at the right to answer.

32. What is the most common race time?

10K Race Times (Minutes)

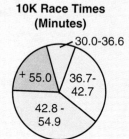

Applications and Problem Solving

33. **Consumer Math** Jackson bought a pennant at the souvenir shop at the top of Pike's Peak. He paid for the $7.59 pennant with a $10 bill. Should he expect about $4.00 or $2.00 in change? *(Lesson 3-7)*

34. **Shoes** The formula that relates shoe size *(s)* and foot length *(f)*, in inches, for men is $s = 3f - 24$. What is the shoe size for a man whose foot is 11 inches long? *(Lesson 3-10)*

Curriculum Connection Projects

- **Current Events** Read and interpret the graph from the front page of a *USA Today* newspaper.
- **Chemistry** Make a list of the atomic masses of the first 36 elements from a Periodic Table. Round each to the nearest whole number, tenth, and hundredth.

Read More About It

Carroll, Lewis. *Alice in Wonderland.*
Reisburg, Ken. *Coin Fun.*
Stokes, Jack. *Let's Make a Kite.*

3 Test

Write each fraction as a decimal.

1. $\dfrac{341}{10,000}$ 2. $\dfrac{7}{100}$ 3. $\dfrac{13}{1,000}$ 4. $\dfrac{2}{100}$ 5. $\dfrac{5}{10}$

6. sixty-nine ten-thousandths

7. two hundred and six thousandths

Measure each line segment to the nearest centimeter.

8. _____   9. _____

Measure each line segment to the nearest millimeter.

10. _____   11. _____

12. **School**   Jennifer's semester math average is 76.7. Her twin sister is close with an average of 76.35. Whose average is higher?

13. Order 17.4, 1.747, 1.8, 17.36, and 17.09 from least to greatest.

14. **Consumer Math**   Sonia and four friends ate at a restaurant. Dividing up the bill, each owed $7.146. What should each have paid?

Round to the underlined place-value position.

15. 901.2̲63   16. 0.9̲9̲9̲

Estimate. Use any method.

17. $7.14 + 7 + 6.7 + 6.9$ 18. $45.9 - 6.12$

Add or subtract.

19. $9.04 + 12.8$ 20. $54.29 - 3.8$ 21. $54 - 1.8$

22. At breakfast, Mel ordered juice for $0.89, scrambled eggs and toast for $3.69, and a glass of milk for $0.59. Did he spend about $6.00 or $4.00?

23. **Geometry**   Art drew a triangle with three equal sides. Each side measured 4.2 inches. Find the perimeter of the triangle.

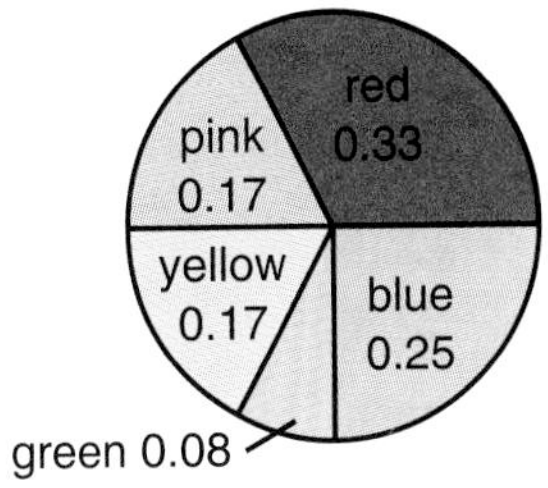

Solve. Use the circle graph.

24. Which color is the most popular?

25. Which two colors together are favored by about half the students?

Student's Favorite Colors

red 0.33
pink 0.17
yellow 0.17
blue 0.25
green 0.08

Bonus   Redraw the graph above to show what might happen if twice as many chose blue.

3 Academic Skills Test

Directions: Choose the best answer. Write A, B, C, or D.

1. There will be 120 people at the music awards banquet. Each table seats 8 people. How can you find the number of tables needed?

   A. Add 8 to 120.
   B. Divide 120 by 8.
   C. Multiply 8 times 120.
   D. none of these.

2. 704 − 299 is about

   A. 400   B. 500
   C. 900   D. 1,000

3. 50 × 120 is

   A. 500   B. 600
   C. 5,000   D. 6,000

4. 350 ÷ 6 is about

   A. 60   B. 50
   C. 6   D. 5

5. What is the value of $8 + 6 \times 10$?

   A. 68   B. 120
   C. 140   D. 480

6. If $n = 4$, $m = 3$, and $x = 12$, what is the value of $x \cdot m + n$?

   A. 19   B. 24
   C. 40   D. 84

7. Which is a true sentence?

   A. 0.5 > 0.48
   B. 0.48 > 0.5
   C. 1.4 > 2.5
   D. 5.01 < 4.08

8. Which average would be best to describe this data?
   195, 188, 70, 185, 190

   A. mean   B. median
   C. mode   D. none of these

9.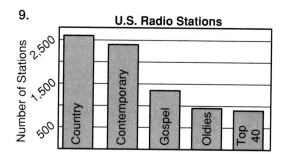

   About how many oldies stations are there?

   A. 550   B. 950
   C. 1,050   D. 2,500

10. A poll was taken about favorite colors. Blue was the favorite of 45 people; green, 9 people; orange, 3 people; purple, 5 people; red, 29 people, and 16 liked yellow the best. Which is the best interval to use to form a scale for the color data?

    A. 1   B. 5
    C. 25   D. 50

11. What is the median of the following test scores?

    80   100   90   60   80   90   100
    90   90   60   70   90   100   80

    A. 70   B. 85
    C. 90   D. 95

12.

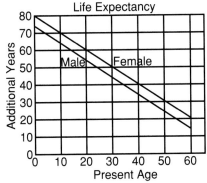

To about what age can a 20-year-old man expect to live?

A. 80   B. 75

C. 60   D. 55

13. The forumla $d = rt$ is used to find the distance (d) traveled when the rate of speed (r) and time traveled (t) are known. How far does a car travel in 2 hours at a rate of 45 miles per hour?

A. 45 miles   B. 47 miles

C. 90 miles   D. 245 miles

14. Which shows the number 3 and 24 hundredths?

A. 0.24   B. 0.324

C. 3.024   D. 3.24

15. _____

The line segment above measures about

A. 6 meters   B. 60 millimeters

C. 60 meters   D. 60 centimeters

16. Which shows the number thirty-five million?

A. 35,000,000,000 B. 305,000,000

C. 350,000,000   D. 35,000,000

17. To the nearest tenth, 3.486 rounds to

A. 3.49   B. 3.48

C. 3.5   D. 3.4

Test-Taking Tip

Most standardized tests have a time limit, so you must use your time wisely. Some questions will be much easier than others. If you cannot answer a question within a few minutes, go on to the next one. If there is still time left when you get to the end of the test, go back to the questions that you skipped.

Take time out after several problems to freshen your mind.

18. One boat is 5.05 meters long and another is 4.8 meters long. How much space is needed to park the boats end-to-end?

A. 9.85 m   B. 9.13 m

C. 5.53 m   D. 0.25 m

19. Elise bought pencils that cost $0.29 each and pens that cost $0.89 each. What do you need to know to find out how much she spent?

A. the cost of each pen and pencil

B. how much money she has

C. how much change she received

D. how many pens and how many pencils she bought

20.

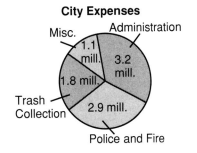

Which item costs about 2 million dollars?

A. administrative   B. police and fire
   costs

C. miscellaneous   D. trash collection

Academic Skills Test

Decimals: Multiplication

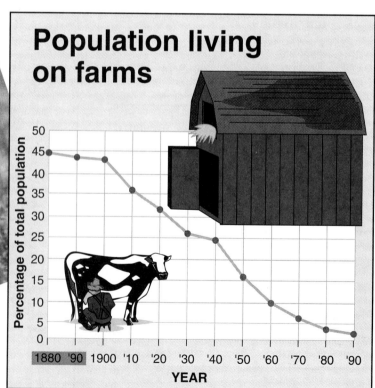

Spotlight on Farmland

Have You Ever Wondered...

- How the number of farms in the United States has changed over the years?

- What percent of the population lives on farms?

Population living on farms

Percentage of total population

YEAR

THE FAR SIDE By GARY LARSON
© Chronicle Features, 1984
CAR!

Looking Ahead

In this chapter, you will see how mathematics can be used to answer the questions about farmland. The major objectives of the chapter are to:

- use powers and exponents in expressions
- solve problems by solving a simpler problem
- multiply decimals
- find perimeter, area, and circumference

Chapter Project

Farmland

Work in a group.

1. Make a list of the foods that you eat that are grown or produced on a farm.

2. Pretend that you are starting a farm. Describe what you need to start your farm.

3. Draw a scale model of your farm. Include the size of the farm and how it would be divided among your different crops.

U.S. Farmland

Year	Number of farms (thousands)	Acres (thousands)	Average acreage
1880	4,009	536,082	134
1890	4,565	623,219	137
1900	5,740	841,202	147
1910	6,366	881,431	139
1920	6,454	958,677	149
1930	6,295	990,112	157
1940	6,102	1,065,114	175
1950	5,388	1,161,420	216
1960	3,962	1,176,946	297
1970	2,954	1,102,769	373
1980	2,432	1,038,885	427
1990	2,143	988,000	461

4-1 Estimating Products

Objective

Estimate products using rounding and compatible numbers.

Are you a *numismatist*? Many teens probably are. A numismatist is a collector of coins, tokens, and money.

Alisa and her family went to Spain for a vacation. In Spain, the basic unit of money is the *peseta*. During the time of their visit, one U.S. dollar was equal to 95.24 pesetas. If they paid $78 per night at the hotel, about how much was that in pesetas?

To find the answer to this question, you need to multiply 78 and 95.24. Since an exact answer is not needed, you can estimate.

One way to estimate a product is to use rounding. First, round each factor to its greatest place-value position. Then multiply. Do not round 1-digit factors.

$$
\begin{array}{rll}
\$78 & \rightarrow & 80 \qquad \textit{Round 78 to 80.} \\
\times 95.24 & \rightarrow & \times\,100 \quad \textit{Round 95.24 to 100.} \\
\hline
& & 8{,}000
\end{array}
$$

The hotel was about 8,000 pesetas per night.

Examples

1 Estimate 377 × 5.

$$
\begin{array}{rll}
377 & \rightarrow & 400 \\
\times 5 & \rightarrow & \times 5 \\
\hline
& & 2{,}000
\end{array}
$$

377 × 5 is about 2,000.

2 Estimate 12.6 × 3.

$$
\begin{array}{rll}
12.6 & \rightarrow & 10 \\
\times 3 & \rightarrow & \times 3 \\
\hline
& & 30
\end{array}
$$

12.6 × 3 is about 30.

Example 3 *Problem Solving*

Publishing On an average day, 125 books are published in the United States. About how many books are published in a year?

To find the answer, you need to multiply 125 by 365 since there are 365 days in a year.

$$
\begin{array}{rll}
365 & \rightarrow & 400 \qquad \textit{Round to the nearest hundred.} \\
\times 125 & \rightarrow & \times\,100 \quad \textit{Round to the nearest hundred.} \\
\hline
& & 40{,}000
\end{array}
$$

There are about 40,000 books published in a year.

You can review compatible numbers on page 18.

Another way to estimate products is to use compatible numbers. In multiplication, compatible numbers are numbers whose product equals some power of 10. It is easy to find these products mentally.

Example 4 *Problem Solving*

Calculators Akiro uses a calculator to find the product of 40.32 and 251. He gets 1012.032 for an answer. Use estimation to check Akiro's answer.

$$40.32 \quad \rightarrow \quad 40 \qquad \textit{Round to the nearest ten.}$$
$$\times\, 251 \quad \rightarrow \quad \times\, 250 \qquad \textit{Round to 250 since it is easy to}$$
$$\textit{multiply 4 and 25 mentally.}$$

Since $4 \times 25 = 100$, then $40 \times 250 = 10{,}000$. Akiro's answer, 1012.032, is not close to the estimate, 10,000. Akiro should enter the problem in the calculator again.

Checking for Understanding

Communicating Mathematics

Read and study the lesson to answer each question.

1. **Tell** how you could use rounding to estimate 26×52.

2. **Write,** in your own words, how you could use compatible numbers to estimate 403×25.

3. **Tell** which is greater: the exact product of 22.3×5 or the estimated product. Explain.

Guided Practice

Write the rounded numbers that could be used to estimate each product.

4. 37×289 5. 2.6×11 6. $52 \times \$1.04$

Write the compatible numbers that could be used to estimate each product.

7. 26×39 8. 48×21 9. $\$12.30 \times 85$

Estimate each product.

10. 26×41 11. 198×71 12. 22×43

13. Estimate the product $\$3.40 \times 76$.

14. Estimate the product 6.8×211.

15. Estimate the product 9×34.

Exercises

Estimate each product by rounding.

16. 66×18 **17.** 31×63 **18.** $42 \times \$57.99$

19. 4.8×77 **20.** 2.4×13.6 **21.** 283×479

Estimate each product by using compatible numbers.

22. $1,280 \times 732$ **23.** 493×209 **24.** $41 \times 2,438$

25. 19×542 **26.** 124×78 **27.** $29 \times \$7.69$

Estimate each product. Use an appropriate strategy.

28. 406×28 **29.** 334×78 **30.** $1,245 \times 83$

31. $2,473 \times 432$ **32.** $\$15.20 \times 9$ **33.** 221.6×3.8

34. Estimate the product of 4 and $28.98.

35. Estimate $17,350 \times 53$.

36. Student Council At Gillian School there are 36 student council representatives as shown on the circle graph at the right. Which grade (both boys and girls) has the most representatives? *(Lesson 3-9)*

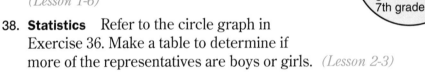

**Student Council
Representatives**

37. Find the value of $11 - 3 \div 3 + 2$. *(Lesson 1-6)*

38. Statistics Refer to the circle graph in Exercise 36. Make a table to determine if more of the representatives are boys or girls. *(Lesson 2-3)*

39. Solve the equation $n = 3.569 + 781.2$. *(Lesson 3-8)*

40. Solve the equation $45 - x = 30$. *(Lesson 1-9)*

41. Museum On an average day, the Smithsonian Institution adds 2,700 items to its collection. About how many items does it add in a week? in a month? in a year?

42. Critical Thinking Deidre buys a dozen doughnuts and two gallons of apple cider for her Halloween party. If the doughnuts cost 27¢ each and the cider costs $2.79 a gallon, estimate how much she spent altogether.

43. Calculators Bill uses a calculator to multiply 63.9 and 5.2 and gets an answer of 3,322.8. Use estimation to check his answer. Do you think Bill's answer is correct?

44. Journal Entry Write one or two sentences to explain why it is important to estimate products when using a calculator.

4-2 Powers and Exponents

Objective

Use powers and exponents in expressions.

Words to Learn

factor
exponent
base
power
squared
cubed

DID YOU KNOW

The smallest fish and also the smallest vertebrate is the dwarf goby, found in fresh water in the Philippines and the Marshall Islands. The longest length of these tiny fish is 1.25 cm.

In 1990, about 1,000,000 people visited the John Shedd Aquarium in Chicago, Illinois. You can express 1,000,000 as the product $10 \cdot 10 \cdot 10 \cdot 10 \cdot 10 \cdot 10$.

When two or more numbers are multiplied, each of these numbers is called a **factor** of the product. For the sentence $2 \times 12 = 24$, 2 and 12 are factors of 24.

When a number is multiplied by itself several times, you can use an exponent to make the notation simpler. The **exponent** indicates the number of times the number, or **base,** is used as a factor.

$$1,000,000 = \underbrace{10 \cdot 10 \cdot 10 \cdot 10 \cdot 10 \cdot 10}_{6 \; factors} = 10^{\overset{\uparrow}{6}}_{\;base} \;\leftarrow\; exponent$$

A number is called a **power** when it is expressed using exponents. The powers 3^2, 4^3, and 6^4 are read as follows.

3^2 three to the second power, or three **squared**
4^3 four to the third power, or four **cubed**
6^4 six to the fourth power

Examples

1 Write $7 \cdot 7 \cdot 7 \cdot 7$ using exponents.

7 is the base. Because 7 is a factor 4 times, the exponent is 4.

$7 \cdot 7 \cdot 7 \cdot 7 = 7^4$

2 Write 5^3 as a product.

The base is 5. The exponent means 5 is a factor 3 times.

$5^3 = 5 \cdot 5 \cdot 5$

3 Evaluate 6^3.

The base is 6. The exponent 3 means 6 is a factor 3 times.

$6^3 = 6 \cdot 6 \cdot 6$
$\quad = 216$

The rules for order of operations can now be expanded to include powers.

Order of Operations	1. Do all powers before other operations.
	2. Multiply and divide in order from left to right.
	3. Add and subtract in order from left to right.

Example 4

Evaluate $4 \cdot 3^2 - 8$.

$$4 \cdot 3^2 - 8 = 4 \cdot 9 - 8 \quad \textit{Evaluate } 3^2 \textit{ first. } 3^2 = 3 \cdot 3 \textit{ or } 9$$
$$= 36 - 8 \quad \textit{Multiply 4 and 9. Then subtract.}$$
$$= 28$$

Example 5 *Connection*

Algebra   Evaluate x^5 if $x = 4$.

$$x^5 = 4^5 \qquad \textit{Replace } x \textit{ with 4.}$$
$$= 4 \cdot 4 \cdot 4 \cdot 4 \cdot 4$$
$$= 1,024$$

You can also use a calculator to find powers.

4 [yˣ] 5 [=] 1024   or   4 [×] 4 [×] 4 [×] 4 [×] 4 [=] 1024

Calculator Hint

• • • • • • • • • • •

Many calculators have a key labeled y^x. This key allows you to compute exponents very quickly. Suppose you want to find 16^5.

16 [yˣ] 5 [=]

1048576

The result appears immediately.

Checking for Understanding

Communicating Mathematics

Read and study the lesson to answer each question.

1. **Write** a paragraph explaining to an absent classmate what 4^3 means and how to evaluate 4^3.

2. **Show** a classmate how to evaluate 5^{10} on a calculator.

3. **Tell** how to find 1^{100}.

Guided Practice

Write each product using exponents.

4. $2 \cdot 2 \cdot 2$
5. $3 \cdot 3 \cdot 3 \cdot 3 \cdot 3$
6. $c \cdot c \cdot c \cdot c$

Write each power as a product.

7. 6^2
8. 10^7
9. 5^5
10. 9^6
11. 11^4

Evaluate each expression.

12. 2^4
13. 5^7
14. 10^8
15. 7^2
16. 1^5

17. Evaluate n^2 if $n = 3$.

Exercises

Independent Practice

Write each product using exponents.

18. $4 \cdot 4 \cdot 4 \cdot 4 \cdot 4 \cdot 4$ 19. $10 \cdot 10 \cdot 10$ 20. $y \cdot y \cdot y \cdot y$

21. $2 \cdot 2 \cdot 3 \cdot 3 \cdot 3 \cdot 3$ 22. $12 \cdot 12 \cdot 7 \cdot 7$ 23. $5 \cdot 5 \cdot 5 \cdot 1 \cdot 1$

Write each power as a product.

24. 15^4 25. 23^5 26. 7^7 27. 9^{10} 28. 6^5

29. 100^2 30. 14^6 31. 277^3 32. 88^8 33. $2{,}100^4$

Evaluate each expression.

34. 10^7 35. 6^8 36. 2^{10}

37. $3^5 \cdot 7^3$ 38. $5^2 \cdot 7^3 \cdot 10^4$ 39. $4^2 \cdot 3^4 \cdot 6^2$

40. $2^3 \cdot 5^2 \cdot 1^{10}$ 41. 5 squared   42. 7 cubed

43. Evaluate x^5 if $x = 6$.

44. If $m = 12$, what is the value of m^2?

Mixed Review

45. **Nutrition** In 1967, the Egyptian people ate more cereals than people in any other country, eating an average of 21.95 ounces per person. About how many ounces of cereals would a family of seven eat? *(Lesson 4-1)*

46. Estimate the quotient $781 \div 7$ using compatible numbers. *(Lesson 1-5)*

47. Solve the equation $33 - a = 30$ mentally. *(Lesson 1-9)*

48. Bob's answering machine can record fifty 30-second messages. How many minutes of tape does his machine have? *(Lesson 1-1)*

49. **Statistics** Is the mean a misleading measure of central tendency for this set of data: 17, 21, 20, 19, 17, 21, 18, 22, 21? Explain. *(Lesson 2-8)*

Problem Solving and Applications

50. **Finance** The money that the government of the United States owes to others is called the federal debt. In early 1992 the federal debt was about 3×10^9 dollars. How much money does this represent?

51. **Critical Thinking** Tell whether the following statement is *true* or *false*. Any power of 2 is an even number. Explain your answer.

52. **Science** A cubic meter of water has a mass of 10^3 kilograms. How much water is this in kilograms?

53. **Data Search** Refer to pages 114 and 115.
    a. How many fewer farms were there in 1980 than in 1880?
    b. How many times as large was the average acreage per farm in 1980 than the average acreage per farm in 1880?

DATA SEARCH

Reviewing Multiplication of Whole Numbers

In this lesson, you will review how to multiply whole numbers.

To multiply whole numbers, align the numbers on the right. Then multiply in each place-value position from right to left and add the products obtained.

Examples

Estimation Hint

● ● ● ● ● ● ● ● ● ●

Before solving a problem, estimate the answer. This way, you can compare the answer with the estimate and see if the answer looks correct.

1 Find 179 × 4.

Estimate: 200 × 4 = 800

Align the numbers on the right.	*Multiply the ones.*	*Multiply the tens.*	*Multiply the hundreds.*
179 × 4	3 179 × 4 6	33 179 × 4 16	33 179 × 4 716

716 is close to the estimate of 800.

2 Multiply 372 and 30.

Estimate: 400 × 30 = 12,000

Align the numbers on the right.		*Multiply by the ones.*	*Multiply by the tens.*
372 × 30	*Zero times any number is zero.*	372 × 30 0	2 372 × 30 11,160

3 Find 455 × 52.

Estimate: 500 × 50 = 25,000

Align the numbers on the right.	*Multiply by the ones.*	*Multiply by the tens.*	*Add.*
455 × 52	11 455 × 52 910	22 11 455 × 52 910 2275	22 11 455 × 52 910 2275 23,660

Exercises

Independent Practice

Multiply.

1. 237
   × 6

2. 569
   × 3

3. 44
   × 61

4. 87
   × 93

5. 459
   × 13

6. 678
   × 26

7. 1,207
   × 248

8. 2,102
   × 319

9. 44×15

10. $8 \times 1,250$

11. 80×900

12. 43×29

13. 142×200

14. $2,312 \times 57$

15. Find the product of 700 and 37.

16. What is 1,290 times 655?

17. What is 3,667 multiplied by 800?

Evaluate each expression if $a = 13$, $b = 48$, and $c = 430$.

18. ab

19. $2c$

20. bc

21. a^2

Problem Solving and Applications

22. **Television** The cost of Doogie's lab coat on the television show *Doogie Howser, M.D.* is $38. He uses six lab coats per season. How much is spent on his lab coats in one season?

23. **Astronomy** Pluto is about 39 times as far away from the sun as Earth is. If Earth's average distance from the sun is 93 million miles, how far away is Pluto?

24. **Critical Thinking** Find a pattern in the products below. How can you use your pattern to multiply mentally by 11?

$11 \times 12 = 132$
$11 \times 23 = 253$
$11 \times 34 = 374$
$11 \times 45 = 495$

Reviewing Multiplication of Whole Numbers **123**

4-3A Multiplying a Decimal by a Whole Number

A Preview of Lesson 4-3

Objective

Multiply decimals by whole numbers using models.

Materials

decimal models
colored pencils

In this lab, you will explore the multiplication of decimals by whole numbers using decimal models. In the activities below, the first factor of the multiplication will be represented by the number of rows on the decimal model. The second factor will be represented by the number of columns on the decimal model.

Activity One

Work in groups of three.

- Use 10-by-10 decimal models like the one shown at the right. The whole decimal model represents 1, and each row or column represents 0.1.

- Multiply 3 and 0.5 using models.

- Place three decimal models on top of each other to represent the factor 3.

- Color five columns green to represent 0.5.

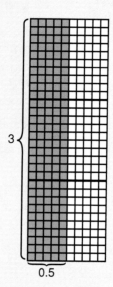

What do you think?

1. How many squares are green?

2. What is the product of 3 and 0.5?

Activity Two

Multiply 0.6 and 4.

- Take four decimal models and line them up as shown below.

- Color six rows green to represent 0.6.

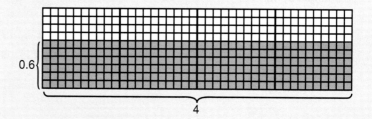

0.6

4

What do you think?

3. How many squares are green?

4. What is the product of 0.6 and 4?

5. What is the relationship between the total number of decimal places in the factors and the number of decimal places in the product?

6. What multiplication sentence does this figure represent? Explain your reasoning.

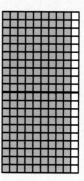

Use decimal models and colored pencils to find each product.

7. 3×0.2 8. 5×0.5 9. 0.1×3

Extension

10. Based on what you have learned in this lab, predict the product of 7 and 0.3. Justify your answer.

4-3 Multiplying Decimals by Whole Numbers

Objective

Multiply decimals by whole numbers.

At 15 years of age, Elizabeth Okino is considered one of the best gymnasts in the United States. She won the 1991 McDonald's American Cup and broke the world record set by Nadia Comaneci in 1976.

To accomplish this, Elizabeth practices 7.5 hours a day, 6 days a week. How many hours a week does she practice?

To find the answer to this question, you need to multiply 7.5 hours by 6 days. When multiplying a decimal by a whole number, multiply as with whole numbers.

One way to determine where to place the decimal point in the product is to use estimation.

Estimate: $8 \times 6 = 48$.

$$\begin{array}{r} 7.5 \\ \times\, 6 \\ \hline 45.0 \end{array}$$ *Multiply as with whole numbers.*

45.0 *Since the estimate is 48, place the decimal point after 5.*

Since multiplication is repeated addition, one way you can check this answer is to add 7.5 six times.

The answer checks. Elizabeth practices 45 hours per week.

$$\begin{array}{r} 7.5 \\ 7.5 \\ 7.5 \\ 7.5 \\ 7.5 \\ +\, 7.5 \\ \hline 45.0 \end{array}\;\checkmark$$

Another way to determine where to place the decimal point in the product is to count the number of decimal places in the decimal factor. The product must have the same number of decimal places. If more decimal places are needed, annex zeros.

Example 1

Multiply 5.02 and 3.

$$\begin{array}{r} 5.02 \\ \times\, 3 \\ \hline 15.06 \end{array}$$

5.02 ← *two decimal places*

15.06 ← *two decimal places*

Check your answer by adding.

$$\begin{array}{r} 5.02 \\ 5.02 \\ +\, 5.02 \\ \hline 15.06 \end{array}\;\checkmark$$

Example 2 *Problem Solving*

Sports Kareem Abdul-Jabbar is the all-time leader in points scored in professional basketball. In his 20 years in the NBA, he averaged 1,919.35 points scored per season. How many points did Abdul-Jabbar score in his career?

Multiply 1,919.35 by 20.

Estimate: 2,000 × 20 = 40,000

$$\begin{array}{r} 1{,}919.35 \\ \times\ 20 \\ \hline 38{,}387.00 \end{array}$$ ← *two decimal places*

← *two decimal places*

Compared to the estimate, the answer is reasonable.

Check with a calculator.

1919.35 20 ⊟ ∃8∃8⅂

Abdul-Jabbar scored 38,387 points in his career.

Checking for Understanding

Communicating Mathematics

Read and study the lesson to answer each question.

1. **Tell** why the product of 1.8×4 should be less than 8.

2. **Show** the multiplication of 2.45 and 6 using decimal models.

Guided Practice

Use estimation to place the decimal point in each product.

3. $0.44 \times 3 = 132$ 4. $1.54 \times 4 = 616$

5. $5.88 \times 5 = 2940$ 6. $4.91 \times 2 = 982$

Multiply.

7. $\begin{array}{r} 0.5 \\ \times\ 2 \\ \hline \end{array}$ 8. $\begin{array}{r} 0.12 \\ \times\ 5 \\ \hline \end{array}$ 9. $\begin{array}{r} 3.42 \\ \times\ 4 \\ \hline \end{array}$ 10. $\begin{array}{r} 11.5 \\ \times\ 7 \\ \hline \end{array}$

11. 6.32×6 12. 2.03×7 13. 4×0.28

14. Find the product of 4.91×5.

15. What is 3 multiplied by 0.07?

Exercises

Multiply.

16. 0.4
    × 8

17. 0.67
    × 75

18. 0.98
    × 112

19. 4.9
    × 328

20. 0.5
    × 452

21. 3.25
    × 802

22. 12.32
    × 1950

23. 160.5
    × 82

24. $2{,}388 \times 1.65$

25. 64×0.005

26. 4.03×112

27. 754×0.0125

28. $1{,}250 \times 2.5$

29. 15.3×39

30. Find the product of 0.36 and 88.
31. What is 2,967 times 0.071?

Solve each equation.

32. $a = 5 \times 0.61$

33. $b = 21 \times 0.81$

34. $p = 107 \times 0.88$

35. $q = 1.56 \times 34$

36. Evaluate the expression $2^3 \cdot 3^2$. *(Lesson 4-2)*
37. Round 345.6278 to the nearest tenth. *(Lesson 3-5)*
38. **Statistics** Order this set of decimals and circle the median: 67.4, 67.13, 7.123, 67.09, 66.98. *(Lesson 3-4)*
39. Multiply $8{,}255 \times 2.13$. *(Lesson 4-3)*
40. **Population** 1988 census data showed Denver with a population of 492,200 people and Atlanta with a population of 420,220 people. How many more people were living in Denver? *(Lesson 1-1)*

41. **Money Exchange** If the Japanese yen (¥) is worth $0.0074, what is the value of ¥3,750?
42. **Astronomy** Pluto, normally the farthest planet from the sun, is also the slowest. Its average speed around the sun is 10,604 miles per hour. Earth, by contrast, travels 6.28 times faster. What is the average speed of Earth?
43. **Number Sense** Copy the boxes at the right. Arrange the digits 0 through 9 in the boxes so that the product is the greatest possible. Use each digit only once.

    □ □,□ □ □
    × □ □,□ □ □

44. **Critical Thinking** Place the decimal point in one of the factors so that the product is correct.

    19235 × 200 = 384.7

45. **Botany** Some types of bamboo, the only food that pandas eat, can grow as much as 3.4 inches per day. At this rate, how much could it grow in a week?

4-4 Using the Distributive Property

Objective

Compute products mentally using the distributive property.

Words to Learn

distributive property

Suppose you are the owner of the Musical Store. You sell rock posters for $2.50 each. If you sell 16 posters on Monday and 14 on Tuesday, how much money did your store make from the sale of posters those two days?

Before you solve this problem, let's introduce another way to show multiplication, parentheses. For example, you can write $2 \times a$ as $2(a)$. Let's include parentheses with the rules for order of operations.

Order of Operations	1. Do all operations within parentheses first.
	2. Do all powers before other operations.
	3. Multiply and divide in order from left to right.
	4. Add and subtract in order from left to right.

Now let's go back to the problem of how much money the Musical Store made on posters Monday and Tuesday. One way to solve it is by multiplying the price by the total number of posters sold.

price of   number of
a poster   posters sold
  ↓   ↓
$2.50(16 + 14) = 2.50(30)$ *Use the order of operations.*
$\qquad\qquad\quad = 75$ *Add inside the parentheses first.*

Another way to solve the problem is by finding how much money the store made each day and then adding.

money earned   money earned
on Monday   on Tuesday
    ↓   ↓
$2.50 \times 16 + 2.50 \times 14 = 40 + 35$ *Use the order of operations.*
$\qquad\qquad\qquad\qquad = 75$ *Do the multiplication first.*

On Monday and Tuesday the store sold $75 worth of rock posters. The solution is the same using both methods shown above. So, the following sentence is true.

$$2.50(16 + 14) = 2.50 \times 16 + 2.50 \times 14$$

This example illustrates the distributive property.

Distributive Property	**Arithmetic:**	$4(2 + 7) = 4 \cdot 2 + 4 \cdot 7$
	Algebra:	For any numbers a, b, and c, $a(b + c) = ab + ac$.

The distributive property allows us to solve problems *in parts*. This makes it easy to solve some multiplication problems mentally.

Example 1

Find 5×13 mentally using the distributive property.

$$5 \times 13 = 5(10 + 3) \qquad \textit{Use 10 + 3 for 13.}$$
$$= 5 \times 10 + 5 \times 3 \quad \textit{THINK: 50 + 15}$$
$$= 50 + 15$$
$$= 65$$

Example 2 *Problem Solving*

Earning Money   Suppose you babysit for the Suber family to earn some extra money. The Subers pay you $3 per hour. If you babysit 4.5 hours, how much will you make?

Estimate: $3 \times 5 = 15$

$$3 \times 4.5 = 3(4 + 0.5) \qquad \textit{Use 4 + 0.5 for 4.5.}$$
$$= 3 \times 4 + 3 \times 0.5 \quad \textit{THINK: 12 + 1.5}$$
$$= 13.5$$

You will make $13.50. Is the answer reasonable?

Checking for Understanding

Communicating Mathematics

Read and study the lesson to answer each question.

1. **Tell** how you could rewrite $12 \times 5 + 12 \times 8$ using the distributive property.
2. **Tell** the order of operations you would use to find $7(3^2 + 5)$.

Guided Practice

Rewrite each expression using the distributive property.

3. $3(10 + 2)$ 4. $12(4 + 0.5)$ 5. $8(50 + 3)$

Find each product mentally. Use the distributive property.

6. 6×16 7. 12×15 8. 2.4×11
9. 4.3×20 10. 3.5×12 11. 5.2×14

Exercises

Independent Practice

Find each product mentally. Use the distributive property.

12. 6×17
13. 104×8
14. 15×14
15. 30×3.1
16. 6×4.2
17. 110×13
18. 7.9×80
19. 8.9×5
20. 10.5×40
21. 0.9×15
22. 20.7×6
23. 20×2.08

24. Find the product of 1.5 and 40 mentally.
25. Find the product of 7 and 30.9 mentally.

Mixed Review

26. State the greater number: 0.091 or 0.1. *(Lesson 3-4)*
27. Write 17^7 as a product. *(Lesson 4-2)*
28. Make a horizontal bar graph for the set of data at the right. *(Lesson 2-5)*

Raul's Earnings				
Mon.	Tues.	Wed.	Thurs.	Fri.
$21.50	13.75	19.15	20.00	25.50

Problem Solving and Applications

29. **Critical Thinking** Use the distributive property to write two expressions for the figure at the right.

30. **Music** Bruce's piano teacher charges $12.50 for a lesson. If Bruce has 5 lessons in one month, how much does he owe his piano teacher? Solve mentally.

31. **Journal Entry** Write a paragraph in your own words explaining when the distributive property can help you solve a problem mentally.

Mid-Chapter Review

1. **Buildings** The World Trade Center in New York is 110 stories tall. If an average story is about 12 feet high, about how tall is the World Trade Center in feet? *(Lesson 4-1)*

Write each power as a product. *(Lesson 4-2)*

2. 4^5
3. 10^1
4. 3^7
5. 25^4
6. 300^3

Solve each equation. *(Lesson 4-3)*

7. $c = 16 \times 2.3$
8. $k = 1.55 \times 30$

9. Find the product of 3.6 and 20 mentally. *(Lesson 4-4)*

10. **Consumer Math** You buy 7 movie tickets for $6.50 each. How much do they cost? Solve mentally using the distributive property. *(Lesson 4-4)*

DECISION MAKING

Organizing a Read-A-Thon

Situation

In order to raise money to keep your local library open, the middle schools in your district have agreed to organize a Read-A-Thon. Students will read books and ask friends and family to sponsor them. The sponsors will pledge a certain amount of money for each book that a student reads. Your class has set a goal of $500. You need to figure out how many books students need to read and how many sponsors they must have in order to collect $500 in pledges.

Hidden Data

How many students will participate?
How many weeks will students have to complete the Read-A-Thon?
Is there a range of pledges that sponsors should be asked to make, perhaps $0.50 - $1.00 per book?

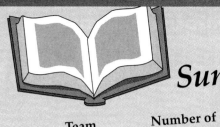

1991 Harris Middle School

Summary

Team (Grade/Number)	Number of Participants	Books Read	Pledges Collected
6-A	108	972	$ 923.40
6-B	124	420	353.00
6-C	133	1,197	1,199.50
7-A	95	665	565.25
7-B	135	897	672.75
7-C	126	504	252.00
8-A	111	675	708.75
8-B	137	1,233	1,243.50
8-C	101	584	423.75
Total	1,070	7,147	6,341.90

READ A THON

Analyzing the Data

1. About how many books were sixth-grade students able to read for the Read-A-Thon?
2. On the average, were the pledges for Team 6-B more or less than $1.00 per book?
3. About how much money did each student at Harris Middle School collect in the 1991 Read-A-Thon?
4. On the average, how much money did sponsors pledge per book for Team 7-A?

Making a Decision

5. **Record** your estimate for each in order to reach your goal of $500.
   a. the number of students that need to participate
   b. the number of books each student will need to read
   c. the pledge amount per book
6. Using your estimates, **revise** the specifics of the Read-A-Thon, such as the length of time allowed for reading or the minimum amount of each pledge so that you can meet your goal.

Making Decisions in the Real World

7. **Investigate** the amount of library donations made by people in your community. Is your goal reasonable?

4-5A Multiplying Decimals

A Preview of Lesson 4-5

Objective

Multiply decimals using models.

Materials

decimal models
colored pencils

On page 124, you used decimal models to multiply a whole number and a decimal. In this lab, you will use decimal models to multiply two decimals.

Activity One

Work with a partner.
Multiply 0.4 and 0.6 using models.

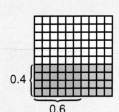

- Use a 10-by-10 decimal model. The whole decimal model represents 1, and each row or column represents 0.1.

- Color 4 rows of the model blue to represent 0.4.

- Color 6 columns of the model yellow to represent 0.6.

What do you think?

1. How many squares are green? What is the product?
2. How many decimal places were there in both factors? How many are in the product? Describe the relationship.

Activity Two

Multiply 1.3 and 0.2 using models.

- Use two decimal models.
- Color 13 rows blue to represent 1.3.
- Color 2 columns yellow to represent 0.2.

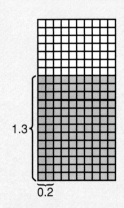

What do you think?

3. How many squares are green? What is the product?
4. How many decimal places were there in both factors? How many are in the product? Describe the relationship.

4-5 Multiplying Decimals

Objective

Multiply decimals.

When Carl was in Germany as an exchange student, he wanted to buy some German deutsch marks (DM), the German currency, to take home. At a bank in Frankfurt, the teller told Carl that $1 was worth 1.71 DM. If Carl had $2.25, how many deutsch marks could he buy?

Economist

Economics focuses on how money is earned and distributed. Economists specialize in areas of study that include public finance, money supply and banking, international trade, and employment.

A college degree and a broad knowledge of finance and statistics are required to become an economist.

For more information contact:
American Economics Association
1313 21st Ave.
Suite 809
Nashville, TN 37212

To answer this question, you need to multiply 1.71 by 2.25. When you multiply decimals, multiply as with whole numbers. One way to place the decimal point is to use estimation.

Estimate: $2 \cdot 2 = 4$ *Round 1.71 to 2 and 2.25 to 2.*

$$\begin{array}{r} 1.71 \\ \times 2.25 \\ \hline 3.8475 \end{array}$$

Multiply as with whole numbers.

Since your estimate is 4, place the decimal point after the 3.

Rounded to the nearest hundredth, Carl could buy 3.85 DM (or 3 DM and 85 pfennigs) with $2.25.

Another way to place the decimal point in the product is by counting the decimal places in each factor. The product will have the same number of decimal places as the sum of the number of decimal places in the factors.

Examples

1   Find $3.4 \cdot 1.2$.   *Estimate: $3 \cdot 1 = 3$*

$$\begin{array}{r} 3.4 \\ \times 1.2 \\ \hline 68 \\ 34 \\ \hline 4.08 \end{array}$$

← *one decimal place*
← *one decimal place*

← *two decimal places*

The product is 4.08. Compared to the estimate, it seems reasonable.

2 Multiply 0.03 and 1.24.   *Estimate: 0 · 1 = 0*

$$
\begin{array}{r}
1.24 \\
\times 0.03 \\
\hline
0.0372
\end{array}
$$

  1.24   ← *two decimal places*
×0.03   ← *two decimal places*
0.0372   ← *To make four decimal places, annex a zero on the left.*

The product is 0.0372.

3 Find 4.32 · 2.3.   *Estimate: 4 · 2 = 8*

  4.32   ← *two decimal places*
× 2.3   ← *one decimal place*
  1296
  864
9.936   ← *three decimal places*

The product is 9.936.

| Example 4 | *Connection* |

Algebra   Evaluate 5.6*n* if *n* = 12.43.

$5.6n = 5.6 \cdot 12.43$

Estimate: 6 · 12 = 72

Use a calculator.   *Replace n with 12.43.*

5.6 ⊠ 12.43 ⊟ 𝟨𝟫.𝟨𝟢𝟪

Checking for Understanding

Communicating Mathematics   Read and study the lesson to answer each question.

1. **Show** how to place the decimal point in the product 0.829 · 31.45 = 2607205 by using estimation.

2. **Write** the multiplication sentence represented by the decimal model at the right.

3. **Write** two different methods you can use to place the decimal point in a product of two decimals.

Guided Practice   Place the decimal point in each product.

4. 0.35 · 1.4 = 0490   5. 5.2 · 0.065 = 03380

6. 3.06 · 4.28 = 130968   7. 0.9 · 0.15 = 0135

Multiply.

8. 2.15 · 3.84   9. 1.3 · 7.3   10. 0.3 · 0.012

Exercises

Independent Practice

Multiply.

11. $9.85 \cdot 1.16$ 12. $2.3 \cdot 89.6$ 13. $0.4 \cdot 3.6$

14. $8.37 \cdot 703.22$ 15. $0.0009 \cdot 0.008$ 16. $0.6 \cdot 0.031$

17. Find the product of 14.3 and 2.864.

Solve each equation.

18. $x = 28.2 \cdot 4.4$ 19. $y = 102.13 \cdot 1.25$

20. $n = 0.002 \cdot 3.9$ 21. $a = 10.356 \cdot 12.21$

Evaluate each expression if $a = 10.6$, $b = 2.02$, and $c = 0.005$.

22. ab 23. $a(b + c)$ 24. abc 25. $bc(a - c)$

Mixed Review

26. Find the product of 6.8 and 40 mentally, using the distributive property. *(Lesson 4-4)*

27. Solve the equation $56 \div n = 8$. *(Lesson 1-9)*

28. **Fishing** Alphonse Bielevich caught an Atlantic cod fish in New Hampshire that weighed 98.75 pounds. Donald Vaughn caught a Pacific cod fish in Alaska that weighed 30 pounds. What is the difference in the weights? *(Lesson 3-8)*

29. Write the number 204.2398 in words. *(Lesson 3-2)*

Problem Solving and Applications

30. **Military Supplies** The graph at the right shows a sample of the average number of supplies given to each United States soldier in the Persian Gulf war in 1991. The number of tubes of lip balm is 3.2 times the number of cans of foot powder per soldier. Complete the graph by finding the average number of tubes of lip balm given to each soldier.

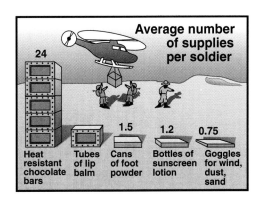

31. **Critical Thinking**
    a. When is the product of two decimals less than both factors?
    b. When is the product of two decimals greater than both factors?

32. **Research** In the business section of your local newspaper, find the foreign currency exchange column. Determine how much of each of the following you could get with your allowance or money earned babysitting: Japanese yen, French franc, Mexican peso, and British pound.

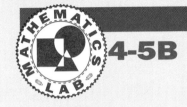

4-5B Spreadsheets

A Follow-Up of Lesson 4-5

Objective

Explore the uses of a computer spreadsheet.

Words to Learn

spreadsheet
cell

A computer **spreadsheet** arranges data and formulas in a column and row format. A spreadsheet is used for organizing and analyzing data.

A spreadsheet is made up of **cells.** A cell can contain data, labels, or formulas. Each cell has a name. The cell name can be found by combining the letter from the top of a particular column with the number found in a particular row. For example, the first cell in the upper left hand corner of a spreadsheet is called cell A1. The word *Item* is stored in cell A1.

The spreadsheet below shows a price list for a retail store.

	A	B	C	D	E
1	ITEM	REGULAR PRICE	RATE OF DISCOUNT	DISCOUNT	SALE PRICE
2	BLOUSE	$35.99	20%	$7.20	$28.79
3	SKIRT	$49.99	25%	$12.50	$37.49
4	JEANS	$29.90	15%	$4.49	$25.41
5	SHORTS	$15.00	20%	$3.00	$12.00
6	SWEATSHIRT	$18.99	35%	$6.65	$12.34
7	T-SHIRT	$19.95	10%	$2.00	$17.95

What do you think?

1. What is stored in cell B5?
2. What is stored in cell D1?
3. Name the cell that holds the data $25.41.
4. Name the cell that holds the label Sale Price.
5. How do you think the numbers in column E are computed?

Extension

6. Make a list of three different kinds of data you would like to put on a spreadsheet.
7. Interview a person that uses spreadsheets on a daily basis. Find out what type of data they use.

4-6 Perimeter

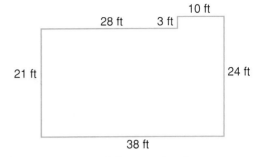

Objective

Find perimeter.

Words to Learn

perimeter
side
length
width

DID YOU KNOW

The word perimeter comes from two Greek words *peri* and *metron*, meaning "to measure around."

Mr. Nguyen wants to put a fence around a section of his back yard so his dog can play. How much fencing will he need?

Mr. Nguyen needs to know the perimeter of the section he wants to fence so he can know how much fencing material to buy. The **perimeter** *(P)* of any closed figure is the distance around the figure. You can find the perimeter by adding the measures of the **sides** of the figure.

$$P = 28 + 3 + 10 + 24 + 38 + 21$$
$$P = 124$$

The perimeter of Mr. Nguyen's dog run is 124 feet. So, Mr. Nguyen needs 124 feet of fencing.

There is an easier way to find the perimeter of a rectangle. Since opposite sides of a rectangle have the same length, you can multiply the **length** by 2 and the **width** by 2. Then add the products.

Perimeter of a Rectangle	**In words:** The perimeter of a rectangle is two times the length (*l*) plus two times the width (*w*).
	In symbols: $P = 2l + 2w$

Example 1 *Problem Solving*

Mental Math Hint

● ● ● ● ● ● ● ● ● ●

You can rewrite the formula for perimeter using the distributive property:
$2l + 2w = 2(l + w)$.

Gardening   Alicia has a rectangular vegetable garden in her back yard that is 6.1 meters long and 2.5 meters wide. She wants to put a border around her garden to keep the rabbits out. How much border will she need?

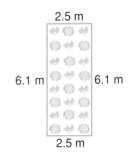

$P = 2l + 2w$
$P = 2 \times 6.1 + 2 \times 2.5$ *Replace l with 6.1*
$P = 12.2 + 5$ *and w with 2.5.*
$P = 17.2$

The perimeter of Alicia's garden is 17.2 meters. So, Alicia needs 17.2 meters of border.

An easy way to find the perimeter of a square is to multiply the length of one side by 4. You can use this formula because each side of a square has the same length.

Perimeter = 4 × length of one side

or

$P = 4s$

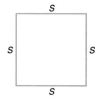

Example 2

Find the perimeter of a square tile whose sides measure 30.48 centimeters.

$P = 4s$
$P = 4 \times 30.48$ *Replace s with 30.48.*

4 ⊠ 30.48 ⊟ l2l.92

30.48 cm

The perimeter of the square tile is 121.92 centimeters.

Checking for Understanding

Communicating Mathematics

Read and study the lesson to answer each question.

1. **Tell** why the rectangle at the right gives you enough information to find its perimeter.

1.5 m
4.5 m

2. **Draw** and label a rectangle that has a length of 3 inches and a width of 2 inches. Find its perimeter.

3. **Draw** and label a square with a side of 1.2 centimeters. Find its perimeter.

Guided Practice

Find the perimeter of each figure.

4.
square
2.5 m

5.
13 in.
9 in.
13 in.

6.
14 m
6 m
6 m
17 m

7.
10 ft
22 ft   22 ft
10 ft

8.
1.5 m   1.5 m
1.5 m   1.5 m
1.5 m

9.
rectangle   1.63 m
4.81 m

Exercises

Independent Practice

Find the perimeter of each figure.

10.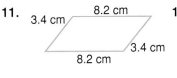

11.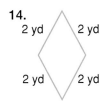

12.

13.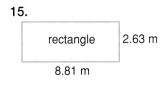

14. 2 yd   2 yd
      2 yd   2 yd

15.

16. Find the perimeter of a triangle if each side is 7.2 inches long.

17. Find the perimeter of the rectangle at the right if $a = 5.3$ and $b = 6.4$.

b cm
a cm

18. Draw and label a 5-sided figure that has a perimeter of 7.5 inches.

19. A square has a perimeter of 20 feet. What is the length of a side of this square?

Mixed Review

20. Multiply 0.0035 by 1.21. *(Lesson 4-5)*

21. Estimate the sum $21.1 + 19 + 20 + 20.3 + 20.1 + 18.8$ using clustering. *(Lesson 3-6)*

22. Data Analysis Explain why the graph at the right is misleading, and tell what false impression might be drawn from it. *(Lesson 2-8)*

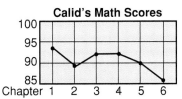

Problem Solving and Applications

23. Sports A football field is a rectangle 360 feet long, including end zones, and 160 feet wide. If a jogger runs completely around the outside of the field, how far will the jogger run?

24. History The Great Pyramid of Cheops, built in about 2600 B.C., has a square base with each side measuring 230.42 meters. What is the perimeter of the base?

25. Critical Thinking Use the perimeter of Mr. Nguyen's dog run and draw three other shapes that have the same perimeter. Do any of them make a better dog run? Why or why not?

4-7 Area of Rectangles and Squares

Objective

Find the area of rectangles and squares.

Words to Learn

area

A new sport was added for the 1992 Olympic games in Barcelona, Spain. The sport is badminton. Suppose the officials of the Department of Parks and Recreation in your community want to form a badminton team. The city decides to build a regulation badminton court that has the measurements at the right.

What is the area of the court?

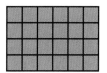

Before we can answer this question, we need to understand the concept of area. **Area** is the number of square units needed to cover a surface.

The rectangle at the right has an area of 24 square units.

Some units of area are the square inch (in²), square centimeter (cm²), and square foot (ft²).

Another way to find the area of a rectangle is to multiply.

Area of a Rectangle	**In words:** The area of a rectangle is the product of its length (l) and width (w).
	In symbols: $A = l \cdot w$

Mental Math Hint

● ● ● ● ● ● ● ● ● ●

You can use the distributive property to find 44 · 20.

44 · 20 = 20 · 44
= 20 (40 + 4)
= 800 + 80
= 880

Now we can find the area of the badminton court.

$A = l \cdot w$
$A = 44 \cdot 20$ *Replace l with 44 and w with 20.*
$A = 880$ The area is 880 square feet.

Example 1

Find the area of a rectangle with a length of 12 inches and a width of 5 inches.

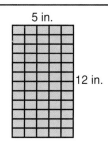

$A = l \cdot w$ *Replace l with 12 and w with 5.*
$A = 12 \cdot 5$
$A = 60$ *Check by counting squares.*

The area of the rectangle is 60 square inches.

Example 2 *Problem Solving*

Gardening   Sandy wants to cover her strawberry garden with nylon net to keep the birds from eating the strawberries. The garden is 12.5 feet long and 7.25 feet wide. How much net does she need to cover her garden?

$A = l \cdot w$
$A = 12.5 \cdot 7.25$ *Replace l with 12.5 and w with 7.25.*

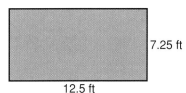
7.25 ft
12.5 ft

12.5 $\boxed{\times}$ 7.25 $\boxed{=}$ *90.625*

Sandy needs 90.625 square feet of nylon net.

Since each side of a square has the same length, you can square the measure of one of its sides to find its area.

Area of a Square	**In words:** The area of a square is the square of the length of one of its sides (s).
	In symbols: $A = s^2$

Example 3

Find the area of the square at the right.

$A = s^2$
$A = 1.5^2$ *Replace s with 1.5.*

1.5 in.

The area of the square is 2.25 square inches.

Checking for Understanding

Communicating Mathematics

Read and study the lesson to answer each question.

1. **Tell** a classmate how to find the area of a rectangle.

2. **Draw** a rectangle on centimeter grid paper that has an area of 20 square centimeters. Compare your answer with those of your classmates. What do you discover?

Guided Practice

Find the area of each rectangle.

3.
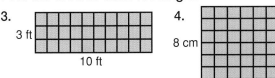
3 ft
10 ft

4.
8 cm

6 cm

5.
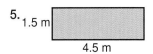
1.5 m
4.5 m

Find the area of each square.

6.

7 yd

7.

2.2 km

8.

0.45 in.

Exercises

Find the area of each rectangle or square described below.

9. *l*, 19 mm; *w*, 12 mm

10. *l*, 6.7 mi; *w*, 0.3 mi

11. *s*, 13 ft

12. *s*, 11.31 m

13.

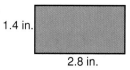

1.4 in.

2.8 in.

14.

1.67 yd

15.

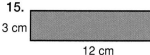

3 cm

12 cm

16.

3.1 m

17.

4.7 cm

0.2 cm

18.

0.09 ft

19. What is the area of a rectangle with a length of 13.2 meters and a width of 6.5 meters?

20. Estimate to determine if 919 is a reasonable answer for 428 + 691. *(Lesson 1-2)*

21. **Babysitting**   If Ellen works Monday through Friday babysitting and makes $10.75 each day she works, how much will she make in three weeks? *(Lesson 4-3)*

22. Find the perimeter of the rectangle at the right. *(Lesson 4-7)*

21 ft

5 ft

23. **Construction**   International soccer fields are rectangular and measure 100 meters by 73 meters. A new soccer field needs to be covered with sod. How many square meters of sod will be needed for the field?

24. **Critical Thinking**   Find the dimensions of all rectangles such that the measures of the length and width are whole numbers and the area is 36 square units.

25. **Remodeling**   A rectangular kitchen that is 11.5 feet by 24.5 feet is to be covered with square-foot tiles. How many tiles will be needed?

26. **Journal Entry**   Describe at least three real-life situations where you would need to find area.

4-7B Area and Perimeter

A Follow-Up of Lesson 4-7

Objective

Find relationships between perimeter and area.

Materials

centimeter grid paper
scissors

In this lab, you will use grid paper to experiment with area and perimeter and to see the relationship between them.

Try this!

Work in groups of three.

- Each member of the group should use a sheet of centimeter grid paper. Draw a rectangular shape with a perimeter of 48, staying on the lines. Examples of rectangular shapes with a perimeter of 14 centimeters are presented below.

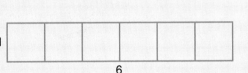

- Cut out your rectangular shape. Find the area by counting the number of squares. Compare with other members of the group.

What do you think?

1. Describe the perimeter of each rectangular shape.
2. Describe the area of each rectangular shape.
3. What can you conclude about the relationship between area and perimeter?

Extension

4. Suppose these shapes represent a dog pen you want to build with 48 feet of fence. Which rectangular shape would you choose? Why?
5. On centimeter grid paper, draw a rectangular shape with the least perimeter possible using exactly 12 squares of the grid. Find the perimeter of the shape. What conclusions can you make about a rectangular shape with the least perimeter?

4-8 Solve a Simpler Problem

Objective
Solve problems by first solving a simpler problem.

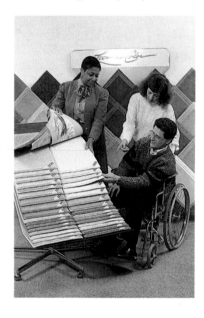

Paula wants to carpet her L-shaped dining room and living room area. How much carpet will she need?

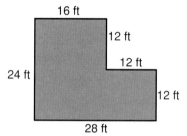

Explore   What do you know?
You know the dimensions of each room by looking at the diagram.
What do you need to find?
You need to find the area of the dining room and living room.

Plan   To find the area of the L-shaped rooms, you can first solve a simpler problem. Divide the L-shape into two regions. Find the area of each region. Then add the area of the regions together to find the total area.

Solve   Find the area of region X.
$A = l \times w$
$A = 16 \times 12$
$A = 192$

Find the area of region Y.
$A = l \times w$
$A = 28 \times 12$
$A = 336$

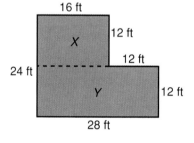

Add to find the total area.
$192 \text{ ft}^2 + 336 \text{ ft}^2 = 528 \text{ ft}^2$

Paula will need 528 square feet of carpet.

Examine   Check your solution by solving the problem another way. Divide the L-shape area differently and find the area.

Two students are making posters for the school band's winter concert. If they make two posters in two hours, how many posters can six students make in eight hours? Assume all the students work at the same rate.

Simplify the problem by finding out how long it takes each student to make one poster.
$2 \div 2 = 1$ Each student makes 1 poster in 2 hours.

Then find how many posters each can make in eight hours and multiply by the number of students, or six.

Divide 8 by 2 since each poster takes 2 hours.
$8 \div 2 = 4$ Each student can make 4 posters in 8 hours.

$6 \times 4 = 24$ Six students can make 24 posters in 8 hours.

Checking for Understanding

Communicating Mathematics

Read and study the lesson to answer each question.

1. **Tell** why solving a simpler problem can sometimes help you solve a more difficult problem.

2. **Write** the sequence of keystrokes you would use if you were using a calculator to solve the problem in the lesson opener.

Guided Practice

Solve by first solving a simpler problem.

3. Find the sum of the whole numbers from 1 to 100.

4. Find the number of minutes in the month of January.

5. Mike's Subs has made a submarine sandwich for fourteen people to share. How many cuts must be made to divide the sandwich equally among the fourteen people?

6. Find the sum of the first 100 odd numbers.

7. The Strand Theater seats 536 people. At last night's movie, there was one empty seat for every three people. How many people were in the movie theater?

Problem Solving

Practice Solve using any strategy.

8. Find the area of the figure below.

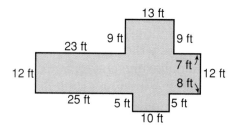

Strategies

• • • • • • • •

Look for a pattern.

Solve a simpler problem.

Act it out.

Guess and check.

Draw a diagram.

Make a chart.

Work backward.

9. A rectangular flower garden 12 feet long and 9 feet wide is divided into equal sections for 6 kinds of flowers. What are three possible perimeters for one section of the garden?

10. **Statistics** The Mad Hatters bowling team had the following scores in a tournament: 138, 149, 175, 134, 189, 203, 148, 120, 109, 156, 155, 145, 189, 176, 143, 109, 190, 133, and 176. Make a frequency table for this data.

11. How many 1-foot square tiles are needed to cover the floor of a kitchen that is 14 feet by 10 feet?

12. **Mathematics and Money** Read the following paragraph.

Coins with animals in their design make an interesting subject for a collection. Many coins carry a picture of an animal which is regarded as a national emblem, such as the kiwi bird of New Zealand. Ireland, an agricultural nation, has had a hen and chick, a pig, and a bull on its coins. Horses, antelope, and mythical beasts are all found on coins.

The smallest known coin ever to have been minted weighed 0.0005 ounce. How much would a stack of 28 of these coins weigh?

4-9 Circles and Circumferences

Objective

Find the circumference of circles.

Words to Learn

circle
center
radius
diameter
circumference

Have you ever used a telescope to look at the stars? Scientists use telescopes to obtain information about space objects. One kind of telescope is the *radio telescope.* This instrument has large dishes to collect data from space. The radio telescope located in Arecibo, Puerto Rico, has the largest dish. The shape of the dish is like a circle.

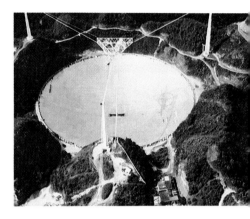

A **circle** is a set of points in a plane, all of which are the same distance from a fixed point in the plane called the **center.**

The distance from the center to any point on the circle is called the **radius** *(r)*. The distance across the circle through its center is called the **diameter** *(d)*. The diameter of a circle is twice the length of its radius. The diameter of the dish of the telescope in Puerto Rico is 305 meters.

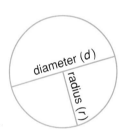

The **circumference** *(C)* is the distance around the circle.

In this Mini-Lab, you will see how circumference and diameter are related.

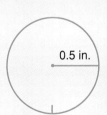

Mini-Lab

Work with a partner.
Materials: compass, scissors, pencil, notebook paper, inch ruler

0.5 in.

- With a compass, draw a circle with a radius of 0.5 inch.

- Cut the circle out and make a pencil mark anywhere on the edge.

- On a sheet of notebook paper, make a pencil mark on one of the lines. Place your circle on the line in such a way that the mark on the circle is on the mark of the line.

- Roll the circle until the mark on the circle touches the line again. Mark this spot on the line of the notebook.

Talk About It

a. What is the distance between the two marks on the paper?

b. What is the diameter of the circle ?

c. How do the diameter of the circle and this distance compare?

d. What can you say about the diameter of the circle and its circumference?

From the Mini-Lab, we can conclude that the circumference of a circle is directly related to its diameter. The circumference of a circle is a little more than three times its diameter. The number of times is called π (pi). π is about 3.14.

| **Circumference** | **In words:** | The circumference of a circle is equal to π times its diameter or π times twice its radius. |
| | **In symbols:** | $C = \pi d$ or $C = 2\pi r$ |

Example 1

Find the circumference of a circle with a radius of 6 inches. Use 3.14 for π.

$C = 2\pi r$
$\approx 2 \cdot 3.14 \cdot 6$ *Replace π with 3.14 and r with 6.*
≈ 37.68

The circumference of the circle is about 37.68 inches.

6 in.

Example 2 *Connection*

Science   Find the circumference of the Arecibo radio dish with a diameter of 305 meters. Use 3.14 for π.

$C = \pi d$ *Estimate: 3 × 300 = 900*
$\approx 3.14 \cdot 305$ *Replace π with 3.14 and d with 305.*

3.14 ⊠ 305 ⊜

The circumference of the Arecibo dish is about 957.7 meters.

Calculator Hint

● ● ● ● ● ● ● ● ● ● ●

π is such an important number in mathematics that it usually has its own key on a calculator. Does your calculator have this key? What value does it display when you press this key?

Checking for Understanding

Communicating Mathematics

Read and study the lesson to answer each question.

1. **Show** how you can use a compass and ruler to find the circumference of the circle at the right. Check your result using the formula for circumference.

1 cm

2. **Write,** in your own words, how to find the circumference of a circle with a diameter of 5 centimeters.

Guided Practice

Find the circumference of each circle. Use 3.14 for π. Round answers to the nearest tenth.

3.
5 mm

4.
3.1 m

5.
8 ft

6.
9.4 cm

7. $d = 4$ in.

8. $d = 2.5$ cm

9. $r = 0.95$ m

10. $r = 8.7$ ft

Exercises

Independent Practice

Find the circumference of each circle. Use 3.14 for π. Round answers to the nearest tenth.

11.
12 cm

12.
2.1 m

13.
10.5 ft

14.
13.6 in.

15. $d = 13.2$ m

16. $r = 0.6$ ft

17. $r = 63.2$ yd

18. $d = 1.25$ cm

19. If a circle has a radius of 12 miles, what is its circumference?

Mixed Review

20. Find the area of the rectangle at the right.
(Lesson 4-8)

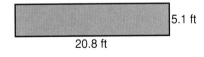
5.1 ft
20.8 ft

21. Find 8.5×4 mentally. Use the distributive property. *(Lesson 4-4)*

Problem Solving and Applications

22. **Critical Thinking** Tell how the circumferences of two circles compare if the diameter of one is twice as long as the diameter of the other.

23. **Measurement** A can is needed to hold tennis balls. The diameter of the circular lid is to be 2.9 inches. A rectangular piece of steel is needed to be bent into a tube to make the can. How wide should the rectangle be? Round your answer to hundredths.

Top
Side of can
◄ ? ►
Bottom

Study Guide and Review

Communicating Mathematics

Choose the letter that best matches each phrase.

1. distance from the center of a circle to any point on the circle
2. distance around any figure
3. twice the radius
4. $a(b+c) = ab+ac$
5. $l \cdot w$
6. $2\pi r$
7. one of the numbers being multiplied
8. number of times a base is used as a factor

   a. factor
   b. area
   c. circumference
   d. radius
   e. exponent
   f. perimeter
   g. diameter
   h. distributive property

9. Tell why you can multiply two decimal numbers and get an answer that is less than the factors you used.
10. Explain the two methods used when multiplying decimals for placing the decimal point correctly. Why is it a good idea to do both?

Skills and Concepts

Objectives and Examples

Upon completing this chapter, you should be able to:

Review Exercises

Use these exercises to review and prepare for the chapter test.

- estimate products using rounding and compatible numbers *(Lesson 4-1)*

  a. $\begin{array}{r} 31.78 \\ \times 4 \end{array}$ $\rightarrow$ $\begin{array}{r} 30 \\ \times 4 \\ \hline 120 \end{array}$ *31.78 × 4 is about 120.*

  b. $\begin{array}{r} 398.1 \\ \times 236 \end{array}$ $\rightarrow$ $\begin{array}{r} 400 \\ \times 250 \\ \hline 100,000 \end{array}$ *398.1 × 236 is about 100,000.*

Estimate each product.

11. 6.9×88
12. 4.2×39.6
13. 38.5×791

Estimate each product.

14. 4.2×240.7
15. $51.2 \times 1,891$
16. 121×7.9

- use powers and exponents in expressions *(Lesson 4-2)*

  $4^3 = 4 \cdot 4 \cdot 4 = 64$
  $6 \cdot 3^2 = 6 \cdot 3 \cdot 3 = 6 \cdot 9 = 54$

Evaluate each expression.

17. 4^6
18. $5^2 \cdot 3^2 \cdot 2^4$
19. x^3 if $x = 5$
20. ten squared

Objectives and Examples

• multiply decimals by whole numbers
  (Lesson 4-3)

  $$\begin{array}{r} 12.32 \\ \times\ 12 \\ \hline 147.84 \end{array}$$ *Estimate: 12 · 12 = 144 or count decimal places (2).*

• compute products mentally using the distributive property   *(Lesson 4-4)*

  $90 \times 3.8 = 90(3 + 0.8)$
  $\qquad\qquad = 90 \times 3 + 90 \times 0.8$
  $\qquad\qquad = 270 + 72$
  $\qquad\qquad = 342$

• multiply decimals   *(Lesson 4-5)*

  $$\begin{array}{r} 38.76 \\ \times\ 4.2 \\ \hline 7752 \\ 15504 \\ \hline 162.792 \end{array}$$ ← *two decimal places*
  ← *one decimal place*

  ← *three decimal places*

• find perimeter   *(Lesson 4-6)*

  6.15 m   3.21 m

  $P = 2l + 2w$
  $\quad = 2 \times 6.15 + 2 \times 3.21$
  $\quad = 12.30 + 6.42$
  $\quad = 18.72\ m$

• find the area of rectangles and squares   *(Lesson 4-7)*

  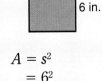

  8.3 m   4 m   6 in.

  $A = l \cdot w$
  $\quad = 8.3 \times 4$
  $\quad = 33.2\ m^2$

  $A = s^2$
  $\quad = 6^2$
  $\quad = 36\ in^2$

Review Exercises

Multiply.

21. $\begin{array}{r} 17.31 \\ \times\ 40 \\ \hline \end{array}$ **22.** $\begin{array}{r} 0.15 \\ \times\ 227 \\ \hline \end{array}$

23. Solve the equation $m = 201 \times 3.94$.

Find each product mentally. Use the distributive property.

24. 40×8.9 **25.** 30×10.76

26. 9×16 **27.** 5×6.8

Multiply.

28. $8.74 \cdot 2.23$ **29.** $0.04 \cdot 5.1$

30. $0.04 \cdot 0.0063$ **31.** $11.089 \cdot 5.6$

32. Evaluate $ca(b - a)$ if $a=2$, $b=5$, and $c=3$.

Find the perimeter of each figure.

33.
2 km   4.3 km   4.9 km

34.

8.1 in.   rectangle   25 in.

35. A square has a perimeter of 100 meters. What is the length of each side?

Find the area of each figure.

36. the rectangle in Exercise 34.

37. the square in Exercise 35.

38. a rectangle with $l = 11.2$ mi and $w = 9.3$ mi

39.
square   7.6 m

Objectives and Examples	Review Exercises

- find the circumference of circles
  (Lesson 4-9)

2.4 m

$c = \pi d = 2\pi r$
$= 2 \cdot 3.14 \cdot 2.4$
$= 15.072$

Find the circumference of each circle. Round to the nearest tenth.

40.
18.2 ft

41.
8.2 cm

Applications and Problem Solving

42. Find the area of the figure at the right. *(Lesson 4-8)*

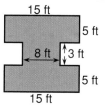

15 ft
5 ft
8 ft 3 ft
5 ft
15 ft

43. **Elevators**   The express elevator to the 60th floor of the "Sunshine 60" building in Tokyo, Japan, operates at the incredible speed of 22.72 miles per hour. If one mile per hour is the same as 88 feet per minute, what is the speed of this elevator in feet per minute? *(Lesson 4-3)*

44. **Fitness**   Janice exercises every morning by running once around a circular path surrounding a park in her neighborhood. The diameter of the circle of the path she follows is 1.7 kilometers. How far does she run? *(Lesson 4-9)*

Curriculum Connection Projects

- **Science**   Find the distance of three of our satellites from Earth. Then find the circumference of the orbit of each. Remember, the surface of Earth is about 6,400 kilometers from its center.

- **Language Arts**   Use a copy of today's newspaper to find, measure, and list the areas of various rectangular and square ads.

Read More About It

Korman, Gordon. *Son of Interflux.*

Rockwell, Thomas. *How To Get Fabulously Rich.*

Wilkinson, Elizabeth. *Making Cents: Every Kid's Guide to Making Money.*

154   **Chapter 4** Study Guide and Review

4 Test

Estimate each product.

1. $23 \cdot 4{,}108$

2. $49 \cdot 6.7$

3. $484 \cdot 319$

4. **Consumer Math**   Mr. Huang has bought a new car. His payments will be $319.50 a month for 48 months. *About* how much will he pay in all?

Evaluate each expression.

5. 2^5

6. n^4 if $n=3$

7. six squared

Multiply.

8. $\begin{array}{r} 0.81 \\ \times\ 22 \\ \hline \end{array}$

9. 22.38×803

10. $\begin{array}{r} 30.98 \\ \times\ 5.6 \\ \hline \end{array}$

Find each product mentally. Use the distributive property.

11. 21.8×30

12. 7×9.3

Solve each equation.

13. $x = 47.3 \cdot 5.6$

14. $20.86 \cdot 4.11 = m$

15. Evaluate $b(a + c)$ if $a = 7.8$, $b = 5.05$, and $c = 0.02$.

Find the perimeter of each figure.

16.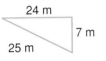

17. a square with sides of length 5.4 inches

18.
17.1 in.
10.5 in.

Find the area of each rectangle.

19. $l = 10.5$ mi; $w = 17.1$ mi

20.
0.04 cm
1.6 cm

21. **Decorating**   Jackie plans to put new wallpaper in her bedroom on one wall. If the wall measures 2.46 meters by 4.3 meters, how many square meters of wallpaper does she need?

Find the circumference of each circle. Round to the nearest tenth.

22. 30.5 ft

23. 1.1 in.

24. The radius is 1.6 km.

25. The diameter is 9.8 cm.

Bonus   The number 1.369 could be a measurement for many uses. Name one use where it might be necessary to be that exact, and one where it would seem silly to be so exact. Explain the difference.

Chapter

5

Decimals: Division

Spotlight on Sports

Have You Ever Wondered...

- In what sporting activities people participate most?
- What spectator sports are disliked by the most adults in the United States?

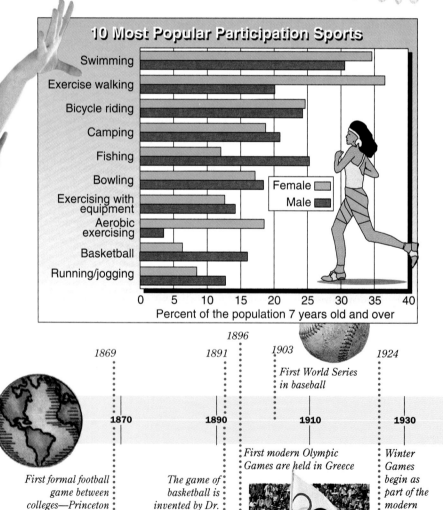

10 Most Popular Participation Sports

Swimming
Exercise walking
Bicycle riding
Camping
Fishing
Bowling
Exercising with equipment
Aerobic exercising
Basketball
Running/jogging

Female
Male

0   5   10   15   20   25   30   35   40
Percent of the population 7 years old and over

1869
1870
First formal football game between colleges—Princeton and Rutgers

1891
1890
The game of basketball is invented by Dr. James Naismith.

1896
First modern Olympic Games are held in Greece

1903
1910
First World Series in baseball

1924
1930
Winter Games begin as part of the modern Olympics

156

Sports

Work in a group.

1. Use a television schedule to make a chart that shows all the sporting events that are shown during one week.

2. Make a list that describes which sports are shown most often and at prime times.

3. From your list, which spectator sports do you think are most popular? Find out which sports your friends and family most enjoy watching to see if their information agrees with your list.

Looking Ahead

In this chapter, you will see how mathematics can be used to answer the questions about sports in the United States.

The major objectives of the chapter are to:

- divide decimals by whole numbers and decimals

- estimate and round quotients

- measure capacity and mass in the metric system

- solve equations by using inverse operations

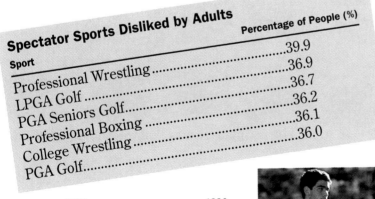

Spectator Sports Disliked by Adults	Percentage of People (%)
Sport	
Professional Wrestling	39.9
LPGA Golf	36.9
PGA Seniors Golf	36.7
Professional Boxing	36.2
College Wrestling	36.1
PGA Golf	36.0

1960
American Football League begins

1990

1950   **1970**   **1990**

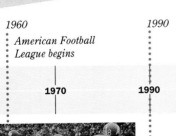

Pete Sampras wins U.S. Open in tennis

157

Reviewing Division of Whole Numbers

Objective

Divide whole numbers.

The next two pages review division of whole numbers.

Examples

Find each quotient.

Estimation Hint
● ● ● ● ● ● ● ● ● ● ●
You can use compatible numbers to help you estimate.

LOOK BACK

You can review compatible numbers on page 18.

---

1   252 ÷ 7   *Estimate: 280 ÷ 7 = 40*

From the estimate, you know that the quotient will have two digits.

$$\begin{array}{r} 3 \\ 7\overline{)252} \\ -21 \\ \hline 4 \end{array} \quad \rightarrow \quad \begin{array}{r} 36 \\ 7\overline{)252} \\ -21 \\ \hline 42 \\ -42 \\ \hline 0 \end{array}$$

Divide in each place-value position from left to right.

252 ÷ 7 = 36. The estimate shows that the answer of 36 is reasonable.

2   3,444 ÷ 42   *Estimate: 3,200 ÷ 40 = 80*

$$\begin{array}{r} 8 \\ 42\overline{)3,444} \\ -336 \\ \hline 8 \end{array} \quad \rightarrow \quad \begin{array}{r} 82 \\ 42\overline{)3,444} \\ -336 \\ \hline 84 \\ -84 \\ \hline 0 \end{array}$$

3,444 ÷ 42 = 82. The estimate shows that the answer of 82 is reasonable.

"When am I ever going to use this?"

Suppose you have been job interviewing and one company offers you $300 per week for 40 hours of work and another company offers you $200 for 32 hours of work. Which is the better job offer if you consider only pay per hour?

---

3   5,237 ÷ 68   *Estimate: 4,900 ÷ 70 = 70*

$$\begin{array}{r} 7 \\ 68\overline{)5,237} \\ -476 \\ \hline 47 \end{array} \rightarrow \begin{array}{r} 77 \\ 68\overline{)5,237} \\ -476 \\ \hline 477 \\ -476 \\ \hline \end{array} \rightarrow \begin{array}{r} 77 \text{ R1} \\ 68\overline{)5,237} \\ -476 \\ \hline 477 \\ -476 \\ \hline 1 \end{array}$$

5,237 ÷ 68 = 77 R1. The estimate shows that the answer is reasonable.

Exercises

Independent Practice

Find each quotient.

1. $8)\overline{248}$
2. $22)\overline{462}$
3. $29)\overline{3,451}$
4. $21)\overline{425}$
5. $29)\overline{812}$
6. $6)\overline{360}$
7. $93)\overline{558}$
8. $67)\overline{429}$
9. $13)\overline{4,550}$
10. $5)\overline{4,630}$
11. $34)\overline{837}$
12. $57)\overline{398}$

13. $728 \div 58$
14. $4,293 \div 105$
15. $847 \div 25$
16. $5,599 \div 509$
17. $6,924 \div 9$
18. $403 \div 3$
19. $769 \div 23$
20. $674 \div 8$
21. $5,124 \div 32$
22. $105 \div 31$
23. $1,054 \div 20$
24. $8,267 \div 65$
25. $131 \div 16$
26. $763 \div 35$

27. Find the quotient $643 \div 37$.
28. What is 750 divided by 5?
29. Find x if $x = 345 \div 4$.
30. Evaluate $r \div s$ if $r = 3,477$ and $s = 43$.

Problem Solving and Applications

31. **Geometry**   The area of a rectangular garden is 450 square feet. The width is 15 feet. What is the length of the garden?

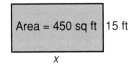

Area = 450 sq ft   15 ft

x

32. **Sales**   The Buckeye Band Boosters sold $8,610 worth of candy in 21 school days. What were their average daily sales?

33. **Critical Thinking**   Using mental math only, determine which quotient is greater, $6,924 \div 98$ or $6,924 \div 99$. Explain.

Reviewing Division of Whole Numbers   **159**

5-1 Dividing Decimals by Whole Numbers

Objective

Divide decimals by whole numbers.

TEEN SCENE

In 1989, *Sassy* was one of the fastest growing teen magazines. *Sassy* had a paid circulation of 450,000 and received more than one letter per subscription.

Rafael wants to subscribe to *Baseball Cards* magazine. The subscription costs $18.00 for 12 issues. How much does each issue cost?

You can estimate the answer.
$$20 \div 10 = 2$$

You can also use the division algorithm. When dividing a decimal by a whole number, the decimal point in the quotient is placed directly above the decimal point in the dividend. Then divide as you do with whole numbers.

$$
\begin{array}{r}
1.50 \\
12\overline{)18.00} \\
-12 \\
\hline
6\,0 \\
-6\,0 \\
\hline
00
\end{array}
$$

Place the decimal point.
Then divide as with whole numbers.
Use your estimate to help you check that the decimal point is placed correctly.

Each issue costs $1.50.

By comparing the actual answer with your estimate, you can see that your actual answer is reasonable.

Examples

Find each quotient.

Mental Math Hint

● ● ● ● ● ● ● ● ● ● ●

When dividing 15.6 by 13, think, "How many groups of 13 are in 15.6?"

1 $15.6 \div 13$

Estimate: $15 \div 15 = 1$

$$
\begin{array}{r}
1.2 \\
13\overline{)15.6} \\
-13 \\
\hline
2\,6 \\
-2\,6 \\
\hline
0
\end{array}
$$

Place the decimal point. Divide as with whole numbers.

$15.6 \div 13 = 1.2$
The estimate shows that the answer is reasonable.

2 $55.6 \div 8$

Estimate: $56 \div 8 = 7$

$$
\begin{array}{r}
6.95 \\
8\overline{)55.60} \\
-48 \\
\hline
7\,6 \\
-7\,2 \\
\hline
40 \\
-40 \\
\hline
0
\end{array}
$$

$55.6 \div 8 = 6.95$
The estimate shows that the answer is reasonable.

Checking for Understanding

Communicating Mathematics

Read and study the lesson to answer each question.

1. **Show** where to place the decimal point in the quotient for $3.588 \div 12$.

2. **Tell** why you can use an estimate to place the decimal point in the quotient for Example 1.

Guided Practice

Find each quotient.

3. $3 \overline{)29.37}$

4. $27 \overline{)5.13}$

5. $18 \overline{)124.92}$

6. $43 \overline{)111.37}$

7. $25 \overline{)43.25}$

8. $9 \overline{)75.6}$

9. $35.2 \div 100$

10. $4.37 \div 23$

11. $78.75 \div 35$

Exercises

Independent Practice

Find each quotient.

12. $61 \overline{)276.33}$

13. $11 \overline{)29.48}$

14. $32 \overline{)475.2}$

15. $42 \overline{)28.56}$

16. $28 \overline{)6.72}$

17. $10 \overline{)8.956}$

18. $34 \overline{)256.36}$

19. $48 \overline{)7.152}$

20. $52 \overline{)479.96}$

21. $6.25 \div 25$

22. $302.5 \div 55$

23. $117.44 \div 16$

24. $784.4 \div 37$

25. $263.25 \div 81$

26. $344.736 \div 76$

27. Find the quotient of 70.59 and 13.

Evaluate each expression if $a = 4.5$, $b = 24$, $c = 65.43$, and $d = 100$.

28. $a \div b$

29. $c \div a$

30. $c \div d$

31. $(a + b) \div d$

32. $(c - a) \div b$

Mixed Review

33. **Statistics** Erica's bowling scores for the month were 119, 134, 135, 125, 143, and 130.
    a. What is the range of her scores?
    b. What would be the best interval to use to graph?
    (Lesson 2-4)

34. **Sports** Allen and Tom finished a one-mile run first and second. Their times were 5.3 minutes and 5.14 minutes. Which decimal represents the least time? *(Lesson 3-4)*

35. Evaluate the expression $z + w$ if $z = 6.45$ and $w = 71.2$. *(Lesson 3-8)*

36. Find the product of 2.91 and 0.067. *(Lesson 4-5)*

37. **Geometry**   If the perimeter of a square is 18.56 cm, how long is one side of the square?

38. **Consumer Math**   It costs about $17.25 to fill up the gasoline tank on an average compact car. If the tank holds 15 gallons, what is the cost per gallon?

39. **Geography**   Long Beach and Laguna Beach in California are about 42.5 kilometers apart. On the Hammond's Road Atlas America, they are about 2 millimeters apart. About how many actual kilometers does a millimeter on the map represent?

40. **Make Up a Problem** that can be solved by dividing a decimal by a whole number.

41. **Critical Thinking**   What coins would you have if you have 35 coins that make $1.15?

CULTURAL KALEIDOSCOPE

Imke Durre

Imke Durre was born with cataracts in 1972. Three years later, her outlook became worse when glaucoma and failed surgery left her completely blind. In November of 1991, Imke was named one of only 13 advanced-placement scholars in the United States by the College Board. She is attending Yale University in Connecticut, where she is studying applied mathematics with an emphasis on weather and climate.

Her father, Karl Durre, developed a computer program that permits Imke to do her work—including mathematical problems—in braille. What she is brailling can be seen at the same time in print on the computer screen. Imke prints out her homework and tests for her teachers to review.

One of her high school math teachers said Imke had an extraordinary ability to visualize mathematical concepts and can work with equations describing three-dimensional objects. Imke would like to become a climatologist, someone who studies weather patterns, storms, and global warming.

5-1B Dividing by Powers of Ten

A Follow-Up of Lesson 5-1

Objective
Divide by powers of ten.

Materials
calculator

You learned about powers of ten in Chapter 4. Did you know you can divide by powers of ten by simply moving the decimal point?

Activity One

Work with a partner.

- Enter 50.25 into your calculator.
- Divide 50.25 by 10.

    50.25 ÷ 10 =

- Record your answer. Press the clear key.

- Now enter 496.25 into your calculator.
- Divide 496.25 by 10.

    496.25 ÷ 10 =

- Record your answer. Press the clear key.

- Enter 16.34 into your calculator.
- Divide 16.34 by 10.

    16.34 ÷ 10 =

- Record your answer. Press the clear key.

What do you think?

1. What happened to the placement of the decimal point each time you divided by 10?
2. How many places did the decimal point move upon dividing by 10?
3. Was the quotient greater or less than the dividend after you divided? Explain how you know.
4. What do you think would happen if you divided 34.68 by 10? Try it.
5. What do you think would happen if you divided 9.1 by 10? Try it.

Mathematics Lab 5-1B Dividing by Powers of Ten   **163**

Activity Two

Work with a partner.

- Enter 50.25 into your calculator.

- Divide 50.25 by 100.

    50.25 $\boxed{\div}$ 100 $\boxed{=}$

- Record your answer. Press the clear key.

- Now enter 496.25 into your calculator.

- Divide 496.25 by 100.

    496.25 $\boxed{\div}$ 100 $\boxed{=}$

- Record your answer. Press the clear key.

- Enter 16.34 into your calculator.

- Divide 16.34 by 100.

    16.34 $\boxed{\div}$ 100 $\boxed{=}$

- Record your answer. Press the clear key.

What do you think?

6. What happened to the placement of the decimal point each time you divided by 100?

7. How many places did the decimal point move upon dividing by 100?

8. Was the quotient greater or less than the dividend after you divided? Explain how you know.

9. Compare dividing by 10 to dividing by 100. How are they alike? How are they different?

10. What do you think would happen if you divided 34.68 by 100? Try it.

11. What do you think would happen if you divided 34.68 by 1,000? Try it.

12. Try dividing 64.19 by 10 and by 100 without using a calculator.

5-2 Rounding Quotients

Objective

Round decimal quotients to a specified place.

Jerry and five friends bought a six-pack of Gatorade® to drink after their baseball game. Each friend wants to pay for his share of the Gatorade®. If the six-pack costs $2.49, how much does each friend owe to the nearest cent?

Estimation Hint

• • • • • • • • • •

When dealing with money, always round the quotient up.

Usually when you divide with decimals the answer does not come out even. You need to round the quotient to a specified place-value position. To find out how much each person owes for the six-pack of Gatorade®, you need to round the quotient to the nearest cent. Always divide to one more place-value position than the place to which you are rounding.

$$\begin{array}{r} 0.415 \\ 6\overline{)2.490} \\ -24 \\ \hline 09 \\ -6 \\ \hline 30 \\ -30 \\ \hline 0 \end{array}$$

Estimate: $2.40 ÷ 6 = $0.40
Place the decimal point.
Annex a zero.
Divide to the thousandths place.

Calculator Hint

• • • • • • • • • •

Always estimate before using a calculator to solve problems. It is a good way to check that you have entered the numbers correctly.

Jerry and his five friends cannot pay $0.415 each. $0.415 rounded to the nearest cent is $0.42. Each person owes $0.42. The estimate shows that the answer is reasonable. To check this amount, multiply $0.42 and 6.

$0.42 × 6 = $2.52   *Why couldn't each person pay $0.41?*

Example

Round 72.89 ÷ 38 to the nearest tenth.

$$\begin{array}{r} 1.91 \\ 38\overline{)72.89} \\ -38 \\ \hline 34\,8 \\ -34\,2 \\ \hline 69 \\ -38 \\ \hline 31 \end{array}$$

Estimate: 80 ÷ 40 = 2
Place the decimal point.
Divide to the hundredths place.

You do not need to divide further.

72.89 ÷ 38 rounded to the nearest tenth is 1.9. The estimate shows the answer is reasonable.

You can solve this same problem using a calculator.

72.89 ⊡÷⊡ 38 ⊡=⊡ 1.9181579

Look at the digit one place to the right of the tenths place. To the nearest tenth, 1.9181579 rounds to 1.9.

Checking for Understanding

Communicating Mathematics

Read and study the lesson to answer each question.
1. **Tell** why you round the quotient up when dividing with money.
2. **Tell** to which place-value position you would divide to find the price of 1 orange if 3 oranges cost $1.00.

Guided Practice

Round each quotient to the nearest tenth.
3. 4.38 ÷ 2
4. 73.29 ÷ 65
5. 745.32 ÷ 52

Round each quotient to the nearest hundredth.
6. 650.23 ÷ 29
7. 55.39 ÷ 7
8. 233.45 ÷ 42

Exercises

Independent Practice

Round each quotient to the nearest tenth.
9. 73.5 ÷ 29
10. 87.5 ÷ 33
11. 327.54 ÷ 19
12. 6.47 ÷ 3
13. 144.47 ÷ 8
14. 77.5 ÷ 16

Round each quotient to the nearest hundredth.
15. 8.075 ÷ 14
16. 597.3 ÷ 62
17. 69.597 ÷ 45
18. 4.567 ÷ 21
19. 414.23 ÷ 79
20. 48.366 ÷ 51

Mixed Review

21. Evaluate the expression $ab - b$ if $a = 7$ and $b = 3$. *(Lesson 1-7)*
22. Write the number 507,269 in words. *(Lesson 2-1)*

Problem Solving and Applications

23. **Medicine** A pharmacist has 84.730 grams of a prescription medicine. He wants to separate it into 48 capsules. To the nearest thousandths of a gram, how many grams will go into each capsule?

24. **Travel** Trish's family traveled 1,345.8 miles on their vacation trip. They used exactly 55 gallons of gasoline. To the nearest tenth, what was their average number of miles per gallon?

25. **Smart Shopping** Which is the better buy, a 305.8-gram can of soup for 42 cents or a 539.4-gram can for 73 cents?

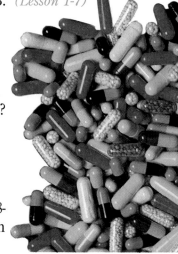

5-3A Dividing by Decimals

A Preview of Lesson 5-3

Objective

Use models to divide decimals by decimals.

Materials

decimal models

In Chapter 3, you used base-ten blocks to model decimals. Now you can use your decimal models to divide decimals by decimals. Just remember,

 represents one and ▯ represents one tenth.

Try this!

Work with a partner.

Use your blocks to find 1.2 ÷ 0.4.

- Place one and two tenths in front of you.

- Trade your ones block for ten tenths.

- Separate the tenths in front of you into groups of four tenths.

What do you think?

1. Why did you need to trade your ones block for tenths?
2. Why did you separate the tenths into groups of four tenths?
3. How many groups of four tenths do you have?
4. What is the quotient of 1.2 ÷ 0.4?
5. Use decimal models to divide 2.6 by 0.2.

5-3 Dividing by Decimals

Objective

Divide decimals by decimals.

Will is saving his allowance to buy a pair of Nike® shoes that cost $68.25. If Will earns $3.25 per week, how many weeks will Will need to save?

First, estimate the number of weeks.
$$69 \div 3 = 23$$
It should take Will about 23 weeks to save the money.

LOOKBACK

You can review multiplying by powers of 10 on page 119.

When dividing decimals by decimals, you usually change the divisor to a whole number. To do this, multiply both the divisor and dividend by a power of 10. Then divide as with whole numbers.

Multiply 3.25 and 68.25 by 100.

$$3.25\overline{)68.25} \quad \rightarrow \quad 3.25\overline{)68.25} \quad \rightarrow \quad 325.\overline{)6825.}$$

$$
\begin{array}{r}
21 \\
325.\overline{)6825.} \\
-650 \\
\hline
325 \\
-325 \\
\hline
0
\end{array}
$$

Multiply to check.

$$
\begin{array}{r}
\$3.25 \\
\times 21 \\
\hline
325 \\
650 \\
\hline
\$68.25 \ \checkmark
\end{array}
$$

Mental Math Hint

• • • • • • • • • •

When multiplying by powers of ten, move the decimal to the right as many places as the number of zeros in the power of ten.

It will take Will 21 weeks to save the money. Checking this answer against the estimate, the answer seems reasonable.

Examples

Find each quotient.

1 $6.425 \div 2.5$ *Estimate: $6 \div 2 = 3$*

$$
2.5\overline{)6.425} \quad \rightarrow \quad
\begin{array}{r}
2.57 \\
25\overline{)64.25} \\
-50 \\
\hline
14\,2 \\
-12\,5 \\
\hline
1\,75 \\
-1\,75 \\
\hline
0
\end{array}
$$

Multiply by 10.
Place the decimal point.
Divide as with whole numbers.

$6.425 \div 2.5 = 2.57$. The estimate shows that the answer is reasonable. Check: $2.5 \times 2.57 = 6.425$ ✓

2 2.736 ÷ 0.048

$$0.048 \overline{)2.736} \quad \rightarrow \quad 48 \overline{)2736.}$$

$$\begin{array}{r} 57.0 \\ 48 \overline{)2736.} \\ -240 \\ \hline 336 \\ -336 \\ \hline 0 \end{array}$$

Multiply by 1,000.
Place the decimal point.
Divide.

$$2.736 \div 0.048 = 57.0 \qquad \text{Check: } 0.048 \times 57 = 2.736 \; \checkmark$$

Checking for Understanding

Communicating Mathematics

Read and study the lesson to answer each question.

1. **Tell** which way you move the decimal point when multiplying by 10.
2. **Tell** why you change the divisor to a whole number when dividing by a decimal.
3. **Tell** whether you multiply the divisor and dividend by 10, 100, or 1,000 when dividing 4.37 by 2.3.

Guided Practice

Write the power of 10 by which each number is multiplied to make a whole number.

4. 3.5
5. 73.35
6. 2.630
7. 0.057
8. 10.05
9. 7.6354
10. 0.00056
11. 5.0093

Estimate first. Then find each quotient.

12. $1.25 \overline{)5.25}$
13. $0.8 \overline{)1.84}$
14. $0.065 \overline{)3.965}$
15. $0.06 \overline{)0.18}$
16. $7.2 \overline{)46.8}$
17. $0.0436 \overline{)0.0654}$

Exercises

Independent Practice

Write the power of 10 by which each number is multiplied to make a whole number.

18. 4.3
19. 16.32
20. 5.677
21. 0.046
22. 47.577
23. 0.084
24. 10.404
25. 0.00439

Estimate first. Then find each quotient.

26. $2.3 \overline{)5.29}$
27. $0.25 \overline{)75.25}$
28. $5.9 \overline{)7.08}$
29. $0.59 \overline{)83.19}$
30. $5.23 \overline{)120.29}$
31. $37.6 \overline{)808.4}$
32. $0.028 \overline{)15.176}$
33. $0.73 \overline{)875.27}$
34. $0.093 \overline{)0.12555}$

35. Evaluate $a \div b$ if $a = 7.502$ and $b = 3.41$.

36. Find the quotient of 967.3 and 1.7.

37. What is 0.117 divided by 0.45?

38. Divide 79.04 by 16. *(Lesson 5-1)*
39. **Statistics** Rain fell for one week at the Contreras' farm. The amounts were 0.5, 1.1, 0.1, 0.2, 0.5, 1.0, and 0.7 inches.
    a. What was the average daily rainfall for the week?
    b. What was the mode? *(Lesson 2-7)*
40. Round 673.018 to the nearest tenth. *(Lesson 3-5)*
41. If 27 students paid $3.75 each to go to the local museum, about how much money was paid in all? *(Lesson 4-1)*
42. Find the value of $3 \cdot 5 - 12 \div 2$. *(Lesson 1-6)*

Problem Solving and Applications

43. The cost of tennis balls in 10 cities is shown in the graph below.
    a. To the nearest tenth, how many times greater is the cost of tennis balls in Copenhagen than in Los Angeles?
    b. To the nearest tenth, how many times greater is the cost of tennis balls in London than in Tokyo?
    c. The cost in Sao Paulo is how many times the average United States cost?

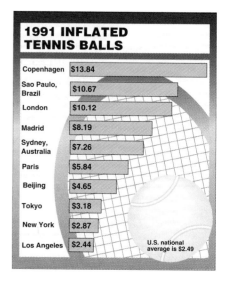

1991 INFLATED TENNIS BALLS

Copenhagen	$13.84
Sao Paulo, Brazil	$10.67
London	$10.12
Madrid	$8.19
Sydney, Australia	$7.26
Paris	$5.84
Beijing	$4.65
Tokyo	$3.18
New York	$2.87
Los Angeles	$2.44

U.S. national average is $2.49

44. **Critical Thinking** Lisa wants to buy a car when she is sixteen. She has saved $800 and can afford to pay $12.50 a week. Her father will sell her the family car for $1,200. How many weeks will she have to pay if she uses all of her savings for a down payment? How many months is this?
45. **Sales** Monica sells bunches of flowers on the corner on Saturday for $2.75 a bunch. At the end of one Saturday, she had collected $57.75. How many bunches had she sold?

COMPUTER CONNECTION

46. **Computer Connection** The spreadsheet at the right shows the unit price for a jar of jelly. To find the unit price, divide the cost of the item by its size.

ITEM	COST	SIZE	UNIT PRICE
JELLY	$1.59	12 OZ	0.1325
CEREAL	$3.35	18 OZ	
BREAD	$1.19	16 OZ	
KETCHUP	$0.89	14 OZ	

Find the unit price for the next three items.
Round to the nearest cent.

5-4 Zeros in the Quotient

Objective

Divide decimals involving special cases of zero in the quotient.

Janet Jackson's *Rhythm Nation* compact disc has 13 tracks. If the entire CD takes 40.25 minutes to play, what is the average time for each track to the nearest hundredth of a minute?

First, estimate the answer.

$$39 \div 13 = 3$$

Each track should average about 3 minutes.

40.25 ÷ 13 = 3.0961538

Suppose you had a blank, 60-minute cassette tape and you wanted to record your favorite songs from your CD to the tape. If each song was about 4.35 minutes long, how many songs could you fit on one side of the tape?

To the nearest hundredth, each track averages 3.10 minutes. The estimate shows the answer is reasonable.

A zero appears in this quotient. Using paper and pencil, you can see why.

```
        3.096
  13)40.250      Divide to the thousandths place to round to the
    −39          nearest hundredth.
     1 2         12 < 13 so a zero is written in the quotient.
     −0
     1 25
    −1 17
        80
       −78
         2
```

Example 1 *Problem Solving*

Consumer Math Shanna needs to buy new batteries for her boom box. Danko's Drug Store sells a package of four batteries for $3.80. How much does one battery cost?

```
     0.95
  4)3.80       Write a zero in the ones place
   −3 6        of the quotient.
     20
    −20
      0
```

One battery costs $0.95.

Example 2 *Connection*

Geometry   The area of a rectangular deck is 51.714 square yards. The length of the deck is 5.07 yards. Find the width of the deck. Since $A = l \cdot w$, divide.

$$5.07\overline{)51.714} \quad \rightarrow \quad 507\overline{)5171.4} \qquad \textit{Estimate: } 50 \div 5 = 10$$

$$
\begin{array}{r}
10.2 \\
507\overline{)5171.4} \\
-507 \\
\hline
101 \\
-0 \\
\hline
101\,4 \\
-101\,4 \\
\hline
0
\end{array}
$$

The width of the deck is 10.2 yards. The estimate shows that the answer is reasonable.

Checking for Understanding

Read and study the lesson to answer each question.

1. **Tell** why a zero is sometimes needed in the quotient.
2. **Tell** how you know $2.5 \div 0.24$ does not equal 1.42.

Find each quotient to the nearest hundredth.

3. $3.98\overline{)4.02}$
4. $47.6\overline{)95.23}$
5. $4.11\overline{)82.35}$
6. $974.2 \div 48.3$
7. $86.3 \div 7.89$
8. $4.8 \div 9.7$
9. $47.56 \div 23.5$
10. $125 \div 24$
11. $20.9 \div 4.23$

Exercises

Find each quotient to the nearest hundredth.

12. $13\overline{)27}$
13. $36\overline{)145}$
14. $61\overline{)2,441}$
15. $72.9 \div 69.8$
16. $1.95 \div 93$
17. $97.2 \div 4.81$
18. $69.2 \div 340.5$
19. $234.4 \div 11.2$
20. $0.0125 \div 1.7$
21. $529 \div 13$
22. $0.625 \div 30.9$
23. $7 \div 145$

24. Evaluate $s \div t$ if $s = 1.43$ and $t = 13$.

25. Evaluate $(p + q) \div r$ if $p = 0.007$, $q = 0.014$, and $r = 2.5$.

Mixed Review

26. Estimate 89.51 ÷ 2.03. Then find the quotient to the nearest tenth. *(Lesson 5-3)*

27. **Geometry**   Find the perimeter of a triangle whose sides are each 23 centimeters long. *(Lesson 4-7)*

28. Divide 2,267.25 by 7.5. *(Lesson 5-3)*

29. **Earning Money**   Martha, DeeAnne, and Sally did odd jobs for people in their neighborhood and earned a total of $65.25. If they split the money evenly, how much would each person earn? *(Lesson 5-1)*

Problem Solving and Applications

30. **Critical Thinking**   If a whole number greater than 1 is divided by a decimal number greater than 1, would the quotient be always greater, sometimes greater, or never greater than 1?

31. **Consumer Math**   The tank of Steve's new car can hold 20 gallons of gasoline. He drove 387 miles on 18.9 gallons. To the nearest tenth, how many miles did his car get for each gallon?

32. **Journal Entry**   Write a letter to a student explaining why 156.14 ÷ 14.8 is not 1.55.

5 Mid-Chapter Review

Find each quotient. *(Lessons 5-1 and 5-4)*

1. 521.4 ÷ 4
2. 145.2 ÷ 11
3. 30.4 ÷ 8
4. 569.5 ÷ 17
5. 8.7 ÷ 6
6. 56.28 ÷ 14
7. 8.034 ÷ 2.6
8. 0.0122 ÷ 1.6
9. 4.08 ÷ 0.2
10. 3.228 ÷ 4
11. 78.26 ÷ 1.3
12. 6.468 ÷ 2.1

Round each quotient to the nearest tenth. *(Lesson 5-2)*

13. 275.9 ÷ 61
14. 29.56 ÷ 11
15. 475.23 ÷ 32
16. 28.59 ÷ 42
17. 27.59 ÷ 3
18. 5.378 ÷ 27

Estimate first. Then find the quotient. *(Lesson 5-3)*

19. 44.8 ÷ 1.6
20. 0.585 ÷ 4.5
21. 1.5 ÷ 0.5
22. 8.8 ÷ 0.004
23. 76.2 ÷ 0.6
24. 9.6 ÷ 1.2

25. Twenty reams of paper cost $19.90. To the nearest cent, find the cost of each ream. *(Lesson 5-4)*

5-5 Capacity and Mass in the Metric System

Objective

Use metric units of capacity and mass.

Words to Learn

gram
kilogram
milligram
liter
milliliter

Oscar de la Hoya is a boxer from Los Angeles who competed in the 1992 Olympics in Barcelona, Spain. Oscar boxes in the lightweight division and needs to keep his mass between 58 and 60 kilograms.

In the metric system, all units are defined in terms of a basic unit. The basic unit of mass in the metric system is the **gram** (g). A **kilogram** (kg) is 1,000 grams. A **milligram** (mg) is 0.001 gram.

Many items you use every day are measured in grams, kilograms, or milligrams.

- A paper clip has a mass of about 1 gram.

- Your math textbook has a mass of about 1 kilogram.

- A grain of salt has a mass of about 1 milligram.

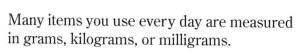

Mini-Lab

Work in groups of four.
Materials: balance, assorted canned goods and food packages

- Find the mass in grams of the contents of each can or package.

- Record each mass next to the name of the item.

- Arrange the cans or packages in order of increasing mass.

Talk About It

a. What conclusions can you draw from this activity?

b. Choose another item. Without finding its mass, place the item where you think it would belong in the group. Find the mass to check.

c. Choose one item in your group. Try to find another item that has the same mass.

Write the unit of mass that you would use to measure each of the following.

1 a nickel

Since a nickel has a small mass like a paper clip, the *gram* is more appropriate.

2 a newborn baby

Since a baby has a mass closer to your textbook, the *kilogram* is more appropriate.

The basic unit of capacity in the metric system is the **liter** (L). A liter is a little more than a quart. The **milliliter** (mL) is 0.001 liter. It takes 1,000 milliliters to make a liter.

- A pitcher has a capacity of about 1 liter.
- An eyedropper has a capacity of about 1 milliliter.

Examples

Write the unit of capacity that you would use to measure each of the following.

3 a tank of gas for a car

Since the tank of gas is a large amount, the *liter* is more appropriate.

4 a teaspoon of cough syrup

Since a teaspoon of cough syrup is a small amount, the *milliliter* is more appropriate.

Checking for Understanding

Communicating Mathematics

Read and study the lesson to answer each question.

1. **Tell** why cereal boxes appear to contain more cereal than they actually do.

2. **Tell** how you decide which metric unit to use when finding the mass of an object.

Write the unit that you would use to measure each of the following.

3. water in a bathtub   4. a pitcher of lemonade

5. the mass of a wrestler   6. the liquid in a dropper

7. a bag of flour   8. a can of soda

9. the mass of a quarter   10. a glass of milk

Exercises

Write the unit that you would use to measure each of the following.

11. a can of evaporated milk   12. a telephone

13. an aspirin   14. the mass of a dog

15. a can of orange juice   16. a large bottle of soda

17. punch in a large bowl   18. vanilla used in a cookie recipe

Mixed Review

19. Divide 1,269.45 by 6.3. *(Lesson 5-4)*

20. **Geometry** Find the area of a square if each side measures 12 kilometers. *(Lesson 4-7)*

21. Find the quotient of 0.506 divided by 25.3. *(Lesson 5-3)*

Problem Solving and Applications

22. **Money** Luke has a jar full of nickels. He knows that a nickel has a mass of about 5 grams. The nickels from his jar have a mass of 2,500 grams. How many nickels does Luke have?

23. **Research** Find examples of several objects of different sizes in a newspaper or magazine. Write the metric units you would use to measure each object.

24. **Mathematics and Sports** Read the following paragraph.

The first great period of boxing popularity began in the 1920s. While boxing at all weights was popular, the heavyweight division was the most popular. Jack Dempsey was the first popular idol. During Joe Louis' long reign, radio broadcasts of championship fights began and the boxing audience grew. Televised fights became popular and remained so through the 1950s. Television brought a decline in local boxing, because audiences preferred to watch nationally-recognized boxers on national television.

A light heavyweight boxer cannot have mass more than 79 kilograms. If a middleweight boxer must have mass no more than 7 kilograms less than a light heavyweight boxer, describe his mass.

5-6 Changing Units in the Metric System

Objective
Change units within the metric system.

At Piggly Wiggly, soft drinks are sold in 2-liter containers as well as in 354-milliliter containers. About how many 354-milliliter containers would equal 2 liters?

From Lesson 5-5, you know that 1 L = 1,000 mL. So, 2 L = 2,000 mL. You need to find 2,000 ÷ 354.

Estimate: 2,000 ÷ 400 = 5
About five 354-mL containers are equal to 2 liters.

To change from one unit to another within the metric system, you either multiply or divide by powers of 10. The chart below shows the relationship between the units in the metric system and the powers of 10.

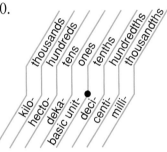

Each place value is ten times the place value to its right.

Mental Math Hint
● ● ● ● ● ● ● ● ● ●
Remember you can multiply or divide by a power of ten by moving the decimal point.

To change from a larger unit to a smaller unit, you need to multiply. To change from a smaller unit to a larger unit, you need to divide.

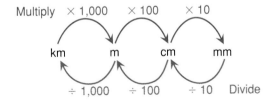

Examples

LOOKBACK
You can review the metric units of length on page 85.

1 0.7 cm = ■ mm
To change from centimeters to millimeters multiply by 10 since 1 cm = 10 mm.

0.7 × 10 = 7
0.7 cm = 7 mm

2 328 mL = ■ L
To change from milliliters to liters, divide by 1,000 since 1 L = 1,000 mL.

328 ÷ 1,000 = 0.328
328 mL = 0.328 L

3   150 mg = ■ g

To change from
milligrams to grams,
divide by 1,000 since
1 g = 1,000 mg.

150 ÷ 1,000 = 0.15
150 mg = 0.15 g

4   5.02 kg = ■ g

To change from
kilograms to grams,
multiply by 1,000 since
1 kg = 1,000 g.

5.02 × 1,000 = 5,020
5.02 kg = 5,020 g

Checking for Understanding

Communicating Mathematics

Read and study the lesson to answer each question.

1. **Tell** how you know whether to multiply or divide when changing from milliliters to liters.

2. **Show** how you would change 3 liters to milliliters.

Guided Practice

Write whether you multiply or divide to change each measurement. Then complete.

3. 150 m = ■ cm
5. 154 g = ■ kg
7. ■ L = 253 mL

4. ■ m = 150 cm
6. 210 mm = ■ cm
8. 2.5 L = ■ mL

Exercises

Independent Practice

Complete.

9. 5.25 kg = ■ g
11. 427 mg = ■ g
13. ■ m = 0.593 km
15. ■ km = 547 m
17. Change 0.6 milliliters to liters.
18. How many grams are in 32.1 kilograms?

10. ■ g = 9.54 mg
12. 3.29 L = ■ mL
14. ■ mg = 5.37 g
16. 43.5 mL = ■ L

Mixed Review

19. To measure the water in a bathtub, would you use liters or milliliters as your units? *(Lesson 5-5)*

20. **Geometry**   Find the circumference of the circle at the right. *(Lesson 4-9)*

1.1 cm

Problem Solving and Applications

21. **Food**   A gallon container holds 3,784 milliliters of milk. How many liters is this?

22. **Critical Thinking**   A chemistry experiment required 12.45 grams of chemical to produce one liter of a gas. If Juanita wanted 24 liters of gas, how many kilograms of the chemical will she need?

5-7 Solving Equations

Objective

Solve equations involving multiplication by using inverse operations.

Misa is shopping for a skateboard. She has checked the prices at Herman's Sporting Goods. She has found that a skateboard at Roush Sporting Goods costs 2 times as much as the same skateboard at Herman's. If Roush's skateboard costs $64, how much does Herman's skateboard cost? The problem says that Roush's skateboard is twice the cost of Herman's. So, 2 times some number is $64.

$$2 \times n = 64$$ You need to find n, the number of dollars the skateboard costs at Herman's.

DID YOU KNOW

The skateboard first appeared in the early 1960s on paved areas along California beaches to keep surfboard fanatics busy when there were no ocean waves.

Multiplication and division are inverse operations. They undo each other.

$$2 \times n = 64 \text{ means the same as } n = 64 \div 2.$$

Now you can use division to solve the problem.
Since $32 = 64 \div 2$, the skateboard at Herman's costs $32.

Many problems can be solved by using equations and inverse operations.

Examples

1 Solve $3r = 12$.

Use division to undo the multiplication.

$3r = 12$ means the same as $r = 12 \div 3$.
$r = 12 \div 3$
$r = 4$

Check: $3 \times 4 = 12$ ✔

2 Solve $6.3 = 2.1m$.

$6.3 = 2.1m$ means the same as $6.3 \div 2.1 = m$.

$6.3 \div 2.1 = m$
$3 = m$

Check: $2.1 \times 3 = 6.3$ ✔

| Example 3 | *Problem Solving* |

Consumer Math Juan bought some pens that cost $0.95 each. If the subtotal before adding tax was $11.40, how many pens did he buy?

Write an equation where p is the number of pens.

$$\$0.95p = \$11.40$$

Use division to undo multiplication.

$$p = 11.40 \div 0.95$$
$$p = 12$$
$$\text{Check: } \$0.95 \times 12 = \$11.40 \ \checkmark$$

Juan bought 12 pens.

Checking for Understanding

Communicating Mathematics Read and study the lesson to answer each question.

1. **Tell** what operation is the inverse of addition.
2. **Write** an equation for the statement: five times some number is 25.

Guided Practice Solve each equation.

3. $3x = 9$
4. $15 = 5z$
5. $0.5y = 15.25$
6. $13.34 = 5.8r$
7. $0.7s = 21$
8. $3.25t = 9.75$
9. $2.9m = 11.89$
10. $48n = 12$
11. $3.75f = 0.225$

Exercises

Independent Practice Solve each equation.

12. $5m = 25$
13. $7t = 28$
14. $75 = 5y$
15. $2.3x = 13.57$
16. $147 = 30p$
17. $24.8 = 1.24a$
18. $15n = 60$
19. $2.5488 = 4.32t$
20. $37.2y = 167.4$
21. $39.9 = 3.5m$
22. $6.5r = 33.8$
23. $3.75 = 0.04s$
24. $\$12.65 = \$0.55d$
25. $2.96w = 5.328$
26. $53.5 = 0.5y$
27. Solve the equation $3.1t = 24.862$.
28. What is the solution of $186.76 = 32.2n$?

Mixed Review 29. Mrs. Sanchez used six, 2.4-meter extension cords to hang holiday lights. How many meters long were the cords in all? *(Lesson 3-7)*

30. Round 8,135 to the nearest thousand. *(Lesson 1-2)*

31. **Statistics** Find the mean for the scores 19, 17, 28, 32, and 23. *(Lesson 2-7)*

32. Evaluate y^3 if $y = 6$. *(Lesson 4-2)*

Problem Solving and Applications

Set up an equation and solve.

33. **Food** The track team has 18 members. They stopped at a restaurant on the way home from a meet. If the total bill was $85.50, what was the average cost of each team member's meal?

34. **Critical Thinking** Four times a number plus 3 is 30. What is the number?

35. **Geography** The number of independent countries has tripled in the last 27 years to over 165 nations. What was the number of independent countries 27 years ago?

36. **Journal Entry** Write an explanation of the following. Given the equation $\frac{x}{3.5} = 7.0$, how do you know that the equation is true if $x = 24.5$?

Save Planet Earth

Hazardous Waste in the Home Are you innocently dumping toxic substances down the drain or into the sewer system? Many people don't know how to dispose of household hazardous waste properly and the result can be serious water contamination.

Hazardous wastes found around the home include: paints and thinners, car battteries, oven and drain cleaners, mothballs, floor and furniture polish, antifreeze, rug and upholstery cleaners, and products used to clean toilets.

How You Can Help

- Keep hazardous materials in their original containers.
- Don't overbuy—the more you buy, the more you have to dispose of. Share with friends and neighbors whenever possible.
- Recycle whenever possible. Motor oil, car batteries, and paint thinners can be recycled.
- Use safer substitutes whenever possible.

5-8 Choose the Method of Computation

Objective

Solve problems by choosing estimation, mental math, paper and pencil, or calculator.

Tracie is sending mail-order catalogs to 4,600 customers. She has a postage budget of $2,000. It costs $0.54 in postage to mail each catalog. Does she have enough money in her budget to mail 4,600 catalogs?

Explore   Look at the information you are given.
The postage budget is $2,000.00.
There are 4,600 catalogs to send.
It costs $0.54 to mail each catalog.

Plan   Decide which method to use to solve the problem.
Answer the questions in the chart to help you decide.

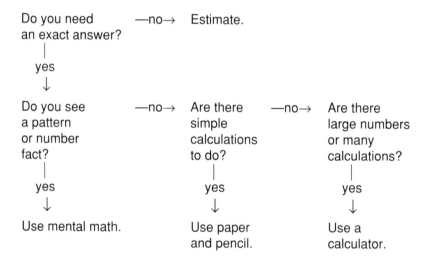

Solve   Since you do not need an exact answer, you can estimate. Divide the budget by the postage to find the number of catalogs that can be sent. Round $0.54 to $0.50.

$$2,000 \div 0.50 = 4,000$$

Tracie does not have enough money to send 4,600 catalogs.

Examine   Which method(s) of computation would you use to find the exact number of catalogs Tracie could mail? Use the exact answer to check your estimate.

Checking for Understanding

Communicating Mathematics

Read and study the lesson to answer each question.

1. **Show** how you would find the exact number of catalogs Tracie could mail using two different methods.

2. **Write** two general rules to follow when you are deciding on a method of computation.

Guided Practice

Choose an appropriate method of computation. Then solve.

3. A *Chorus Line* was one of the longest running Broadway plays with 6,137 performances. If there were 9 performances per week, how many weeks had the show run at that time?

4. In 1989, the United States' oil consumption was 6,323.6 millions of barrels while Japan's was 1,818.1 millions of barrels. How many more barrels did the United States consume than Japan?

5. Seventy-five students tried out for a dance troupe and 40 of them were accepted. How many students did not make the troupe?

Problem Solving

Practice

Solve using any strategy.

6. The Chang sisters have a combined age of 60 years. Lin is the youngest, Tara is 4 years older than Lin, and Olin is 8 years older than Lin. How old is each sister?

Strategies

• • • • • • •

Look for a pattern.
Solve a simpler problem.
Act it out.
Guess and check.
Draw a diagram.
Make a chart.
Work backward.

7. Marco wants to fence in his tomato garden that is 8 feet by 6 feet. How much fencing does he need?

8. The hiking trails in Highline Park are 3.8, 7.5, 10.2, 5.9, and 3.7 miles long. Brianna hiked two of the trails for a total of 9.7 miles. What were the lengths of the trails she hiked?

9. Brad had $100. He spent half of his money at the game store. Then he spent half of the money he had left at the sporting goods store. How much money does Brad have left?

10. Lawana bowled a 125, a 148, and a 162 during league competition. If her previous average was 128 pins, how many pins over her average did she bowl?

11. **Data Search**   Refer to pages 156 and 157.
    a. Do a greater percentage of females ride bicycles or go bowling?
    b. Do a greater percentage of males play basketball or exercise with equipment?

DATA SEARCH

5 Study Guide and Review

Communicating Mathematics

Fill in the blanks with the appropriate term(s).

1. The answer in a division problem is called the __?__.
2. When rounding quotients, you need to divide to __?__ more place-value position than you are rounding.
3. To estimate answers to division problems, use __?__ and __?__ __?__.
4. The __?__ is the basic unit of capacity in the metric system.
5. Multiplication and division are __?__ operations.
6. Explain why the answer to $160.16 \div 8$ is 20.02, not 20.2 or 2.2.

Skills and Concepts

Objectives and Examples

Upon completing this chapter, you should be able to:

- divide decimals by whole numbers
  (Lesson 5-1)

  $96.9 \div 51 = ?$

  $$\begin{array}{r} 1.9 \\ 51\overline{)96.9} \\ -51 \\ \hline 45\,9 \\ -45\,9 \\ \hline 0 \end{array}$$ ← *decimal point directly over the decimal in the dividend.*

- round decimal quotients to a specified place *(Lesson 5-2)*

  Round $6.21 \div 3$ to the nearest tenth.

  $$\begin{array}{r} 2.07 \\ 3\overline{)6.21} \end{array}$$ *Divide to one more place-value position than the place you are rounding.*

  $6.21 \div 3$ rounded to the nearest tenth is 2.1.

Review Exercises

Use these exercises to review and prepare for the chapter test.

Find each quotient.

7. $12.24 \div 36$
8. $32\overline{)203.84}$
9. $1000\overline{)17.97}$
10. $5.0175 \div 10$

Evaluate each expression.

11. $a \div b$ if $a = 10.89$, $b = 25$
12. $d \div c$ if $c = 100$, $d = 7.64$

Round each quotient to the nearest tenth.

13. $35.56 \div 7$
14. $84.36 \div 12$

Round each quotient to the nearest hundredth.

15. $16.349 \div 9$
16. $249.77 \div 13$

Objectives and Examples

- **divide decimals by decimals**
  (Lesson 5-3)

  $166.14 \div 21.3 = ?$

  $21.3\overline{)166.14} \quad \rightarrow \quad 213\overline{)1661.4}^{\,7.8}$

  Multiply the divisor and dividend by 10.

- **divide decimals involving special cases of zero in the quotient**
  (Lesson 5-4)

  $571.2 \div 56 = ?$

  $$
  \begin{array}{r}
  10.2 \\
  56\overline{)571.2} \\
  -56 \\
  \hline
  11 \\
  -0 \\
  \hline
  11\,2 \\
  -11\,2 \\
  \hline
  0
  \end{array}
  $$

  11 < 56 so zero is written in the quotient.

- **use metric units of capacity and mass**
  (Lesson 5-5)

  What unit of mass would you use to measure an egg?
  Since an egg has a small mass, use the *gram.*
  What unit of capacity would you use to measure a bucket of water?
  Since a bucket of water is more than a quart, use the *liter.*

- **change units within the metric system** *(Lesson 5-6)*

  Multiply to change a larger unit to a smaller unit. $9.2\text{ g} = 9{,}200\text{ mg}$
  Divide to change a smaller unit to a larger unit. $120\text{ mL} = 0.12\text{ L}$

Review Exercises

Estimate first. Then find each quotient.

17. $569.2 \div 1.6$

18. $0.045\overline{)6.345}$

19. Evaluate $e \div f$ if $e = 1.56$ and $f = 3.2$.

Find each quotient.

20. $7{,}380 \div 36$

21. $24.3\overline{)2.43}$

22. $6.5\overline{)3510}$

23. $8.134 \div 98$

24. $3{,}190.95 \div 45$

Write the unit that you would use to measure each of the following.

25. a candy apple

26. a stack of books

27. a pitcher of lemonade

28. a dose of cough medicine

Complete.

29. $300\text{ mL} = \blacksquare\text{ L}$

30. $\blacksquare\text{ g} = 1\text{ mg}$

31. $\blacksquare\text{ m} = 0.75\text{ km}$

32. $387\text{ g} = \blacksquare\text{ kg}$

Objectives and Examples

Review Exercises

- solve equations involving multiplication by using inverse operations
  (*Lesson 5-7*)

  Solve $1.9x = 77.9$.
  Use division to undo the multiplication.
  $1.9x = 77.9$ means the same as
  $x = 77.9 \div 1.9$.
  $x = 41$

Solve each equation.

33. $25y = 100$

34. $25m = 40$

35. $24.16 = 8x$

36. $1.1p = 66$

37. $33.64 = 1.45h$

Applications and Problem Solving

38. **Car Loans**   Ellen's older brother, Drew, bought a new car and financed it over a 48-month period. The total of all the payments he will make is $15,336.96. About how much is each car payment? (*Lesson 5-1*)

39. **Weather**   In 1985, Indianapolis received a total of 46.98 inches of precipitation. To the nearest tenth, what was the average precipitation per month? (*Lesson 5-2*)

40. **Charity**   Marisela collected $131.25 to buy gifts for the children of a family whose home burned. If on average, each person gave Marisela $1.25, how many people contributed? (*Lesson 5-4*)

41. **Fundraising**   The Pep Club makes a profit of $4 for each $7 homecoming mum they sell. If the club made a profit of $2,160, how many mums did they sell? (*Lesson 5-8*)

Curriculum Connection Projects

- **Sports**   Find the average running speed of the Olympic gold-medal winners in the 100-, 200-, and 400-meter events by dividing each distance by the winner's time.

- **Consumer Awareness**   Change the units of five grocery items at home from grams to kilograms and milligrams; and, five more from liters to kiloliters and milliliters.

Read More About It

Belden, Willane Schneider. *Mind-Find.*
Cooper, Susan. *Over Sea Under Stone.*
Foder, R.V. *Nickels, Dimes and Dollars: How Currency Works.*

5 Test

Round each quotient to the nearest tenth.

1. $818.4 \div 5$

2. $279.8 \div 8$

3. $524.88 \div 4$

Estimate first. Then find each quotient.

4. $19.36 \div 44$

5. $21.6 \overline{)49.68}$

Find each quotient mentally.

6. $100 \overline{)829.3}$

7. $45.1 \div 10$

Evaluate each expression.

8. $a \div b$ if $a = 7.260$ and $b = 30$

9. $c \div d$ if $c = 108.9$ and $d = 3.3$

Find each quotient.

10. $58.305 \div 11.5$

11. $1{,}274.7 \div 2.1$

Write the unit that you would use to measure each of the following.

12. a bottle of windshield-washer fluid

13. a potato

14. Mr. Ricci bought a new hot water bottle that holds 1,500 milliliters of water. How many liters is that?

Complete.

15. $739 \text{ mL} = \blacksquare \text{ L}$

16. $\blacksquare \text{ mg} = 2.1 \text{ kg}$

Solve each equation.

17. $3x = 17.7$

18. $43.95 = 1.5y$

19. **Baking** Wu-Lun is making eggnog for a holiday party. His recipe calls for 5.25 liters of whipping cream to serve 42 people. How much whipping cream should he use if he plans to serve six people?

20. **Transportation** Danielle drove 11.5 km each day for 5 days. How far did she travel in all?

Bonus The label on a bag of fertilizer states that it will cover 3 m². To cover a rectangular yard with dimensions of 7.3 m by 5.2 m, how many bags of fertilizer are needed?

Patterns and Number Sense

Spotlight on Water

Have You Ever Wondered...

- What it means when newspapers report that only about $\frac{3}{100}$ of the world's water is usable?

- How much water is used per person per day?

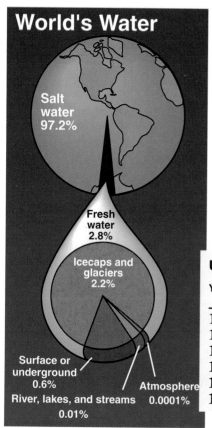

World's Water

Salt water 97.2%

Fresh water 2.8%

Icecaps and glaciers 2.2%

Surface or underground 0.6%

Atmosphere 0.0001%

River, lakes, and streams 0.01%

U.S. Water Use Per Day

Year	Total (bil. gal.)	Per capita (gal.)	Irrigation (bil. gal.)	Rural (bil. gal.)	Industrial and misc. (bil. gal.)	Steam electric utilities (bil. gal.)
1960	61	339	52	2.8	3.0	0.2
1965	77	403	66	3.2	3.4	0.4
1970	87	427	73	3.4	4.1	0.8
1975	96	451	80	3.4	4.2	1.9
1980	100	440	83	3.9	5.0	3.2
1985	92	380	74	9.2	6.1	6.2

Clouds are so interesting. I love to find shapes in them

There's a chicken cloud, and a hamburger cloud, and a bicycle cloud

And I do believe that one's a rain cloud

10-19   JIM DAVIS   © 1984 United Feature Syndicate, Inc.

Chapter Project

Water

Work in a group.

1. Keep track of the water you use each day for one week.

2. Make a graph showing about how much you use for each activity and the average for each day.

3. List several ways in which water can be saved. Show the information on a poster.

Looking Ahead

In this chapter, you will see how mathematics can be used to answer questions about the water supply. The major objectives of the chapter are to:

- find the prime factorization of a composite number

- solve problems by making an organized list

- find the greatest common factor and least common multiple of two or more numbers

- express mixed numbers as improper fractions and vice versa

- write decimals as fractions and vice versa

6-1 Divisibility Patterns

Objective

Use divisibility rules for 2, 3, 5, 6, 9, and 10.

Have you ever ridden a roller coaster? It is a lot of fun according to Don Helbing from Cincinnati, Ohio. Every summer from 1981 to 1990, Don rode the Racer roller coaster at Kings Island amusement park. His goal was to complete 10,000 rides. He spent 665 days visiting the park and 500 hours riding the coaster to reach his goal. About how many rides did he complete per hour?

To solve this problem, you need to divide 10,000 by 500. "Is 500 a factor of 10,000?" is another way of asking this question. The factors of a whole number divide that number with a remainder of zero. So, let's find out if 500 is a factor of 10,000.

$$10,000 \;\boxed{\div}\; 500 \;\boxed{=}\; 20$$

Since the quotient is a whole number, the remainder is zero, and 500 is a factor of 10,000. You can also say that 10,000 is *divisible* by 500. Don completed about 20 rides per hour.

You can find the factors of large numbers mentally by using some of the divisibility patterns for integers. The rules you can use to determine if a number is divisible by 2, 3, 5, 6, 9, or 10 are as follows.

A number is divisible by:
- 2 if the ones digit is divisible by 2.
- 3 if the sum of the digits is divisible by 3.
- 5 if the ones digit is 0 or 5.
- 6 if the number is divisible by 2 and 3.
- 9 if the sum of the digits is divisible by 9.
- 10 if the ones digit is zero.

Examples

1 Is 26 divisible by 2?

The ones digit is 6. 6 is divisible by 2.

Check by using a calculator.

$26 \;\boxed{\div}\; 2 \;\boxed{=}\; 13$

26 is divisible by 2.

2   Is 457 divisible by 3?

$4 + 5 + 7 = 16$ *Find the sum of the digits.*

16 is not divisible by 3. So, 457 is not divisible by 3.

Check by using a calculator.

457 ÷ 3 = ।52.33333   *Since the result is not a whole number, the remainder is not zero.*

457 is not divisible by 3.

3   Determine whether 128 is divisible by 2, 3, 5, 6, 9, or 10.

- Is the ones digit divisible by 2?   Yes

- Is the sum of the digits divisible by 3?
  $1 + 2 + 8 = 11$
  11 is not divisible by 3. So 128 is not divisible by 3.

- Is the ones digit a 0 or a 5?   No

- Is the number divisible by 2 and 3?   No

- Is the sum of the digits divisible by 9?
  $1 + 2 + 8 = 11$
  11 is not divisible by 9. So 128 is not divisible by 9.

- Is the ones digit 0?   No

Therefore, 128 is divisible by 2 only.

Checking for Understanding

Communicating Mathematics   Read and study the lesson to answer each question.
1. **Tell** how you can determine whether 68 is divisible by 2.
2. **Write** two factors of 45 that are greater than 1 and less than 45.

Guided Practice   State whether each number is divisible by 2, 3, 5, 6, 9, or 10.

3. 17	4. 1,032	5. 780	6. 102
7. 382	8. 576	9. 1,002	10. 215
11. 104	12. 203	13. 882	14. 2,736

Exercises

Determine if the first number is divisible by the second number.

15. 225; 5   16. 1,040; 3   17. 280; 10   18. 405; 9

19. 426; 6   20. 1,081; 2   21. 153; 3   22. 888; 5

23. 945; 9   24. 2,002; 6   25. 4,155; 10   26. 11,112; 3

27. Is 439 divisible by 9?   28. Is 2,680 divisible by 5?

Use mental math skills, paper and pencil, or a calculator to find a number that satisfies the given conditions.

29. a number divisible by both 2 and 5

30. a three-digit number divisible by 3 and 6

31. a four-digit number divisible by 2, 5, and 10

32. a five-digit number divisible by 3 and 6

33. **Algebra** Solve the equation $5.2m = 130$. *(Lesson 5-7)*

34. **Decorating** Megan plans to make 25 bows to decorate the school gym for the graduation party. She has 64 feet (or 768 inches) of ribbon to use. How many inches, to the nearest tenth, does that give her for each bow? *(Lesson 5-4)*

35. Find the quotient $7,651.5 \div 100$ mentally. *(Lesson 5-1)*

36. **Geometry** Find the perimeter of the rectangle at the right. *(Lesson 4-6)*

3.4 m

11.75 m

37. **Entertainment** A new theater of 1,000 seats is being planned. Each row will have the same number of seats. The builder wants to put 30 seats per row. Will the plan work? Explain.

38. **Critical Thinking** What is the greatest three-digit number that is not divisible by 3 or 5?

39. **Card Games** To play the card game *War,* you need to deal out all the cards. If there are 52 cards in a deck and there are 3 people playing, will each person get the same number of cards?

COMPUTER
CONNECTION

40. **Computer Connection** You can use the BASIC program at the right to find all of the factors of a given number. The computer divides the input number, N, by each number less than or equal to N, and then prints the divisor and quotient. If the quotient is a whole number, the divisor is a factor of N. Use the program to find all of the factors of 100.

```
10   INPUT N
20   FOR D = 1 TO N
30   PRINT D, N/D
40   NEXT D
```

6-2A **Rectangular Arrays**

A Preview of Lesson 6-2

Objective

Recognize prime and composite numbers based on rectangular arrays.

Materials

colored tiles
grid paper

You can review area of rectangles on page 142.

In this lab, you will use tiles to help you find as many different shaped rectangles as you can for a certain number of square units.

Try this!

Work with a partner.

- Use 8 square tiles. Build as many different-shaped rectangles as you can. Draw each shape on grid paper as a record.

- Use a chart like the one below to record your findings.

Area	Possible Lengths and Widths	Drawings	Number of Arrangements
8	1 × 8 2 × 4		2

What do you think?

1. How many rectangles could you make with 8 tiles?
2. Could you use a length of 3 with your 8 tiles and make a rectangle? Why or why not?
3. What numbers were used as possible lengths or widths for your 8 tiles?
4. Repeat the process with areas of 2 square units through 12 square units.
5. Which numbers of square tiles had only one arrangement?
6. Which numbers of square tiles had more than one arrangement?
7. Is there a relationship between the multiplication table and the number of arrangements you have? If so, describe it.

Extension

8. Make a guess about which numbers between 12 and 25 square units can have more than one rectangular shape. Explain why you selected those numbers. Then try it.

6-2 Prime Factorization

Objective

Find the prime factorization of a composite number.

Words to Learn

composite
prime
factor tree
prime factorization

LOOK BACK

You can review factors on page 119.

Divisibility rules help you determine easily whether 2, 3, or 5 are factors.

Clara's little brother has a pegboard with 30 pegs. He is learning how to stack the pegs. Clara suggests that he use all the pegs to make stacks of 2 pegs. How many stacks will they have? If they make stacks of other factors of 30, how many stacks will they have? To answer this question, find all the factors of 30.

Divisible by	Test	He could make
1?	Yes, 1 is a factor of every number.	30 stacks of 1.
2?	Yes, 30 ends in 0.	15 stacks of 2.
3?	Yes, 3 + 0 = 3. 3 is divisible by 3.	10 stacks of 3.
4?	30 ÷ 4 = 7.5 No, the quotient is not a whole number.	
5?	Yes, 30 ends in 0.	6 stacks of 5.
6?	You don't need to check 6 since you already know that 6 is a factor of 30.	

The numbers 1, 2, 3, 5, 6, 10, 15, and 30 are factors of 30. 30 is a composite number. A **composite** number has more than two factors. In the previous mathematics lab, you found that there was only one rectangle you could make with an area of 3 square units. The lengths of the sides were 1 unit and 3 units. So, 3 is a prime number. A **prime** number has exactly two factors, 1 and the number itself.

Example 1

Determine whether 12 is a prime or a composite number. Draw a picture to explain your answer.

1×12 2×6 3×4

12 has more than two factors. Therefore, it is a composite number.

The numbers 0 and 1 are neither prime nor composite. Zero has an endless number of factors. The number 1 has only one factor, itself.

The diagrams below show two different ways to find the factors of 12. These diagrams are called **factor trees.**

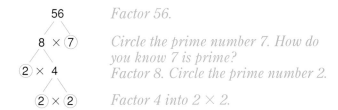

The factors are in a different order, but the result is the same.

The prime factorization of 12 is $2 \times 2 \times 3$.

When a number is expressed as a product of factors that are all prime, the expression is called the **prime factorization** of the number.

Example 2

Find the prime factorization of 56.

Method 1:   Use a factor tree.

$$56$$
$$8 \times \textcircled{7}$$
$$\textcircled{2} \times 4$$
$$\textcircled{2} \times \textcircled{2}$$

Factor 56.

Circle the prime number 7. How do you know 7 is prime?
Factor 8. Circle the prime number 2.

Factor 4 into 2×2.

The prime factorization of 56 is $2 \times 2 \times 2 \times 7$.

Calculator Hint

When you use a calculator to find prime factors, it is helpful to record them on paper as you find each one.

Method 2:   A calculator may also be used.

$56 \; \boxed{\div} \; 2 \; \boxed{=} \; 28 \; \boxed{\div} \; 2 \; \boxed{=} \; 14 \; \boxed{\div} \; 2 \; \boxed{=} \; 7$

Write the prime numbers in order from least to greatest. Check to see if the product is 56.

$2 \times 2 \times 2 \times 7 = 56$ ✓

Checking for Understanding

Communicating Mathematics

Read and study the lesson to answer each question.

1. **Draw** a rectangular shape to illustrate whether 15 is a prime or a composite number.
2. **Tell** how to find the prime factorization of 20.
3. **Tell** what the only even prime number is.

Draw a picture of all the different shaped rectangles that have the given area.

4. 18　　　　5. 13　　　　6. 21　　　　7. 27　　　　8. 36

Tell whether each number is prime, composite, or neither.

9. 42　　　10. 1　　　11. 56　　　12. 97　　　13. 112

List all factors of each number.

14. 15　　　15. 18　　　16. 27　　　17. 39　　　18. 42

Exercises

Independent Practice　Find the prime factorization of each number.

19. 49　　　　20. 65　　　　21. 48　　　　22. 14

23. 28　　　　24. 19　　　　25. 32　　　　26. 81

27. 90　　　　28. 102　　　29. 124　　　30. 87

31. Find the least prime number that is greater than 70.

Mixed Review　32. State whether 708 is divisible by 2, 3, 5, 6, 9, or 10 using divisibility rules. *(Lesson 6-1)*

33. Which estimate is a better measurement of a large bag of garbage, 4.5 kilograms or 4.5 grams? *(Lesson 5-5)*

34. **Algebra**　Evaluate the expression zwx if $w = 0.02$, $x = 20.7$, and $z = 3.001$. *(Lesson 4-5)*

35. **Statistics**　Refer to the circle graph of students' favorite sports on page 104. How much more of the students' votes were for swimming than tennis? *(Lesson 3-9)*

Problem Solving and Applications　36. **Games**　There are 64 squares on a checkerboard.

a. Is 64 a prime or composite number?

b. Draw a picture of a different way that the 64 squares could be arranged.

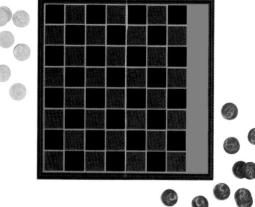

37. **Critical Thinking**　What type of number has an odd number of factors? Hint: Use your diagrams.

38. **History**　In 1742, Christian Goldbach of Russia suggested that any even number greater than 2 could be expressed as the sum of two prime numbers. Some numbers can be expressed in several ways. For example, 40 = 17 + 23 or 40 = 29 + 11. Find two prime numbers whose sum is the given number.

a. 50　　　　　　b. 38　　　　　　c. 76

6-3 **Make a List**

Objective
Solve problems by making an organized list.

Molly's Muffin Shop sells bran, corn, apple cinnamon, and blueberry muffins in three sizes: mini, regular, and large. How many different options do you have if you want to buy one muffin?

Explore What do you know?
There are 4 varieties of muffins.
There are 3 sizes available.

What do you need to find out?
You need to find out how many different options you have when buying one muffin.

Plan Make a list of all the possible options. Then count the options to find the total.

Solve

bran - mini	corn - mini
bran - regular	corn - regular
bran - large	corn - large
apple cinnamon - mini	blueberry - mini
apple cinnamon - regular	blueberry - regular
apple cinnamon - large	blueberry - large

There are 12 options.

Examine Look at the relationship between the number of varieties (4) and the number of sizes (3). The product of these numbers (4×3) is equal to the number of options (12).

Example

Twin primes are two prime numbers that are consecutive odd integers such as 3 and 5, 5 and 7, and 11 and 13. What is the next pair of twin primes?

Make a list of numbers from 1 to 30.

1 ② ③ 4 ⑤ 6 ⑦ 8 9 10
⑪ 12 ⑬ 14 15 16 ⑰ 18 ⑲ 20
21 22 ㉓ 24 25 26 27 28 ㉙ 30

Circle all the prime numbers. Then look for twin primes. The next pair of twin primes is 17 and 19.

Checking for Understanding

Communicating Mathematics

Read and study the lesson to answer the question.

1. **Tell** the advantages of an organized list over a random list.

Guided Practice

Solve using the strategy make a list.

2. A manufacturer offers 5 different styles of jeans in 5 different colors. How many combinations of style and color are possible?

3. How many 2-digit numbers can you make using the digits 1, 3, 5, and 7?

4. Keela went to dinner and had to decide what to order. She had a choice of soup or salad, chicken or pasta, and fruit or pie. How many different dinners are possible?

Problem Solving

Practice

Solve. Use any strategy.

5. What is the first pair of twin primes after 100?

6. Justin is 4 years older than Kelly. The sum of their ages is 42. How old is each?

Strategies

• • • • • • •

Look for a pattern.

Solve a simpler problem.

Act it out.

Guess and check.

Draw a diagram.

Make a chart.

Work backward.

7. Lita saved $125 in June and $175 in July. How much does she need to save in August to average $150 per month in savings?

8. The Hobson's vacation lasted 12 days. They budgeted $55 per night for lodging and $75 per day for food. How much did they budget for these two items for the whole trip?

9. Colby is planning to travel to Seattle, San Francisco, and Salt Lake City. He cannot decide in which order to schedule his visits. How many choices does he have?

10. How many license plate combinations can you make with the letters A, B, Z and the numbers 2, 4, 6 if each license plate starts with three different letters and ends with three different numbers?

6-4 Greatest Common Factor

Objective

Find the greatest common factor of two or more numbers.

Words to Learn

greatest common factor (GCF)

Suppose that 24 sixth graders and 32 seventh graders from your school are selected to participate in the Flag Day parade. The parade director wants the students to form a pattern with rows that have the same number of people. Sixth graders and seventh graders will be in different rows. What is the greatest number of students that can be in each row? How many rows will there be?

Problem Solving Hint

• • • • • • • • • •

To solve this problem, you can make a list.

To help answer these questions, you can find the factors of 24 and 32. Make a list of all factors of each number from least to greatest.

factors of 24:   1, 2, 3, 4, 6, 8, 12, 24

factors of 32:   1, 2, 4, 8, 16, 32

Notice that 1, 2, 4, and 8 are common factors of both numbers, 24 and 32. The greatest of the common factors is 8. The greatest of the common factors of two or more numbers is called the **greatest common factor (GCF)** of the numbers.

The greatest number of students that can be in each row for the Flag Day parade is 8. Since $3 \times 8 = 24$, there are 3 rows of sixth graders. Since $4 \times 8 = 32$, there are 4 rows of seventh graders. There are a total of 7 rows of sixth and seventh graders.

Example 1

Find the greatest common factor of 27 and 45.

List all the factors of each number.

factors of 27:   1, 3, 9, 27

factors of 45:   1, 3, 5, 9, 15, 45

The common factors are 1, 3, and 9.

The greatest common factor of 27 and 45 is 9.

Example 2 *Problem Solving*

School There are 42 apples and 77 oranges to be sold as part of the country fair at your school. The organizer wants the same number of items of fruit in each bag. Apples and oranges should not be mixed. What is the greatest number of items that can be placed in a bag?

factors of 42: 1, 2, 3, 6, 7, 14, 21, 42

factors of 77: 1, 7, 11, 77

The greatest common factor of 42 and 77 is 7. Thus, 7 items should be placed in each bag.

You can also use prime factorization to find the GCF of numbers. First, find the prime factorization of each number, and then find the products of their common factors.

Example 3

Use prime factorization to find the GCF of 20 and 30.

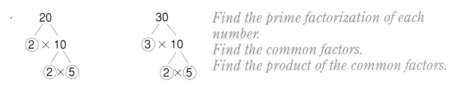

Find the prime factorization of each number.
Find the common factors.
Find the product of the common factors.

The greatest common factor of 20 and 30 is 2×5 or 10.

Checking for Understanding

Communicating Mathematics Read and study the lesson to answer each question.

1. **Write** in your own words what is meant by the term *greatest common factor.*
2. **Tell** how to find the greatest common factor of 8 and 12.
3. **Tell** how to find all the common factors of 9 and 12.

Guided Practice List all the factors of each number from least to greatest.

4. 14   5. 25   6. 28   7. 38

Find all common factors of each pair of numbers.

8. 9, 15   9. 12, 18   10. 8, 24

Find the GCF of each pair of numbers by listing the factors of the numbers.

11. 24, 28   **12.** 32, 24   **13.** 60, 36

Find the GCF of each pair of numbers by using prime factorization.

14. 45, 54   **15.** 84, 56   **16.** 64, 72

Exercises

Independent Practice

Find the GCF of each pair of numbers.

17. 8, 30   **18.** 18, 36   **19.** 30, 75

20. 25, 35   **21.** 28, 49   **22.** 32, 48

23. 44, 55   **24.** 90, 36   **25.** 72, 28

26. Find the GCF of 6, 18, and 42.

27. Find the GCF of 8, 28, and 52.

Mixed Review

28. Find the prime factorization of 120. *(Lesson 6-2)*

29. Geometry Find the perimeter of the figure at the right. *(Lesson 4-6)*

30. Pets Jason's fish tank holds 46.2 liters of water. If there should be 2.2 liters of water per fish, how many fish should Jason put in the tank? *(Lesson 5-3)*

31. Statistics How do you know that the graph at the right is misleading? Why might the tutors have prepared the graph this way? *(Lesson 2-8)*

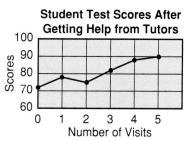

Problem Solving and Applications

32. Critical Thinking Give an example of two composite numbers whose greatest common factor is 1.

33. Critical Thinking Write two numbers that have a GCF of 14.

34. Algebra Eric and his sister Nola are teenagers. The GCF of their ages is 3. If Eric is 15, how old is Nola?

35. Decorating Marcos has two rolls of streamers to use in decorating the gym for a school dance. One roll is 126 feet long. The other is 48 feet long. If he wants to make all the streamers the same length, what is the longest length each streamer can be?

6-5A Understanding Fractions

A Preview of Lesson 6-5

Objective

Use models to represent fractions. Find a fraction equal to another fraction.

Materials

paper
pencil
ruler

In previous lessons, you have used models to help you understand decimals, multiplication, and division. In this lab, you will use models to help you understand fractions.

Activity One

Work with a partner.

Represent $\frac{1}{6}$ using a model.

- On a sheet of paper, draw a rectangle like the one shown below.

- Separate the rectangle into six equal parts. Shade one part.

What do you think?

1. What fraction was represented in the activity?
2. How would you represent $\frac{2}{6}$ using the same model?
3. If you shade five parts of the rectangle, what fraction would that be?
4. What other model can you think of to represent the fraction in Activity One?
5. Are there other ways to show $\frac{1}{6}$ on the rectangle above? If so, draw an example.

Activity Two

Compare $\frac{2}{6}$ and $\frac{4}{12}$.

- Draw two identical rectangles as shown below.

- Separate the upper rectangle into six equal parts. Separate the bottom rectangle into twelve equal parts.

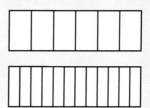

- Shade two parts of the upper rectangle, and four of the bottom rectangle.

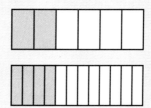

What do you think?

6. What fractions are represented by each rectangle?

7. Do these fractions name the same number? Explain why.

8. Are there other fractions that name the same number? If so, draw shaded rectangles to show your example.

Write the pair of fractions represented by each model.

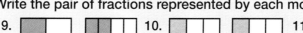

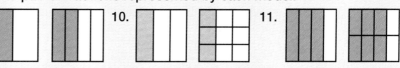

Draw models for each pair of fractions. Tell whether each pair represents the same number.

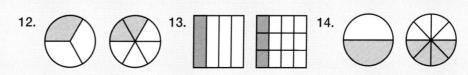

15. $\frac{1}{2}, \frac{4}{8}$

16. $\frac{1}{3}, \frac{2}{6}$

17. $\frac{2}{8}, \frac{1}{4}$

Simplifying Fractions

Objective

Express fractions in simplest form.

Words to Learn

equivalent fractions
simplest form

What is your favorite color of M&M's®? What fraction of that color do you think a package would contain? A $1\frac{1}{2}$-ounce package of M&M's® contains about 56 M&M's®. Usually, there are more brown pieces than any other color. Suppose you open a package that has 16 brown M&M's®. The fraction that represents the part of the M&M's® that are brown is $\frac{16}{56}$.

The fraction $\frac{16}{56}$ can also be written as $\frac{8}{28}$, $\frac{4}{14}$, and $\frac{2}{7}$. All these fractions name the same number. Fractions that name the same number are called **equivalent fractions.**

The rectangles below are the same size. The same part or fraction of each rectangle is shaded. So, the fractions $\frac{2}{7}$ and $\frac{4}{14}$ are equivalent. That is, $\frac{2}{7} = \frac{4}{14}$.

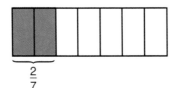

$$\frac{2}{7}$$

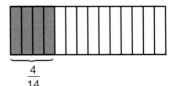

$$\frac{4}{14}$$

You can multiply or divide the numerator and denominator of a fraction by the same nonzero number to find an equivalent fraction.

Examples

Replace each ■ with a number so that the fractions are equivalent.

1 $\frac{4}{7} = \frac{■}{28}$

Multiply the numerator and denominator by 4.

$$\overset{\times 4}{\frac{4}{7}} = \frac{■}{28}, \text{ so } \frac{4}{7} = \frac{16}{28}.$$
$$\underset{\times 4}{}$$

2 $\frac{18}{81} = \frac{■}{9}$

Divide the numerator and denominator by 9.

$$\overset{\div 9}{\frac{18}{81}} = \frac{■}{9}, \text{ so } \frac{18}{81} = \frac{2}{9}.$$
$$\underset{\div 9}{}$$

In previous lessons, you have learned about factors and the greatest common factor. You can use factors to write a fraction in simplest form. A fraction is in **simplest form** when the GCF of the numerator and denominator is 1. To write a fraction in simplest form, find the GCF of the numerator and denominator first. Then divide both the numerator and denominator by the GCF.

Calculator Hint

● ● ● ● ● ● ● ● ● ● ●

You can use a fraction calculator to simplify fractions. For example, to simplify $\frac{4}{16}$, enter

4 [÷] 16 [Simp] [=].

Repeat [Simp] [=] until the N/D → n/d does not appear on the screen.

Examples

Write each fraction in simplest form.

3 $\frac{4}{12}$

factors of 4: 1, 2, 4
factors of 12: 1, 2, 3, 4, 6, 12

$$\frac{4}{12} = \frac{1}{3}$$

Divide the numerator and denominator by the GCF, 4.

Since the GCF of 1 and 3 is 1, the fraction $\frac{1}{3}$ is in simplest form.

4 $\frac{28}{77}$

factors of 28: 1, 2, 4, 7, 14, 28
factors of 77: 1, 7, 11, 77

$$\frac{28}{77} = \frac{4}{11}$$

Divide the numerator and denominator by the GCF, 7.

Since the GCF of 4 and 11 is 1, the fraction $\frac{4}{11}$ is in simplest form.

Checking for Understanding

Communicating Mathematics

Read and study the lesson to answer each question.

1. **Draw** a picture to show $\frac{12}{16}$ and the fraction in simplest form.

2. **Tell** what fractions are represented by the figures at the right. Are they equivalent?

Guided Practice

3. Refer to the figure below.

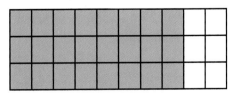

a. What fraction does the shaded part of the figure describe?
b. What is the GCF of the numerator and denominator of the fraction?
c. Write the fraction in simplest form.
d. Draw a picture of the fraction in simplest form.

State whether each fraction is in simplest form. If not, write it in simplest form.

4. $\dfrac{4}{9}$ 5. $\dfrac{3}{27}$ 6. $\dfrac{12}{48}$ 7. $\dfrac{18}{21}$

8. $\dfrac{51}{66}$ 9. $\dfrac{35}{49}$ 10. $\dfrac{41}{85}$ 11. $\dfrac{81}{100}$

Exercises

Independent Practice

12. Refer to the figure at the right.

    a. What fraction does the shaded part of the figure describe?
    b. What is the GCF of the numerator and denominator of the fraction?
    c. Write the fraction in simplest form.
    d. Draw a picture of the fraction in simplified form.

State whether each fraction is in simplest form. If not, write it in simplest form.

13. $\dfrac{8}{25}$ 14. $\dfrac{9}{24}$ 15. $\dfrac{24}{30}$ 16. $\dfrac{12}{96}$ 17. $\dfrac{35}{63}$

18. $\dfrac{57}{60}$ 19. $\dfrac{49}{84}$ 20. $\dfrac{32}{40}$ 21. $\dfrac{50}{63}$ 22. $\dfrac{13}{78}$

23. $\dfrac{8}{100}$ 24. $\dfrac{75}{81}$ 25. $\dfrac{18}{144}$ 26. $\dfrac{27}{54}$ 27. $\dfrac{49}{101}$

Mixed Review

28. Use the distributive property to mentally solve $(5.7 \times 11) + (0.3 \times 11)$. *(Lesson 4-4)*

29. Estimate the sum $4.231 + 3.98 + 4 + 4.197 + 3.76$. *(Lesson 3-6)*

30. List all common factors of 20 and 50. *(Lesson 6-4)*

31. **Animals**   Isra recorded the weights of guinea pigs in the pet store where she worked. The weights were 1.8, 1.754, 2.09, 1.91, and 2.1 pounds. List these weights in order from least to greatest. *(Lesson 3-4)*

Problem Solving and Applications

32. **Critical Thinking**   Copy the figure at the right and then show $\dfrac{3}{5}$.

33. **Sports**   Yang hit the ball 12 times out of 15 times at bat. Write a fraction in simplest form to express his batting success.

34. **Statistics**   Approximately 80 of every 100 American households have cable TV. Write this as a fraction in simplest form.

35. **Journal Entry**   Write one or two sentences on how you know whether a fraction is in simplest form.

Mixed Numbers and Improper Fractions

Objective

Express mixed numbers as improper fractions and vice versa.

Words to Learn

mixed numbers
improper fraction

To win the U.S. Triple Crown, a horse must win these three classic races.

- Kentucky Derby, a $1\frac{1}{4}$ mile race
- Preakness Stakes, a $1\frac{3}{16}$ mile race
- Belmont Stakes, a $1\frac{1}{2}$ mile race

Eleven horses have won the Triple Crown. The last horse to win the Triple Crown was *Affirmed* in 1978.

The numbers $1\frac{1}{4}$, $1\frac{3}{16}$, and $1\frac{1}{2}$ are called mixed numbers.

Mixed numbers show the sum of a whole number and a fraction. Mixed numbers can also be written as fractions.

Mini-Lab

Work with a partner.

Materials: paper, pencil, ruler

Draw a model for $1\frac{1}{4}$.

- Draw a rectangle like the one shown below. Shade the rectangle to represent 1.

- Draw an identical rectangle beside the first one. Separate the rectangle on the right into four equal parts to show fourths. Shade one part to represent $\frac{1}{4}$.

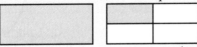

- Separate the whole number portion into $\frac{1}{4}$'s.

Talk About It

a. How many shaded $\frac{1}{4}$'s are there?

b. What fraction is equivalent to $1\frac{1}{4}$?

DID YOU KNOW

The first horse was an animal no larger than a small dog. It is known as *Euhippus* or the "dawn horse." It had four toes on its front feet and three toes on its back feet. Gradually, over millions of years, the horse lost all its toes except one.

A fraction, like $\frac{7}{3}$ or $\frac{4}{4}$, with a numerator that is greater than or equal to the denominator is called an **improper fraction.**

From the Mini-Lab, you can conclude that you can express a mixed number as an improper fraction.

Example 1

Express $2\frac{1}{2}$ as an improper fraction.

Find the number of parts in the whole numbers. Then add the fraction.

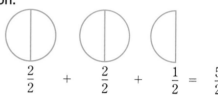

$$\frac{2}{2} \quad + \quad \frac{2}{2} \quad + \quad \frac{1}{2} \quad = \quad \frac{5}{2}$$

A short-cut is to multiply the whole number by the denominator and add the numerator. Then write this sum over the denominator.

$$2\frac{1}{2} = \frac{(2 \times 2) + 1}{2} = \frac{5}{2}$$

We can also express an improper fraction as a mixed number. To do this, divide the numerator by the denominator.

Example 2

Express $\frac{9}{4}$ as a mixed number. Draw a model to illustrate.

Divide 9 by 4.

$$4\overline{\smash{\big)}9} \quad \overset{2}{}\ \text{R1}$$

$$\frac{9}{4} = 2\frac{1}{4}$$

Write the remainder as a fraction that has the remainder as the numerator and the divisor as the denominator.

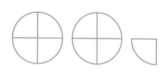

Checking for Understanding

Communicating Mathematics

Read and study the lesson to answer each question.

1. **Tell,** in your own words, what is meant by a mixed number.

2. **Write** a mixed number and an improper fraction for the figure at the right.

Guided Practice

Draw a model and express each mixed number as an improper fraction.

3. $1\frac{1}{3}$ 4. $1\frac{3}{4}$ 5. $2\frac{2}{3}$ 6. $3\frac{1}{2}$

Draw a model and express each fraction as a mixed number.

7. $\frac{7}{3}$ 8. $\frac{9}{5}$ 9. $\frac{7}{6}$ 10. $\frac{5}{3}$

Exercises

Independent Practice

Express each mixed number as an improper fraction.

11. $1\frac{2}{5}$ 12. $1\frac{5}{6}$ 13. $3\frac{1}{8}$ 14. $5\frac{2}{3}$ 15. $7\frac{3}{4}$

16. $2\frac{5}{8}$ 17. $1\frac{4}{9}$ 18. $4\frac{1}{5}$ 19. $9\frac{3}{7}$ 20. $11\frac{1}{3}$

Express each fraction as a mixed number.

21. $\frac{11}{8}$ 22. $\frac{13}{6}$ 23. $\frac{7}{2}$ 24. $\frac{18}{3}$ 25. $\frac{14}{9}$

26. $\frac{23}{4}$ 27. $\frac{17}{3}$ 28. $\frac{28}{7}$ 29. $\frac{25}{8}$ 30. $\frac{39}{11}$

31. What improper fraction represents the mixed number three and one-fourth?

Mixed Review

32. **Astronomy** The distance from Earth to the sun is close to 10^8 miles. How many miles is this? *(Lesson 4-2)*

33. Change 12,237 milliliters to liters. *(Lesson 5-6)*

34. Use the figure at the right to state what fraction the shaded part describes. *(Lesson 6-5)*

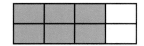

35. Evaluate $14 - 2 \times 5$. *(Lesson 1-6)*

Problem Solving and Applications

36. **Critical Thinking** Explain, in your own words, how you can determine whether a fraction is less than, equal to, or greater than 1.

37. **Cooking** Juana's biscuit recipe calls for $\frac{1}{2}$ cup of milk. If she triples the recipe, she will need $\frac{3}{2}$ cups of milk. Write $\frac{3}{2}$ as a mixed number.

6 Mid-Chapter Review

State whether each number is divisible by 2, 3, 5, 6, 9, or 10. *(Lesson 6-1)*

1. 222 2. 353 3. 1,050 4. 525

Find the prime factorization of each number. *(Lesson 6-2)*

5. 106 6. 77 7. 56 8. 99

9. What is the next pair of twin primes after 17 and 19? *(Lesson 6-3)*

10. Find the GCF of 45 and 90. *(Lesson 6-4)*

11. Write $\frac{12}{15}$ in simplest form. *(Lesson 6-5)*

12. Write a mixed number and an improper fraction for the figure at the right. *(Lesson 6-6)*

6-7 Length in the Customary System

Objective

Measure line segments with a ruler divided in fourths and in eighths.

Words to Learn

inch
foot
yard
mile

For many people, the length of their arm is about 8 times the length of their index finger.

Mini-Lab

Work with a partner.

Materials: a ruler or tape measure, paper, pencil

- Draw a long line segment and mark off eight segments using the length of your index finger.
  Measure this length to the nearest inch. Record.
- Measure the length of your arm from your shoulder to the end of your middle finger to the nearest inch. Record.

Talk About It

a. Is the total length of the eight line segments close to the actual length of your arm?

b. Divide the actual length of your arm by eight to get an estimate of the length of your index finger. Check by measuring your finger.

The **inch** is one of the most commonly used customary units of length. Others are the **foot, yard,** and **mile.**

$$1 \text{ foot (ft)} = 12 \text{ inches (in.)}$$
$$1 \text{ yard (yd)} = 3 \text{ feet or } 36 \text{ inches}$$
$$1 \text{ mile (mi)} = 5{,}280 \text{ feet or } 1{,}760 \text{ yards}$$

Sometimes you have to measure objects using units shorter than an inch. Most rulers are separated into fourths or eighths.

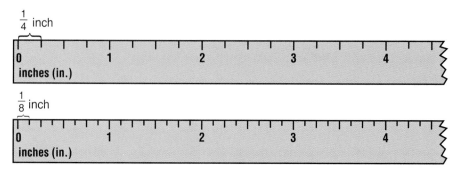

$\frac{1}{4}$ inch

$\frac{1}{8}$ inch

1 Draw a line segment measuring $1\frac{1}{4}$ inches.

Use a ruler separated into fourths or into eighths.

Find $1\frac{1}{4}$ on the ruler.

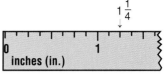

Draw a line segment from 0 to $1\frac{1}{4}$.

2 Find the length of the nail to the nearest eighth inch.

Line up the nail with a ruler separated into eighths.

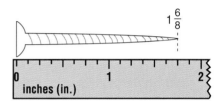

The length of the nail is $1\frac{6}{8}$ or $1\frac{3}{4}$ inches.

Checking for Understanding

Communicating Mathematics

Read and study the lesson to answer each question.

1. **Tell** why you can use either a ruler separated into fourths or a ruler separated into eighths to measure an object to the nearest fourth inch.

2. **Tell** the shortest length you can measure using the ruler at the right.

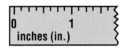

Guided Practice

Draw a line segment for each length.

3. $3\frac{1}{4}$ inches

4. $\frac{7}{8}$ inch

5. $2\frac{1}{8}$ inches

Find the length of each line segment to the nearest fourth inch.

6. ———

7. ———

8. ————————

Exercises

Independent Practice

Draw a line segment for each length.

9. $3\frac{3}{4}$ inch

10. $2\frac{1}{2}$ inch

11. $\frac{5}{8}$ inch

12. $1\frac{7}{8}$ inch

13. $4\frac{3}{4}$ inch

14. $2\frac{1}{8}$ inch

Find the length of each object to the nearest eighth inch.

15.

16.

17.

18.

19. Find the width of your desk to the nearest fourth inch.

20. Find the width of your math book to the nearest eighth inch.

Mixed Review

21. **Algebra** Evaluate the expression $a + b \div c$ if $a = 5$, $b = 12$, and $c = 4$. *(Lesson 1-6, 1-7)*

22. Express the fraction $\frac{30}{7}$ as a mixed number. *(Lesson 6-6)*

23. Divide 200.568 by 1.22 *(Lesson 5-3)*

Problem Solving and Applications

24. **Auto Mechanics** Mr. Flores uses wrenches of different sizes when working on cars. He needs to choose a wrench that will loosen the bolt at the right. What size wrench will he need: $\frac{3}{8}$ inch, $\frac{5}{8}$ inch, or $\frac{3}{4}$ inch?

25. **Critical Thinking**
    a. How many eighth inches are in a foot?
    b. How many fourth inches are in a yard?

26. **Mathematics and Social Studies** Read the following paragraph.

Up until the fifteenth century, Europeans had only a limited view of the world. Then, in 1492, a navigator named Christopher Columbus began sailing west from Spain and accidentally discovered the West Indies. Further Spanish expeditions followed, leading to the discovery of the South American mainland. At first, people believed this new land was part of Asia. Later, they realized that for them an entire new continent had been discovered.

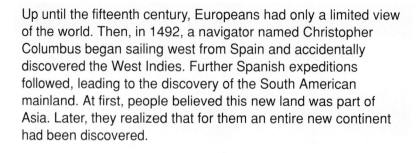

Suppose two of the crew members on Columbus' voyage were responsible for loading crates of supplies onto the ship. One member made a row of crates that was 30 feet long and the other made a row that was 21 feet long. What is the largest size crate they might have loaded if all the crates were the same size?

6-8 Least Common Multiple

Objective

Find the least common multiple of two or more numbers.

Words to Learn

multiples
common multiples
least common multiple
(LCM)

Problem Solving Hint

● ● ● ● ● ● ● ● ● ●

You can make a list to find multiples.

Regina has soccer practice every other school day. She has band practice every third school day. What is the first day that she will have to bring both her trumpet and her soccer equipment to school?

To answer this question, you can use multiples. A **multiple** of a number is the product of that number and any whole number. Regina has soccer practice every second day, or multiples of 2. Band practice is every third day, or multiples of 3.

The multiples of 2 and 3 are shown below.
multiples of 2: 0, 2, 4, 6, 8, 10, . . .
multiples of 3: 0, 3, 6, 9, . . .

Remember that the three dots mean "and so on."

Notice that 0 and 6 are multiples of both 2 and 3. They are called **common multiples.** The least of the common multiples of two or more numbers, other than zero, is called the **least common multiple (LCM).** The least common multiple of 2 and 3 is 6.

Regina would have to bring both her trumpet and her soccer equipment to school on the sixth school day.

Mini-Lab

Work with a partner.
Materials: scissors, ruler, paper, and pencil

- Cut eight strips of paper 3 inches long by 1 inch wide.
- Cut six strips of paper 4 inches long by 1 inch wide.
- Place the three-inch strips of paper end-to-end to make a train.
- Repeat the process with the four-inch strips of paper. Place this train under the three-inch strips train.

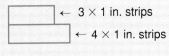

← 3 × 1 in. strips
← 4 × 1 in. strips

Talk About It

a. Sketch a diagram of the trains.

b. Describe where the ends of the strips of the trains are lined up. What is that length?

c. What is the least common multiple of 3 and 4?

1 List the first five multiples of 3.

multiples of 3: 0, 3, 6, 9, 12

Calculator Hint

● ● ● ● ● ● ● ● ● ●

You can use a calculator to find the LCM. Divide multiples of the greater number by the lesser number until you get a whole number quotient.

Find the LCM for each pair of numbers.

2 4 and 6
multiples of 4:
0, 4, 8, 12, 16, . . .
multiples of 6:
0, 6, 12, 18, 24, . . .

The LCM is 12.

3 6 and 8
multiples of 6:
0, 6, 12, 18, 24, 30, . . .
multiples of 8:
0, 8, 16, 24, 32, 40, . . .

The LCM is 24.

Example 4 *Problem Solving*

Remodeling Marbella is replacing her kitchen floor. She has placed a row of tiles that are 3 inches long along the wall. The next row has tiles 4 inches long, and the third row has tiles 5 inches long. At what point will the ends of the tiles be lined up?

Find the multiples of 3, 4, and 5.

multiples of 3: 0, 3, 6, 9, . . . , 57, 60, 63, . . .
multiples of 4: 0, 4, 8, 12, . . . , 56, 60, 64, . . .
multiples of 5: 0, 5, 10, 15, . . . , 55, 60, . . .

The LCM of 3, 4, and 5 is 60. The ends of the tiles will be lined up at 60 inches.

Checking for Understanding

Communicating Mathematics

Read and study the lesson to answer each question.

1. **Tell** what is meant by the term LCM.
2. **Make a model,** using paper strips, to find the LCM of 2 and 5.
3. **Tell,** in your own words, how to find the LCM of 6 and 9.

Guided Practice

List the first five multiples of each number.

4. 2   5. 7   6. 12   7. 10

Find the LCM for each set of numbers.

8. 4, 8   9. 9, 12   10. 3, 8   11. 2, 7

Exercises

Determine whether the first number is a multiple of the second number.

12. 35; 5   **13.** 28; 7   **14.** 188; 8   **15.** 48; 6

Find the LCM for each set of numbers.

16. 8, 12   **17.** 9, 15   **18.** 24, 30   **19.** 15, 24

20. 12, 16   **21.** 14, 21   **22.** 13, 16   **23.** 28, 32

24. 10, 35, 40   **25.** 9, 12, 15   **26.** 15, 25, 75   **27.** 9, 18, 3

28. Find the LCM of your age and one of your parents' ages.

Mixed Review

29. State whether 405 is divisible by 2, 3, 5, 6, 9, or 10. *(Lesson 6-1)*

30. Find the width of your pencil to the nearest eighth inch. *(Lesson 6-7)*

31. **Geometry** The length of a rectangle is 5 inches longer than it is wide. What are the length and width if the perimeter is 50 inches? *(Lesson 1-8)*

32. How many milliliters are equal to 2.14 liters? *(Lesson 5-6)*

Problem Solving and Applications

33. **Critical Thinking** The LCM of two numbers is $2^3 \cdot 3^2$. Find two pairs of numbers that fit this description.

34. **Family** Maria and Nicole see each other when they visit their grandparents who live next door to each other. Maria visits her grandparents every two weeks. Nicole visits hers every three weeks. They are both at their grandparents this weekend.

a. How many weeks will it be before they are at their grandparents at the same time?

b. If they are together on December 12, on what date will they next see each other?

Save Planet Earth

Holiday Waste Holiday celebrations can actually hurt the environment by making additional waste, harming animals, and using some of Earth's threatened resources.

How You Can Help

- Buy a live tree that can be planted in your backyard after the holiday season.

- Use an alternative to traditional wrapping paper. Be creative: use old maps, posters, or comics.

- For additional information, look for the book *The New Green Christmas* by the Evergreen Alliance.

Comparing and Ordering Fractions

Objective

Compare and order fractions.

Words to Learn

least common denominator (LCD)

DID YOU KNOW

Nielsen Media Research provides the networks with the ratings for each TV program. The company uses the viewing habits of 1,200 American families to determine its ratings.

Television is the most powerful source of information available to millions of people. To determine such things as what shows to keep on the air and what kinds of advertisement to place at certain times, the networks obtain data from Nielsen Media Research.

According to Nielsen Media Research, during an average week from 8:00 P.M. until 11:00 P.M., male teens spend $\frac{1}{3}$ of this time watching TV and males over 18 years of age spend $\frac{3}{7}$ of this time watching TV. Who spends more time watching TV in the evening, male teens or males over 18?

The answer to this question can be obtained by comparing the fractions $\frac{1}{3}$ and $\frac{3}{7}$. One way to compare these fractions is to express them as fractions with the same denominator. Any common denominator could be used. But using the least common denominator makes the computation easier.

The **least common denominator (LCD)** is the LCM of the denominators. To find the LCD of $\frac{1}{3}$ and $\frac{3}{7}$, you need to find the LCM of 3 and 7.

multiples of 3: 0, 3, 6, 9, 12, 15, 18, 21, 24, . . .
multiples of 7: 0, 7, 14, 21, 28, . . .
The LCM of the denominators, 3 and 7, is 21.

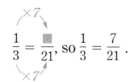

$$\frac{1}{3} = \frac{\blacksquare}{21}, \text{ so } \frac{1}{3} = \frac{7}{21}. \qquad \frac{3}{7} = \frac{\blacksquare}{21}, \text{ so } \frac{3}{7} = \frac{9}{21}.$$

Now compare the numerators of the two fractions. You can use models to help you do this.

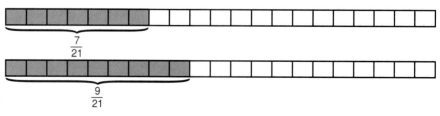

$\frac{7}{21}$

$\frac{9}{21}$

Since $9 > 7$, $\frac{9}{21} > \frac{7}{21}$. Therefore, $\frac{3}{7} > \frac{1}{3}$. So, males over 18 spend more time watching TV in the evening.

Replace each ● with <, >, or = to make a true sentence.

1 $\dfrac{5}{6}$ ● $\dfrac{7}{8}$

The LCM of 6 and 8 is 24. Express $\dfrac{5}{6}$ and $\dfrac{7}{8}$ as fractions with a denominator of 24.

$\dfrac{5}{6} = \dfrac{\blacksquare}{24}$, so $\dfrac{5}{6} = \dfrac{20}{24}$. $\dfrac{7}{8} = \dfrac{\blacksquare}{24}$, so $\dfrac{7}{8} = \dfrac{21}{24}$.

Since $20 < 21$, $\dfrac{20}{24} < \dfrac{21}{24}$. Therefore $\dfrac{5}{6} < \dfrac{7}{8}$.

2 $\dfrac{5}{12}$ ● $\dfrac{3}{4}$

The LCM of 12 and 4 is 12. Express $\dfrac{5}{12}$ and $\dfrac{3}{4}$ as fractions with a denominator of 12.

$\dfrac{5}{12} = \dfrac{5}{12}$ $\dfrac{3}{4} = \dfrac{\blacksquare}{12}$, so $\dfrac{3}{4} = \dfrac{9}{12}$.

Since $5 < 9$, $\dfrac{5}{12} < \dfrac{9}{12}$. Therefore $\dfrac{5}{12} < \dfrac{3}{4}$

Checking for Understanding

Communicating Mathematics

Read and study the lesson to answer each question.

1. **Tell,** in your own words, how the LCM is used to compare fractions.

2. **Write** the steps you would take to rewrite $\dfrac{3}{8}$ as $\dfrac{\blacksquare}{32}$.

3. **Make a model** to compare $\dfrac{2}{3}$ and $\dfrac{3}{4}$.

Guided Practice

Find the LCD for each pair of fractions.

4. $\dfrac{1}{4}, \dfrac{2}{3}$ 5. $\dfrac{3}{5}, \dfrac{3}{4}$ 6. $\dfrac{1}{2}, \dfrac{3}{8}$

Replace each ● with <, >, or = to make a true sentence.

7. $\dfrac{3}{5}$ ● $\dfrac{2}{3}$ 8. $\dfrac{2}{9}$ ● $\dfrac{1}{3}$ 9. $\dfrac{6}{9}$ ● $\dfrac{2}{3}$

10. $\dfrac{1}{4}$ ● $\dfrac{2}{3}$ 11. $\dfrac{3}{5}$ ● $\dfrac{3}{4}$ 12. $\dfrac{1}{2}$ ● $\dfrac{3}{8}$

13. $\dfrac{5}{8}$ ● $\dfrac{10}{16}$ 14. $\dfrac{1}{8}$ ● $\dfrac{1}{12}$ 15. $\dfrac{5}{6}$ ● $\dfrac{9}{10}$

Exercises

Independent
Practice

Find the LCD for each pair of fractions.

16. $\frac{1}{6}, \frac{2}{9}$

17. $\frac{3}{4}, \frac{1}{6}$

18. $\frac{1}{2}, \frac{2}{3}$

Replace each ⬤ with <, >, or = to make a true sentence.

19. $\frac{3}{7}$ ⬤ $\frac{2}{5}$

20. $\frac{3}{10}$ ⬤ $\frac{1}{4}$

21. $\frac{5}{6}$ ⬤ $\frac{1}{9}$

22. $\frac{9}{15}$ ⬤ $\frac{11}{15}$

23. $\frac{5}{7}$ ⬤ $\frac{4}{9}$

24. $\frac{5}{14}$ ⬤ $\frac{9}{28}$

25. Which is greater, $\frac{5}{8}$ or $\frac{7}{12}$?

Order the following fractions from least to greatest.

26. $\frac{1}{2}, \frac{3}{5}, \frac{5}{6}, \frac{2}{3}$

27. $\frac{2}{5}, \frac{4}{7}, \frac{5}{11}, \frac{4}{9}$

28. $\frac{1}{3}, \frac{1}{4}, \frac{1}{2}, \frac{1}{7}$

29. $\frac{1}{6}, \frac{2}{5}, \frac{3}{7}, \frac{3}{5}$

Mixed Review

30. Find the product of 17.241 and 16. *(Lesson 4-3)*

31. **Algebra** Solve the equation $0.012s = 9.36$. *(Lesson 5-7)*

32. Find the least common multiple of 7 and 21. *(Lesson 6-8)*

33. **Animals** If a rabbit has a mass of 800 grams, what is its mass in kilograms? *(Lesson 5-6)*

Problem Solving
and
Applications

Statistics The chart at the right shows the statistics for the votes for Supreme Court nominees. Use the chart to answer Exercises 34-36.

34. Write a fraction to show the votes for John J. Parker compared to the total number of votes on his confirmation.

35. Write a fraction to show the votes for Robert Bork compared to the total number of votes on his confirmation. Write an equivalent fraction in simplest form.

36. Write a fraction to show what part of the total votes were in support of Clarence Thomas. Write an equivalent fraction in simplest form.

JUSTICE VOTES

Closest confirmation votes for Supreme Court nominees :

NOMINEE	For	Against
J. Parker	39	41
G. Carswell	45	51
C. Haynsworth Jr.	45	55
R. Bork	42	58
M. Pitney	50	26
C. Thomas	52	48

37. **Critical Thinking** I am a fraction in simplest form. My numerator and denominator are twin primes. Twin primes are prime numbers that have a difference of 2. The sum of my numerator and denominator is equal to a dozen. Who am I?

38. **Journal Entry** Write, in your own words, how to determine which of two fractions is greater.

6-10 Writing Decimals as Fractions

Objective

Express terminating decimals as fractions in simplest form.

Words to Learn

terminating decimal

To compete in today's market and to meet government requirements, car makers have to build cars that are more fuel efficient. The average gas mileage for cars in the United States almost doubled from 14.2 miles per gallon in 1974 to 28.2 miles per gallon in 1991.

The number 14.2 is an example of a **terminating decimal.** A terminating decimal can be written as a fraction with a denominator of 10, 100, 1,000, and so on. For example, 14.2 can be written as the mixed number $14\frac{2}{10}$. In simplest form, $14\frac{2}{10}$ is $14\frac{1}{5}$.

Mental Math Hint

• • • • • • • • • • •

Here are some commonly used decimal-fraction equivalencies. It is helpful to know them by memory.

$0.5 = \frac{1}{2}$

$0.25 = \frac{1}{4}$

$0.2 = \frac{1}{5}$

$0.125 = \frac{1}{8}$

Examples

Express each decimal as a fraction in simplest form.

1 0.6

$0.6 = \frac{6}{10}$ *Write the decimal as a fraction. Use a denominator of 10 since 0.6 is 6 tenths.*

$= \frac{\overset{3}{\cancel{6}}}{\underset{5}{\cancel{10}}}$ *Simplify. The GCF of 6 and 10 is 2.*

$= \frac{3}{5}$

2 4.375

$4.375 = 4\frac{375}{1,000}$ *Write the decimal as a mixed number. Use a denominator of 1,000 since 0.375 is 375 thousandths.*

$= 4\frac{\overset{3}{\cancel{375}}}{\underset{8}{\cancel{1,000}}}$ *Simplify. The GCF of 375 and 1,000 is 125.*

$= 4\frac{3}{8}$

Example 3 *Connection*

Geometry   0.64 of the figure at the right is shaded. What fraction of the figure is shaded?

$0.64 = \frac{64}{100}$

$= \frac{16}{25}$ $\frac{16}{25}$ of the figure is shaded.

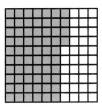

TEEN SCENE

Did you know your parent's insurance rates will go up an average of $800 a year when you are added to their policy? Rates differ depending on the number of vehicles and whether you are male or female. Many insurance companies offer a good student discount if your grades are good.

Checking for Understanding

Communicating Mathematics

Read and study the lesson to answer each question.

1. **Tell** how you would express 0.9 as a fraction.

2. **Draw a model** of 0.5. Rewrite the decimal as a fraction in simplest form.

Guided Practice

Complete.

3. $0.3 = \dfrac{\blacksquare}{10}$

4. $0.1 = \dfrac{\blacksquare}{10}$

5. $0.8 = \dfrac{\blacksquare}{10}$

6. $0.25 = \dfrac{\blacksquare}{100} = \dfrac{\blacksquare}{4}$

Express each decimal as a fraction in simplest form.

7. 2.4 8. 0.16 9. 5.45 10. 3.625

11. Write *eighty-two hundredths* as a decimal and as a fraction in simplest form.

Exercises

Independent Practice

Express each decimal as a fraction in simplest form.

12. 0.6 13. 0.03 14. 0.012 15. 3.36

16. 4.4 17. 7.88 18. 11.002 19. 2.101

20. 6.65 21. 1.97 22. 4.456 23. 22.2

24. Write *nine thousandths* as a decimal and as a fraction in simplest form.

25. Write *thirteen and two tenths* as a decimal and as a fraction in simplest form.

Mixed Review

26. **Algebra** Evaluate $a - b$ if $a = 7.3$ and $b = 2.94$. *(Lesson 3-8)*

27. Find the greatest common factor of 45 and 75. *(Lesson 6-4)*

28. Order the fractions $\dfrac{3}{5}, \dfrac{5}{8}$, and $\dfrac{4}{7}$ from least to greatest. *(Lesson 6-9)*

Problem Solving and Applications

29. **Critical Thinking** A factory requires that its employees complete at least $\dfrac{9}{10}$ of their work within three days after it is assigned. Mr. Williams completed 0.93 of his work. Did he meet the factory standard? Explain.

30. **Statistics** A 1990 survey showed that Americans are driving more miles with fewer passengers in each vehicle. Write each decimal as a fraction in simplest form.

     a. The members of an average household drive 41.4 miles per day.

     b. There are 1.8 vehicles per household.

     c. The average length of a trip to work is 10.9 miles.

31. **Journal Entry** Find three decimals in a magazine or newspaper. Write a sentence explaining the use of each decimal. Express each decimal as a fraction in simplest form.

6-11 Writing Fractions as Decimals

Objective

Express fractions as terminating and repeating decimals.

Words to Learn

repeating decimal

Enrique and his mother went to the store to buy some ground beef. They asked for $\frac{3}{4}$ of a pound of ground beef. The scale read 0.72 pounds. Did they have enough ground beef?

To solve this problem, you need to find out if $\frac{3}{4}$ is equal to 0.72. To do this, remember that a fraction is another way of writing a division problem. So, $\frac{3}{4}$ means $3 \div 4$.

$$3 \; \boxed{\div} \; 4 \; \boxed{=} \; 0.75$$

Since $\frac{3}{4} = 0.75$ and the scale read 0.72, Enrique and his mother did not have enough ground beef.

To express a fraction as a decimal, simply divide the numerator by the denominator. If the division ends or terminates with a remainder of zero, then the result is a terminating decimal.

Stock Broker

Selling stocks is one way companies raise money. If you buy stock, you own a share of the company.

A stockbroker is a person who makes a living buying and selling stocks. Every time a stock is bought or sold, the broker receives a fee.

In order to work with stock prices, you need to be able to express fractions as decimals.

For more information, contact:
Professional Salespersons of America
3801 Monaco NE
Albuquerque, NM 87111

Examples

Express each fraction as a decimal.

1 $\frac{3}{8}$

Use a calculator.

$$3 \; \boxed{\div} \; 8 \; \boxed{=} \; 0.375$$

$$\frac{3}{8} = 0.375$$

Use paper and pencil.

$$\frac{3}{8} \rightarrow 8 \overline{)3.000} = 0.375$$

```
        0.375
   8 ) 3.000
      -2 4
        60
       -56
         40
        -40
          0
```

Annex zeros to the numerator.
3 = 3.000

2 $160\frac{3}{4}$

Use a calculator.

$$160\frac{3}{4} = 160 + \frac{3}{4}$$

$$160 \; \boxed{+} \; 3 \; \boxed{\div} \; 4 \; \boxed{=} \; 160.75$$

$$160\frac{3}{4} = 160.75$$

Use paper and pencil.

$$\frac{3}{4} \rightarrow$$

```
        0.75
   4 ) 3.00
      -2 8
        20
       -20
         0
```

$$160 + 0.75 = 160.75$$

Not all fractions can be expressed as terminating decimals. When dividing the numerator by the denominator of a fraction, if a remainder of zero cannot be obtained, the digits in the quotient repeat. This quotient is then called a **repeating decimal.** An example of a repeating decimal is 0.33333. . . . It can be written as $0.\overline{3}$. The bar above 3 means that the digit 3 repeats.

Example 3

Write the first ten decimal places of $0.1\overline{6}$.

The bar above 6 means that 6 repeats.
$$0.1\overline{6} = 0.\underbrace{1666666666}_{ten\ decimal\ places}$$

Example 4

Calculator Hint

• • • • • • • • • • •

Some calculators round answers and others truncate answers. Truncate means to cut off at at a certain place-value position, ignoring the digits that follow. Does your calculator round or truncate?

Express $\dfrac{5}{11}$ as a decimal. Use bar notation to show a repeating decimal.

Use a calculator.

5 ÷ 11 = 0.45454545

$\dfrac{5}{11} = 0.45454545\ldots$
$\qquad = 0.\overline{45}$

Use paper and pencil.

$$\dfrac{5}{11} \rightarrow \begin{array}{r} .4545 \\ 11\overline{)5.0000} \\ -44 \\ \hline 60 \\ -55 \\ \hline 50 \\ -44 \\ \hline 60 \\ -55 \\ \hline 5 \end{array} \rightarrow 0.\overline{45}$$

Example 5 *Problem Solving*

School American students spend 180 days a year in school. German students spend $1\frac{1}{3}$ times as many days. How many days a year do German students spend in school?

Multiply 180 and $1\frac{1}{3}$. First, use your calculator to change $1\frac{1}{3}$ to a decimal.

1 + 1 ÷ 3 = 1.333333

Keeping 1.333333 on your calculator display, multiply by 180.

1.333333 × 180 = 240

German students spend 240 days a year in school.

Checking for Understanding

Communicating Mathematics Read and study the lesson to answer each question.

1. **Tell,** in your own words, how to express a fraction as a decimal.

2. **Show** an example of a fraction that can be expressed as a terminating decimal and one that can be expressed as a repeating decimal.

Guided Practice Express each decimal using bar notation.

3. $0.66666\ldots$ 4. $0.83333\ldots$ 5. $0.121212\ldots$

Write the first ten decimal places of each decimal.

6. $0.\overline{47}$ 7. $0.1\overline{67}$ 8. $0.5\overline{67}$ 9. $0.0\overline{321}$

Express each fraction or mixed number as a decimal. Use bar notation to show a repeating decimal.

10. $\dfrac{2}{3}$ 11. $\dfrac{5}{8}$ 12. $3\dfrac{1}{11}$ 13. $\dfrac{5}{12}$

Exercises

Independent Practice Express each fraction or mixed number as a decimal. Use bar notation to show a repeating decimal.

14. $\dfrac{7}{12}$ 15. $4\dfrac{3}{11}$ 16. $1\dfrac{4}{5}$ 17. $2\dfrac{7}{8}$

18. $\dfrac{11}{12}$ 19. $\dfrac{3}{10}$ 20. $\dfrac{2}{9}$ 21. $3\dfrac{1}{3}$

22. $7\dfrac{3}{8}$ 23. $\dfrac{4}{7}$ 24. $2\dfrac{2}{3}$ 25. $\dfrac{9}{12}$

26. Express *five and two-thirds* as a decimal.

Mixed Review 27. Express 0.74 as a fraction in simplest form. *(Lesson 6-10)*

28. Find the prime factorization of 300. *(Lesson 6-2)*

29. **Weather**   The average annual precipitation in Georgia is 48.61 inches. About how many inches of precipitation does Georgia average each month? *(Lesson 5-1)*

30. **Statistics**   Using the guinea pig weights given in Exercise 31 on page 206, construct a horizontal bar graph. *(Lesson 2-5)*

Problem Solving and Applications 31. **Consumer Awareness**   According to the Census Bureau, in 1989, Americans' consumption of sugar was $62\dfrac{1}{5}$ pounds per person. If the average price of sugar is $0.32 per pound, about how much did each American spend on sugar in 1989?

32. **Critical Thinking**   Make an educated guess about how you can determine whether a fraction in simplest form is expressed as a terminating or repeating decimal by examining the denominator.

33. **Data Search**   Refer to pages 188 and 189. What fraction of Earth's water is found in rivers, lakes, and streams?

6 Study Guide and Review

Communicating Mathematics

State whether each sentence is *true* or *false.* If false, replace the underlined words or numbers to make a true sentence.

1. A number is divisible by <u>7</u> if the sum of the digits is divisible by <u>7</u>.
2. The <u>LCM</u> of 10 and 15 is <u>5</u>.
3. A composite number has <u>more than</u> two factors.
4. The number 29 is a <u>prime</u> number.
5. The fraction $\frac{3}{7} \le \frac{5}{14}$.
6. The greatest common factor of 14 and 21 is <u>7</u>.
7. The fraction $\frac{8}{12}$ is equivalent to the fraction $\frac{4}{6}$.
8. The mixed number $3\frac{2}{3}$ is equivalent to the improper fraction $\frac{13}{3}$.
9. Write in your own words how to write a fraction in simplest form.
10. Tell why the number 6 will never appear in the prime factorization of any number.

Skills and Concepts

Objectives and Examples

Upon completing this chapter, you should be able to:

Review Exercises

Use these exercises to review and prepare for the chapter test.

- use divisibility rules for 2, 3, 5, 6, 9, and 10 *(Lesson 6-1)*

  The number 630 is divisible by 2, 3, 5, 6, 9, and 10.

State whether each number is divisible by 2, 3, 5, 6, 9, and 10.

11. 51
12. 300
13. 423

- find the prime factorization of a composite number *(Lesson 6-2)*

  Find the prime factorization of 12.

  Factor 12.

  $12 = 2 \times 2 \times 3$

  12
  6 × ②
  ② × ③

Find the prime factorization of each number.

14. 54
15. 75
16. 124

Objectives and Examples

- find the greatest common factor (GCF) of two or more numbers
  (Lesson 6-4)

  The GCF of 12 and 18 is 6.

  factors of 12: 1, 2, 3, 4, 6, 12
  factors of 18: 1, 2, 3, 6, 9, 18

- express fractions in simplest form
  (Lesson 6-5)

  $\dfrac{42}{70}$ is not in simplest form.

  factors of 42: 1, 2, 3, 6, 7, 14, 21, 42
  factors of 70: 1, 2, 5, 7, 10, 14, 35, 70
  $\dfrac{42}{70} = \dfrac{3}{5}$

- express mixed numbers as improper fractions and vice versa *(Lesson 6-6)*

  $\dfrac{27}{5} \rightarrow \overset{\;\;\;5\;\;\text{R2}}{5\overline{)27}}$ $\qquad \dfrac{27}{5} = 5\dfrac{2}{5}$

  $1\dfrac{2}{3} = \dfrac{(1 \times 3) + 2}{3} = \dfrac{5}{3}$

- measure line segments with a ruler separated into fourths or in eighths
  (Lesson 6-7)

  Measure to the nearest eighth inch.

  $1\dfrac{3}{8}$ inch

  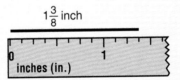

- find the least common multiple of two or more numbers *(Lesson 6-8)*

  The LCM of 8 and 12 is 24.

  multiples of 8: 0, 8, 16, 24, 32, . . .
  multiples of 12: 0, 12, 24, 36, 48, . . .

- compare and order fractions
  (Lesson 6-9)

  $\dfrac{3}{7}$ ● $\dfrac{4}{9}$ *LCM of 7 and 9 is 63*

  $\dfrac{3}{7} = \dfrac{27}{63}$ $\dfrac{4}{9} = \dfrac{28}{63}$ So, $\dfrac{3}{7} < \dfrac{4}{9}$.

Review Exercises

17. List all of the factors of 30.

Find the GCF of each pair of numbers.

18. 30, 36

19. 39, 26

20. 18, 28

State whether each fraction is in simplest form. If not, write it in simplest form.

21. $\dfrac{15}{18}$

22. $\dfrac{2}{9}$

23. $\dfrac{24}{28}$

Express each fraction as a mixed number.

24. $\dfrac{18}{11}$ 25. $\dfrac{30}{7}$

Express each mixed number as an improper fraction.

26. $3\dfrac{4}{5}$ 27. $5\dfrac{1}{7}$

Find the length of each line segment to the nearest eighth inch.

28. ———   29. ——————

30. ————————

31. Draw a line segment that is $4\dfrac{1}{4}$ inches long.

Find the LCM for each set of numbers.

32. 15, 25

33. 15, 20, 12

34. 28, 35

Replace each ● with <, >, or = to make a true sentence.

35. $\dfrac{3}{5}$ ● $\dfrac{7}{10}$ 36. $\dfrac{5}{9}$ ● $\dfrac{6}{11}$

Objectives and Examples

Review Exercises

- express terminating decimals as fractions in simplest form
  (Lesson 6-10)

  $1.72 = 1\frac{72}{100} = 1\frac{18}{25}$

Express each decimal as a fraction in simplest form.

37. 0.8

38. 0.04

39. 3.65

40. 0.36

- express fractions as terminating and repeating decimals *(Lesson 6-11)*

  Express $\frac{5}{16}$ as a decimal.

  $5 \boxed{\div} 16 \boxed{=}$ 0.3125;

  $\frac{5}{16} = 0.3125$

  Express $2\frac{5}{6}$ as a decimal.

  $2 \boxed{+} 5 \boxed{\div} 6 \boxed{=}$ 2.8333333

  $2\frac{5}{6} = 2.83333\ldots$

  $= 2.8\overline{3}$

Express each fraction or mixed number as a decimal. Use bar notation to show a repeating decimal.

41. $\frac{5}{8}$

42. $3\frac{1}{11}$

43. $\frac{7}{9}$

44. $5\frac{3}{16}$

Applications and Problem Solving

45. **Weather**   When the office opens late on snow days, Renee has to call Jim, David, Risa, and Isabella and let them know about the delayed opening. In how many different orders can she make her phone calls? *(Lesson 6-3)*

46. **Jobs**   Zu-wang worked $6\frac{2}{5}$ hours last week. If he earns $4.75 per hour, how much did he earn last week? *(Lesson 6-11)*

47. **Jobs**   Refer to Exercise 46. If Ellen worked $6\frac{3}{7}$ hours, who worked longer, Ellen or Zu-wang? *(Lesson 6-9)*

Curriculum Connection Projects

- **Meteorology**   Find the GCF and LCM for yesterday's high and low temperatures for at least three cities listed in today's newspaper.
- **Sports**   From the sports section of today's newspaper, find the final scores for at least three games. Compare the scores of both teams as a proper fraction, as an improper fraction, and as a mixed number. Then simplify.

Read More About It

Cooper, Susan. *The Dark is Rising.*
Lamb, Geoffrey. *Card Tricks.*
White, Laurence B. Jr. and Ray Broekel. *Math-a-Magic: Number Tricks for Magicians.*

6 Test

State whether each number is divisible by 2, 3, 5, 6, 9, or 10.

1. 75

2. 864

3. **Games** In one card game, all 52 cards are to be dealt out to the players. If there are six players, will all players have the same number of cards?

4. How many 3-digit numbers can you make using the digits 2, 4, 6, 8?

Find the prime factorization of each number.

5. 36

6. 108

7. 52

8. List all the factors of 20.

Find the GCF of each pair of numbers.

9. 27, 45

10. 18, 48

11. 21, 38

State whether each fraction is in simplest form. If not, write it in simplest form.

12. $\dfrac{40}{100}$

13. $\dfrac{6}{25}$

14. $\dfrac{45}{50}$

Express each fraction as a mixed number.

15. $\dfrac{20}{7}$

16. $\dfrac{20}{19}$

17. $\dfrac{7}{3}$

Express each mixed number as an improper fraction.

18. $2\dfrac{5}{7}$

19. $4\dfrac{1}{3}$

20. $100\dfrac{1}{2}$

Find the length of each line segment to the nearest eighth inch.

21. ━━━━━━━━━━

22. ━━━━━

Find the LCM for each set of numbers.

23. 6, 16

24. 18, 12

25. 7, 12

Replace each ● with <, >, or = to make a true sentence.

26. $\dfrac{5}{6}$ ● $\dfrac{7}{9}$

27. $\dfrac{6}{11}$ ● $\dfrac{4}{9}$

28. $\dfrac{8}{12}$ ● $\dfrac{12}{18}$

Express each decimal as a fraction in simplest form.

29. 0.05

30. 5.1

31. 0.42

Express each fraction or mixed number as a decimal. Use bar notation to show a repeating decimal.

32. $6\dfrac{5}{6}$

33. $\dfrac{7}{16}$

Bonus Find the greatest common factor of $16a$ and $24ab$.

Chapter 6 Academic Skills Test

Directions: Choose the best answer. Write A, B, C, or D.

1. In the school auditorium, there are 428 seats on the main floor and 136 in the balcony. About how many seats are there?

   A. 400 B. 450

   C. 550 D. 650

2. If $n = 12$, what is the value of $3n + 8$?

   A. 23 B. 36

   C. 38 D. 44

3. If $x + 30 = 65$, what is the value of x?

   A. 25 B. 30

   C. 35 D. 65

4. How do you read the number 4,056,000,000?

   A. four billion, fifty-six million
   B. four trillion, fifty-six billion
   C. four trillion, fifty-six million
   D. four hundred fifty-six million

5. Grace has one quarter, two nickels, and four pennies. How many different amounts can she make using one or more of the coins?

   A. 24 B. 29

   C. 39 D. none of these

6. The class scores on a history quiz were: 8, 9, 8, 8, 7, 9, 10, 7, 10, 8, 8, 9, 9, 10, 9, and 8. What was the mode?

   A. 8 B. 8.5

   C. 8.56 D. 8 and 9

7. Which shows the number five and eighty-six thousandths?

   A. 5.086 B. 5.860

   C. 5.0086 D. 5.86

8. Which is the most reasonable estimate of $8.96 + 10.98$?

   A. 10 B. 20

   C. 80 D. 180

9. 8×3.21 is about

   A. 24 B. 32

   C. 240 D. 320

10. $2.5 \times 5.4 =$

    A. 0.135 B. 1.35

    C. 13.5 D. 1,350

11. What is the perimeter of a rectangular bulletin board that is 6 feet wide and 4 feet high?

    A. 10 feet
    B. 20 feet
    C. 24 feet
    D. none of these

12. A dog is on a 20-foot leash attached to a stake in the ground. If the dog walks in a circle at the end of the leash, how far can he walk before returning to where he started? Round to the nearest foot.

A. 13 ft   B. 20 ft
C. 63 ft   D. 126 ft

13. For their fall fund-raiser, the 82 members of the band sold a total of 720 bags of apples. Estimate the average number sold by each band member.

A. 9 bags   B. 80 bags
C. 90 bags   D. 400 bags

14. $26 \div 3.25 =$

A. 0.008   B. 0.08
C. 0.8   D. 8

15. Which is the best estimate for the weight of a 1-year-old?

A. 8 g   B. 8 kg
C. 60 kg   D. 60 mg

16. If $36 = 1.8y$, what is the value of y?

A. 2   B. 6.48
C. 20   D. 64.8

17. Which numbers are factors of 75?

A. 3, 5, and 6
B. 3, 5, and 9
C. 3 and 5
D. 5 and 10

Test-Taking Tip

Although most of the basic formulas that you need to answer the questions on a standardized test are usually given to you in the test booklet, it is a good idea to review the formulas beforehand.

The perimeter of a rectangle is twice the sum of its length and width.

The perimeter of a square is four times the length of a side.

The circumference of a circle with radius r is $2\pi r$ or πd.

You may be able to eliminate all or most of the answer choices by estimating. Also, look to see which answer choices are not reasonable for the information given in the problem.

18. Which fraction is in simplest form?

A. $\dfrac{3}{18}$ B. $\dfrac{5}{12}$

C. $\dfrac{6}{10}$ D. $\dfrac{12}{12}$

19. What is the least common multiple of 8 and 12?

A. 2   B. 4
C. 24   D. 48

20. Ricco is using a calculator to find $\dfrac{2}{5}$ of a number. What equivalent decimal should he enter for $\dfrac{2}{5}$?

A. 0.2   B. 0.25
C. 0.4   D. 2.5

Fractions: Addition and Subtraction

Spotlight on Snow

Have You Ever Wondered. . .

- Why snowflakes have different shapes?
- What the record is for the greatest amount of snowfall in one place?

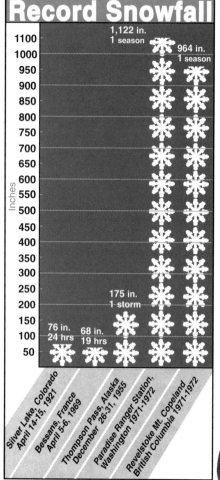

TEMPERATURE GIVES SNOW ITS SHAPE

Snow shapes are formed by temperature.

Hollow columns
21°F to 14°F
and
below -8°F

Sector plates
14°F to 10°F
and
3°F to -8°F

Dendrites
10°F to 3°F

Needles
25°F to 21°F

Thin plates
32°F to 25°F

Looking Ahead

In this chapter, you will see how mathematics can be used to answer questions about weather. The major objectives of the chapter are to:

- estimate fraction sums and differences
- add and subtract fractions and mixed numbers
- add and subtract measures of time
- solve problems by eliminating possibilities

Chapter Project

Snow

Work in a group.

1. Cut out the United States weather map from a newspaper for two weeks.

2. Note which areas on the map have low or high temperatures and which have snow or rain.

3. Compare the weather predicted with the actual weather in your area. Make a graph from this data.

4. Read the weather predictions for each day in the *Farmer's Almanac*.

5. Make a table to compare the predictions and the actual weather.

GARFIELD, TAKE A LOOK OUTSIDE AND SEE WHAT THE WEATHER IS LIKE

DID IT SNOW LAST NIGHT?

YES, IT DID

© 1991 United Feature Syndicate, Inc.

JIM DAVIS 1-7-92

231

7-1 Rounding Fractions and Mixed Numbers

Objective

Round fractions and mixed numbers.

Mabet is going to bake cookies for a bake sale at school. To decide what type of cookies to bake, Mabet surveyed her classmates. They were allowed to vote for more than one type of cookie. Here are the results of the survey:

Type of cookie	Fraction of People Who Liked Them
chocolate chip	$\frac{4}{5}$
oatmeal	$\frac{7}{15}$
raisin	$\frac{1}{5}$

Mabet concluded from her survey that almost all her classmates like chocolate chip cookies, about half like oatmeal cookies, and only a few like raisin cookies. Mabet was actually rounding the fractions.

To round fractions and mixed numbers to the nearest one-half unit, you can use the following guidelines. A number line can help you decide how to round.

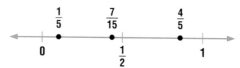

When am I ever going to use this?

Suppose you were mailing letters to two friends. One letter weighed $2\frac{1}{4}$ ounces and the other weighed $1\frac{1}{8}$ ounces. Mailing charges are 29 cents for the first ounce and 25 cents for any additional fraction of an ounce. How much is the postage?

- If the numerator is almost as large as the denominator, round the number up to the next whole number.
$\frac{4}{5}$ rounds to 1. *4 is almost as large as 5.*

- If the numerator is about half of the denominator, round the fraction to $\frac{1}{2}$.
$\frac{7}{15}$ rounds to $\frac{1}{2}$. *7 is about half of 15.*

- If the numerator is much smaller than the denominator, round the number down to the next whole number.
$\frac{1}{5}$ rounds to 0. *1 is much smaller than 5.*

1 Round $\frac{5}{8}$ to the nearest half.

$\frac{5}{8}$ *is closer to* $\frac{1}{2}$ *than 1.*

The numerator is about half of the denominator.
$\frac{5}{8}$ rounds to $\frac{1}{2}$.

2 Round $1\frac{3}{16}$ to the nearest half.

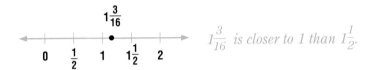

$1\frac{3}{16}$ *is closer to 1 than* $1\frac{1}{2}$.

The numerator, 3, is much smaller than its denominator, 16.
$1\frac{3}{16}$ rounds to 1.

You often need to round fractions when measuring.

Example 3 *Connection*

Measurement Find the length of the line segment to the nearest one-half inch.

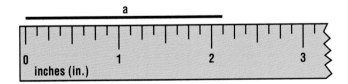

$2\frac{1}{8}$ is closer to 2 inches than $2\frac{1}{2}$ inches. Therefore, to the nearest one-half inch, the length of the line segment is 2 inches.

As shown in the following example, you should round a number up when it is better for a measure to be too large than too small.

Example 4 *Problem Solving*

Cooking   A recipe to make tacos calls for $1\frac{1}{2}$ pounds of ground beef. Should you buy a 2-pound package or a $1\frac{1}{8}$-pound package? You should buy the 2-pound package to have enough ground beef to make the tacos.

You should round a number down when it is better for a measure to be too small than too large as shown in the following example.

Example 5 *Problem Solving*

Cooking   You are not sure how much salt to put in a pot of chili. Should you put one-half teaspoon or one teaspoon of salt in the pot?

To avoid ruining the chili, you should round down to one-half teaspoon of salt. You can always add more.

Checking for Understanding

Communicating Mathematics

Read and study the lesson to answer each question.

1. **Tell** when it is better for a measure to be too large than too small. Give an example.

2. **Tell** if $\frac{4}{9}$ is closer to 0, $\frac{1}{2}$, or 1?

3. **Show** what number on the number line should be rounded to 3 if you were rounding the number to the nearest half.

Guided Practice

Tell whether each number should be rounded up, down, or to the nearest half unit.

4. your height, accurate to $\frac{1}{2}$ inch

5. a patch for a $3\frac{1}{8}$-inch tear

Round each number to the nearest half.

6. $2\frac{9}{10}$ 7. $\frac{3}{8}$ 8. $\frac{3}{16}$ 9. $4\frac{9}{16}$ 10. $7\frac{1}{10}$

Exercises

Independent Practice

Tell whether each number should be rounded up, down, or to the nearest half unit.

11. the width of mini-blinds to fit in a window $30\frac{1}{4}$ inches wide

12. the weight of a fish you caught, accurate to $\frac{1}{2}$ pound

13. the capacity of a pitcher needed to hold $\frac{7}{8}$ gallon of lemonade

Round each number to the nearest half.

14. $\frac{7}{16}$ 15. $2\frac{1}{8}$ 16. $12\frac{7}{8}$ 17. $7\frac{9}{10}$

18. $4\frac{5}{8}$ 19. $\frac{2}{16}$ 20. $10\frac{15}{16}$ 21. $3\frac{1}{6}$

22. $\frac{1}{3}$ 23. $7\frac{2}{5}$ 24. $6\frac{4}{5}$ 25. $\frac{3}{4}$

26. $7\frac{2}{3}$ 27. $\frac{5}{6}$ 28. $\frac{5}{9}$ 29. $5\frac{3}{7}$

Find the length of each line segment to the nearest one-half inch.

30. —————   31. ————————————

32. ——————————   33. ————————————

34. Name three fractions that round to $2\frac{1}{2}$.

Mixed Review

35. Express $\frac{9}{11}$ as a decimal using bar notation. *(Lesson 6-11)*

36. **Geometry**   Find the perimeter of the figure shown at the right. *(Lesson 4-6)*

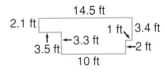

37. Change 3,400 centimeters to meters. *(Lesson 5-6)*

38. Find the prime factorization of 204. *(Lesson 6-2)*

39. Express $\frac{35}{42}$ in simplest form. *(Lesson 6-5)*

Problem Solving and Applications

40. **Critical Thinking**   Round $5\frac{5}{6}$ to the nearest one-fourth.

41. **Carpentry**   Mr. Day needs to buy a board to cut a $6\frac{1}{2}$-foot closet shelf. Should he buy a 6-foot or 8-foot board?

42. **Smart Shopping**   Candy bars are priced at 2 for 99¢. What will one candy bar cost?

43. **Journal Entry**   Write, in your own words, how you know when to round a fraction to 0, $\frac{1}{2}$, or 1 when rounding to the nearest half.

Estimating Sums and Differences

Objective

Estimate sums and differences of fractions and mixed numbers.

DID YOU KNOW

It is estimated that by the year 2030, the Hispanic population of the United States will more than double and the Asian population will more than quadruple.

The United States is considered a melting pot because immigrants come to the United States from many different places. In the early 1900s, most immigrants came from Europe. By 1989, about $\frac{8}{20}$ of the U.S. immigrants came from Latin America, and $\frac{3}{20}$ came from Canada and Europe. Asians make up the other part of the immigrant population. About what fraction are from Asia?

Round fractions to the nearest half to estimate their sum or difference.

Examples

1 Estimate to find the fraction of immigrants from Asia.

First estimate the fraction from Latin America, Canada, and Europe. The portion of U.S. immigrants that are from Latin America, Canada, and Europe is $\frac{8}{20} + \frac{3}{20}$.

$\frac{8}{20}$ is closer to $\frac{1}{2}$ than 0.

$\frac{3}{20}$ is closer to 0 than $\frac{1}{2}$.

Add: $\frac{1}{2} + 0 = \frac{1}{2}$

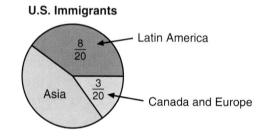

U.S. Immigrants

Latin America — $\frac{8}{20}$

$\frac{3}{20}$ — Canada and Europe

Asia

About $\frac{1}{2}$ of the immigrants are from Latin America, Canada, and Europe. That means about $1 - \frac{1}{2}$ or $\frac{1}{2}$ are from Asia.

2 Estimate $\frac{7}{8} - \frac{5}{16}$.

$\frac{7}{8}$ is closer to 1 than $\frac{1}{2}$. $\frac{5}{16}$ is closer to $\frac{1}{2}$ than 0.

Subtract: $1 - \frac{1}{2} = \frac{1}{2}$

$\frac{7}{8} - \frac{5}{16}$ is about $\frac{1}{2}$.

To estimate sums and differences of mixed numbers, round each number to the nearest whole number.

Examples

Estimate.

3 $5\frac{7}{12} + 9\frac{5}{6}$

$5\frac{7}{12}$ is closer to 6 than 5.

$9\frac{5}{6}$ is closer to 10 than 9.

Add: $6 + 10 = 16$

$5\frac{7}{12} + 9\frac{5}{6}$ is about 16.

4 $7\frac{1}{4} - 2\frac{1}{2}$

$7\frac{1}{4}$ is closer to 7 than 8.

$2\frac{1}{2}$ is between 2 and 3.

Round $2\frac{1}{2}$ to 2.

Subtract: $7 - 2 = 5$

$7\frac{1}{4} - 2\frac{1}{2}$ is about 5.

Sometimes when estimating sums and differences of fractions and mixed numbers, you need to round all fractions up.

Example 5   *Problem Solving*

Cooking   Ani is preparing a meal for her friends. She wants to make three desserts – a cake, some cookies, and an apple pie. To make the cake, she needs $1\frac{1}{2}$ cups of flour. For the cookies, she needs $1\frac{1}{4}$ cups of flour. For the pie, she needs $2\frac{3}{4}$ cups of flour. About how much flour does she need to make the three desserts?

Ani wants to make sure that she has enough flour to make the desserts. She rounds up.

Estimate: $2 + 2 + 3 = 7$

$1\frac{1}{2} + 1\frac{1}{4} + 2\frac{3}{4}$ is about 7. Therefore, Ani needs about 7 cups of flour to make the desserts.

Example 6   *Connection*

Geometry   Mrs. Coe wants to buy fence to put around her flower bed. About how much fence will she need?

Mrs. Coe wants to be sure she buys enough fence to enclose the bed. She needs to round up.

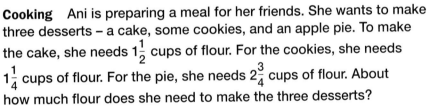

Estimate: $14 + 9 + 14 + 9 = 46$

Mrs. Coe will need about 46 feet of fence.

Checking for Understanding

Communicating Mathematics

Read and study the lesson to answer each question.

1. **Draw** a number line like the one below. Show where $2\frac{3}{8}$ would be on the line.

2. **Tell** what would happen if you rounded $2\frac{1}{2}$ to 3 in Example 4.

Guided Practice

Round each fraction or mixed number to the nearest whole number.

3. $4\frac{5}{8}$ 4. $6\frac{1}{9}$ 5. $\frac{8}{9}$ 6. $7\frac{11}{12}$ 7. $\frac{1}{7}$

Estimate.

8. $3\frac{1}{3} + 7\frac{4}{5}$ 9. $\frac{3}{5} - \frac{1}{10}$ 10. $9\frac{1}{4} - 4\frac{1}{16}$

11. $3\frac{7}{8} - \frac{7}{16}$ 12. $\frac{2}{5} + 1\frac{4}{10}$ 13. $5\frac{1}{2} - \frac{1}{4}$

Exercises

Independent Practice

Estimate.

14. $2\frac{3}{4} - 1\frac{5}{8}$ 15. $\frac{4}{5} - \frac{1}{3}$ 16. $8\frac{1}{3} + \frac{1}{2}$

17. $8\frac{3}{16} + 10\frac{1}{4}$ 18. $\frac{1}{2} + \frac{9}{10}$ 19. $5\frac{1}{8} - \frac{5}{6}$

20. $\frac{3}{20} - \frac{1}{12}$ 21. $10\frac{5}{6} + \frac{1}{8}$ 22. $\frac{11}{22} - \frac{3}{8}$

23. $21\frac{5}{8} + 4\frac{3}{4}$ 24. $8\frac{1}{2} - 2\frac{15}{18}$ 25. $1\frac{8}{15} - \frac{4}{5}$

26. $2\frac{12}{20} - 1\frac{3}{10}$ 27. $8\frac{5}{6} + \frac{1}{12}$ 28. $11\frac{2}{3} - \frac{5}{9}$

29. Estimate the sum of $\frac{2}{3}$ and $6\frac{1}{4}$.

30. Estimate the difference of $2\frac{3}{5}$ and $\frac{3}{4}$.

Mixed Review

31. Round $\frac{8}{11}$ to the nearest half. *(Lesson 7-1)*

32. The distance from Pluto to Earth is 4,644,000,000 miles at its maximum. Write this number in words. *(Lesson 2-1)*

33. Tell whether the number 462 is divisible by 6 and/or 9 using divisibility patterns. *(Lesson 6-1)*

34. Divide 2,087.8 by 26. *(Lesson 5-4)*

Problem Solving and Applications

35. **Home Economics** A 14-pound turkey should roast about $4\frac{2}{3}$ hours. If it has been roasting $2\frac{3}{4}$ hours, about how much longer should the turkey roast?

36. **Music** The Morse Middle School band is practicing 3 pieces to play for this month's PTA meeting. The band director estimates that the length of the pieces are $3\frac{1}{2}$ minutes, 3 minutes, and $4\frac{1}{2}$ minutes. About how long will the band play?

37. **Critical Thinking** When estimating $3\frac{5}{8} - 1\frac{1}{2}$, $3\frac{5}{8}$ rounds to 4. Should $1\frac{1}{2}$ be rounded to 1 or 2 to give you the better estimate?

38. **Crafts** Mrs. Roso needs $3\frac{1}{2}$ yards of ribbon to make a large bow and $1\frac{3}{4}$ yards to make a smaller bow. About how much ribbon should she buy to make sure she has enough ribbon to make a large bow and a small bow?

39. **Geometry** Estimate the perimeter of the rectangle at the right.

$2\frac{4}{5}$ in.

$12\frac{1}{8}$ in.

40. **Data Search** Refer to pages 230 and 231.
If snow fell at the same rate throughout the entire storm, which location would have more snow, Silver Lake, Colorado, in 1921 or Bessans, France, in 1969?

7-3 Adding and Subtracting Fractions: Like Denominators

Objective

Add and subtract fractions with like denominators.

Thad and Omar had a pizza party on Friday evening. They ordered a large supreme pizza. The pizza was cut into 8 slices. While they were waiting for their two guests to arrive, Thad ate $\frac{1}{8}$ of the pizza and Omar ate $\frac{3}{8}$ of it. How much pizza did Thad and Omar eat before the guests arrived?

To help us find the answer to this question, let's do the following Mini-Lab.

Mini-Lab

Work with a partner.

Materials: grid paper, colored pencils

Add $\frac{1}{8}$ and $\frac{3}{8}$.

- On your grid paper, draw a rectangle like the one shown below. This rectangle shows 8 eighths.

- With a colored pencil, color one section of the rectangle to represent $\frac{1}{8}$.

- With a different colored pencil, color three more sections of the rectangle to represent $\frac{3}{8}$.

Talk About It

a. How many sections of the rectangle are colored?

b. What fraction represents the number of colored sections of the rectangle?

c. If you color two more sections of the rectangle, what fraction would that represent?

You add and subtract the numerators of fractions the same way you add and subtract whole numbers. The denominator of the fraction names the units being added or subtracted.

Estimation Hint

• • • • • • • • • • •

Use rounding to estimate your answer before adding or subtracting. Then compare your answer to your estimate to see if it is reasonable.

From the Mini-Lab, we know that

1 eighth   plus   3 eighths   equals   4 eighths.

$$\frac{1}{8} \quad + \quad \frac{3}{8} \quad = \quad \frac{4}{8} \quad \text{or} \quad \frac{1}{2}$$

Thad and Omar ate $\frac{1}{2}$ of the pizza.

Adding Like Fractions	To add fractions with like denominators, add the numerators.

Example 1

Find the sum of $\frac{3}{5}$ and $\frac{4}{5}$. *Estimate:* $\frac{1}{2} + 1 = 1\frac{1}{2}$

$$\frac{3}{5} + \frac{4}{5} = \frac{3+4}{5}$$
$$= \frac{7}{5}$$
$$= 1\frac{2}{5}$$

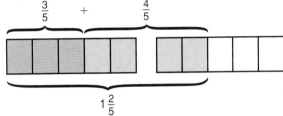

Subtracting Like Fractions	To subtract fractions with like denominators, subtract the numerators.

Subtraction has three meanings. Each of these is shown in one of the following examples.

- to take away part of a set
- to find a missing addend
- to compare the size of two sets

Example 2

Tony found $\frac{5}{8}$ of a cherry pie in the refrigerator. He ate $\frac{1}{8}$ of it. How much pie is left?

$$\frac{5}{8} - \frac{1}{8} = \frac{5-1}{8}$$
$$= \frac{4}{8} \text{ or } \frac{1}{2}$$

$\frac{1}{2}$ of the pie is left.

Example 3 *Connection*

Algebra The tank of Juan's lawn mower holds $\frac{5}{6}$ gallons of gas. It took $\frac{3}{6}$ gallons to fill the tank. How much gasoline was already in the tank?

You can use mental math to solve this problem.

$\blacksquare + \frac{3}{6} = \frac{5}{6}$ *Think: What plus $\frac{3}{6}$ is $\frac{5}{6}$?*

$\frac{2}{6} + \frac{3}{6} = \frac{5}{6}$, so the tank already contained $\frac{2}{6}$, or $\frac{1}{3}$, gallon.

Example 4

Reed lives $\frac{7}{10}$ mile from school. Tasha lives $\frac{3}{10}$ mile from school. How much farther from school does Reed live than Tasha?

$$\frac{7}{10} - \frac{3}{10} = \frac{7-3}{10}$$

$$= \frac{4}{10} \text{ or } \frac{2}{5}$$

Reed lives $\frac{2}{5}$ mile farther from school than Tasha.

Checking for Understanding

Communicating Mathematics

Read and study the lesson to answer each question.

1. **Write** a fraction problem that uses the missing addend meaning of subtraction.
2. **Show** a model like the one used in the Mini-Lab to represent Example 2.

Guided Practice

Add or subtract. Write each answer in simplest form.

3. $\frac{3}{4} - \frac{3}{4}$ 4. $\frac{1}{5} + \frac{2}{5}$ 5. $\frac{7}{9} - \frac{1}{9}$ 6. $\frac{17}{20} - \frac{3}{20}$

Tell whether you would add or subtract to solve. Then solve.

7. Bill bought two apples at Yarnell's Fruit Stand. The Red Delicious apple weighed $\frac{3}{4}$ pound. The Granny Smith apple weighed $\frac{2}{4}$ pound. How much more did the Red Delicious weigh?

Exercises

Independent Practice

Add or subtract. Write each answer in simplest form.

8. $\frac{3}{11} + \frac{7}{11}$

9. $\frac{6}{7} - \frac{2}{7}$

10. $\frac{3}{8} + \frac{5}{8}$

11. $\frac{4}{9} - \frac{4}{9}$

12. $\frac{11}{12} - \frac{5}{12}$

13. $\frac{7}{15} + \frac{3}{15}$

14. $\frac{5}{9} + \frac{7}{9}$

15. $\frac{3}{6} + \frac{5}{6}$

16. $\frac{7}{12} + \frac{8}{12}$

17. $\frac{7}{15} + \frac{8}{15}$

18. $\frac{12}{16} - \frac{9}{16}$

19. $\frac{3}{10} + \frac{5}{10}$

20. Find the sum of $\frac{6}{7}$ and $\frac{4}{7}$.

21. Find the difference of $\frac{8}{9}$ and $\frac{5}{9}$.

Tell whether you would add or subtract to solve. Then solve.

22. Sherri's aunt bought $\frac{2}{3}$ yard of terry cloth to make hand towels. She used $\frac{1}{3}$ yard. How much material does she have left?

23. Angie is making a punch mixture that calls for $\frac{3}{4}$ quart of pineapple juice, $\frac{2}{4}$ quart of orange juice, and $\frac{2}{4}$ quart of lemonade. How much punch does the recipe make?

Mixed Review

24. **Statistics**   The local flower shop took orders for the December Dance as follows: 120 carnation corsages, 75 rose corsages, 95 carnation and rose mixed, 40 orchid, and 30 other. Construct a vertical bar graph to show this data. *(Lesson 2-5)*

25. Find the value of the expression $2^4 \cdot 5^2$. *(Lesson 4-2)*

26. Find the least common multiple (LCM) of 24 and 30. *(Lesson 6-8)*

27. Express 0.84 as a fraction in simplest form. *(Lesson 6-10)*

Problem Solving and Applications

28. **Science**   Mr. Chan added a granulated sugar solution to $\frac{4}{8}$ pint of water before heating it to make $\frac{5}{8}$ pint of hummingbird nectar. How much sugar solution did he add?

29. **Geography**   According to the 1990 census, about $\frac{11}{100}$ of the population of the United States lives in California. Another $\frac{8}{100}$ of the population lives in New York. What part of the population lives in the two states?

30. **Upholstery**   Mr. Talbert estimates that it will take $\frac{7}{8}$ yard of material to upholster the seat of a chair and $\frac{5}{8}$ yard to upholster the back of the chair. About how many yards of material will be needed to upholster the chair?

31. **Critical Thinking**   Find the sum of $\frac{1}{20} + \frac{19}{20} + \frac{2}{20} + \frac{18}{20} + \frac{3}{20} + \frac{17}{20} + \cdots \frac{10}{20}$. Look for a pattern to help you.

DECISION MAKING

Planning a Party

Situation

Your class has decided to have a going-away party for a foreign exchange student at your school. A committee has been appointed to plan the party and make the final decision on the food and drink items. Your committee has been given a $100 budget. You must determine how to stay within the $100 budget and have enough food that 25 people will enjoy.

Hidden Data

Does the price of food from the Catering Service include items such as paper plates, cups, and plastic utensils and condiments such as ketchup, mustard, and mayonnaise?
Is there a delivery charge?
Does your state charge sales tax on food?

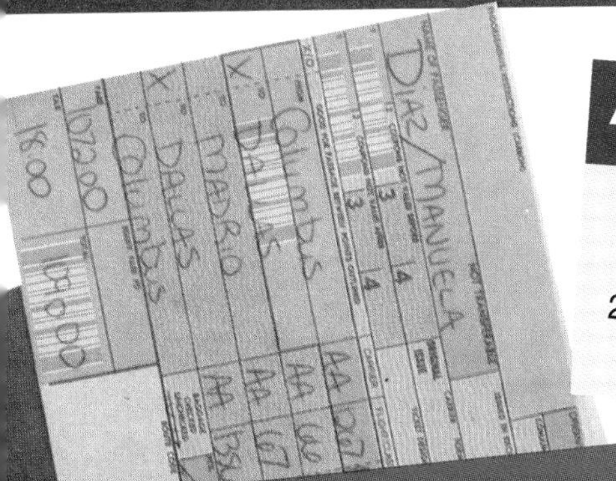
DIAZ/MANUELA
Columbus
DALLAS
MADRID
Columbus

Analyzing the Data

1. What is the total cost of a tray of 20 hamburgers and a deluxe meat and cheese tray that serves 10 to 12?
2. What is the cost for 25 soft drinks?

Making a Decision

3. **How much** juice would you need to serve 25 people?
4. **Is it** more convenient to purchase party platters or ready-made sandwiches?
5. **Will everyone** enjoy the ready-made sandwiches?
6. **What are** the advantages of buying ice cream sandwiches over ice cream in half-gallon containers?
7. **Should you** buy less food, assuming not everyone will eat the same quantity, or buy more in case that is not enough for everyone?

ESPAÑA
PASAPORTE
PASSPORT
Apellido / Surname
DIAZ
MANUELA
14 NOV
Sexo / Sex
F
MADRID

Specialty Catering Services
Ice Cream
Sandwiches......$0.50 each
Half Gallon......$6.75
Refreshments
Iced tea...........$0.50 each
Soft drinks.....$0.75 each
Juice................$3.00/gal
Pizza
Small w/ 2 toppings..........$8.95
Medium w/ 3 toppings...$12.95
Large with 3 toppings..... $15.95
Hamburgers
Tray of 20 hamburgers.....$69.95
Tray of 30 hamburgers.....$79.95
Deluxe Meat and Cheese Tray
Roast Beef, Ham, or Corned Beef, Turkey Breast, American, and Swiss Cheese served with white, wheat, or rye bread
Serves 10 to 12....$19.95
Serves 15 to 20....$39.95
Serves 25 to 30....$59.95
Small Sandwiches
Choice of Tuna, Chicken or Ham Salad, Turkey Breast, Roast Beef on freshly baked rolls
Tray of 20.....$19.95
Tray of 40.....$39.95
Tray of 60.....$59.95

Making Decisions in the Real World

8. **Investigate** the cost of planning a party using items from your local grocery store. Did you follow the same steps?

7-4 Solving Equations

Objective

Solve equations involving addition and subtraction of fractions.

The highest mountain in the world is __?__ . This sentence is neither true nor false until you substitute the name of a mountain for __?__ . If you choose Mount Mitchell, the sentence is false. If you choose Mount Everest, the sentence is true.

LOOKBACK

You can review equations on page 32.

As you learned in Chapter 1, a sentence such as $x + 2 = 7$ is called an equation. Any number that makes the equation true is called a solution.

In Chapter 1, the solutions to equations were whole numbers. Fractions can also be solutions to equations.

Examples

Solve each equation.

1 $x = \dfrac{1}{6} + \dfrac{3}{6}.$ *Since the denominators are the same, add the numerators.*

$x = \dfrac{4}{6}$ *Simplify $\dfrac{4}{6}$.*

$x = \dfrac{2}{3}$

2 $\dfrac{10}{12} - \dfrac{4}{12} = f$ *Subtract the numerators.*

$\dfrac{6}{12} = f$ *Simplify $\dfrac{6}{12}$.*

$\dfrac{1}{2} = f$

In Chapter 5, you learned about inverse operations. You can also use inverse operations to solve equations involving addition and subtraction of fractions.

Problem Solving Hint

● ● ● ● ● ● ● ● ● ● ●

You may want to use the guess-and-check strategy to help you solve these equations.

Example 3

Solve $\frac{1}{9} + s = \frac{8}{9}$.

Use subtraction to undo the addition.

$\frac{1}{9} + s = \frac{8}{9}$ means the same as $s = \frac{8}{9} - \frac{1}{9}$.

$$s = \frac{8}{9} - \frac{1}{9}$$

$$s = \frac{8-1}{9}$$

$$s = \frac{7}{9}$$

Checking for Understanding

Communicating Mathematics

Read and study the lesson to answer each question.

1. **Write,** in your own words, what a solution to an equation is.

2. **Write** an equation with fractions.

Guided Practice

Name the number that is a solution of the given equation.

3. $\frac{2}{8} + \frac{3}{8} = r; \frac{1}{8}, \frac{3}{8}, \frac{5}{8}$

4. $y = \frac{8}{10} - \frac{1}{10}; \frac{7}{10}, \frac{8}{10}, \frac{9}{10}$

Solve each equation.

5. $\frac{6}{18} + r = \frac{13}{18}$

6. $w + \frac{1}{10} = \frac{9}{10}$

7. $c - \frac{1}{3} = \frac{2}{3}$

8. $t - \frac{5}{8} = \frac{2}{8}$

9. $\frac{1}{12} + m = \frac{6}{12}$

10. $u - \frac{2}{5} = \frac{1}{5}$

Exercises

Independent Practice

Name the number that is a solution of the given equation.

11. $y + \frac{9}{16} = \frac{14}{16}; \frac{0}{16}, \frac{5}{16}, \frac{9}{16}$

12. $p - \frac{11}{18} = \frac{4}{18}; \frac{7}{18}, \frac{9}{18}, \frac{15}{18}$

13. $\frac{8}{9} = g - \frac{4}{9}; \frac{4}{9}, \frac{8}{9}, \frac{12}{9}$

14. $\frac{9}{12} = \frac{2}{12} + z; \frac{7}{12}, \frac{8}{12}, \frac{9}{12}$

15. $r - \frac{2}{7} = \frac{7}{7}; \frac{5}{7}, \frac{7}{7}, \frac{9}{7}$

16. $z - \frac{9}{15} = \frac{4}{15}; \frac{5}{15}, \frac{11}{15}, \frac{13}{15}$

Solve each equation.

17. $\frac{7}{9} + d = \frac{8}{9}$

18. $\frac{10}{11} = v - \frac{9}{11}$

19. $\frac{6}{7} = \frac{1}{7} + t$

20. $m - \frac{3}{6} = \frac{5}{6}$

21. $\frac{8}{13} = \frac{4}{13} + a$

22. $r - \frac{4}{5} = \frac{2}{5}$

23. $\frac{8}{8} = z + \frac{3}{8}$

24. $v + \frac{9}{12} = \frac{11}{12}$

25. $\frac{2}{14} = w - \frac{12}{14}$

Mixed Review

26. Write the number thirty-seven thousandths as a decimal. *(Lesson 3-2)*

27. Replace ● with <, > or = to make a true sentence. $1\frac{3}{4}$ ● $1\frac{7}{9}$
(Lesson 6-9)

28. Add $\frac{4}{11} + \frac{5}{11}$. *(Lesson 7-3)*

29. **Geometry** Find the area of a square whose sides are each 1.1 meters long. *(Lesson 4-8)*

Problem Solving and Applications

30. **Carpentry** Jerrod is planing a door to fit an opening. He needs to plane $\frac{3}{16}$ inch off the door. He has already planed $\frac{2}{16}$ inch. How much more of the door does he need to plane?

31. **Critical Thinking** Write a problem that can be solved using the equation $\frac{2}{8} + x = \frac{3}{8}$.

32. **Geometry** The perimeter of the triangle shown at the right is $\frac{7}{5}$ inches. What is the value of *x*?

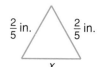

$\frac{2}{5}$ in. $\frac{2}{5}$ in.   *x*

33. **Cooking** Max is making a cake from a recipe that calls for $\frac{3}{4}$ cup of oil. He has $\frac{1}{4}$ cup of oil in his cupboard. How much more oil does Max need?

CULTURAL KALEIDOSCOPE

Bessie Smith

Bessie Smith was born in Chattanooga, Tennessee, in 1894 and was one of seven children. She loved music and when she was only nine years old she began singing on street corners for nickels and dimes. At 18, she got a job in a show where she met the great Gertrude "Ma" Rainey, a well-known blues singer, who quickly spotted her talent.

By 1920, Bessie Smith had become one of the most popular and highest-paid blues singers in the country. She worked with many of the best jazz musicians in the business, including Louis Armstrong, Charlie Green, and Benny Goodman. Her singing influenced later artists such as Mahalia Jackson and Billie Holiday. In 1923, she made her first record, "Down Hearted Blues." It was an instant hit, selling more than two million copies.

7-5A Renaming Sums

A Preview of Lesson 7-5

Objective

Find a common unit name for uncommon objects.

Materials

4 pennies
3 nickels
2 pencils
3 pens

What would you do if your teacher asked you to add 4 apples and 6 oranges? You might say "You can't add apples and oranges!"

If you have 4 apples and 6 oranges and you want to tell someone how many you have all together, you need to find a common unit name for them. You could name them 10 things, 10 objects, 10 round objects, or 10 fruits. If you want to find the best unit name, you would probably say you have 10 fruits.

Try this!

Work with a partner.

- Choose one person to be the recorder.
- Put 4 pennies and 3 nickels together. Write as many unit names as you can think of to describe the sum of the pennies and nickels.

- Look at your list. Choose the best unit name for the pennies and nickels.
- Repeat the steps using pencils and pens.

What do you think?

1. Explain why you need a common unit name to find the sum.
2. Did you find that some unit names fit better than others? Why or why not?

Extension

3. Make a class graph showing different unit names for the pennies and nickels. Which name appeared most often? Explain why.
4. Bring in different objects that could have a common unit name.

Mathematics Lab 7-5A Renaming Sums   **249**

7-5 Adding and Subtracting Fractions: Unlike Denominators

Objective

Add and subtract fractions with unlike denominators.

The average American teenage boy spends about $\frac{1}{4}$ of his weekly allowance on food, $\frac{1}{7}$ on clothes, and $\frac{1}{9}$ on entertainment. What is the fraction of money spent per week on food and clothes? To find the fraction, you must add $\frac{1}{4}$ and $\frac{1}{7}$.

To find a sum or difference, you need a common unit name. In the preview lab, you came up with common unit names for different objects. When you work with fractions with different, or unlike, denominators you do the same thing.

LOOKBACK

You can review LCM on page 213.

To find the sum or difference of two fractions with unlike denominators, rename the fractions using the least common denominator. Then add or subtract and simplify.

1 How much does the average American teenage boy spend on food and clothes?

Add $\frac{1}{4}$ and $\frac{1}{7}$.

Find the LCM of 4 and 7.
The LCM of 4 and 7 is 28.

$$\frac{1}{4} + \frac{1}{7} = \frac{7}{28} + \frac{4}{28}$$

Rename $\frac{1}{4}$ as $\frac{7}{28}$ and $\frac{1}{7}$ as $\frac{4}{28}$.

$$= \frac{7+4}{28}$$

$$= \frac{11}{28}$$

The average American teenage boy spends about $\frac{11}{28}$ of his allowance on food and clothes.

Calculator Hint

• • • • • • • • • • •

You can use a fraction calculator to solve problems. For example, to solve $\frac{5}{6} + \frac{1}{6}$, enter the following.

5 $\boxed{/}$ 6 $\boxed{+}$

1 $\boxed{/}$ 6 $\boxed{=}$ $\boxed{\text{Simp}}$

Repeat $\boxed{\text{Simp}}$ $\boxed{=}$ until the N/D $\rightarrow$ n/d does not appear on the screen.

2 Find $\frac{5}{6} - \frac{3}{4}$.

$$\frac{5}{6} - \frac{3}{4} = \frac{10}{12} - \frac{9}{12}$$

The LCM of 6 and 4 is 12.

$$= \frac{10-9}{12}$$

Rename $\frac{5}{6}$ as $\frac{10}{12}$ and $\frac{3}{4}$ as $\frac{9}{12}$.

$$= \frac{1}{12}$$

3 Solve $y = \frac{3}{10} + \frac{9}{20}$.

$$y = \frac{3}{10} + \frac{9}{20}$$

The LCM of 10 and 20 is 20.

$$y = \frac{6}{20} + \frac{9}{20}$$

Rename $\frac{3}{10}$ as $\frac{6}{20}$.

$$y = \frac{6+9}{20}$$

$$y = \frac{15}{20} \text{ or } \frac{3}{4}$$

Simplify.

Checking for Understanding

Communicating Mathematics

Read and study the lesson to answer each question.

1. **Tell** the first step in adding fractions with unlike denominators.

2. **Tell** how to find the LCD for $\frac{7}{10}$ and $\frac{3}{4}$.

Guided Practice

Find the LCD of each pair of fractions.

3. $\frac{1}{2}, \frac{1}{6}$ 4. $\frac{3}{8}, \frac{5}{6}$ 5. $\frac{1}{9}, \frac{1}{6}$ 6. $\frac{4}{15}, \frac{5}{6}$

Add or subtract. Write each answer in simplest form.

7. $\frac{2}{3} - \frac{1}{4}$ 8. $\frac{9}{10} + \frac{3}{4}$ 9. $\frac{7}{8} - \frac{1}{6}$ 10. $\frac{5}{12} - \frac{1}{9}$

11. $\frac{3}{7} + \frac{5}{6}$ 12. $\frac{1}{3} + \frac{4}{5}$ 13. $\frac{4}{15} + \frac{7}{9} - \frac{1}{3}$

Exercises

Find the LCD of each pair of fractions.

14. $\frac{1}{3}, \frac{1}{9}$

15. $\frac{3}{7}, \frac{5}{14}$

16. $\frac{8}{15}, \frac{2}{3}$

17. $\frac{1}{4}, \frac{3}{10}$

18. $\frac{4}{5}, \frac{7}{15}$

19. $\frac{5}{12}, \frac{7}{18}$

20. $\frac{2}{6}, \frac{3}{7}$

21. $\frac{11}{20}, \frac{7}{12}$

Add or subtract. Write each answer in simplest form.

22. $\frac{1}{4} + \frac{3}{8}$

23. $\frac{9}{10} - \frac{2}{5}$

24. $\frac{2}{3} - \frac{1}{6}$

25. $\frac{4}{15} + \frac{2}{5}$

26. $\frac{7}{8} - \frac{3}{6}$

27. $\frac{4}{5} - \frac{1}{2}$

28. $\frac{7}{9} + \frac{5}{6}$

29. $\frac{9}{10} + \frac{7}{15}$

30. $\frac{3}{7} - \frac{1}{6}$

31. $\frac{3}{8} + \frac{5}{6}$

32. $\frac{5}{18} - \frac{1}{12}$

33. $\frac{3}{4} - \frac{1}{5}$

34. $\frac{1}{3} + \frac{5}{6} + \frac{3}{4}$

35. $\frac{1}{2} + \frac{4}{5} - \frac{7}{10}$

36. What is the sum of the fractions $\frac{3}{7}, \frac{8}{12}$, and $\frac{9}{12}$?

Solve each equation.

37. $d = \frac{5}{6} - \frac{1}{3}$

38. $\frac{7}{8} - \frac{1}{2} = p$

39. $z = \frac{1}{4} + \frac{2}{3}$

40. **Algebra**   Find the solution in simplest form for the equation $y - \frac{7}{12} = \frac{5}{12}$. *(Lesson 7-4)*

41. Round $4\frac{2}{15}$ to the nearest half. *(Lesson 7-1)*

42. Express $4\frac{2}{15}$ as an improper fraction. *(Lesson 6-6)*

43. **Algebra**   Solve the equation $2.1g = 142.8$. *(Lesson 5-7)*

44. Express $2\frac{3}{8}$ as a decimal. *(Lesson 6-11)*

45. **Sports**   Mark plays Pony League baseball. He pitched $\frac{1}{3}$ of a game Monday and $\frac{7}{9}$ of a game on Saturday. How much did he pitch this week?

46. **Statistics**   Todd made a circle graph to show the class how he spends his day.
    a. What part of the day does he spend sleeping, eating, and attending school?
    b. What part of the day does Todd have for other things?

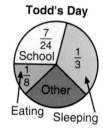

Todd's Day

47. **Geography**   The continent of Asia has about $\frac{3}{5}$ of the world's population. North America has about $\frac{1}{20}$ of the world's population. How much more of the world's population does Asia have than North America?

48. **Critical Thinking**   A piece of wire was cut into thirds. One-third was used. One-fifth of the remaining wire was used. The piece left is 16 feet long. What was the original length of the wire?

7-6 Adding and Subtracting Mixed Numbers

Objective
Add and subtract mixed numbers.

Felix, Sarah, and their father walked down the beach to a fishing pier. On the way back they stopped to help a friend build a sand castle. If it took them $1\frac{1}{4}$ hours to walk to the pier and $2\frac{3}{4}$ hours more to get back, how many hours were they gone? You need to add $1\frac{1}{4}$ hours and $2\frac{3}{4}$ hours.

Mini-Lab

Work with a partner.
Materials: paper plates, scissors

- Place one paper plate in front of you to show the whole number 1 in the mixed number $1\frac{1}{4}$.

- Cut another paper plate into fourths. Place $\frac{1}{4}$ of the plate in front of you.

- Use more paper plates to show $2\frac{3}{4}$.

- Combine the pieces to make as many whole paper plates as you can.

Talk About It

a. How many whole paper plates do you have?
b. How many pieces of paper plates do you have?
c. Did you combine any of the pieces to form a whole paper plate?
d. How long were Felix, Sarah, and their father gone?

TEEN SCENE

The highest waves for surfing are thought to be those at Makaha Beach, Hawaii. They sometimes reach a height of 30 to 33 feet, which is the highest that people can surf safely.

<table>
<tr><td>

Adding and Subtracting Mixed Numbers

</td><td>

1. Add or subtract the fractions.
2. Then add or subtract the whole numbers.
3. Rename and simplify if necessary.

</td></tr>
</table>

Estimation Hint

• • • • • • • • • • •

One way to estimate with fractions and mixed numbers is to use rounding. If the fraction is $\frac{1}{2}$ or greater, round up to the next whole number. If the fraction is less than $\frac{1}{2}$, the whole number remains the same.

Examples

1 Find $7\frac{5}{8} - 3\frac{1}{8}$.   *Estimate: $8 - 3 = 5$*

$$\begin{array}{r} 7\frac{5}{8} \\ -3\frac{1}{8} \\ \hline \frac{4}{8} \end{array}$$ ➡ $$\begin{array}{r} 7\frac{5}{8} \\ -3\frac{1}{8} \\ \hline 4\frac{4}{8} = 4\frac{1}{2} \end{array}$$

2 Find $13\frac{1}{3} + 5\frac{3}{4}$.   *Estimate: $13 + 6 = 19$*

$$13\frac{1}{3} \Rightarrow 13\frac{4}{12}$$ *The LCM of 3 and 4 is 12.*

$$+5\frac{3}{4} \Rightarrow +5\frac{9}{12}$$

$$\phantom{+5\frac{3}{4} \Rightarrow} 18\frac{13}{12}$$ *Rename $\frac{13}{12}$ as $1\frac{1}{12}$.*

$$18 + 1\frac{1}{12} = 19\frac{1}{12}$$

Example 3 *Connection*

Algebra   Solve $8\frac{9}{10} - 6\frac{1}{4} = n$.   *Estimate: $9 - 6 = 3$*

$$8\frac{9}{10} \Rightarrow 8\frac{18}{20}$$ *The LCM of 4 and 10 is 20.*

$$-6\frac{1}{4} \Rightarrow -6\frac{5}{20}$$

$$\phantom{-6\frac{1}{4} \Rightarrow} 2\frac{13}{20}$$

$$2\frac{13}{20} = n$$

Checking for Understanding

Read and study the lesson to answer each question.

1. **Make a model** to show $2\frac{2}{4} + 1\frac{3}{4}$.

2. **Tell** what you would do first when adding $3\frac{2}{8}$ and $1\frac{3}{4}$.

Guided Practice

Add or subtract. Write each answer in simplest form.

3. $3\frac{2}{9} + 1\frac{2}{3}$

4. $7\frac{3}{4} + 10\frac{5}{12}$

5. $13\frac{5}{8} - 5\frac{1}{6}$

6. $5\frac{3}{4} - 3\frac{1}{2}$

7. $17\frac{3}{10} + 9\frac{1}{4}$

8. $14\frac{6}{7} + \frac{1}{2}$

Exercises

Independent Practice

Add or subtract. Write each answer in simplest form.

9. $5\frac{1}{3} + 8\frac{1}{3}$

10. $3\frac{7}{8} - 1\frac{3}{8}$

11. $9\frac{9}{10} - 4\frac{3}{10}$

12. $6\frac{5}{9} + 10\frac{7}{9}$

13. $11\frac{1}{3} - 8\frac{1}{6}$

14. $7\frac{4}{5} + \frac{8}{15}$

15. $16\frac{3}{4} + \frac{2}{3}$

16. $21\frac{2}{3} - 7\frac{2}{3}$

17. $19\frac{5}{6} - 5\frac{5}{8}$

18. $13\frac{7}{15} + 6\frac{8}{15}$

19. $11\frac{3}{4} - 3\frac{1}{6}$

20. $18\frac{5}{9} - 18\frac{1}{6}$

21. $15\frac{3}{4} + 7\frac{7}{10}$

22. $23\frac{9}{16} + 19\frac{3}{4}$

23. $35\frac{5}{7} - 4\frac{2}{5}$

24. What is the sum of $13\frac{2}{7}$ and $6\frac{4}{5}$?

25. Solve the equation $y = 4\frac{8}{9} - 3\frac{3}{8}$.

Algebra   Solve each equation. Write each solution in simplest form.

26. $s = 8\frac{1}{3} - 2\frac{1}{6}$

27. $b = 5\frac{1}{4} + 6\frac{1}{6}$

28. $11\frac{3}{4} + 9\frac{3}{5} = z$

29. $24\frac{7}{9} - 9\frac{1}{6} = g$

Mixed Review

30. Find $\frac{7}{12} - \frac{1}{3}$. Write the answer in simplest form. *(Lesson 7-5)*

31. Estimate $4\frac{1}{5} + 8\frac{8}{9} - 1\frac{7}{9}$. *(Lesson 7-2)*

32. Evaluate the expression $1.1 \times 20 - 18.6 \div 3.1$. *(Lesson 1-7)*

Problem Solving and Applications

33. **Travel**   Mrs. Hobbs is taking a flight from Atlanta to Dallas. Travel time is about 2 hours. When Mrs. Hobbs reports to the gate, they tell her the flight is $1\frac{1}{2}$ hours late. About how long will it be before she arrives in Dallas?

34. **Health**   Kim's baby brother weighed $7\frac{1}{2}$ pounds at birth. He weighed $7\frac{3}{4}$ pounds at his two-week checkup. How much weight had the baby gained?

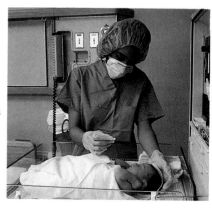

35. **Critical Thinking**   I am a mixed number in simplest form. My denominator is twice my whole number. When I am written as an improper fraction, my numerator is 5 more than 3 times my denominator. Who am I?

36. **Critical Thinking**   Two and seven-tenths plus one and three-fifths is equal to four and three-tenths. Is this a true statement? Explain.

37. **Geometry**   Find the perimeter of the triangle shown at the right.

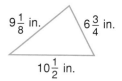

$9\frac{1}{8}$ in. $6\frac{3}{4}$ in. $10\frac{1}{2}$ in.

Mid-Chapter Review

Round each number to the nearest half.   *(Lesson 7-1)*

1. $3\frac{1}{5}$

2. $11\frac{6}{7}$

3. $41\frac{7}{8}$

4. $2\frac{1}{3}$

Estimate.   *(Lesson 7-2)*

5. $4\frac{6}{7} - 2\frac{4}{9}$

6. $22\frac{3}{8} - 14\frac{8}{9}$

7. $9\frac{1}{2} + 5\frac{3}{5}$

Add or subtract. Write each answer in simplest form.   *(Lessons 7-3, 7-5, and 7-6)*

8. $\frac{14}{15} - \frac{7}{15}$

9. $\frac{13}{20} + \frac{5}{20}$

10. $\frac{3}{2} + \frac{7}{2}$

11. $\frac{2}{5} - \frac{1}{6}$

12. $\frac{9}{12} + \frac{7}{20}$

13. $\frac{8}{15} - \frac{4}{9}$

14. $13\frac{2}{3} + 6\frac{1}{2}$

15. $11\frac{5}{8} - 9\frac{2}{5}$

16. $16\frac{7}{10} + 3\frac{2}{7}$

Solve each equation.   *(Lesson 7-4)*

17. $m + \frac{3}{5} = \frac{4}{5}$

18. $z - \frac{8}{11} = \frac{4}{11}$

19. $\frac{14}{13} = d + \frac{5}{13}$

7-7 Subtracting Mixed Numbers: Renaming

Objective

Subtract mixed numbers involving renaming.

DID YOU KNOW

Thomas Saint of England patented the first sewing machine in 1790. It was foot-pedal driven. In 1889, the first electric sewing machine was manufactured by the Singer Company.

Karen is learning how to sew. She is making a skirt for herself and one for her little sister Katie. Karen has a $30\frac{1}{4}$-inch waist. Katie has a $24\frac{3}{4}$-inch waist. How much shorter should Katie's waistband be than Karen's? You need to subtract to compare the waistbands.

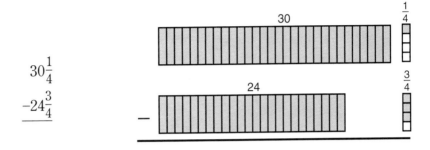

$$\begin{array}{r} 30\frac{1}{4} \\ -24\frac{3}{4} \\ \hline \end{array}$$

Notice that you cannot subtract $\frac{3}{4}$ from $\frac{1}{4}$. Rename $30\frac{1}{4}$ as $29\frac{5}{4}$.

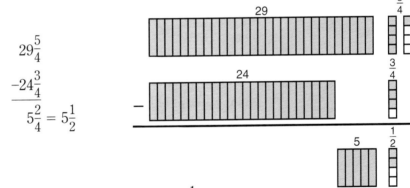

$$\begin{array}{r} 29\frac{5}{4} \\ -24\frac{3}{4} \\ \hline 5\frac{2}{4} = 5\frac{1}{2} \end{array}$$

Katie's waistband should be $5\frac{1}{2}$ inches shorter than Karen's.

As shown above, sometimes it is necessary to rename the fraction of a mixed number as an improper fraction in order to subtract.

LOOK BACK

You can review renaming mixed numbers as improper fractions on page 207.

Example 1

Find $13 - 8\frac{7}{9}$. *Estimate: 13 − 9 = 4*

$$\begin{array}{rcl} 13 & \Rightarrow & 12\frac{9}{9} \\ -8\frac{7}{9} & \Rightarrow & -8\frac{7}{9} \\ \hline & & 4\frac{2}{9} \end{array}$$

Rename 13 as $12\frac{9}{9}$.

$13 - 8\frac{7}{9} = 4\frac{2}{9}$

Example 2

Find $7\frac{1}{3} - 2\frac{3}{4}$. *Estimate: 7 − 3 = 4*

Step 1 $7\frac{1}{3}$ ➡ $7\frac{4}{12}$ *The LCM of 3 and 4 is 12.*

$-2\frac{3}{4}$ ➡ $-2\frac{9}{12}$

Step 2 $7\frac{4}{12}$ ➡ $6\frac{16}{12}$ *Since you cannot subtract $\frac{9}{12}$*

$\phantom{7\frac{4}{12}}\quad -2\frac{9}{12}\qquad -2\frac{9}{12}$ *from $\frac{4}{12}$, you must rename*

$ 4\frac{7}{12}$ *$7\frac{4}{12}$ as $6\frac{16}{12}$.*

$7\frac{1}{3} - 2\frac{3}{4} = 4\frac{7}{12}$

Example 3 *Problem Solving*

Sewing   Mrs. Young is making drapes for her living room. She needs $12\frac{3}{4}$ yards of material for a larger window and $7\frac{1}{2}$ yards for a smaller window. If Mrs. Young bought 30 yards of material, how much will she have left after making the drapes?

First, you need to add to find out the total number of yards that Mrs. Young needs. Add $12\frac{3}{4}$ and $7\frac{1}{2}$.

$12\frac{3}{4}$ ➡ $12\frac{3}{4}$ *Estimate: 13 + 8 = 21*
   The LCM of 4 and 2 is 4.

$+7\frac{1}{2}$ ➡ $+7\frac{2}{4}$

$ 19\frac{5}{4}$ *Rename $\frac{5}{4}$ as $1\frac{1}{4}$..*

$19 + 1\frac{1}{4} = 20\frac{1}{4}$

Now, subtract $20\frac{1}{4}$ from 30 to find how much material will be left.

$30 29\frac{4}{4}$ *Estimate: 30 − 20 = 10*

$-20\frac{1}{4}$ ➡ $-20\frac{1}{4}$ *Rename 30 as $29\frac{4}{4}$.*

$ 9\frac{3}{4}$

Mrs. Young will have $9\frac{3}{4}$ yards of material left.

Checking for Understanding

Communicating Mathematics

Read and study the lesson to answer each question.

1. **Write**, in your own words, how you know when you need to rename before subtracting mixed numbers.

2. **Tell** how to rename $6\frac{1}{4}$.

Guided Practice

Complete.

3. $3\frac{5}{8} = 2\frac{\blacksquare}{8}$

4. $6\frac{3}{10} = \blacksquare \frac{13}{10}$

5. $14\frac{8}{15} = \blacksquare \frac{23}{15}$

Subtract. Write each answer in simplest form.

6. $\begin{array}{r} 5\frac{2}{5} \\ -1\frac{6}{10} \\ \hline \end{array}$

7. $\begin{array}{r} 17\frac{1}{3} \\ -12\frac{5}{6} \\ \hline \end{array}$

8. $\begin{array}{r} 4\frac{1}{2} \\ -1\frac{6}{7} \\ \hline \end{array}$

9. $\begin{array}{r} 8\frac{1}{14} \\ -5\frac{4}{7} \\ \hline \end{array}$

10. $\begin{array}{r} 18\frac{3}{8} \\ -5\frac{3}{4} \\ \hline \end{array}$

11. $10 - 7\frac{6}{7}$

12. $16 - 7\frac{5}{8}$

13. $9\frac{3}{4} - 6\frac{2}{5}$

Exercises

Independent Practice

Complete.

14. $9\frac{2}{5} = \blacksquare \frac{7}{5}$

15. $6\frac{2}{9} = 5\frac{\blacksquare}{9}$

16. $2\frac{3}{4} = 1\frac{\blacksquare}{4}$

17. $13\frac{5}{8} = \blacksquare \frac{13}{8}$

18. $4\frac{1}{8} = 3\frac{\blacksquare}{8}$

19. $15\frac{2}{3} = \blacksquare \frac{5}{3}$

20. $7\frac{13}{16} = 6\frac{\blacksquare}{16}$

21. $8\frac{11}{12} = 7\frac{\blacksquare}{12}$

22. $7\frac{5}{6} = 6\frac{\blacksquare}{6}$

Subtract. Write each answer in simplest form.

23. $\begin{array}{r} 3\frac{1}{4} \\ -1\frac{3}{4} \\ \hline \end{array}$

24. $\begin{array}{r} 8\frac{3}{5} \\ -6\frac{4}{5} \\ \hline \end{array}$

25. $\begin{array}{r} 9\frac{3}{8} \\ -4\frac{5}{8} \\ \hline \end{array}$

26. $\begin{array}{r} 3\frac{1}{5} \\ -1\frac{7}{10} \\ \hline \end{array}$

27. $\begin{array}{r} 16\frac{1}{3} \\ -2\frac{3}{4} \\ \hline \end{array}$

28. $\begin{array}{r} 15\frac{3}{8} \\ -9\frac{5}{6} \\ \hline \end{array}$

29. $\begin{array}{r} 7\frac{1}{6} \\ -3\frac{1}{3} \\ \hline \end{array}$

30. $\begin{array}{r} 12\frac{3}{15} \\ -\frac{2}{5} \\ \hline \end{array}$

31. $7 - 1\frac{1}{3}$

32. $18\frac{7}{16} - 8\frac{3}{4}$

33. $14\frac{4}{9} - 6\frac{5}{6}$

34. If you subtract $11\frac{5}{7}$ from $15\frac{3}{4}$, what is the result?

35. Find the difference of $7\frac{4}{7}$ and $3\frac{3}{5}$.

Solve each equation. Write each solution in simplest form.

36. $7\frac{1}{6} - 2\frac{5}{6} = y$ 37. $9\frac{2}{7} - 8\frac{5}{7} = d$ 38. $t = 12\frac{9}{10} - 10\frac{2}{3}$

Mixed Review   39. Estimate $18\frac{3}{16} + 4\frac{1}{9}$. *(Lesson 7-2)*

40. Divide 64.16 by 0.32. *(Lesson 5-3)*

41. Order the fractions $\frac{2}{7}, \frac{1}{3}, \frac{2}{9}, \frac{2}{5}, \frac{4}{1}$ from greatest to least. *(Lesson 6-9)*

42. Find $7\frac{6}{7} + 2\frac{3}{5}$. Write the answer in simplest form. *(Lesson 7-6)*

43. Find the least common multiple of 15, 20, and 25. *(Lesson 6-8)*

Problem Solving and Applications

44. **Critical Thinking**   Which difference is greater, $28\frac{1}{7} - 6\frac{3}{4}$ or $30\frac{1}{8} - 8\frac{3}{4}$?

45. **Sports**   Some of the differences between Olympic softball and Olympic baseball are listed in the chart below. Answer each question.
    a. Which sport has the larger ball? How much larger is it compared to the other sport's ball?
    b. Which ball is heavier? How much heavier is it?

	Olympic Softball	Olympic Baseball
Size	$11\frac{7}{8}$ to $12\frac{1}{8}$ inches in circumference	9 to $9\frac{1}{4}$ inches in circumference
Weight	$6\frac{1}{4}$ to 7 ounces	5 to $5\frac{1}{4}$ ounces

46. **Statistics**   The graph below shows the number of vehicles per household according to a 1990 survey conducted by the U.S. Transportation Department. Use the graph to answer each question.

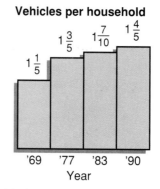

Vehicles per household

    a. How many more vehicles per household did Americans have in 1990 than in 1983?
    b. In what year did Americans have the fewest vehicles per household?

7-8 Adding and Subtracting Measures of Time

Objective

Add and subtract measures of time.

Spiridon Loues of Greece won the first modern Summer Olympic Games marathon in 1896. His time was 2 hours, 58 minutes, 50 seconds. In the 1988 Summer Olympics, Gelindo Bordin of Italy won the marathon with a time of 2 hours, 10 minutes, 47 seconds. How much faster was Bordin's time?

DID YOU KNOW

In early times, Romans used the water clock. They equated time with water just like we equate time with money. For example, if a speaker in the Senate talked too much, his colleagues would shout that his water should be taken away.

Loues' time:	2 h 58 min 50 s
Bordin's time:	−2 h 10 min 47 s
	48 min 3 s

Bordin ran the marathon 48 minutes, 3 seconds faster than Loues.

To add or subtract measures of time:
1. Add or subtract the seconds.
2. Add or subtract the minutes.
3. Add or subtract the hours.
Rename if necessary in each step.

1 **hour** (h) = 60 **minutes** (min)
1 minute (min) = 60 **seconds** (s)

Example 1 *Problem Solving*

Tennis In 1991, Jimmy Connors defeated Aaron Krickstein in a qualifying round for the men's singles at the U.S. Open Tennis Tournament. The match lasted 4 hours and 41 minutes. Suppose Connors played another match later in the same week that lasted 3 hours and 35 minutes. How long did Connors play tennis that week?

   4 h 41 min *There are no seconds, so*
+3 h 35 min *add the minutes. Then*
   7 h 76 min *add the hours.*

Now rename 76 min as 1 h and 16 min.

7h + 1 h 16 min = 8 h 16 min
Connors played tennis 8 hours and 16 minutes that week.

Example 2 **Problem Solving**

School Rob starts school at 7:50 A.M. and ends his day at 2:35 P.M. How long is Rob at school?

To answer this question, you need to find how much time has elapsed. Think of a clock.

7:50 A.M. to 12:00 noon is 4 h 10 min.

12:00 noon to 2:35 P.M. is 2 h 35 min.

The elapsed time is 4 h 10 min + 2 h 35 min or 6 h 45 min. Rob is in school for 6 hours and 45 minutes each day.

Checking for Understanding

Communicating Mathematics

Read and study the lesson to answer each question.

1. **Write** the number of minutes in 1 h 15 min.
2. **Write** the answer to complete the statement 95 s = __?__ min __?__ s.
3. **Show** the elapsed time from 5:15 P.M. to 7:45 P.M. using a clock.

Guided Practice

Complete.

4. 7 h 6 min = 6 h __?__ min
5. 17 min 75 s = __?__ min 15 s
6. 3 h 85 min 105 s = 4 h __?__ min __?__ s
7. 6 h 8 min 25 s = 5 h 67 min __?__ s

Add or subtract. Rename if necessary.

8.   5 min 25 s
  +10 min 16 s

9.   7 h 39 min
  −1 h 16 min

10.   8 h 15 min
  −6 h 40 min

11.   8 h   10 s
 −3 h 20 min 25 s

12.   3 h 35 min 50 s
 +6 h 50 min 30 s

Find the elapsed time.

13. 10: 30 A.M. to 2:15 P.M.

14. 8:15 P.M. to 6:45 A.M.

Exercises

Complete.

15. 5 h 18 min = 4 h __?__ min

16. 2 min 85 s = __?__ min 25 s

17. 7 min 21 s = __?__ min 81 s

18. 9 h 93 min 8 s = __?__ h 33 min 8 s

19. 1 h 28 min 59 s = __?__ min 59 s

20. 7 h 20 min 10 s = 6 h __?__ min 70 s

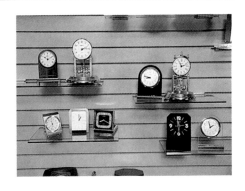

Add or subtract.

21. 5 h 34 min
    −2 h 9 min

22.   3 min 25 s
    +7 min 15 s

23.   8 h 48 s
    +7 h 11 s

24. 9 h 20 min
    −5 h 35 min

25.   4 h 50 min
    +10 h 30 min

26. 12 min 30 s
    −5 min 58 s

27. 48 min 25 s
    +25 min 35 s

28. 4 h
    +12 h 55 min

29. 9 h 45 min 2 s
    −2 h 10 min 55 s

30. 4 h 18 min 15 s
    +12 h 55 min 28 s

31. 9 h 27 min 45 s
    +2 h 50 min 45 s

32. 7 h 30 min 41 s
    −3 h 47 min 52 s

33. 8 h
    −3 h 20 min 15 s

Find the elapsed time.

34. 8:05 A.M. to 11:45 A.M.

35. 7:00 A.M. to 6:55 P.M.

36. 5:30 P.M. to 3:10 A.M.

37. 8:30 P.M. to 10:40 A.M.

38. Add 8 h 12 min and 5 h 15 s.

39. Subtract 7 h 25 min 10 s from 13 h 55 min.

40. **Astronomy**   Mercury moves in its orbit at a speed of 29.75 miles per second. Write this number as a mixed number in simplest form. *(Lesson 6-10)*

41. Find $\frac{7}{12} + \frac{11}{12}$ in simplest form. *(Lesson 7-3)*

42. Find $7\frac{1}{6} - 3\frac{3}{4}$ in simplest form. *(Lesson 7-7)*

43. Express 0.75 as a fraction in simplest form. *(Lesson 6-10)*

44. **Space**   The countdown to the launching of a spacecraft was stopped at 12 h 7 min 36 s before launch. The next stop was at 1 h 28 min 10 s before launch. How far did the countdown progress between stops?

45. **Travel**   According to a road atlas, it takes 7 hours 5 minutes to travel by car from Indianapolis to St. Louis, Missouri, 4 hours 35 minutes from St. Louis to Springfield, Missouri, and 3 hours 50 minutes from Springfield to Tulsa, Oklahoma. How long is the drive from Indianapolis to Tulsa?

46. **Critical Thinking**   Regina is flying from New York City to Los Angeles to visit her grandmother. Her flight leaves at 9:15 A.M. The non-stop flight takes about 6 hours. About what time will it be in Los Angeles when Regina arrives? *(Hint: Remember that there is a time difference between New York and Los Angeles.)*

47. **Journal Entry**   Write in your own words how adding measures of time is similar to adding mixed numbers.

48. **Mathematics and Time**   Read the following paragraphs.

In 1929, Canadian W. Marrison produced the first quartz crystal clock. The mineral quartz has a piezoelectric effect. This means that if it receives electric charges it will vibrate at a fixed frequency and act like a pendulum. The vibrations are used to control the speed of an electric motor, which drives the clock hands. A quartz clock can keep extremely accurate time.

In 1967, the Seiko Company produced the first quartz wristwatch. The battery in the watch produces electric current, which makes the crystal vibrate. A circuit controls the vibration rate to give one pulse per second. It then converts one-second pulses into numbers in a digital display.

a. A game starts in Los Angeles at 2:00 P.M. and ends at 5:10 P.M. If the game is shown live on TV in New Jersey, what time does the game begin and end?

b. The light on an answering machine blinks 5 times every 6 seconds. How many times does it blink in 24 hours?

7-9 **Eliminate Possibilities**

Objective

Solve problems by using estimation to eliminate possibilities.

Kyle's mother works at Williams Insurance Company. Her benefits package offers a choice of three savings plans. The first plan pays 8 cents on every dollar of her annual salary of $17,500. The second plan pays 5 cents on every dollar for the first $10,000 she earns. The third plan pays a flat rate of $1,000 per employee. If Kyle's mother wants to join the plan that pays the most, which plan should she choose?

Explore   What do you know?
Kyle's mother earns $17,500 a year.
One plan pays 8 cents on every dollar of her annual salary.
One plan pays 5 cents on every dollar of the first $10,000 she earns.
One plan pays a flat rate of $1,000 per employee.

What do you need to find out?
You need to find out which plan pays the most.

Plan   Estimate how much each plan pays per year by multiplying the company investment by the amount Kyle's mother earns per year.
Eliminate the lowest paying plans.

Solve   Plan 1: Round 8 cents to 10 cents.
Multiply mentally by $17,500.

$$\$0.10 \times \$17,500 = \$1,750$$

Plan 2: Multiply 5 cents by $10,000.

$$\$0.05 \times \$10,000 = \$500$$

Plan 3: This plan pays a flat rate of $1,000 per employee.

Look at your three choices.
Solve by eliminating two of the choices.

Kyle's mother should choose the first plan.

Check your estimate by using a calculator.

$$0.08 \; \boxed{\times} \; 17500 \; \boxed{=} \; 1400$$

The first plan pays the most.

Example

Estimate the following product. Choose the correct answer by eliminating possibilities.

$$1{,}890 \times 21$$

a. 396,900   b. 30,000   c. 39,690   d. 3,000

You can estimate by using compatible numbers. 1,890 rounds to 2,000. 21 rounds to 20.

$$2{,}000 \times 20 = 40{,}000$$

Choice **a** can be eliminated because 396,900 is much greater than 40,000. Choice **d** can be eliminated because 3,000 is much less than 40,000.

Round again to get a closer estimate. Round 1,890 to 1,900. $1{,}900 \times 20 = 38{,}000$.

Choice **b** can be eliminated because 30,000 is much less than 38,000. By eliminating possibilities, you can see that choice **c** must be the correct answer. Check your answer by dividing.

$$39690 \; \boxed{\div} \; 21 \; \boxed{=} \; 1890 \; \checkmark$$

Checking for Understanding

Communicating
Mathematics

Read and study the lesson to answer each question.

1. **Tell** how the eliminate possibilities strategy would help you on a multiple choice test.
2. **Tell** how eliminating possibilities could help you make a good choice when choosing between several sale items to buy.

Solve by eliminating possibilities.

3. Last year, Fred's car odometer read 45,500.4. This year the odometer reads 57,200.9. Find about how many miles Fred has driven over the past year.

    a. 100   **b.** 1,100   **c.** 11,000   **d.** 111,000

4. Yankee Stadium has a seating capacity of 57,545. If all tickets sell for $10.95, the gross receipts for a sold-out, three-day homestand would be:

    a. $1,890,353.30   **b.** $189,353.30   **c.** $18,903.53

5. Norm used his calculator to divide 435,800 by 49.8. He should expect the quotient to be about:

    a. 87   **b.** 870   **c.** 807   **d.** 8,700

Problem Solving

Solve. Use any strategy.

6. What is the greatest number you can multiply by 78 to get a product between 2,000 and 2,500?

7. Because of erosion, the height of some mountains is worn down about $3\frac{1}{2}$ inches every 1,000 years. At this rate, about how long would it take a mountain as tall as Mount Kilimanjaro, which is 19,340 feet, to wear down?

    a. 6,000 years   **b.** 660,000 years   **c.** 66,000,000 years

8. Is the mean of 123.9, 43.6, 120.89, 502.9, 12.7, and 72.34 about 150 or 15?

9. A manufacturer offers four different styles of tennis shoes in white canvas, white leather, black canvas, black leather, and blue canvas. How many combinations of style and color are possible?

10. What pages would you need to open in a book so that the product of the facing page numbers is 1,190?

11. At an intersection, a flashing yellow light blinks 25 times per minute. In one day, the number of times the light blinks is about:

    a. 360   **b.** 3,600   **c.** 36,000   **d.** 360,000

Strategies
● ● ● ● ● ● ●
Look for a pattern.
Solve a simpler problem.
Act it out.
Guess and check.
Draw a diagram.
Make a chart.
Work backward.

Study Guide and Review

Communicating Mathematics

Choose the letter that best matches each phrase.

1. the number of minutes in one hour and twenty minutes
2. the number of liters (rounded) of lemonade to prepare for use in a $2\frac{7}{8}$-liter pitcher
3. an estimate for $\frac{1}{11} + 4\frac{5}{8} - 1\frac{4}{9}$
4. the solution for $m = \frac{6}{13} + \frac{7}{13}$
5. these must be alike for addition or subtraction
6. these get added when fractions with like denominators are added
7. the least common denominator is the __?__ of the unlike denominators
8. the difference $\frac{11}{13} - \frac{7}{10}$
9. Explain why $4\frac{3}{7} - 1\frac{5}{7}$ doesn't become $3\frac{13}{7} - 1\frac{5}{7}$ when you rename or "borrow" for subtraction.
10. Tell what the three meanings of subtraction are.

a. one
b. numerators
c. denominators
d. GCF
e. 80
f. LCM
g. $\frac{4}{3}$
h. 120
i. $2\frac{1}{2}$
j. 3
k. $\frac{19}{130}$

Skills and Examples

Objectives and Examples
Upon completing this chapter, you should be able to:

- round fractions and mixed numbers
  (Lesson 7-1)
  Round to the nearest half:
  $\frac{13}{16} \rightarrow 1$ *13 is close to 16.*
  $\frac{5}{9} \rightarrow \frac{1}{2}$ *5 is about half of 9.*
  $\frac{2}{11} \rightarrow 0$ *2 is much smaller than 11.*

Review Exercises
Use these exercises to review and prepare for the chapter test.

Tell whether each number should be rounded up, down, or to the nearest half unit.

11. the diameter of a plant to put in an $11\frac{7}{8}$-inch pot
12. the weight of a dog

Round to the nearest half.

13. $\frac{9}{16}$ 14. $9\frac{2}{9}$

15. $11\frac{4}{7}$ 16. $\frac{9}{11}$

Objectives and Examples

- estimate sums and differences of fractions *(Lesson 7-2)*
Estimate.

$$\frac{13}{16} - \frac{5}{11} \rightarrow 1 - \frac{1}{2} = \frac{1}{2}$$

$$\frac{8}{9} + 8\frac{1}{5} \rightarrow 1 + 8 = 9$$

- add and subtract fractions with like denominators *(Lesson 7-3)*

$$\frac{7}{8} + \frac{5}{8} = \frac{7+5}{8} = \frac{12}{8} = 1\frac{4}{8} = 1\frac{1}{2}$$

- solve equations involving addition and subtraction of fractions *(Lesson 7-4)*

Solve $\frac{6}{7} - h = \frac{2}{7}$.

Since $6 - 4 = 2$, $h = \frac{4}{7}$.

- add and subtract fractions with unlike denominators *(Lesson 7-5)*

$$\frac{7}{12} - \frac{11}{24} = \frac{14}{24} - \frac{11}{24} = \frac{14-11}{24} = \frac{3}{24} = \frac{1}{8}$$

- add and subtract mixed numbers *(Lesson 7-6)*

$$
\begin{aligned}
11\frac{2}{3} &\rightarrow 11\frac{16}{24} \\
+2\frac{3}{8} &\rightarrow 2\frac{9}{24} \\
\hline
&\quad 13\frac{25}{24} = 13 + 1\frac{1}{24} = 14\frac{1}{24}
\end{aligned}
$$

- subtract mixed numbers involving renaming *(Lesson 7-7)*

$$
\begin{aligned}
4\frac{1}{4} - \frac{2}{3} &= 4\frac{1}{4} \rightarrow 4\frac{3}{12} \rightarrow 3\frac{15}{12} \\
&\qquad\quad -\frac{2}{3} \qquad -\frac{8}{12} \qquad -\frac{8}{12} \\
&\qquad\qquad\qquad\qquad\qquad\qquad 3\frac{7}{12}
\end{aligned}
$$

Review Exercises

Estimate.

17. $8\frac{3}{16} - 4\frac{1}{4}$

18. $\frac{5}{8} + \frac{17}{18}$

19. $\frac{7}{8} - \frac{7}{15}$

20. $\frac{1}{8} + 4\frac{2}{3}$

Add or subtract. Write each answer in simplest form.

21. $\frac{8}{15} + \frac{11}{15}$ 22. $\frac{4}{7} - \frac{1}{7}$

23. $\frac{7}{8} - \frac{1}{8}$ 24. $\frac{4}{6} + \frac{2}{6}$

Solve each equation.

25. $\frac{7}{10} - m = \frac{3}{10}$

26. $k - \frac{3}{5} = \frac{4}{5}$

27. $\frac{9}{14} = \frac{3}{14} + b$

Add or subtract. Write each answer in simplest form.

28. $\frac{2}{3} - \frac{3}{5}$ 29. $\frac{4}{9} + \frac{5}{6}$

30. $\frac{7}{10} - \frac{1}{5}$ 31. $\frac{2}{3} + \frac{1}{4}$

Add or subtract. Write each answer in simplest form.

32. $7\frac{1}{7} + 2\frac{3}{7}$ 33. $11\frac{5}{8} - 3\frac{1}{6}$

Solve each equation. Write each solution in simplest form.

34. $f = 6\frac{1}{4} - \frac{1}{8}$ 35. $5\frac{2}{5} + 1\frac{3}{5} = h$

Subtract. Write each answer in simplest form.

36. $\quad 8\frac{1}{8}$
    $-3\frac{1}{4}$
    $\overline{}$

37. $\quad 8$
    $-2\frac{2}{3}$
    $\overline{}$

Objectives and Examples

- add and subtract measures of time
  (Lesson 7-8)

  3 h 50 min
  +2 h 15 min
  _____
  65 min = 1 h 5 min
  5 h + 1 h 5 min = 6 h 5 min

Objectives and Examples

Add or subtract. Rename if necessary.

38.   5 h 20 min   39.   7 h 45 min
      +2 h 16 min   −4 h 32 min

40.   2 h 35 min   41.   9 h 7 min
      +6 h 41 min   −8 h 7 min 8 s

Applications and Problem Solving

42. **Sewing** Melinda and Dolores both decided to make their dresses for graduation. Melinda's pattern calls for $6\frac{1}{3}$ yards of fabric and Dolores' pattern calls for $5\frac{5}{6}$ yards. How much more fabric does Melinda need to buy? *(Lesson 7-6)*

43. **Smart Shopping** How much money should Danny get from his wallet to pay for a tape that costs $8.10; $8 or $8.50? *(Lesson 7-1)*

44. **Distance** Mieko walks $\frac{1}{2}$ mile to school in the morning, then walks $\frac{2}{3}$ mile to work after school. She gets a ride home at night from her dad. How many miles does she walk each day? *(Lesson 7-5)*

45. **Jobs** Gina began working at 9:15 A.M. and worked until 3:30 P.M. How long did she work? *(Lesson 7-8)*

46. A credit card company pays 5¢ to a local charity for every $2.00 charged. If a cardholder charged $1,400 worth of merchandise, how much money went to charity? *(Lesson 7-9)*
    a. $3.50   b. $7   c. $35   d. $350

Curriculum Connection Projects

- **Recreation** Use two pairs of differently-colored number cubes, one color for numerators, one for denominators. You and a friend each roll a pair and add or subtract your two fractions.

- **Music** Add all the notes, as fractions, from one bar on a sheet of your favorite music. For example, three half notes plus three eighth notes equals $\frac{3}{2} + \frac{3}{8}$ or $1\frac{7}{8}$.

Read More About It

Babbit, Natalie. *Tuck Everlasting.*
Greer, Gary and Ruddick, Bob. *Max and Me and the Time Machine.*
Jesperson, James. *Time and Clocks for the Space Age.*

Round to the nearest half unit.

1. $\dfrac{7}{15}$

2. $7\dfrac{2}{11}$

3. $\dfrac{11}{14}$

4. **Decorating**   One wall space in Gustavo's room is $24\dfrac{3}{8}$ inches wide. When looking for a poster to fit in that space, should he round up or down or to the nearest half inch?

Estimate.

5. $4\dfrac{7}{10} - 2\dfrac{1}{5}$

6. $\dfrac{7}{8} + \dfrac{1}{6}$

7. $\dfrac{15}{16} - \dfrac{4}{9}$

Add or subtract. Write each answer in simplest form.

8. $\dfrac{3}{17} + \dfrac{16}{17}$

9. $\dfrac{7}{8} + \dfrac{1}{6}$

10. $6\dfrac{3}{10} - 2\dfrac{9}{10}$

11. $4\dfrac{2}{9} + 5\dfrac{2}{9}$

12. $12 - 7\dfrac{3}{8}$

13. $3\dfrac{1}{8} - 1\dfrac{1}{16}$

14. $\dfrac{15}{16} - \dfrac{7}{16}$

15. $\dfrac{2}{15} + \dfrac{3}{10} - \dfrac{1}{5}$

16. $5\dfrac{7}{8} + 1\dfrac{5}{6}$

Solve each equation. Write each solution in simplest form.

17. $m = \dfrac{15}{16} - \dfrac{11}{16}$

18. $y = \dfrac{4}{5} + \dfrac{3}{5}$

19. $p = 7\dfrac{3}{5} + 2\dfrac{1}{5}$

20. $15\dfrac{7}{12} - 8\dfrac{3}{8} = d$

21. **Gardening**   Mr. O'Dowd had $\dfrac{1}{2}$ ton of black dirt for his garden delivered on Tuesday. On Thursday, another $\dfrac{1}{4}$ ton was delivered in error. How much dirt did he have in all?

22. **Tickets**   Kwasi got to TicketMaster at 4:52 A.M. to wait in line for concert tickets. If TicketMaster opened at 7:30 A.M., how long did he wait?

Add or subtract.

23.   10 h 12 min
  − 4 h 50 min

24.   3 h 4 min 10 s
  +9 h 58 min 28 s

25.   6 h 55 min
  −5 h 49 min

Bonus   The sum of three mixed numbers each greater than zero is always greater than what number?

Fractions: Multiplication and Division

Spotlight on Energy and Petroleum

Have You Ever Wondered...

- How much gasoline is used by vehicles in the United States?

- What fraction of United States energy consumption consists of petroleum products such as gasoline?

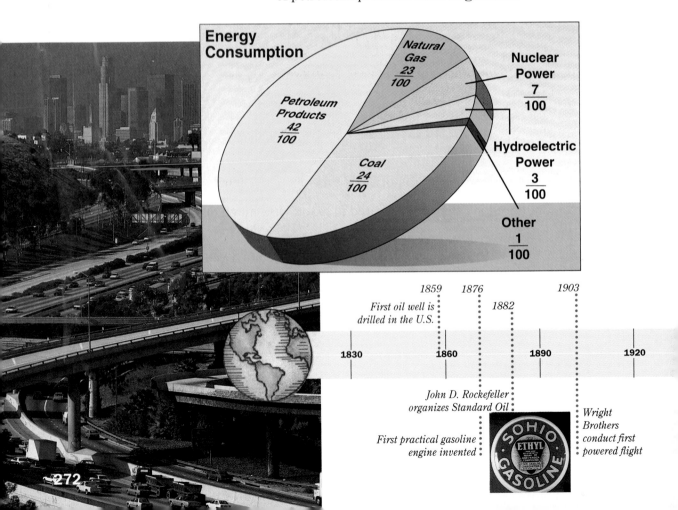

Energy Consumption

Natural Gas $\frac{23}{100}$

Nuclear Power $\frac{7}{100}$

Petroleum Products $\frac{42}{100}$

Coal $\frac{24}{100}$

Hydroelectric Power $\frac{3}{100}$

Other $\frac{1}{100}$

1859
First oil well is drilled in the U.S.

1876

1882

1903

1830   1860   1890   1920

John D. Rockefeller organizes Standard Oil

First practical gasoline engine invented

SOHIO · ETHYL · GASOLINE

Wright Brothers conduct first powered flight

Chapter Project

Energy and Petroleum

Work in a group.

1. Ask an adult who drives a vehicle to keep track of his/her gasoline consumption for one month.

2. Each time gasoline is purchased, the number of gallons, the mileage, the date, and the price should be noted.

3. Make a chart showing the data.

4. Calculate how many gallons of gasoline the person would consume in one year at the same rate.

5. Make a chart to compare the results from each group.

Looking Ahead

In this chapter, you will see how mathematics can be used to answer questions about energy and petroleum. The major objectives of the chapter are to:

- multiply and divide fractions and mixed numbers

- recognize and extend sequences

- change units in the customary system

U.S. Motor Vehicle Fuel Consumption (1988 estimate)

Type or vehicle	Number of registered vehicles	Average miles traveled per vehicle	Fuel consumed (thousand gallons)	Average fuel consumption per vehicle (gallons)
All passenger vehicles	146,451,648	9,865	72,774,695	497
All buses	615,669	8,877	920,056	1,494
All cargo vehicles	42,529,368	13,656	57,111,185	1,343
All motor vehicles	188,981,016	10,718	129,885,880	687

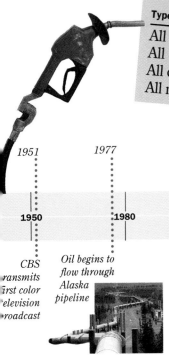

1951
1977
1989
1991

1950
1980
2010

CBS transmits first color television broadcast

Oil begins to flow through Alaska pipeline

Exxon Valdez spills oil into Alaska's Prince William Sound

Iraqi army dumps oil into Persian Gulf

8-1 Estimating Products

Objective

Estimate fraction products using compatible numbers.

LOOKBACK

You can review compatible numbers on page 18.

There are about 7 million small pleasure boats in the United States. About $\frac{2}{3}$ of these boats are motorboats.

About how many motorboats are there in the United States?

To find out, estimate the product of $\frac{2}{3}$ and 7 million by using compatible numbers.

$\frac{2}{3} \times 7 = ?$ *$\frac{2}{3} \times 7$ means $\frac{2}{3}$ of 7.*

$\frac{1}{3} \times 6 = 2$ *Think: The nearest multiple of 3 is 6. 3 and 6 are compatible numbers.*

$\frac{2}{3} \times 6 = 4$ *Since $\frac{1}{3} \times 6 = 2$, it follows that $\frac{2}{3} \times 6 = 2 \times 2$ or 4.*

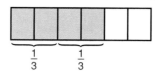

The product of $\frac{2}{3}$ and 7 is about 4. There are about 4 million pleasure motorboats in the United States.

Remember, there is a special symbol, $\approx$, that means *is about equal to*.

$$\frac{2}{3} \times 7 \approx 4$$

Example 1

Estimate the product $\frac{1}{4} \times 9$.

$\frac{1}{4} \times 8 = 2$ *Think: The nearest multiple of 4 is 8.*

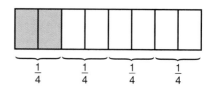

$\frac{1}{4} \times 9 \approx 2$ *The product of $\frac{1}{4}$ and 9 "is about equal to" 2.*

Example 2

Estimate $\frac{5}{6} \times 23$.

$\frac{1}{6} \times 24 = 4$ *Think: The nearest multiple of 6 is 24.*

$\frac{5}{6} \times 24 = 20$ *Since $\frac{1}{6} \times 24 = 4$, it follows that $\frac{5}{6} \times 24 = 5 \times 4$ or 20.*

$\frac{5}{6} \times 23 \approx 20$

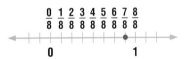

LOOKBACK

You can review rounding fractions on page 232.

You can also estimate products by rounding fractions to 0, $\frac{1}{2}$, or 1.

Examples

Estimate each product.

3 $\frac{2}{5} \times \frac{7}{8}$ *Think: $\frac{7}{8}$ is nearest to 1.*

$$\frac{0}{8}\ \frac{1}{8}\ \frac{2}{8}\ \frac{3}{8}\ \frac{4}{8}\ \frac{5}{8}\ \frac{6}{8}\ \frac{7}{8}\ \frac{8}{8}$$

0   1

$\frac{2}{5} \times 1 = \frac{2}{5}$ *Round $\frac{7}{8}$ to 1.*

$\frac{2}{5} \times \frac{7}{8} \approx \frac{2}{5}$

4 $\frac{4}{7} \times \frac{1}{3}$ *Think: $\frac{4}{7}$ is nearest to $\frac{1}{2}$.*

$$\frac{0}{7}\ \frac{1}{7}\ \frac{2}{7}\ \frac{3}{7}\ \frac{4}{7}\ \frac{5}{7}\ \frac{6}{7}\ \frac{7}{7}$$

0       $\frac{1}{2}$ 1

$\frac{1}{2} \times \frac{1}{3} = \frac{1}{6}$ *Round $\frac{4}{7}$ to $\frac{1}{2}$.*

$\frac{4}{7} \times \frac{1}{3} \approx \frac{1}{6}$

You can also estimate the product of mixed numbers. Round each mixed number to the nearest whole number and then multiply.

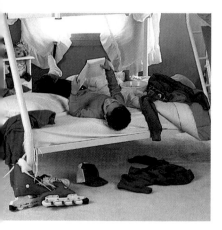

Example 5 *Connection*

Geometry Kal is getting new carpet for his bedroom. The room measures $10\frac{1}{2}$ feet by $9\frac{1}{4}$ feet. Estimate the area of the room.

You need to estimate $10\frac{1}{2} \times 9\frac{1}{4}$.

$11 \times 9 = 99$ *Round $10\frac{1}{2}$ to 11. Round $9\frac{1}{4}$ to 9.*

$10\frac{1}{2} \times 9\frac{1}{4} \approx 99$

The area of Kal's bedroom is about 99 square feet.

Checking for Understanding

Communicating Mathematics

Read and study the lesson to answer each question.

1. **Tell** how you would use rounding to estimate $\frac{1}{4} \times \frac{9}{10}$.

2. **Tell** how the model shows the use of compatible numbers to estimate $\frac{3}{5} \times 9$.

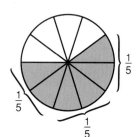

Guided Practice

Round each fraction to 0, $\frac{1}{2}$, or 1.

3. $\frac{2}{3}$ 4. $\frac{1}{8}$ 5. $\frac{7}{9}$ 6. $\frac{3}{8}$ 7. $\frac{1}{4}$

Round each mixed number to the nearest whole number.

8. $1\frac{7}{9}$ 9. $13\frac{1}{4}$ 10. $7\frac{2}{5}$ 11. $16\frac{4}{7}$ 12. $8\frac{1}{2}$

Estimate.

13. $\frac{3}{4} \times 7$ 14. $\frac{6}{13} \times 20$ 15. $4\frac{1}{3} \times 9\frac{7}{8}$ 16. $\frac{1}{6} \times \frac{3}{8}$

Exercises

Independent Practice

Round each fraction to 0, $\frac{1}{2}$, or 1.

17. $\frac{11}{12}$ 18. $\frac{5}{9}$ 19. $\frac{1}{10}$ 20. $\frac{3}{7}$ 21. $\frac{3}{4}$

Estimate.

22. $\frac{1}{3} \times 20$

23. $8\frac{2}{3} \times 4\frac{1}{6}$

24. $\frac{3}{7} \times \frac{9}{10}$

25. $\frac{7}{12} \times \frac{4}{15}$

26. $16\frac{5}{8} \times 5\frac{3}{4}$

27. $\frac{1}{7} \times 33$

28. $3\frac{1}{2} \times 4\frac{1}{6}$

29. $\frac{3}{5} \times \frac{1}{2}$

30. Estimate the product of $\frac{7}{9}$ and $2\frac{1}{3}$.

31. Estimate the product of $5\frac{7}{9}$ and $12\frac{3}{4}$.

Mixed Review 32. Estimate the quotient $410 \div 67$ using compatible numbers. *(Lesson 1-5)*

33. Find the quotient when 68.52 is divided by 12. *(Lesson 5-2)*

34. Express $\frac{120}{150}$ in simplest form. *(Lesson 6-5)*

35. Add 5 hours, 45 minutes, 40 seconds to 11 hours, 20 minutes, 5 seconds. *(Lesson 7-8)*

Problem Solving and Applications

36. **Geometry** Use estimation to find the area of the garden at the right.

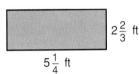

$2\frac{2}{3}$ ft

$5\frac{1}{4}$ ft

37. **Science** The average office worker breathes in about 400 cubic feet of air each day. Dry air is about $\frac{1}{5}$ oxygen. About how many cubic feet of oxygen does an office worker breathe in per day?

38. **Number Sense** Is $\frac{1}{4}$ of $\frac{2}{3}$ less than or greater than $\frac{2}{3}$? Explain.

39. **Critical Thinking** Which point shown on the number line could be the graph of the product of the numbers graphed at *C* and *D*?

40. **Statistics** The graph at the right shows how often people eat salad according to a 1991 poll. Suppose 1,000 people participated in the survey.

   a. About how many people eat salad once a week?

   b. About how many people eat salad less than once a month?

   c. About how many people eat salad several times a week?

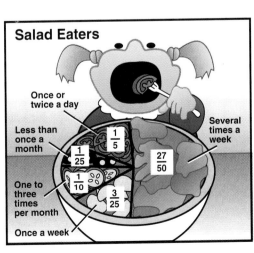

Salad Eaters

Once or twice a day

Several times a week

Less than once a month $\frac{1}{25}$

$\frac{1}{5}$

$\frac{27}{50}$

One to three times per month $\frac{1}{10}$

$\frac{3}{25}$

Once a week

8-2A Multiplying Fractions

A Preview of Lesson 8-2

Objective

Explore multiplying fractions.

Materials

geoboard
geobands

Did you know you can use a geoboard to multiply fractions? Work with a partner on the activity below to find $\frac{1}{4} \times \frac{1}{2}$ and see for yourself!

Try this!

- Place a geoboard and six geobands in front of you.
- Place one geoband in a straight line along the bottom of the geoboard to show fourths.

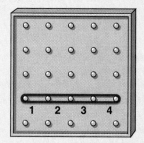

- Place a geoband perpendicular to the first geoband to make a corner. Stretch the geoband to show halves.

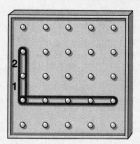

- Use geobands to connect the lines and make a rectangle.

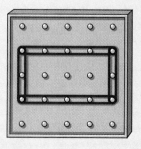

● Starting at the lower left-hand corner of your rectangle, count across one space to show $\frac{1}{4}$. Place a geoband on this peg.

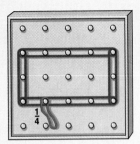

● Starting at the lower left-hand corner again, count up one space to show $\frac{1}{2}$. Place a geoband on this peg.

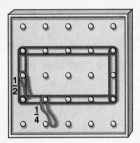

● Connect the two geobands to show a small rectangle.

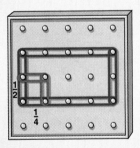

What do you think?

1. If the small rectangle has an area of 1 square unit, what is the area of the large rectangle?
2. What fraction of the area of the large rectangle is the area of your small rectangle?
3. What is $\frac{1}{4} \times \frac{1}{2}$? Explain your reasoning.

Extension

4. Use your geoboard to show $\frac{3}{4} \times \frac{1}{3}$.

8-2 Multiplying Fractions

Objective
Multiply fractions.

DID YOU KNOW

1 in 4 American teenagers volunteer to help charities and social service agencies. Some of the organizations where teenagers offer their services are hospitals, nursing homes, and the Children's Wish Foundation.

Suppose your classroom was collecting money to donate to different charities in your community. $\frac{3}{4}$ of the students donated money. $\frac{1}{3}$ of these students wanted the money to go to a charity that would provide toys for homeless children during the holiday season. About what part of the class was this?

You need to find $\frac{1}{3}$ of $\frac{3}{4}$. $\frac{1}{3}$ of $\frac{3}{4}$ means the same as $\frac{1}{3} \times \frac{3}{4}$.

A model can be used to help you solve the problem. The rectangle is separated into fourths. $\frac{3}{4}$ of the rectangle is colored red.

The rectangle is then separated into thirds. $\frac{1}{3}$ of the rectangle is colored blue.

The portion that overlaps the two colors shows the product.

You can find the same answer if you multiply the numerators and multiply the denominators.

$$\frac{1}{3} \times \frac{3}{4} = \frac{1 \cdot 3}{3 \cdot 4} \qquad \textit{Multiply the numerators.}$$
$$\textit{Multiply the denominators.}$$

$$= \frac{3}{12} \text{ or } \frac{1}{4} \qquad \textit{Simplify.}$$

About $\frac{1}{4}$ of the students who donated wanted the money to buy toys for homeless children during the holidays.

Multiplying Fractions	To multiply fractions, multiply the numerators. Then multiply the denominators. Simplify if necessary.

Calculator Hint

• • • • • • • • • •

You can use a
fraction calculator
to solve problems.
For example, to find
$\frac{1}{2} \times \frac{2}{3}$, enter 1 ⏥ 2
⏥ 2 ⏥ 3 ⏥
⏥ . Repeat ⏥
⏥ until the N/D →
n/d does not appear
on the screen.

Examples

1 Find $\frac{1}{2} \times \frac{2}{3}$.

$$\frac{1}{2} \times \frac{2}{3} = \frac{1 \cdot 2}{2 \cdot 3} \qquad \textit{Multiply the numerators.}$$
$$\textit{Multiply the denominators.}$$

$$= \frac{2}{6} \text{ or } \frac{1}{3} \qquad \textit{Simplify.}$$

2 Find $\frac{1}{4} \times 5$. $\qquad$ *Estimate: $\frac{1}{4}$ of 4 is 1.*

$$\frac{1}{4} \times 5 = \frac{1}{4} \times \frac{5}{1} \qquad \textit{Express 5 as an improper fraction, } \frac{5}{1}.$$

$$= \frac{1 \cdot 5}{4 \cdot 1} \qquad \textit{Multiply the numerators.}$$
$$\textit{Multiply the denominators.}$$

$$= \frac{5}{4} \text{ or } 1\frac{1}{4} \qquad \textit{Simplify.}$$

If the numerator of one fraction and the denominator of another
fraction have a common factor, you can simplify *before* you multiply.

LOOK BACK

You can review
GCF on page 199.

Example 3

Find $\frac{3}{8} \times \frac{4}{5}$.

Simplify before multiplying.

$$\frac{3}{8} \times \frac{4}{5} = \frac{3 \cdot \overset{1}{4}}{\underset{2}{8} \cdot 5} \qquad \textit{Divide both the numerator}$$
$$\textit{and the denominator by 4.}$$
$$\textit{4 is the GCF of 4 and 8.}$$

$$= \frac{3}{10}$$

Example 4 *Connection*

Algebra Solve $\frac{2}{5} \times \frac{15}{16} = n$.

$$\frac{2}{5} \times \frac{15}{16} = n$$

$$\frac{\overset{1}{2} \cdot \overset{3}{15}}{\underset{1}{5} \cdot \underset{8}{16}} = n \qquad \textit{Divide both the numerator and}$$
$$\textit{the denominator by 2 and then by 5.}$$
$$\textit{2 is the GCF of 2 and 16.}$$
$$\textit{5 is the GCF of 15 and 5.}$$

$$\frac{3}{8} = n$$

The solution is $\frac{3}{8}$.

Checking for Understanding

Communicating
Mathematics

Read and study the lesson to answer each question.

1. **Draw** a rectangle to show that $\frac{1}{2} \times \frac{1}{6} = \frac{1}{12}$.

2. **Tell** how the model below shows the product of $\frac{1}{4}$ and $\frac{2}{3}$.

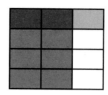

Guided Practice

Find each product. Write in simplest form.

3. $\frac{1}{4} \times \frac{1}{2}$

4. $\frac{2}{3} \times \frac{1}{5}$

5. $\frac{1}{3} \times \frac{3}{5}$

Solve each equation.

6. $k = 12 \times \frac{3}{4}$

7. $b = \frac{2}{3} \times \frac{3}{8}$

8. $\frac{7}{8} \times \frac{2}{9} = r$

Exercises

Independent
Practice

Find each product. Write in simplest form.

9. $\frac{1}{3} \times \frac{2}{5}$

10. $\frac{7}{8} \times \frac{1}{2}$

11. $\frac{5}{9} \times \frac{2}{3}$

12. $\frac{3}{4} \times 7$

13. $\frac{1}{9} \times \frac{3}{4}$

14. $\frac{2}{5} \times \frac{1}{4}$

15. $8 \times \frac{5}{6}$

16. $\frac{3}{4} \times \frac{8}{9}$

17. $\frac{1}{2} \times \frac{4}{5}$

18. Find the product of 6 and $\frac{2}{3}$.

19. Evaluate ab if $a = \frac{1}{3}$ and $b = \frac{6}{7}$.

20. Evaluate cd if $c = \frac{1}{4}$ and $d = \frac{6}{8}$.

21. Solve the equation three-fifths times two thirds equals x.

Solve each equation. Write the solution in simplest form.

22. $y = \frac{3}{8} \times \frac{4}{9}$

23. $u = \frac{7}{10} \times \frac{5}{7}$

24. $\frac{7}{12} \times \frac{4}{5} = x$

25. $z = \frac{9}{10} \times \frac{5}{12}$

26. $20 \times \frac{5}{6} = a$

27. $d = \frac{5}{6} \times \frac{9}{14}$

28. $j = \frac{14}{15} \times \frac{1}{2}$

29. $\frac{8}{9} \times \frac{3}{16} = x$

30. $\frac{4}{15} \times \frac{3}{8} = n$

31. Estimate $\frac{4}{9} \times \frac{1}{5}$. *(Lesson 8-1)*

32. **Science**   One cubic meter of carbon dioxide has a mass of 1.977 kilograms. Round this number to the nearest tenth. *(Lesson 3-5)*

33. **Algebra**   Find the solution in simplest form for the equation $x - \frac{5}{12} = \frac{5}{12}$. *(Lesson 7-4)*

34. Write $1\frac{5}{18}$ as a decimal using bar notation. *(Lesson 6-11)*

35. **Transportation**   If a truck has a $\frac{3}{4}$-ton load, how many pounds is it hauling? Hint: 1 ton = 2,000 pounds.

36. **Critical Thinking**   Find the product $\frac{1}{2} \times \frac{2}{3} \times \frac{3}{4} \times \frac{4}{5} \cdots \frac{99}{100}$.

37. **Science**   About $\frac{7}{10}$ of the human body is water. If a person weighs 160 pounds, about how many pounds are water?

38. **Journal Entry**   Write, in your own words, how to find the product of fractions.

39. **Mathematics and Advertising**   Read the following paragraphs.

Throughout the 20th century, advertising has expanded along with the economy. In 1950, American business spent $5.7 billion to advertise its goods and services; by 1960, that figure would double, and then almost double again by 1970. Between 1970 and 1988, as the Baby Boom generation entered the marketplace, advertising spending grew at a rapid rate, reaching $125 billion by 1989.

Most advertising dollars are spent by about 6,000 advertising agencies, who create the ads and buy the space or time on television, radio, or in magazines and newspapers and so on.

Try our newest sandwiches!

A network sells 24 minutes of a certain program time to advertisers. A company buys $\frac{5}{6}$ of the ad time. How many minutes did the company buy?

Multiplying Mixed Numbers

Objective
Multiply mixed numbers.

Brenna has a baseball album. She has enough baseball cards to fill $8\frac{1}{3}$ pages. $\frac{1}{2}$ of her cards are made by Topps®. If Brenna puts all of her Topps® cards in consecutive pages, how many pages will the cards fill?

You need to find $\frac{1}{2}$ of $8\frac{1}{3}$, or $\frac{1}{2} \times 8\frac{1}{3}$.

LOOKBACK

You can review improper fractions on page 207.

To multiply mixed numbers, express the mixed number as an improper fraction.

$$8\frac{1}{3} = \frac{(8 \times 3) + 1}{3} = \frac{25}{3}$$

Then multiply as with fractions.

$$\frac{1}{2} \times \frac{25}{3} = \frac{25}{6} \quad \textit{Multiply the numerators.}$$
$$\textit{Multiply the denominators.}$$
$$= 4\frac{1}{6} \quad \textit{Simplify.}$$

The Topps® cards will fill $4\frac{1}{6}$ pages.

You can also show this using models.

Mental Math Hint
• • • • • • • • • •
You can compute products like $8 \times 2\frac{1}{2}$ mentally. Think:
$8 \times 2 = 16$
$8 \times \frac{1}{2} = 4$
$16 + 4 = 20$

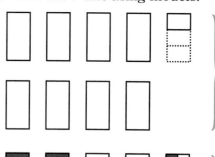

$8\frac{1}{3}$ pages

$\frac{1}{2}$ of $8\frac{1}{3}$ pages

$\frac{1}{2}$ of 8 is 4, and $\frac{1}{2}$ of $\frac{1}{3}$ is $\frac{1}{6}$. So, $\frac{1}{2} \times 8\frac{1}{3}$ is $4 + \frac{1}{6}$ or $4\frac{1}{6}$.

TEEN SCENE

Little League has expanded to include programs for 13- to 18-year-olds. By the 1980s, there were over 14,000 leagues, with 145,000 teams and about 2.5 million participants in over 30 countries with half of the teams and participants in the U.S.

Example 1

Find $10 \times 2\frac{3}{4}$.

Estimate: $10 \times 3 = 30$.

$$10 \times 2\frac{3}{4} = \frac{10}{1} \times \frac{11}{4}$$

Express 10 as an improper fraction.
Express $2\frac{3}{4}$ as an improper fraction.

$$= \frac{\overset{5}{\cancel{10}}}{1} \times \frac{11}{\underset{2}{\cancel{4}}}$$

Divide 10 and 4 by the GCF, 2.

$$= \frac{55}{2} \text{ or } 27\frac{1}{2}$$

Check against your estimate.

Example 2 *Connection*

Algebra   Solve $x = 3\frac{1}{8} \times 2\frac{4}{5}$.

Estimate: $3 \times 3 = 9$.

$$x = 3\frac{1}{8} \times 2\frac{4}{5}$$

$$x = \frac{25}{8} \times \frac{14}{5}$$

Express mixed numbers as improper fractions.

$$x = \frac{\overset{5}{\cancel{25}}}{\underset{4}{\cancel{8}}} \times \frac{\overset{7}{\cancel{14}}}{\underset{1}{\cancel{5}}}$$

Divide 25 and 5 by the GCF, 5.
Divide 14 and 8 by the GCF, 2.

Check against your estimate.

$$x = \frac{35}{4} \text{ or } 8\frac{3}{4}$$

The solution is $8\frac{3}{4}$.

Checking for Understanding

Communicating Mathematics

Read and study the lesson to answer each question.

1. **Tell** how to find the product in Example 1 without finding the GCF.
2. **Tell** how to multiply $\frac{3}{8}$ and $2\frac{1}{2}$.
3. **Write** the product represented by the following.

Guided Practice

Express each mixed number as an improper fraction.

4. $7\frac{1}{2}$ 5. $3\frac{1}{6}$ 6. $9\frac{2}{3}$ 7. $1\frac{7}{8}$ 8. $4\frac{5}{7}$

Find each product. Write in simplest form.

9. $\frac{1}{2} \times 2\frac{1}{3}$ 10. $4\frac{1}{3} \times \frac{1}{4}$ 11. $8 \times 2\frac{5}{6}$ 12. $1\frac{3}{7} \times 2\frac{4}{5}$

Solve each equation. Write each solution in simplest form.

13. $5\frac{1}{4} \times 10 = g$ 14. $12 \times 1\frac{1}{9} = x$ 15. $q = 7\frac{1}{2} \times \frac{3}{10}$

Lesson 8-3 Multiplying Mixed Numbers   **285**

Exercises

Express each mixed number as an improper fraction.

16. $4\frac{2}{3}$ 17. $2\frac{5}{8}$ 18. $5\frac{1}{10}$ 19. $9\frac{1}{3}$ 20. $3\frac{5}{7}$

Find each product. Write in simplest form.

21. $\frac{1}{3} \times 2\frac{1}{4}$ 22. $4\frac{1}{2} \times \frac{1}{5}$ 23. $3 \times 2\frac{3}{4}$ 24. $4\frac{1}{2} \times 1\frac{1}{3}$

25. $1\frac{4}{5} \times 4$ 26. $2\frac{3}{4} \times 2\frac{2}{3}$ 27. $\frac{3}{8} \times 1\frac{5}{9}$ 28. $10 \times 3\frac{1}{5}$

Solve each equation. Write each solution in simplest form.

29. $x = 2\frac{4}{7} \times 5\frac{1}{6}$ 30. $a = \frac{1}{2} \times 3\frac{5}{7}$ 31. $8\frac{1}{3} \times \frac{9}{10} = w$

32. $3\frac{1}{2} \times 5\frac{2}{3} = g$ 33. $f = \frac{5}{12} \times 1\frac{4}{5}$ 34. $2\frac{3}{16} \times \frac{3}{7} = i$

35. Find the product of two and one-half and four and three-fourths.

36. Multiply $\frac{5}{7}$ and $\frac{2}{3}$. *(Lesson 8-2)*

37. Express $\frac{100}{6}$ as a mixed number. *(Lesson 6-6)*

38. Multiply 6.45 and 0.04. *(Lesson 4-5)*

39. **Sewing** Ramon needs $\frac{7}{8}$ yard of blue felt to trim his costume. In the attic he found $\frac{2}{3}$ yard. How much more felt does he need? *(Lesson 7-5)*

40. **Statistics** Refer to Raul's earnings from Exercise 28 on page 131. Find the median of Raul's earnings. *(Lesson 2-7)*

41. **Sewing** Lauren bought $3\frac{3}{4}$ yards of material to make a dress. At $4 per yard, what was the cost of the material?

42. **Music** A dot following a note means that the note gets $1\frac{1}{2}$ times as many beats as the same note without a dot. If a whole note gets four beats, how many beats would a dotted whole note get?

43. **Geometry** Find the area of a room $5\frac{1}{2}$ yards long and $4\frac{2}{3}$ yards wide.

44. **Critical Thinking** Is $2\frac{2}{3} \times 4\frac{1}{2}$ more or less than 10? Explain how you know without actually multiplying.

45. **Measurement** Emily picked $4\frac{1}{2}$ buckets of strawberries at the farm market. Each bucket holds $2\frac{1}{2}$ quarts. How many quarts did Emily pick?

46. **Crafts** Tad's mother is knitting a sweater. She knits $3\frac{1}{2}$ rows each hour. How many rows will she knit in $4\frac{1}{2}$ hours?

8-4 Find a Pattern

Objective

Solve problems by finding and extending a pattern.

Ahmad eats half of a tuna fish sandwich at 1:00 P.M. At 3:00 P.M., he eats half of what was left of his sandwich from lunch. At 5:00 P.M. he eats half of what was left from his after school snack. If Ahmad continues eating at this rate for four more hours, how much of the sandwich has he eaten? How much is left? Will he ever eat the entire sandwich?

Explore   What do you know? Ahmad starts with one sandwich. He eats half at 1:00 P.M. He eats half of what was left at 3:00 P.M. He eats half of what was left of this at 5:00 P.M.

You need to find out how much of the sandwich he has eaten, how much is left, and if the entire sandwich will ever be eaten.

Plan   Draw a model of the sandwich. Shade how much of the whole sandwich is eaten. Record the time of day and the amount eaten and the amount not eaten.

Solve   Make a model for the amount eaten.

1:00 P.M.	3:00 P.M.	5:00 P.M.	7:00 P.M.	9:00 P.M.

Record the data in a chart.

What pattern do you see?

Time of day	1:00	3:00	5:00	7:00	9:00
Amount eaten	$\frac{1}{2}$	$\frac{3}{4}$	$\frac{7}{8}$	$\frac{15}{16}$	$\frac{31}{32}$
Amount left	$\frac{1}{2}$	$\frac{1}{4}$	$\frac{1}{8}$	$\frac{1}{16}$	$\frac{1}{32}$

Amount eaten	Sum	Amount left
$\frac{1}{2}$	$\frac{1}{2}$	$\frac{1}{2}$
$\frac{1}{2} + \frac{1}{4}$	$\frac{3}{4}$	$\frac{1}{4}$
$\frac{1}{2} + \frac{1}{4} + \frac{1}{8}$	$\frac{7}{8}$	$\frac{1}{8}$
$\frac{1}{2} + \frac{1}{4} + \frac{1}{8} + \frac{1}{16}$	$\frac{15}{16}$	$\frac{1}{16}$
$\frac{1}{2} + \frac{1}{4} + \frac{1}{8} + \frac{1}{16} + \frac{1}{32}$	$\frac{31}{32}$	$\frac{1}{32}$

After eight hours, Ahmad had eaten $\frac{31}{32}$ of the tuna fish sandwich and $\frac{1}{32}$ was left. If the pattern continues, Ahmad will not finish the sandwich, but the amount left will be so small that it will be like he did eat it all.

Examine If you were to continue the pattern for the next two-hour interval, the amount left would be $\frac{1}{2} \times \frac{1}{32}$ or $\frac{1}{64}$.

Checking for Understanding

Read and study the lesson to answer the question.

Communicating Mathematics

1. **Show** how multiplication can be used to show one half of one fourth of a sandwich.

Guided Practice

Solve by finding a pattern.

2. Grace and Victoria were hired at the same company in 1992. Victoria earned $13,000 per year and Grace earned $17,000. Each year Victoria received a $1,500 raise, and Grace received a $1,000 raise.
   a. In what year will they earn the same amount of money?
   b. What will be their annual salary in that year?

3. Kevin's age is $\frac{1}{4}$ of his aunt's age. If Kevin is 10 years old, how old will both he and his aunt be when Kevin's age is $\frac{1}{2}$ his aunt's age?

4. Eve saves $1 the first week, $2 the second, $4 the third, $7 the fourth, and $11 the fifth week.
   a. If this pattern continues, how much money would Eve save during the tenth week?
   b. What would be her total savings during the ten weeks?

Problem Solving

Practice

Solve. Use any strategy.

5. Libby is choosing an outfit to wear for a date. She has black, blue, and stonewashed jeans, and red, white, and purple sweaters. How many different combinations of jeans and a sweater are possible?

Strategies

• • • • • • • •

Look for a pattern.

Solve a simpler problem.

Act it out.

Guess and check.

Draw a diagram.

Make a chart.

Work backward.

6. What is the next number in the pattern?
   7, 13, 19, 25 ...

7. Find the area of the figure at the right.

8. What is the next fraction in the pattern?
   $\frac{1}{8}$, $\frac{1}{24}$, $\frac{1}{72}$ ...

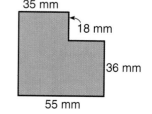

9. At Kibourne Middle School, the bell rings at 7:55, 8:40, 9:25, and 10:10 each morning. If this pattern continues, when would the next three bells ring?

8-5 Sequences

Objective
Recognize and extend sequences.

Words to Learn
sequence

When am I ever going to use this?

Record Industry
"Single specialists" make sure single records are in the stores once they are added to the play lists at radio stations. They also work with entertainers when they go on tour and get involved with merchandising.

These positions require organizational and problem-solving skills.

For more information contact:
Sales and Marketing Executives
Statler Office Tower #458
Cleveland, OH 44115

What type of person is a music composer? The image that probably comes to mind is a white-haired gentleman. But not anymore! The Chicago Symphony Orchestra and its *Young Composers Project* have given teenagers and children from the Chicago area the opportunity to write and compose music. In 1991, the Chicago Symphony presented the works of Rich Carte, who was 7 years old, John Orfe, 14, and Jeff Letterly, 18. These three composers based their works on melodies written by students from Howland School of the Arts.

When composers write music, they need to indicate how long each note should be held. Look at the notes below.

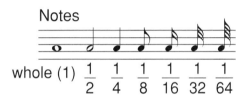

Notes

whole (1) $\frac{1}{2}$ $\frac{1}{4}$ $\frac{1}{8}$ $\frac{1}{16}$ $\frac{1}{32}$ $\frac{1}{64}$

The numbers $1, \frac{1}{2}, \frac{1}{4}, \frac{1}{8}, \frac{1}{16}, \frac{1}{32}$, and $\frac{1}{64}$ form a **sequence.** A sequence of numbers is a list in a specific order.

Notice that each number in the sequence, or pattern, is multiplied by $\frac{1}{2}$ to get the next number.

$$1, \quad \frac{1}{2}, \quad \frac{1}{4}, \quad \frac{1}{8}, \quad \frac{1}{16}, \quad \frac{1}{32}, \quad \frac{1}{64}, \quad \cdots$$

$\times\frac{1}{2}$ $\times\frac{1}{2}$ $\times\frac{1}{2}$ $\times\frac{1}{2}$ $\times\frac{1}{2}$ $\times\frac{1}{2}$

Examples

Find the next number in each sequence.

1   2,   6,   18,   54
$\times 3$ $\times 3$ $\times 3$

Each number in the sequence is multiplied by 3. The next number is 54×3, or 162.

2   125,   25,   5,   1
$\times\frac{1}{5}$ $\times\frac{1}{5}$ $\times\frac{1}{5}$

Each number in the sequence is multiplied by $\frac{1}{5}$. The next number is $1 \times \frac{1}{5}$, or $\frac{1}{5}$.

Another type of sequence is one where the same number is added or subtracted from each number in the sequence.

<table>
<tr><td>

Problem Solving Hint

● ● ● ● ● ● ● ● ● ●

You may want to use the find a pattern strategy to help you with these sequences.

</td><td>

Examples

Find the next number in each sequence.

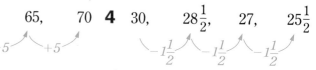

3 55, 60, 65, 70   **4** 30, $28\frac{1}{2}$, 27, $25\frac{1}{2}$

5 is added to each number. The next number in the sequence is 70 + 5, or 75.

$1\frac{1}{2}$ is subtracted from each number. The next number in the sequence is $25\frac{1}{2} - 1\frac{1}{2}$, or 24.

</td></tr>
</table>

Checking for Understanding

Communicating Mathematics

Read and study the lesson to answer each question.

1. **Tell** the sixth number in the sequence in Example 1.
2. **Write** how the numbers are related in the sequence 9, 3, 1, $\frac{1}{3}$.

Guided Practice

Find the next two numbers in each sequence.

3. 1, 3, 5, 7
4. 50, 46, 42, 38
5. 6, 12, 24, 48
6. 12.5, 12, 11.5, 11
7. 18, 6, 2, $\frac{2}{3}$
8. $\frac{1}{2}$, 2, 8, 32

Exercises

Independent Practice

Find the next two numbers in each sequence.

9. 25, 50, 75, 100
10. 5, 15, 45, 135
11. 3, 30, 300, 3,000
12. 12, 6, 3, $1\frac{1}{2}$
13. 10, 50, 250, 1,250
14. 270, 27, 2.7, 0.27
15. 48, 40, 32, 24
16. 3.5, 7, 10.5, 14
17. 64, 16, 4, 1
18. 5, 7.5, 10, 12.5
19. 1, $\frac{3}{4}$, $\frac{9}{16}$, $\frac{27}{64}$
20. 5, $15\frac{1}{2}$, 26, $36\frac{1}{2}$

21. The first number of a sequence is 1. Each successive number is found by multiplying the preceding number by $\frac{3}{8}$. Find the first four numbers of this sequence.

22. **Statistics**   Find the mean and the median for the following set of low temperatures for a city: 45, 41, 20, 42, 44, 40, 40. *(Lesson 2-8)*

23. Find the GCF of 120 and 150. *(Lesson 6-4)*

24. Find $1\frac{5}{7} \times 2\frac{5}{8}$ in simplest form. *(Lesson 8-3)*

Problem Solving and Applications

25. **Science**   If gravity is the only force acting on a falling object, its speed will increase 32.16 feet per second each second.

    a. Copy and complete this table.

    b. Explain what happens to the speed of the ball as each second passes.

Seconds after ball is dropped	1	2	3	4
Speed of ball (ft per second)				

    c. If 60 mph = 88 ft/s, find the speed of the ball in miles per hour (mph) after 4 seconds.

26. **Critical Thinking**   The square at the right represents 1.
    a. Write the first ten numbers of the sequence represented by the model.
    b. What do you think the sum of the first ten numbers is close to? Do not actually add.

27. **Geometry**   Draw the next shape in the sequence.

, ,

28. **Journal Entry**   Make up a sequence and describe the patterns.

Mid-Chapter Review

Estimate. *(Lesson 8-1)*

1. $1\frac{6}{7} \times 21$ 2. $3\frac{1}{3} \times 1\frac{7}{8}$ 3. $5\frac{2}{3} \times 1\frac{1}{4}$ 4. $\frac{6}{10} \times \frac{1}{3}$

Find each product. Write in simplest form. *(Lessons 8-2 and 8-3)*

5. $\frac{1}{2} \times \frac{3}{4}$ 6. $\frac{2}{3} \times 1\frac{5}{7}$ 7. $3\frac{4}{5} \times \frac{2}{3}$ 8. $6\frac{1}{2} \times 4\frac{2}{3}$

Solve by finding a pattern. *(Lesson 8-4)*

9. Marlene's mother earned a total of $140 over 3 days. Each day she earned twice as much as she did the day before. How much did she earn each day?

Find the next two numbers in each sequence. *(Lesson 8-5)*

10. 3, 7, 11, 15   11. 124, 62, 31, 15.5

8-6A Dividing Fractions

A Preview of Lesson 8-6

Objective

Explore dividing fractions.

Materials

geoboard
geobands

In Lesson 8-2A, you learned how to use a geoboard to multiply fractions. You can use your geoboard to divide fractions too.

Try this!

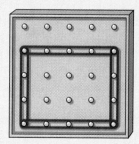

- Look at the geoboard at the right. This rectangle represents 1.

- Use your geobands to make a rectangle to match the one above.
- Use geobands to show thirds.

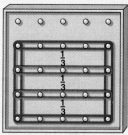

- How many $\frac{1}{3}$'s are there in 1?
- Now separate your rectangle into sixths using geobands.

- How many $\frac{1}{6}$'s are there in 1? How many $\frac{1}{6}$'s are there in $\frac{1}{3}$?

What do you think?

1. To find out how many $\frac{1}{6}$'s are in $\frac{1}{3}$, we write $\frac{1}{3} \div \frac{1}{6}$. What is $\frac{1}{3} \div \frac{1}{6}$?

Dividing Fractions

Objective

Divide fractions.

Words to Learn

reciprocal

Mr. Wagner is going to make a broccoli casserole. The recipe calls for 3 cups of cheddar cheese. Mr. Wagner stopped at the convenience store near his apartment to buy cheese. The only cheddar cheese package they sell contains $\frac{1}{2}$ cup of cheese. How many packages does he need to buy to make the casserole?

You need to find the number of $\frac{1}{2}$ cups in 3 cups. Look at the drawing below that shows $3 \div \frac{1}{2}$.

The dashed lines shows that each cup holds two $\frac{1}{2}$ cup packages of cheese. Three cups hold 3 times 2 or 6 packages of cheese.

$$3 \div \frac{1}{2} = 6 \quad \rightarrow \quad 3 \times 2 = 6$$

The numbers $\frac{1}{2}$ and 2 have a special relationship. The product of $\frac{1}{2}$ and 2 is 1.

$$\frac{1}{2} \times 2 = 1$$

Any two numbers whose product is 1 are called **reciprocals.**

Mental Math Hint

• • • • • • • • • •

The reciprocal of a fraction is found by "inverting" the fraction. That is, the numerator and denominator are interchanged.

Examples

Find the reciprocal of each number.

1 $\frac{4}{15}$

Since $\frac{4}{15} \times \frac{15}{4} = 1$, the reciprocal of $\frac{4}{15}$ is $\frac{15}{4}$.

2 5

Since $5 \times \frac{1}{5} = 1$, the reciprocal of 5 is $\frac{1}{5}$.

You can use reciprocals to divide fractions.

Dividing Fractions	To divide by a fraction, multiply by its reciprocal.

Example 3

Find $\dfrac{5}{9} \div \dfrac{2}{3}$.

$\dfrac{5}{9} \div \dfrac{2}{3} = \dfrac{5}{9} \times \dfrac{3}{2}$ *Multiply by the reciprocal of $\dfrac{2}{3}$.*

$= \dfrac{5}{\underset{3}{9}} \times \dfrac{\overset{1}{3}}{2}$ *Divide 3 and 9 by the GCF, 3.*

$= \dfrac{5}{3} \times \dfrac{1}{2}$ *Multiply the numerators.*
Multiply the denominators.

$= \dfrac{5}{6}$

LOOKBACK

You can review inverse operations on page 179.

Remember that division is related to multiplication. Sometimes it is helpful to write a multiplication sentence first and then write the related division sentence.

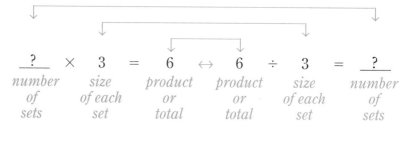

$\underline{\ ?\ }$	×	3	=	6	↔	6	÷	3	=	$\underline{\ ?\ }$
number of sets		*size of each set*		*product or total*		*product or total*		*size of each set*		*number of sets*

Example 4   *Problem Solving*

Cooking   Ms. Newsome made a cheesecake. If each serving is $\dfrac{1}{12}$ of a cheesecake, how many servings does she have if $\dfrac{3}{4}$ of the cheesecake is left?

Explore   You know the size of each set, $\dfrac{1}{12}$, and the total amount, $\dfrac{3}{4}$. You need to find the number of sets.

Plan

$\underline{\ ?\ }$	×	$\dfrac{1}{12}$	=	$\dfrac{3}{4}$	↔	$\dfrac{3}{4}$	÷	$\dfrac{1}{12}$	=	$\underline{\ ?\ }$
number of servings		*size of each serving*		*total*		*total*		*size of each serving*		*number of servings*

Solve $\dfrac{3}{4} \div \dfrac{1}{12} = \dfrac{3}{4} \times \dfrac{\overset{3}{12}}{\underset{1}{1}}$ *Multiply by the reciprocal of $\dfrac{1}{12}$.*

Divide by the GCF of 4 and 12, 4.

$$= 9$$

Ms. Newsome has 9 servings of cheesecake left.

Example 5 *Problem Solving*

Cooking   Mr. Heustess made a pecan pie for his family. He had $\dfrac{3}{4}$ of it left. He decided to divide it into 9 same-sized parts to serve to dinner guests. What part of a whole pie did each dinner guest get?

$$
\begin{array}{ccccccccc}
9 & \times & \underline{\ ?\ } & = & \dfrac{3}{4} & \leftrightarrow & \dfrac{3}{4} & \div & 9 & = & \underline{\ ?\ }
\end{array}
$$

*number   size   product   product   number   size
of   of   or   or   of   of
servings   each   total   total   servings   each
            serving   serving*

$$\dfrac{3}{4} \div 9 = \dfrac{\overset{1}{3}}{4} \times \dfrac{1}{\underset{3}{9}}$$

$$= \dfrac{1}{12}$$

Each serving is $\dfrac{1}{12}$ of a whole pie.

Checking for Understanding

Communicating Mathematics

Read and study the lesson to answer each question.

1. **Tell** how you would use a reciprocal to find $\dfrac{3}{4} \div \dfrac{1}{2}$.

2. **Tell** how this model shows $4 \div \dfrac{1}{3} = 12$.

Guided Practice

Name the reciprocal of each number.

3. $\dfrac{1}{3}$ 4. $\dfrac{5}{7}$ 5. $\dfrac{3}{8}$ 6. 6 7. $\dfrac{5}{12}$ 8. $\dfrac{9}{16}$

Find each quotient. Write in simplest form.

9. $\dfrac{1}{6} \div \dfrac{1}{5}$ 10. $\dfrac{3}{4} \div 6$ 11. $\dfrac{7}{9} \div \dfrac{14}{21}$ 12. $12 \div \dfrac{3}{4}$

Solve each equation. Write the solution in simplest form.

13. $\dfrac{1}{8} \div \dfrac{1}{2} = p$ 14. $\dfrac{5}{6} \div \dfrac{3}{4} = k$ 15. $s = \dfrac{6}{7} \div 8$

Exercises

Independent Practice

Name the reciprocal of each number.

16. $\dfrac{2}{3}$ 17. $\dfrac{1}{9}$ 18. $\dfrac{7}{8}$ 19. $\dfrac{4}{5}$ 20. 9 21. $\dfrac{3}{7}$

22. 5 23. $\dfrac{1}{10}$ 24. 3 25. $\dfrac{7}{16}$ 26. $\dfrac{5}{24}$ 27. $\dfrac{1}{16}$

Find each quotient. Write in simplest form.

28. $\dfrac{1}{3} \div \dfrac{1}{2}$ 29. $\dfrac{3}{5} \div \dfrac{1}{3}$ 30. $\dfrac{2}{3} \div \dfrac{3}{4}$ 31. $\dfrac{1}{8} \div \dfrac{1}{4}$

32. $\dfrac{3}{5} \div \dfrac{3}{4}$ 33. $\dfrac{1}{5} \div \dfrac{3}{10}$ 34. $4 \div \dfrac{2}{3}$ 35. $\dfrac{7}{9} \div \dfrac{2}{3}$

36. $\dfrac{5}{8} \div 2$ 37. $\dfrac{5}{6} \div \dfrac{5}{8}$ 38. $\dfrac{3}{10} \div \dfrac{2}{5}$ 39. $\dfrac{8}{9} \div 4$

40. Divide $\dfrac{5}{8}$ by 2.

41. How many $\dfrac{1}{2}$'s are in $\dfrac{1}{3}$?

Solve each equation. Write the solution in simplest form.

42. $\dfrac{2}{3} \div \dfrac{8}{9} = z$ 43. $d = \dfrac{15}{16} \div \dfrac{5}{8}$ 44. $g = \dfrac{5}{24} \div \dfrac{1}{16}$

Evaluate each expression.

45. $a \div b$, if $a = \dfrac{2}{3}$ and $b = \dfrac{1}{4}$.

46. $m \div p$, if $m = \dfrac{5}{12}$ and $p = 15$.

Mixed Review

47. **Patterns**   Find the next two numbers in the sequence 160, 80, 40, 20. *(Lesson 8-5)*

48. **Cooking**   Toshi needs $2\dfrac{1}{4}$ cups of flour for making cookies, $1\dfrac{2}{3}$ cups for almond bars, and $3\dfrac{1}{2}$ cups for cinnamon rolls. How much flour does she need in all? *(Lesson 7-6)*

49. Find $586.1 \div 0.58$ to the nearest tenth. *(Lesson 5-4)*

50. Find 6.2×50 mentally using the distributive property. *(Lesson 4-4)*

51. Estimate $6.291 + 234.38$. *(Lesson 3-6)*

Problem Solving and Applications

52. **Parking**   Mrs. Gibson parked her car at a 3-hour parking meter. If each half-hour of parking costs 25¢, how many quarters does she need for 3 hours of parking?

53. **Critical Thinking**   If $\dfrac{14}{15} \div \dfrac{\blacksquare}{9} = \dfrac{3}{10}$, find $\blacksquare$.

8-7 Dividing Mixed Numbers

Objective
Divide mixed numbers.

Teri is preparing a speech about the differences and similarities among people from different countries. She wants to include quotes from people of various nationalities. She is going to place the quotes in a column on a $17\frac{1}{2}$-inch long poster board. Each quote will be on an index card that is $2\frac{3}{4}$ inches long. How many quotes can Teri put on the poster board?

Calculator Hint
• • • • • • • • • • •

You can express fractions as decimals then divide. For example express $17\frac{1}{2}$ as 17.5 and $2\frac{3}{4}$ as 2.75. Then enter 17.5 $\div$ 2.75 $=$ to solve.

You need to find out how many $2\frac{3}{4}$ inches are in $17\frac{1}{2}$ inches.

$17\frac{1}{2} \div 2\frac{3}{4} = \frac{35}{2} \div \frac{11}{4}$ *Express each mixed number as an improper fraction.*

$= \frac{35}{2} \times \frac{4}{11}$ *Multiply by the reciprocal.*

$= \frac{35}{\underset{1}{2}} \times \frac{\overset{2}{4}}{11}$ *Divide 2 and 4 by the GCF, 2.*

$= \frac{70}{11}$ or $6\frac{4}{11}$ *Simplify.*

Teri can put 6 quotes on the poster board.

Example 1

Find $3\frac{3}{4} \div \frac{5}{6}$. *Estimate: $4 \div 1 = 4$*

$3\frac{3}{4} \div \frac{5}{6} = \frac{15}{4} \div \frac{5}{6}$ *Express the mixed number as an improper fraction.*

$= \frac{15}{4} \times \frac{6}{5}$ *Multiply by the reciprocal.*

$= \frac{\overset{3}{15}}{\underset{2}{4}} \times \frac{\overset{3}{6}}{\underset{1}{5}}$ *Divide 15 and 5 by the GCF, 5.*
   Divide 6 and 4 by the GCF, 2.

$= \frac{9}{2}$ or $4\frac{1}{2}$ *Simplify.*

$4\frac{1}{2}$ is close to the estimate. So, the answer is reasonable.

Example 2 *Connection*

Algebra   Solve $x = 14 \div 2\frac{2}{3}$.   *Estimate: $15 \div 3 = 5$*

$x = \dfrac{14}{1} \div \dfrac{8}{3}$ *Express each number as an improper fraction.*

$x = \dfrac{\overset{7}{14}}{1} \times \dfrac{3}{\underset{4}{8}}$ *Multiply by the reciprocal. Divide 14 and 8 by the GCF, 2.*

$x = \dfrac{21}{4}$ or $5\frac{1}{4}$ *Simplify.*

The solution $5\frac{1}{4}$ is close to the estimate. So, it is a reasonable solution.

Checking for Understanding

Communicating Mathematics

Read and study the lesson to answer each question.

1. **Tell** why, in the lesson introduction, Teri can put only 6 quotes on the poster board when the answer was greater than 6.

2. **Write,** in your own words, the steps to follow when finding $5\frac{5}{6} \div 8\frac{1}{8}$.

Guided Practice

Write each mixed number as an improper fraction.

3. $3\frac{1}{2}$ 4. $6\frac{2}{3}$ 5. $10\frac{2}{3}$ 6. $5\frac{5}{9}$ 7. $1\frac{7}{18}$

Find each quotient. Write in simplest form.

8. $3 \div 2\frac{1}{2}$ 9. $4\frac{2}{3} \div 5\frac{1}{3}$ 10. $\frac{7}{8} \div 3\frac{3}{4}$

11. $7\frac{1}{3} \div 6$ 12. $5\frac{1}{4} \div 3\frac{1}{2}$ 13. $8 \div 3\frac{1}{2}$

14. If you divide $4\frac{2}{5}$ by $3\frac{1}{4}$, what is the quotient?

Exercises

Independent Practice

Find each quotient. Write in simplest form.

15. $2 \div 3\frac{1}{2}$ 16. $3\frac{1}{4} \div \frac{1}{2}$ 17. $2\frac{3}{4} \div 1\frac{3}{8}$ 18. $6\frac{1}{2} \div 4\frac{1}{6}$

19. $2\frac{1}{7} \div 10$ 20. $2\frac{4}{5} \div 5\frac{3}{5}$ 21. $\frac{15}{16} \div 3\frac{1}{8}$ 22. $1\frac{5}{9} \div 2\frac{1}{3}$

23. Divide 12 by $7\frac{1}{3}$.

Solve each equation. Write the solution in simplest form.

24. $x = 5\frac{3}{4} \div 2\frac{1}{2}$ 25. $3\frac{1}{2} \div 1\frac{1}{2} = w$

26. $10\frac{1}{2} \div \frac{7}{8} = t$ 27. $m = 2\frac{1}{12} \div 1\frac{1}{4}$

Evaluate each expression.

28. $r \div s$, if $r = 2\frac{3}{16}$ and $s = 3\frac{1}{2}$.

29. $t \div u$, if $t = 2\frac{1}{20}$ and $u = \frac{9}{10}$.

Mixed Review

30. Find $\frac{3}{7} \div \frac{3}{5}$ in simplest form. *(Lesson 8-6)*

31. Find $\frac{15}{16} \times \frac{3}{5}$ in simplest form. *(Lesson 8-2)*

32. Find $6\frac{2}{5} - 4\frac{3}{5}$ in simplest form. *(Lesson 7-7)*

33. **Measurement** How many grams are there in 0.05 kilograms? *(Lesson 5-6)*

Problem Solving and Applications

Refer to the graph at the right for Exercises 34 through 37.

34. If there are about $4\frac{3}{4}$ million students in California, what would be the cost per student for each day the school year is lengthened?

35. If there are about $\frac{6}{10}$ million students in South Carolina, how much would it cost per student to lengthen the school year by one day?

36. How many times as much per day would it cost in Texas to lengthen the school year than it would in Nebraska?

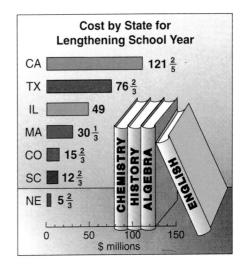

37. How many times as much per day would it cost in Massachusetts to lengthen the school year than it would in Nebraska?

38. **Critical Thinking** Tell whether $\frac{8}{10} \div \frac{2}{3}$ is greater than or less than $\frac{8}{10} \div \frac{3}{4}$ without solving. Explain your reasoning.

39. **Work** Ms. Peterson is setting up a work schedule for her employees. The store is open $7\frac{1}{2}$ hours each day. If employees work $1\frac{1}{2}$-hour shifts, how many shifts will Ms. Peterson need to fill?

8-8 Changing Units in the Customary System

Objective

Change units within the customary system.

Words to Learn

fluid ounce
cup
pint
quart
gallon
ounce
pound
ton

To make lemonade from a 12-ounce can of concentrate, Michael needs to mix the contents of the can with 36 ounces of cold water. How many cups of lemonade can be served from this mix?

Add to find the total amount of lemonade.

$$\begin{array}{r} 12 \text{ fluid ounces} \\ + 36 \text{ fluid ounces} \\ \hline 48 \text{ fluid ounces} \end{array}$$

To find out how many cups 48 fluid ounces makes, change fluid ounces to cups. It takes 8 fluid ounces to make 1 cup. You need to find out how many sets of 8 fluid ounces there are in 48 fluid ounces. So, divide:

$$48 \div 8 = 6$$

A 12-ounce can of concentrate makes 6 cups of lemonade.

DID YOU KNOW

The lemon is a type of berry called a *hesperidium*.

The most commonly used customary units of capacity are the **fluid ounce, cup, pint, quart,** and **gallon.**

1 cup (c) = 8 fluid ounces (fl oz)
1 pint (pt) = 2 cups
1 quart (qt) = 2 pints
1 gallon (gal) = 4 quarts

The most commonly used customary units of weight are **ounce, pound,** and **ton.**

1 pound (lb) = 16 ounces (oz)
1 ton (T) = 2,000 pounds

To change customary units of capacity and weight,

1. Decide if you are changing from smaller to larger units or from larger to smaller units.

2. Divide to change from smaller to larger units. Multiply to change from larger to smaller units.

Example 1

4 lb = ___ oz   *Think: Each pound equals 16 ounces.*

$4 \times 16 = 64$ *Multiply to change from a larger unit (lb) to a smaller unit (oz).*

4 pounds equals 64 ounces.

2 12 qt = ___ gal   *Think: It takes 4 quarts to make 1 gallon.*

$12 \div 4 = 3$ *Divide to change from a smaller unit (qt) to a larger unit (gal).*

12 quarts equals 3 gallons.

3 56,000 oz = ___ T   *Think: There are 16 ounces in 1 pound. There are 2,000 pounds in 1 ton. You need to divide twice.*

56000 ⌹÷⌡ 16 ⌹=⌡ 3500   *Divide to change from ounces to pounds.*

3500 ⌹÷⌡ 2000 ⌹=⌡ 1.75   *Divide to change from pounds to tons.*

56,000 ounces equals 1.75 or $1\frac{3}{4}$ tons.

Example 4 *Problem Solving*

Food A box of cereal weighs 14 ounces. How many pounds does a case of 24 boxes weigh?

24 ⌹×⌡ 14 ⌹=⌡ 336   *First, find the total number of ounces.*

336 ⌹÷⌡ 16 ⌹=⌡ 21   *Think: Each 16 ounces equals 1 pound. Divide to change from ounces to pounds.*

A case of cereal weighs 21 pounds.

Checking for Understanding

Communicating Mathematics   Read and study the lesson to answer each question.

1. **Tell** which unit is larger, quart or pint.
2. **Write** how you change five quarts to pints.
3. **Tell** which unit is lighter, ounce or ton.
4. **Write** how you change 32 ounces to pounds.

Guided Practice   Complete.

5. 48 oz = ___ lb

6. 6 qt = ___ pt

7. $2\frac{1}{2}$ T = ___ lb

8. 22 qt = ___ gal

9. 3 T = ___ oz

10. 16 c = ___ gal

11. How many pounds are in $2\frac{3}{5}$ tons?

Exercises

Independent Practice

Complete.

12. $3 \text{ c} = \_\_ \text{ fl oz}$

13. $10 \text{ lb} = \_\_ \text{ oz}$

14. $6,000 \text{ lb} = \_\_ \text{ T}$

15. $6 \text{ pt} = \_\_ \text{ qt}$

16. $24 \text{ fl oz} = \_\_ \text{ c}$

17. $5 \text{ T} = \_\_ \text{ lb}$

18. $12 \text{ c} = \_\_ \text{ fl oz}$

19. $80 \text{ oz} = \_\_ \text{ lb}$

20. $7 \text{ gal} = \_\_ \text{ qt}$

21. $20 \text{ fl oz} = \_\_ \text{ c}$

22. $1\frac{1}{2} \text{ lb} = \_\_ \text{ oz}$

23. $3\frac{1}{2} \text{ pt} = \_\_ \text{ c}$

24. $500 \text{ lb} = \_\_ \text{ T}$

25. $31 \text{ qt} = \_\_ \text{ gal}$

26. $3.2 \text{ T} = \_\_ \text{ lb}$

27. How many fluid ounces are in 3 quarts?

28. Change 5 gallons to cups.

Mixed Review

29. Divide $5\frac{4}{5}$ by $2\frac{3}{10}$ and write the answer in simplest form. *(Lesson 8-7)*

30. **Time** If Mr. Barriga works 8 hours 30 minutes, eats lunch for 1 hour 15 minutes, and spends a total of 1 hour 20 minutes getting to and from work, how much time is spent away from home? *(Lesson 7-8)*

31. **Algebra** Solve the equation $6.5m = 2,600$. *(Lesson 5-7)*

Problem Solving and Applications

32. a. **Catering** About how many cups of banana punch will the recipe make?

    · Banana Punch ·

    6 c water
    4 c sugar
    4 c bananas, mashed
    ¼ c lemon juice
    46 fl oz pineapple juice
    6 fl oz orange juice
    4 qt ginger ale

    Mix water and sugar. Boil for 6 mins. Let cool. Add remaining ingredients except ginger ale. Freeze. Take out about 2 hrs. before serving. Add ginger ale.

    b. If you double the recipe, how many quarts of water will you need?

    c. How many cups of orange juice will be needed if you quadruple the recipe?

33. **Lawn Care** Mrs. Senter mixes oil and gasoline for her lawn mower's 2-cycle engine. If $\frac{1}{24}$ of a 3-gallon mixture is oil, how many fluid ounces of the mixture is oil?

34. **Safety** A civil engineer is planning to build a bridge. The engineer figures the bridge can hold a maximum of 20 cars. If the average weight of a car is 3,000 pounds and she builds the bridge twice as strong as necessary, how many tons should the bridge be built to hold?

DATA SEARCH

35. **Data Search** Refer to pages 272 and 273. On average, how many pints of gasoline are consumed per passenger vehicle in the United States?

8-8B Measurement

A Follow-Up of Lesson 8-8

Objective

Compare weights.

Materials

measuring cups
containers
water
paper
pencil

Have you ever dreamed of visiting another plant or the moon? You would find many things to be different on another planet, even the weight of water or any other item, including yourself!

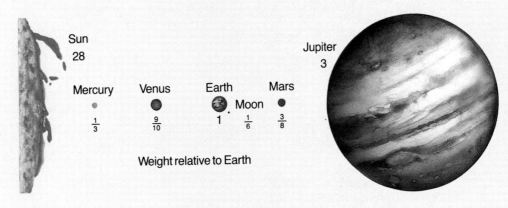

Sun
28

Jupiter
3

Mercury   Venus   Earth   Mars
 $\frac{1}{3}$ $\frac{9}{10}$ Moon $\frac{3}{8}$
                             1      $\frac{1}{6}$

Weight relative to Earth

Try this!

Work with a partner. Choose a work station.

- Record the name of the planet found at your work station. Then record the planet's weight relative to Earth.

- Fill a measuring cup with a cup of water to represent the weight of water on Earth. Pour this amount into the container labeled *Earth*.

- Use the planet's weight relative to Earth to fill as many measuring cups with water as needed to represent the weight of water on this planet. For example, Jupiter's weight is 3 times that of Earth. Empty three cups of water into the container labeled *Jupiter*.

What do you think?

1. Which container weighs more? Why?
2. Move through each work station. Repeat the steps. Does the water on this planet weigh more or less than water on Earth? More or less than on Jupiter? How do you know?
3. How much would a 12-pound baby weigh on the moon?
4. How much would you weigh on the moon?

Study Guide and Review

Communicating Mathematics

Choose the best response for each question.

1. Estimating $\frac{2}{5} \times 14$ by using $\frac{2}{5} \times 15 = 6$ is an example of using

   a. compatible numbers.   b. front-end estimation.   c. rounding.

2. When the word "of" is used in mathematics, you should usually

   a. add.   b. multiply.   c. divide.

3. To multiply $\frac{7}{8}$ by $\frac{3}{8}$,

   a. find only $7 \cdot 3$ since 8 is a common denominator.

   b. find both $7 \cdot 3$ and $8 \cdot 8$.

   c. it depends upon how the problem is worded.

4. When multiplying fractions, you can simplify

   a. before multiplying.   b. after multiplying.   c. before or after multiplying.

5. A sequence of numbers is

   a. a list in a specific order.   b. a set of musical notes.

6. To divide $\frac{2}{3}$ by $\frac{5}{7}$,

   a. find $\frac{3}{2} \times \frac{7}{5}$.   b. find $\frac{3}{2} \times \frac{5}{7}$.   c. find $\frac{2}{3} \times \frac{7}{5}$.

7. The numbers $\frac{8}{9}$ and $\frac{9}{8}$ are called

   a. reciprocals.   b. opposites.   c. improper fractions.

8. Under what circumstances can you multiply two numbers and get a product which is less than either of the two original numbers? Explain.

Skills and Concepts

Objectives and Examples	Review Exercises
Upon completing this chapter, you should be able to:	*Use these exercises to review and prepare for the chapter test.*
• estimate fraction products using compatible numbers *(Lesson 8-1)* $\frac{4}{5} \times 11 = ?$ *Think* $\frac{4}{5} \times 10 = 8$ *since 10 is the nearest multiple of 5.*	Estimate. 9. $\frac{4}{7} \times 22$ 10. $\frac{3}{6} \times 12$ 11. $\frac{1}{7} \times 44$ 12. $10 \times \frac{3}{4}$

Objectives and Examples

- multiply fractions *(Lesson 8-2)*

$$\frac{5}{9} \times \frac{3}{10} = \frac{5 \cdot 3}{9 \cdot 10}$$
$$= \frac{15}{90} \text{ or } \frac{1}{6}$$

- multiply mixed numbers *(Lesson 8-3)*

$$3\frac{1}{2} \times 4\frac{2}{3} = \frac{7}{\cancel{2}_1} \times \frac{\cancel{14}^7}{3}$$
$$= \frac{49}{3} \text{ or } 16\frac{1}{3}$$

- recognize and extend sequences *(Lesson 8-5)*

Find the next number in the sequence $\frac{2}{5}$, 2, 10, 50.
Multiply by 5. $50 \cdot 5 = 250$.

- divide fractions *(Lesson 8-6)*

$$\frac{3}{11} \div \frac{4}{7} = \frac{3}{11} \times \frac{7}{4}$$
$$= \frac{21}{44}$$

- divide mixed numbers *(Lesson 8-7)*

$$5\frac{1}{2} \div 1\frac{5}{6} = \frac{11}{2} \div \frac{11}{6}$$
$$= \frac{\cancel{11}^1}{\cancel{2}_1} \times \frac{\cancel{6}^3}{\cancel{11}_1}$$
$$= \frac{3}{1} \text{ or } 3$$

- change units within the customary system *(Lesson 8-8)*
  a. $56 \text{ oz} = \underline{\quad} \text{ lb}$ *16 oz = 1 lb*
   smaller to larger → divide
   $56 \div 16 = 3\frac{1}{2} \text{ lb}$

  b. $5 \text{ qt} = \underline{\quad} \text{ pt}$ *2 pt = 1 qt*
   larger to smaller → multiply
   $5 \times 2 = 10 \text{ pt}$

Review Exercises

Find each product. Write in simplest form.

13. $\frac{3}{4} \times \frac{5}{6}$ 14. $\frac{7}{8} \times \frac{4}{5}$

15. $\frac{2}{3} \times \frac{3}{4}$ 16. $\frac{14}{15} \times \frac{10}{21}$

Find each product. Write in simplest form.

17. $1\frac{7}{8} \times 7\frac{1}{5}$ 18. $6 \times 3\frac{2}{3}$

19. $2\frac{2}{5} \times 3\frac{1}{8}$ 20. $3\frac{3}{4} \times 1\frac{1}{5}$

Find the next two numbers in each sequence.

21. $\frac{1}{4}, \frac{1}{2}, 1, 2, 4$

22. 625, 125, 25, 5

23. $6, 7\frac{1}{3}, 8\frac{2}{3}, 10$

Find each quotient. Write in simplest form.

24. $\frac{5}{6} \div \frac{3}{5}$ 25. $14 \div \frac{7}{11}$

26. $\frac{4}{9} \div 8$ 27. $\frac{5}{11} \div \frac{10}{11}$

Find each quotient. Write in simplest form.

28. $1\frac{7}{8} \div 7\frac{1}{2}$ 29. $3\frac{1}{3} \div 10$

30. $6 \div 3\frac{4}{5}$ 31. $2\frac{1}{4} \div 4\frac{2}{7}$

Complete.

32. $45 \text{ c} = \underline{\quad} \text{ qt}$

33. $5\frac{1}{2} \text{ T} = \underline{\quad} \text{ lb}$

34. $2\frac{3}{4} \text{ gal} = \underline{\quad} \text{ pt}$

35. $3 \text{ pt} = \underline{\quad} \text{ fl oz}$

36. $12 \text{ oz} = \underline{\quad} \text{ lb}$

Applications and Problem Solving

37. **School**   Mitch is typing a term paper on his home computer. He types 7 pages the first day, 12 pages the second, and 17 pages the third day. He finishes typing his term paper on the fourth day. If he continues this pattern, how many pages long is the term paper? *(Lesson 8-4)*

38. **School**   Seven-eighths of the eighth-grade students will be going through the graduation ceremony. If $\frac{4}{5}$ of the graduates are girls, what portion of the eighth graders are girls who will be going through the ceremony? *(Lesson 8-2)*

39. **Food**   Mrs. Westerman has $4\frac{1}{2}$ cups of her special pudding. How many $\frac{3}{4}$-cup servings will she be able to serve? *(Lesson 8-6)*

40. **Cooking**   Anne's fudge recipe calls for $1\frac{3}{4}$ cups of sugar for 1 batch of fudge. She plans to make 6 batches. *(Lesson 8-8)*
    a. How many cups of sugar does she need?
    b. How many pints is that?

Curriculum Connection Projects

- **Home Economics**   Find the number of $\frac{1}{2}$-ounce servings in a box of your favorite mini-cookies or mini-crackers.

- **Stocks**   From the business section of a newspaper, select ten stocks. Add the high and low values of each. Then multiply each by $\frac{1}{2}$ to find the average for the day.

Read More About It

Borghese, Anita. *The Down to Earth Cookbook.*
Kline, Suzy. *Orp and the Chop Suey Burgers.*
Miller, Marvin. *You Be the Jury: Courtroom II.*

Study Guide and Review

Estimate using compatible numbers or rounding.

1. $\dfrac{2}{9} \times 28$

2. $\dfrac{5}{8} \times \dfrac{9}{10}$

3. $9\dfrac{1}{4} \times 5\dfrac{6}{7}$

Find each product. Write in simplest form.

4. $\dfrac{3}{5} \times 10$

5. $\dfrac{5}{8} \times \dfrac{9}{10}$

6. $\dfrac{9}{11} \times \dfrac{5}{6}$

7. $4\dfrac{1}{2} \times \dfrac{7}{9}$

8. $1\dfrac{1}{4} \times 3\dfrac{1}{5}$

9. $2\dfrac{2}{3} \times 1\dfrac{4}{5}$

10. Civics In Homeland, U. S. A., $\dfrac{4}{5}$ of the adult residents are registered to vote. In the last election, $\dfrac{3}{4}$ of those registered voted. What portion of the adult residents voted?

11. Cooking Alexandra's punch recipe calls for $3\dfrac{1}{2}$ quarts of ginger ale. If she makes half of the recipe, how much ginger ale will she need?

12. Consumer Math Marissa started a savings plan. The first month she deposited $5 into her savings account. Each month she doubled her deposit.
   a. How much will she deposit the ninth month?
   b. How much has she saved altogether?

Find the next two numbers in each sequence.

13. $4, 6\dfrac{1}{2}, 9, 11\dfrac{1}{2}, 14$

14. $810, 270, 90, 30$

Find each quotient. Write in simplest form.

15. $\dfrac{3}{5} \div \dfrac{9}{10}$

16. $\dfrac{2}{5} \div \dfrac{5}{7}$

17. $\dfrac{11}{12} \div \dfrac{2}{3}$

18. $2\dfrac{2}{3} \div 1\dfrac{5}{9}$

19. $2\dfrac{1}{2} \div 1\dfrac{2}{3}$

20. $4\dfrac{3}{5} \div 7$

Solve each equation. Write the solution in simplest form.

21. $x = 6 \div 1\dfrac{4}{5}$

22. $3\dfrac{2}{3} \times 2\dfrac{1}{3} = k$

Complete.

23. $7,000 \text{ lb} = \underline{\ \ } \text{ T}$

24. $5 \text{ c} = \underline{\ \ } \text{ fl oz}$

25. $2\dfrac{1}{2} \text{ qt} = \underline{\ \ } \text{ c}$

Bonus The two sequences below have something in common. As each sequence goes on and on, the numbers in each sequence get closer and closer to the same number. What is that number? Will either sequence ever reach that number? Explain.

$750, 75, 7.5, 0.75, 0.075, \ldots$ 　　　 $8, 4, 2, 1, \dfrac{1}{2}, \dfrac{1}{4}, \ldots$

Investigations in Geometry

Spotlight on Maps

Have You Ever Wondered...

- What time it is in California when it is twelve noon in New York?

- What it means when someone says that South America is 60 degrees west of the prime meridian?

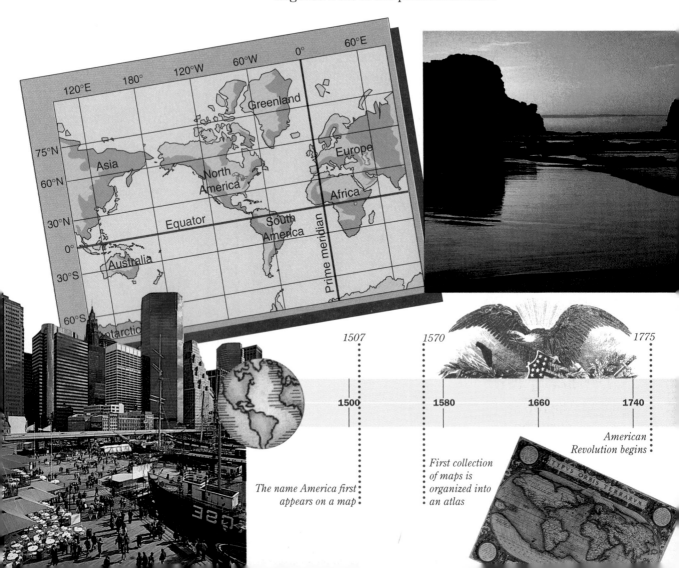

1507

1500

The name America first appears on a map

1570

1580

First collection of maps is organized into an atlas

1660

1775

1740

American Revolution begins

Chapter Project

Maps

Work in a group.

1. Select ten specific places on Earth. Using a map or globe, determine the approximate latitude and longitude of each place.

2. Write down ten random combinations of latitude and longitude. Refer to the map or globe to find the corresponding locations.

3. Record your findings in two charts.

Latitude and Longitude of U.S. Cities (time corresponding to 12:00 noon, eastern standard time)			
City	Lat. n ° '	Long. w. ° '	Time
Atlanta	33 45	84 23	12:00 noon
Baltimore	39 18	76 38	12:00 noon
Boston	42 21	71 5	12:00 noon
Chicago	41 50	87 37	11:00 a.m.
Cleveland	41 28	81 37	12:00 noon
Dallas	32 46	96 46	11:00 a.m.
Denver	39 45	105 0	10:00 a.m.
Detroit	42 20	83 3	12:00 noon
El Paso	31 46	106 29	10:00 a.m.
Honolulu	21 18	157 50	7:00 a.m.
Los Angeles	34 3	118 15	9:00 a.m.
Minneapolis	44 59	93 14	11:00 a.m.
New Orleans	29 57	90 4	11:00 a.m.
New York	40 47	73 58	12:00 noon
Oklahoma City	35 26	97 28	11:00 a.m.
Pittsburgh	40 27	79 57	12:00 noon
St. Louis	38 35	90 12	11:00 a.m.
San Antonio	29 23	98 33	11:00 a.m.
San Francisco	37 47	122 26	9:00 a.m.
Seattle	47 37	122 20	9:00 a.m.

Looking Ahead

In this chapter, you will see how mathematics can be used to answer questions about maps. The major objectives of the chapter are to:

- describe and measure angles

- construct congruent segments and angles

- bisect segments and angles

- name two-dimensional figures

- determine congruence and similarity

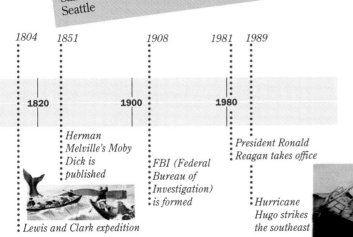

1804   *1851*   *1908*   *1981*   *1989*

1820   **1900**   **1980**

Herman Melville's Moby Dick is published

FBI (Federal Bureau of Investigation) is formed

President Ronald Reagan takes office

Hurricane Hugo strikes the southeast

Lewis and Clark expedition

9-1 Angles

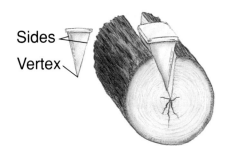
Sides
Vertex

Objective
Classify and measure angles.

Words to Learn
sides
vertex
angle
degree
protractor
acute angle
obtuse angle
right angle

When the Hamptons built their new home, an oak tree on the lot was cut down. Mr. Hampton wishes to cut and split the oak logs to burn in their fireplace. He will use a wedge to split the logs.

From the front, the **sides** of a wedge look like two lines that meet in a point called the **vertex.** Other examples of wedges are needles and ski jumps.

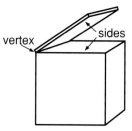
vertex   sides

Notice the arrows on the sides of the angle. These tell you that you can extend the sides.

The vertex and sides of the wedge form an **angle.** The sides of an angle may be opened like a box lid. An angle is large or small according to the amount of openness.

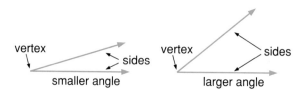
vertex   sides   vertex   sides
smaller angle   larger angle

The most common unit used in measuring angles is the **degree.** Imagine a circle cut into 360 equal-sized parts. Each part would make up a one-degree (1°) angle as shown.

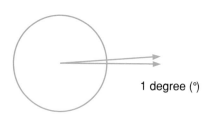
1 degree (°)

You can use a **protractor** to measure angles.
1. Place the center of the protractor on the vertex *(B)* of the angle with the straightedge along one side.
2. Use the scale that begins with 0° on the right side of the angle. Read the angle measure where the other side crosses the same scale. Extend the sides if needed.

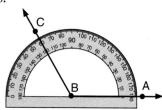
The angle measures 120°.

Angles can be classified according to their measure.

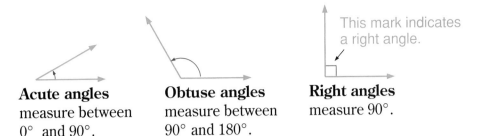

Acute angles
measure between
$0°$ and $90°$.

Obtuse angles
measure between
$90°$ and $180°$.

Right angles
measure $90°$.

This mark indicates
a right angle.

Mini-Lab

Work with a partner.

Materials: round paper plate, scissors, protractor

- Find the center of the plate by folding it in half twice.
- Cut a right-angle wedge along the fold lines.
- Cut an acute wedge from the plate.
- Cut an obtuse wedge from the plate.
- Use a protractor to measure the angle formed by each wedge.

Talk About It

How could you use the right angle wedge to determine if an angle is acute or obtuse?

Examples

Use a protractor to find the measure of each angle.
Classify each angle as acute, right, or obtuse.

1

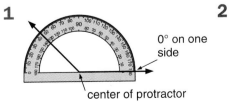

0° on one side

center of protractor

The angle measures
$135°$. It is an obtuse
angle.

2

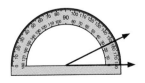

The angle measures
$25°$. It is an acute
angle.

Example 3 *Problem Solving*

Construction   The figure at the right shows the plan for a new apartment community and a new sidewalk. Will the angles formed by the existing sidewalks and the new sidewalk be acute, right or obtuse?

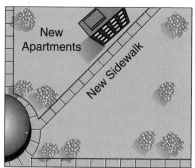

The angle formed by the two existing sidewalks measures 90°. The angles formed by the existing sidewalk and the new sidewalk measure less than 90°. Therefore, they are acute angles.

Checking for Understanding

Communicating Mathematics   Read and study the lesson to answer each question.

1. **Show** the vertex and sides of an angle you see in your classroom.
2. **Show** how to use a protractor to measure an angle.
3. **Tell** the measure of an obtuse angle.
4. **Draw** an acute angle.

Guided Practice   Use a protractor to find the measure of each angle.

5.
6.
7.

Classify each angle as acute, right, or obtuse.

8.
9.
10. a 90° angle
11. a 115° angle

Exercises

Independent Practice   Use a protractor to find the measure of each angle.

12.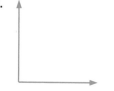
13.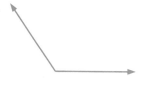
14.

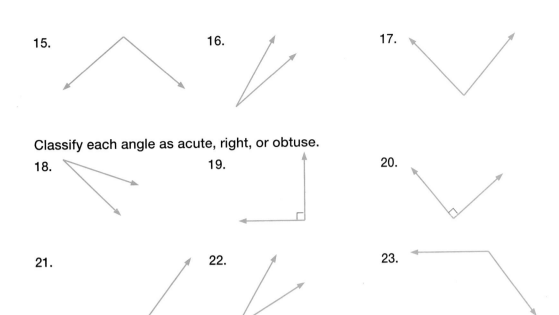

15.   **16.**   **17.**

Classify each angle as acute, right, or obtuse.

18.   **19.**   **20.**

21.   **22.**   **23.**

Classify angles having each measure as acute, right, or obtuse.

24. $38°$ **25.** $149°$ **26.** $85°$ **27.** $90°$

28. $175°$ **29.** $2°$ **30.** $98.5°$ **31.** $161.9°$

32. An angle measures $90.5°$. Is it an obtuse angle or a right angle?

Mixed Review   **33.** **Waste** If every person in the U.S. is responsible for about 2.9 pounds of waste that is disposed of per day, how many ounces is that? *(Lesson 8-8)*

34. Change 4 kilometers to centimeters. *(Lesson 5-6)*

35. Estimate the product of 26.98 and 11 by rounding. *(Lesson 4-1)*

36. Which is the better estimate for the capacity of a glass of milk, 360 liters or 360 milliliters? *(Lesson 5-5)*

Problem Solving and Applications

37. **Carpentry** A carpenter needs to cut two boards as shown to form a right angle. At what angle should the carpenter cut the boards?

38. **Lawn Care** Mr. Greene has installed a sprinkler in each corner of his yard. Should the spray pattern in corner A be acute, right, or obtuse? in corner C? in corner D? in corner B?

39. **Critical Thinking** Students complain that a ramp into the school is too steep.

   a. How would you construct a ramp that is not so steep?

   b. Would the obtuse angle with the ground become larger or smaller?

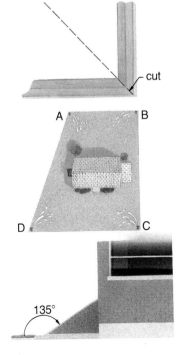

9-2 Using Angle Measures

Objective

Draw angles and estimate measures of angles.

Words to Learn

straightedge

A compass is a device used to determine direction. The needle of a compass points to magnetic north. The eight main points, or directions, on a compass are north (N), northeast (NE), east (E), southeast (SE), south (S), southwest (SW), west (W), and northwest (NW). The directional angle is measured from magnetic north.

If a ship traveling northeast increases its directional angle by 45°, in what direction is it now sailing?

Problem Solving Hint

• • • • • • • • • • •

You can use the strategy draw a diagram to help you with angle measures.

Northeast (NE) has a directional angle of 45° measured from the north (N). The new directional angle is 45° + 45° or 90°. The ship is now sailing east.

You can use the following steps to draw an angle with a certain measure.

1. Draw one side of the angle.

2. Mark the vertex on one end and draw an arrow on the other end.

3. Place the protractor as you would to measure an angle. With a pencil, mark a point beside the mark on the protractor that indicates the number of degrees needed for the angle you are drawing.

4. With a straightedge, draw the side that connects the vertex with the pencil mark. A **straightedge** is a ruler or any object with a straight side, which can be used to draw a line. Draw an arrow on the end of the side.

Example 1

Draw a 150° angle.

Draw one side.
Mark the vertex and
draw an arrow.

Find 150° on the
appropriate scale.
Make a pencil mark.

Draw the side that
connects the vertex
and the pencil mark.

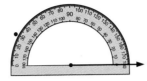

You can estimate the measure of an angle by comparing it to an
angle whose measure you know.

Mini-Lab

Work with a partner.

Materials: paper plates, straightedge, protractor, scissors

- Find and mark the center of two paper plates by folding
  them in half twice.
- Measure and cut wedges of 150°, 90°, 60°, and 30° from
  one plate. Cut wedges of 135°, 120°, and 45° from the other
  plate. Write the measure on each wedge.
- Have one partner draw any angle. The other partner uses
  the wedges to estimate the measure of the angle.
- Repeat for several different angles.

Talk About It

How can you estimate the measure of an angle without using
wedges?

Estimation Hint

• • • • • • • • • •

Lay one corner of
your notebook
paper on top of an
angle to determine
whether the angle is
acute, obtuse, or
right.

Example 2

Tell whether the measure of the angle shown below is greater than,
less than, or about equal to 130°.

130° is greater than a
right angle.
The angle shown is an
acute angle.

Therefore, the measure of the angle is less than 130°.

Example 3 *Connection*

LOOKBACK

You can review circle graphs on page 103.

Statistics The circle graph at the right shows the results of a survey of pizza preference among sixth-graders at Walnut Springs Middle School. Use the measure of the angles to order the types of pizza from most preferred to least preferred.

Pizza Preferences

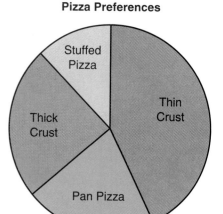

Thin Crust is shown by an obtuse angle. It is obviously the most preferred. Stuffed Pizza is shown by a small, acute angle. It is obviously the least preferred.

Measure the remaining two angles with a protractor. The angle for Thick Crust measures 85°; the angle for Pan Pizza measures 75°.

Therefore, the sixth-graders prefer Thin Crust, Thick Crust, Pan Pizza, and Stuffed Pizza, in that order.

Checking for Understanding

Communicating Mathematics

Read and study the lesson to answer each question.

1. **Tell** how to draw a 125° angle.
2. **Show** an angle of about 45° using two pencils as the sides.

Guided Practice

Use a protractor to draw angles having the following measurements.

3. 140°   4. 15°   5. 80°   6. 128°   7. 37°

Tell whether the measure of each angle is greater than, less than, or about equal to the measurement given.

8. 95°   9. 100°   10. 48°

Exercises

Independent Practice
Use a protractor to draw angles having the following measurements.

11. $75°$ **12.** $130°$ **13.** $5°$ **14.** $90°$ **15.** $155°$

16. $87°$ **17.** $35°$ **18.** $148°$ **19.** $172°$ **20.** $95°$

Tell whether the measure of each angle is greater than, less than, or about equal to the measurement given.

21. $150°$ **22.** $58°$ **23.** $25°$

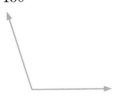 

24. $95°$ **25.** $160°$ **26.** $21°$

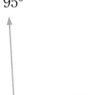 

Without a protractor, draw your best estimate of an angle having each measurement given. Check your estimate with a protractor.

27. $88°$ **28.** $10°$ **29.** $165°$ **30.** $43°$ **31.** $110°$

Mixed Review
32. Determine whether 786 is divisible by 2, 3, 5, 6, 9, or 10. *(Lesson 6-1)*

33. Change 498 grams to kilograms. *(Lesson 5-6)*

34. **Carpeting** Allison will be buying new carpet for her bedroom, which measures 5.3 yards by 4.7 yards. How many square yards will she need? *(Lesson 4-7)*

35. Subtract $1\frac{3}{4}$ from $7\frac{1}{2}$. *(Lesson 7-7)*

Problem Solving and Applications
36. **Horticulture** The branches on young trees should be spread to form angles of at least $60°$ with the tree trunk. This strengthens branches and allows for more air circulation and light. Which branches on the tree at the right need to be spread?

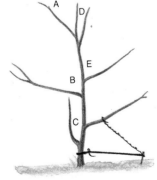

37. **Critical Thinking** Through what angle does the minute hand of a clock turn in 5 minutes?

38. **Journal Entry** Find a circle graph in a newspaper or magazine, and copy it into your Journal. Write one or two sentences that explain the data presented in the graph.

9-3 The Tools of Geometry

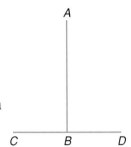

Objective
Construct congruent segments and angles.

Words to Learn
line segment
congruent segments
congruent angles
ray

An *optical illusion* is a misleading image as perceived by the human eye. The figure at the right is an optical illusion that is made up of two line segments, segment *AB* and segment *CD*. A **line segment** is a straight path between two points. To indicate line segment *AB*, write $\overline{AB}$.

Which segment do you think is longer, $\overline{CD}$ or $\overline{AB}$? If you use a ruler to measure both segments, you will discover that both segments have the same length. Segments having the same length are **congruent segments.** To construct congruent segments you need a straightedge and a compass. A compass is used to draw circles or circular arcs and to measure distances.

When am I ever going to use this?

To make sense out of conflicting images or illusions you must sharpen your sense of perception. Perception is relating what you see to the many other occasions on which you have seen the same or similar objects.

Mini-Lab

Work with a partner.
Materials: straightedge, compass, ruler

Construct a line segment congruent to $\overline{AB}$.

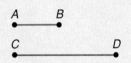

- Use a straightedge to draw a line segment longer than $\overline{AB}$. Label it $\overline{CD}$.

- Place the compass point at *A* and adjust the compass setting so that the pencil on the compass is at *B*. The compass setting equals the length of $\overline{AB}$.

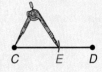

- Use this compass setting and place the compass point at *C*. Draw an arc that intersects $\overline{CD}$ at *E*.

Talk About It
Measure $\overline{AB}$ and $\overline{CE}$. Are these segments congruent?

A compass and a straightedge can also be used to construct congruent angles. **Congruent angles** are angles that have the same measure.

Look at the angle at the right. Remember, the sides of the angle meet in a point called the vertex. The sides of the angle are two rays, ray *SR* and ray *ST*. A **ray** is a path that extends endlessly from one point in a certain direction. The arrow at the end of the ray indicates that the ray is endless.

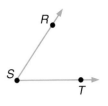

The symbol ∠ means angle.

The angle shown above can be named in two ways, angle *RST* or angle *TSR*. The middle letter is always the vertex. Another way to indicate angle *RST* is ∠ *RST*.

Mini-Lab

Work with a partner.

Materials: straightedge, compass, protractor

Construct an angle congruent to ∠ *ABC*.

- Beside ∠ *ABC*, use a straightedge to draw ray *EF*.

- Go back to ∠ *ABC*. With the compass point at *B*, draw an arc that intersects the sides of ∠ *ABC*. Label the intersections *M* and *N*.

- With the same compass setting, move to ray *EF*. Place the compass point at *E* and draw an arc that intersects ray *EF*. Label this intersection *P*.

- On ∠ *ABC*, set the compass at points *M* and *N* as shown. With that setting, go to ray *EF*. Place the compass point at *P*, and draw an arc that intersects the larger arc you drew before. Label the intersection *Q*.

- Use a straightedge to draw ray *EQ*.

Talk About It
With a protractor, measure ∠ *ABC* and ∠ *QEP*. Are these angles congruent?

Checking for Understanding

Communicating Mathematics

Read and study the lesson to answer each question.

1. **Tell** how to name the angle at the right.

2. **Draw** and label a line segment and an angle.

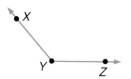

Guided Practice

Trace each segment or angle. Then construct a segment or angle congruent to it.

3. A ●————————● B

4. K ●————● L

5. X ●————————————● Y

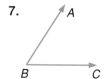

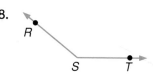

Exercises

Independent Practice

Trace the drawing at the right. Then construct a line segment or angle congruent to each segment or angle named.

9. $\overline{AB}$ 10. $\angle BDC$ 11. $\angle DCB$
12. $\overline{AD}$ 13. $\overline{BC}$ 14. $\angle ADC$
15. $\overline{DB}$ 16. $\angle ABD$ 17. $\angle ADB$

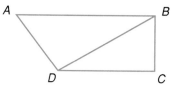

Mixed Review

18. Use a protractor and draw an angle that measures 70°. *(Lesson 9-2)*

19. Measure the line segment ———————————— to the nearest $\frac{1}{8}$ of an inch. *(Lesson 6-7)*

20. Solve the equation $17 - m = 4$ mentally. *(Lesson 1-9)*

Problem Solving and Applications

21. **Optical Illusion**   Which sement is longer, $\overline{MN}$ or $\overline{NQ}$? Use a compass.

22. **Critical Thinking**   Trace $\angle RST$ and $\angle DEF$. Construct an angle whose measure is equal to the sum of the measures of $\angle RST$ and $\angle DEF$.

23. **Journal Entry**   Write one or two sentences explaining why you can use a compass to measure distances.

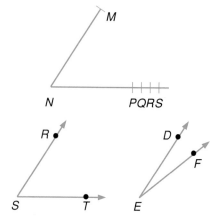

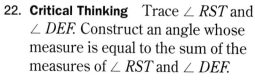

9-4 Finding Bisectors

Objective
Bisect line segments and angles.

Words to Learn
bisect

Origami is the art of folding paper into decorative objects such as birds, flowers, and fish. Originating in China, it is now a part of the Japanese culture.

You can use some paper-folding methods to do several constructions in geometry. To **bisect** something means to separate it into two congruent parts.

Mini-Lab

Work with a partner.

Materials: compass, straightedge, paper

Bisect a line segment using paper folding.

- Use a straightedge to draw $\overline{MN}$.

- Fold point N onto point M and make a crease as shown. The crease bisects $\overline{MN}$. Label the intersection point L.

Talk About It
Measure $\overline{ML}$ and $\overline{LN}$. What can you say about point L?

Bisect a line segment using a straightedge and compass.

- Use a straightedge to draw $\overline{RS}$.

- Place the compass point at R. Set the compass to more than half the length of $\overline{RS}$. Draw two arcs as shown.

- With the same compass setting, place the compass point at S and draw two arcs as shown. These arcs should intersect the first arcs at P and Q.

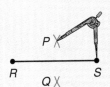

- With a straightedge, draw $\overline{PQ}$. $\overline{PQ}$ bisects $\overline{RS}$ at T.

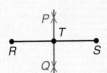

Talk About It
Measure $\overline{RT}$ and $\overline{TS}$. What can you say about point T?

DID YOU KNOW

There are about 100 traditional origami figures. Most origami is folded from an uncut, square piece of paper.

You can also use paper-folding methods to bisect an angle.

Mini-Lab

Work with a partner.
Materials: straightedge, compass, protractor, paper

Bisect an angle using paper folding.

- Draw ∠ *RST*.

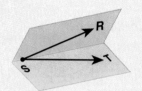

- Fold ray *ST* onto ray *SR* and make a crease. The crease bisects ∠ *RST*.

Talk About It
What can you say about the two angles formed by the crease?

Bisect an angle using a straightedge and compass.

- Use a straightedge to draw ∠ *ABC*.

- Place the compass point at *B* and draw an arc that intersects both sides of the angle. Label these points *D* and *E*.

- Place the compass point at *D* and draw an arc as shown.

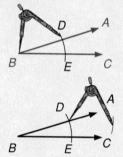

- With the same compass setting, place the compass point at *E* and draw an arc that intersects the one drawn in the previous step. Label the intersection *F*.

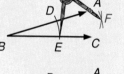

- Using a straightedge, draw ray *BF*. Ray *BF* bisects ∠ *ABC*.

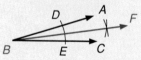

Talk About It
Measure ∠ *ABF* and ∠ *CBF*. What can you say about ray *BF*?

Checking for Understanding

Communicating Mathematics

Read and study the lesson to answer each question.

1. **Tell** how to bisect a line segment using a compass and a straightedge.
2. **Show** how to bisect an angle using paper folding.

Draw the angle or line segment with the given measurement. Then use a straightedge and compass to bisect each angle or line segment.

    3. $45°$ **4.** 3 in.   **5.** 4 cm   **6.** $130°$ **7.** $85°$

    8. $20°$ **9.** 3 cm   **10.** 6 in.   **11.** $115°$ **12.** 35 mm

Exercises

Draw the angle or line segment with the given measurement. Then use a straightedge and compass to bisect each angle or line segment.

    13. 53 mm   **14.** $53°$ **15.** $65°$ **16.** 2 in.   **17.** $135°$

    18. 4 in.   **19.** $28°$ **20.** $105°$ **21.** 6 cm   **22.** 5 in.

23. Trace the angle at the right. Then construct an angle congruent to it. *(Lesson 9-3)*

24. Find the circumference of a circle with a radius that measures 4 meters. *(Lesson 4-9)*

25. Find the sum of $5\frac{3}{8}$ and $8\frac{5}{8}$. *(Lesson 7-6)*

26. **Design** Mrs. Parker is designing a 6-inch diameter wheel with eight spokes similar to the one at the right. Draw a circle that is 6 inches in diameter. Draw a diameter. Construct a second diameter that bisects the first one. Construct two more diameters that bisect the angles of intersection.

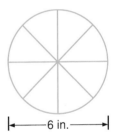

27. **Critical Thinking** Draw a 5-inch line segment. Then use a compass and a straightedge to separate the segment into four congruent parts. How can you tell that the four parts are congruent?

28. **Data Search** Refer to pages 308 and 309. About how many degrees are there between the equator and the southernmost tip of South America?

DATA SEARCH

9-5A Triangles and Quadrilaterals

A Preview of Lesson 9-5

Objective

Sort triangles and quadrilaterals.

Words to Learn

polygon
sides
vertices
triangle
quadrilateral

Materials

geoboard
dot paper
scissors

Simple closed shapes formed by line segments are called **polygons.** The line segments, called **sides,** intersect only at their end points. These points of intersection are called **vertices** (plural for vertex).

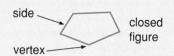

In this lab, you will investigate triangles and quadrilaterals. A **triangle** is a polygon with three sides. A **quadrilateral** is a polygon with four sides.

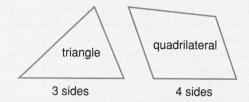

triangle
3 sides

quadrilateral
4 sides

Activity One

Work with a partner.

- One student should make a triangle on the geoboard. A sample is shown at the right.
- Have the partner draw the triangle on dot paper and cut out the triangle.
- Continue this activity until you have ten different triangles. Then switch roles and continue until you have a total of 20 triangles. Try to make a variety of triangles.

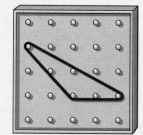

What do you think?

1. One way to name triangles is by their angles. At least two angles of every triangle are acute. Sort your triangles into three groups, based on the third angle.
2. Name the groups *right triangles, obtuse triangles,* or *acute triangles.*
3. Write a definition for each kind of triangle.

Activity Two

Work with a partner.

- Repeat Activity One using quadrilaterals instead of triangles. Several samples are shown below.

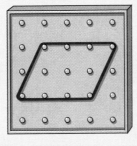 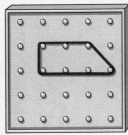 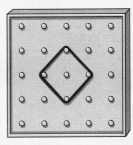

What do you think?

4. Sort your quadrilaterals into two groups, based on any characteristic you choose. Write a description of the quadrilaterals in each group.

5. Sort the quadrilaterals into three groups, based on any characteristic you choose. Write a description of the quadrilaterals in each group.

Extension

6. Write a sentence that describes one characteristic of each polygon. Then draw another polygon that has the same characteristic.

a. 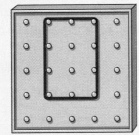   b.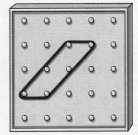

c. 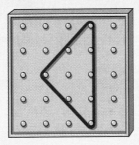   d.

9-5 Two-Dimensional Figures

Objective

Name two-dimensional figures.

Words to Learn

parallel
parallelogram
pentagon
hexagon
octagon
decagon
equilateral triangle

If you've played the game Tetris™, you cover the screen with figures like the ones shown below. Each figure is made up of squares.

A square is a quadrilateral with some special characteristics.

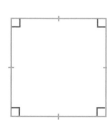

- First, all of the sides are congruent. This is shown by the matching blue marks on the sides of the figure.
- Second, a square has four right angles. This is shown by the red right angle symbols.
- Finally, pairs of opposite sides are **parallel.** You can extend the sides, and the opposite sides will never intersect.

Other types of quadrilaterals share some of the characteristics described above.

Example 1

In what ways are a square and rectangle alike? In what ways are they different?

square | rectangle

A square and rectangle are alike because they
- have four sides.
- have opposite sides parallel.
- have four right angles.

A square and rectangle are different because a square must have four congruent sides and a rectangle may not have four congruent sides.

If a quadrilateral has both pairs of opposite sides parallel, it is called a **parallelogram.**

Example 2

In what ways are a square and parallelogram alike? In what ways are they different?

A square and parallelogram are alike because they
- have four sides.
- have opposite sides parallel.

A square and the parallelogram are different because a square has four congruent sides and the parallelogram may not. Also, a square has four right angles and a parallelogram may not.

In addition to triangles and quadrilaterals, there are other kinds of polygons. A polygon is named according to the number of its sides. Some examples of common polygons are shown below.

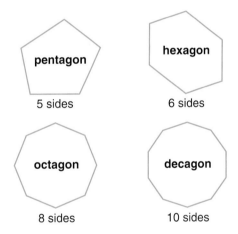

Example 3

In what ways are the triangle and hexagon alike? In what ways are they different?

The triangle and hexagon are alike because they have congruent sides. A triangle with three congruent sides is called an **equilateral triangle.**

They are different because they have a different number of sides.

Checking for Understanding

Communicating Mathematics

Read and study the lesson to answer each question.

1. **Show** a quadrilateral that you see in the classroom.

2. **Draw** a quadrilateral that has exactly one pair of opposite sides parallel.

3. **Tell** as many characteristics as you can about the figure at the right.

4. **Tell** the name that is given to a ten-sided polygon.

Guided Practice

Name each polygon.

5.   6.   7.

Explain how each pair of figures is alike and how each pair is different.

8.    9.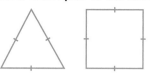

10. Draw an example of an octagon.

Exercises

Independent Practice

Name each polygon.

11.   12.   13.

Explain how each pair of figures is alike and how each pair is different.

14.    15.

Draw an example of each polygon. Mark any congruent sides and right angles.

16. equilateral triangle   17. parallelogram

18. pentagon   19. hexagon

Mixed Review

20. **Sales** Ke Min bought 12 T-shirts. If the total sale before tax was $43.44, what was the price of each T-shirt? *(Lesson 5-7)*

21. Evaluate $6^2 \div 3^2 + 2^3$. *(Lesson 4-2)*

Problem Solving and Applications

22. **Furniture Design** A design for a conference table that will seat six people with equal comfort and work space is desired. What shape should the table be?

23. **Critical Thinking** *True* or *False:* All squares are parallelograms. Explain.

COMPUTER

CONNECTION

24. **Computer Connection** The LOGO program at the right will draw a square on the screen. The FD 50 command draws a line that is 50 units long. The RT 90 command indicates a right angle. Modify the program so that it will draw a rectangle.

```
FD 50 RT 90
FD 50 RT 90
FD 50 RT 90
FD 50 RT 90
```

Mid-Chapter Review

Use a protractor to find the measure of each angle. *(Lesson 9-1)*

1.   2.   3.   4.

Tell whether the measure of each angle is greater than, less than, or about equal to the measurement given. *(Lesson 9-2)*

5. $80°$ 6. $110°$  7. $100°$ 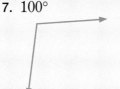 8. $30°$

Trace each segment or angle. Then construct a segment or angle congruent to it. *(Lesson 9-3)*

9.   10.   11.   12.

13. Draw an angle that measures $29°$. Then use a straightedge and compass to bisect the angle. *(Lesson 9-4)*

14. Draw an example of a square. Mark congruent sides and right angles. *(Lesson 9-5)*

9-5B **Wrapping a Box**

A Follow-Up of Lesson 9-5

Objective

Use a net to wrap a box.

Materials

scissors
paper
small open box
small closed box

In this lab, you will make a figure called a net, and use it to wrap an open box. The faces of the box are shaped like squares.

Activity One

Work with a partner.

- Count the number of faces of the open box.
- Place the open box on the paper as shown.
- Draw around the base of the box as many times as necessary to make a figure like the one shown at the right. This figure is called a net.
- Cut the figure out and try wrapping the open box.

What do you think?

1. Did the net cover the box?

Activity Two

- Follow the procedure in Activity One. But this time, draw a net like the one shown at the right.
- Cut the figure out and try wrapping the open box.

What do you think?

2. Did the net cover the open box?
3. How many different nets of five squares can you find that will cover the open box? Look for a pattern.

Extension

4. Count the number of faces of a closed box with faces shaped like squares. How many different nets can you find that will cover the closed box?

9-6 **Using Logical Reasoning**

Objective

Solve problems by using logical reasoning.

Of the 150 students in the sixth grade at Hastings Middle School, 45 are in the orchestra, 65 are in concert band, and 20 are in both the orchestra and concert band. How many students are only in the orchestra? How many students are only in concert band? How many are in either the orchestra or concert band? How many sixth grade students are not in the orchestra or concert band?

Explore   What do you know?
There are 150 students in the sixth grade.
45 students are in the orchestra.
65 students are in concert band.
20 students are in both the orchestra and concert band.

What do you need to find out?
You need to find out the number of students who participate in only one, either one, or neither music group.

Plan   You can draw a Venn diagram to show the relationship between the students in the orchestra and concert band. First, draw a rectangle to represent the number of students in the sixth grade.

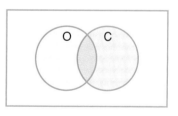

Then draw two overlapping circles inside the rectangle. Label one circle O to represent the orchestra and the other circle C to represent the concert band.

Solve   Find the number of students in the orchestra only.

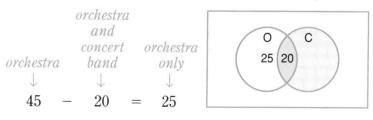

$$\begin{array}{ccccc} orchestra & & \begin{array}{c}orchestra\\and\\concert\\band\end{array} & & \begin{array}{c}orchestra\\only\end{array}\\ \downarrow & & \downarrow & & \downarrow \\ 45 & - & 20 & = & 25 \end{array}$$

Write 25 in the circle labeled O.

Write 20 where the circles overlap to show the number of students that are in both music groups.

Find the number of students in the concert band only.

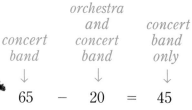

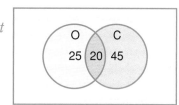

$$
\begin{array}{ccccc}
65 & - & 20 & = & 45
\end{array}
$$

Write 45 in the circle labeled C.

Find the number of students in either the orchestra or concert band.

Add the numbers in each circle to 20, the number of students in both the orchestra and concert band.

$$25 + 45 + 20 = 90$$

You can use this information to find the number of students who are neither in the orchestra or in concert band. Subtract 90 from 150, the number of sixth graders.

$$150 - 90 = 60$$

There are 60 students who are not in the orchestra or in concert band. Write 60 inside the rectangle but outside of the circles.

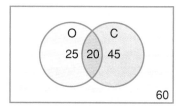

Examine   Check to see that all of the numbers in the Venn diagram add up to 150, the total number of sixth graders at Hastings Middle School.

Example

Marisa entered a supermarket contest. She must correctly draw the next figure in the pattern below. If she draws it correctly, her name will be put into a drawing for a grand prize. Was her entry correct?

Marisa's entry

Marisa's entry was not correct. You can see that the polygons in the pattern have one additional side each time. The next figure should be a hexagon.

Checking for Understanding

Communicating Mathematics

Read and study the lesson to answer each question.

1. **Tell** how a Venn diagram shows relationships among groups.

2. **Draw** a correct fifth figure in the pattern in the Example.

Guided Practice

Solve using logical reasoning.

3. In a line of 24 people waiting for a city bus, 12 have umbrellas, 10 have raincoats, and 7 have both umbrellas and raincoats. The rest have neither.

   a. How many people only have umbrellas?

   b. How many have neither a raincoat nor an umbrella?

4. What are the next two figures in the pattern?

1	2
4	6

1	2
4	6

1	2
4	6

1	2
4	6

Problem Solving

Practice

Solve using any strategy.

5. In a troop of 75 scouts, 24 girls have their camping badge, 39 have their volunteer badge, and 18 have both their camping and volunteer badges.

   a. How many scouts only have their camping badge?

   b. How many scouts have their volunteer badge but not their camping badge?

Strategies

• • • • • • •

Look for a pattern.
Solve a simpler problem.
Act it out.
Guess and check.
Draw a diagram.
Make a chart.
Work backward.

6. Sheila was on the phone for 32 minutes at a rate of $0.16 per minute. Did she spend about $3.00 or $5.00 for the call?

7. Byron, Verneeda, Glenn, and Angela are all doctors. Their specialties are pediatrics, optometry, family medicine, and dermatology. The family doctor and Byron played tennis together on Tuesday. Glenn is married to the dermatologist. Who is the family doctor?

8. What is the next number in this pattern?
   2,456; 4,562; 5,624; ___

9. Of the 54 players trying out for the Wildcat's baseball team, 26 are outfielders, 31 are infielders, and 12 are either infielders or outfielders. How many were neither infielders nor outfielders?

9-7 Line Symmetry

Objective

Describe and define lines of symmetry.

Words to Learn

line of symmetry

Did you know that there are more than 15,000 different species of butterflies? The bright colors and attractive patterns make the butterfly one of the most beautiful insects.

If you draw a line down the middle of a butterfly, the two halves match. When this happens, the line is called a **line of symmetry.** One way to find lines of symmetry is to cut out a shape and observe whether the sides and patterns match exactly when the shape is folded.

Examples

Draw all lines of symmetry for each figure.

1

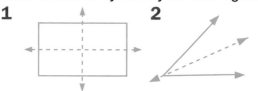

2

3

| A rectangle has two lines of symmetry. | An angle has one line of symmetry. | A rattail comb has no lines of symmetry. |

DID YOU KNOW

The smallest butterfly is the western pygmy blue, which is found in North America. Its wingspan is only about $\frac{3}{8}$ inch.

You can also use a MIRA® to draw figures that have lines of symmetry.

Mini-Lab

Work with a partner.

Materials: MIRA®, paper and pencil

- Trace each figure below. Place the MIRA® on the dashed line of each figure and draw its reflection.

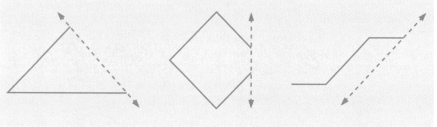

Talk About It
a. How did the reflection look on the other side of the MIRA®?
b. What would you call the dashed lines with respect to the figure and its reflection?

Checking for Understanding

Communicating Mathematics

Read and study the lesson to answer each question.

1. **Tell** in your own words what a line of symmetry is.
2. **Show** how to use a MIRA® to find a line of symmetry.
3. **Draw** a design that has at least two lines of symmetry.

Guided Practice

Tell whether the dashed line is a line of symmetry. Write *yes* or *no*.

4. 　　5. 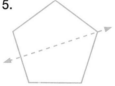　　6. 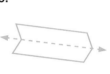　　7.

Trace each figure. Draw all lines of symmetry.

8. 　　9. 　　10. 　　11.

Exercises

Independent Practice

Tell whether the dashed line is a line of symmetry. Write *yes* or *no*.

12. 　　13. 　　14. 　　15.

16. 　　17. 　　18. 　　19.

Trace each figure. Draw all lines of symmetry.

20.

21.

22.

23.

24. How many lines of symmetry does an equilateral triangle have?

Mixed Review

25. Draw an example of a parallelogram. *(Lesson 9-5)*

26. **Weather** During two weeks of torrential rains, Amber measured the rainfall. She recorded 11.33 centimeters in the first week and 15.75 centimeters in the second week. About how much rain is this in the two weeks? *(Lesson 3-6)*

27. Find $\frac{7}{8} - \frac{5}{12}$. *(Lesson 7-5)*

28. Multiply $7\frac{1}{5}$ and $6\frac{1}{4}$. *(Lesson 8-3)*

Problem Solving and Applications

29. **Landscaping** Trace the house and shrubs and draw in additional shrubs to make the landscaping symmetrical with respect to the front door.

30. **Geometry** Draw a parallelogram, a rectangle, and a square. Which figure has the most lines of symmetry?

31. **Critical Thinking** Copy the figure at the right. Shade enough squares so that the figure has a diagonal line of symmetry.

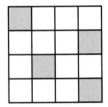

32. **Science** How many lines of symmetry can you find in the picture below?

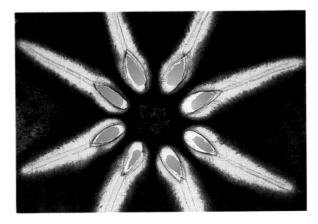

9-8 About Size and Shape

Objective
Determine congruence and similarity.

Words to Learn
congruent figures
similar figures

A photocopier can make copies that are the same size and shape as the original. Figures that are the same size and shape are called **congruent figures.** The symbol ≅ means *is congruent to.*

triangle ABC   ≅   triangle GHI

Some photocopiers can also produce reduced or enlarged copies. Figures that have the same shape, but different size are called **similar figures.** The symbol ~ means *is similar to.*

triangle ABC   ~   triangle DEF

In the figures above, notice that the triangles are named by using the letters that correspond to each vertex. We can also use the symbol Δ to replace the word *triangle* when naming a triangle. For example, triangle *ABC* can be written as Δ*ABC*.

Mini-Lab

Work with a partner.
Materials: scissors, tracing paper

● Trace each pair of figures below and cut them out.

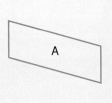 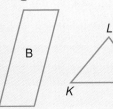 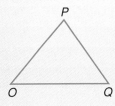

● Try to match each pair of figures exactly.

Talk About It

a. Did either pair of figures fit exactly? Name these figures.

b. What did you find when comparing the angles and segments of each pair of figures?

c. If the figures did not fit exactly, would you be able to fit one figure inside another with equal space separating all corresponding sides? Name these figures.

Tell whether each triangle below is congruent or similar to the triangle at the right.

1

The triangles are the same size and shape. They are congruent.

2

The triangles are the same shape, but not the same size. They are similar.

3

The triangles are not the same shape or size. The triangles are neither congruent nor similar.

Checking for Understanding

Communicating Mathematics

Read and study the lesson to answer each question.

1. **Tell** in your own words when two polygons are similar.

2. **Write** a sentence explaining the difference between congruence and similarity.

3. **Tell** whether the figures at the right are similar.

4. **Draw** two congruent triangles.

Guided Practice

Tell whether each pair of polygons is congruent, similar, or neither.

5.   6.   7.

8.   9.   10.

Exercises

Independent
Practice

Tell whether each pair of polygons is congruent, similar, or neither.

11.

12.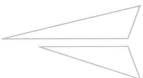

13.

14.

15.

16.

Mixed Review

17. Tell how many lines of symmetry there are in a square. *(Lesson 9-7)*

18. **Cooking**   Millie needs $1\frac{3}{4}$ gallons of sweet cream. The grocery store sells sweet cream in pint containers only. How many pints should she buy? *(Lesson 8-8)*

19. Find the prime factorization of 300. *(Lesson 6-2)*

Problem Solving
and
Applications

20. **Sewing**   The pattern pieces for a pullover blouse are shown below. Trace and cut out the pattern pieces. Then use a folded sheet of paper as cloth to cut the pieces of the blouse. The arrows indicate that a side of the pattern is to be placed on the fold.

21. **Critical Thinking**   The parallelogram shown at the right can be separated into two or more congruent parts. Find at least four different ways in which the figure can be separated into congruent parts.

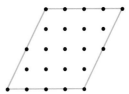

22. **Journal Entry**   Write a few sentences describing a situation from your life in which you use congruent figures or similar figures.

9-8B Translations and Escher Drawings

A Follow-Up of Lesson 9-8

Objective

Create Escher-like drawings using translations.

Words to Learn

translation

Materials

centimeter grid paper
colored pencils

Maurits Cornelis Escher was a Dutch artist who lived from 1898 to 1972. He is famous for his unusual art. Some of his most famous pieces are formed by congruent figures that fit together to cover the entire surface. They often look like birds, reptiles, and fish. You can make an Escher-like drawing by using **translations.** When you do a translation, you slide a figure from one location to another.

Activity One

Work with a partner.

- Draw a square. Now change the top side to look like a shark fin.

- Translate, or slide, the fin to the opposite side of the square. The pattern for our Escher-like drawing is formed.

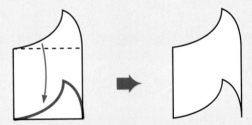

- To make a small Escher-like drawing, start by drawing a 3×3 unit square on a piece of centimeter grid paper.
- Change each square using the fin design you made above. Then use colored pencils to create the pattern.

©M.C. Escher/Cordon Art—Baarn—Holland
Collection Haags
Gemeentemuseum—The Hauge

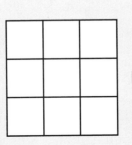 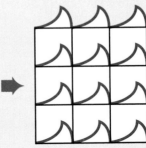 

What do you think?

1. Why do the pieces fit together exactly?

Activity Two

You can make more complex patterns by using two translations. Make your pattern square first.

- Draw a square. Make a bump on the top and translate the bump to the bottom.

- Now make a bump on the left side and translate the bump to the right side.

- Use the steps in your new design to create a large pattern using centimeter grid paper.

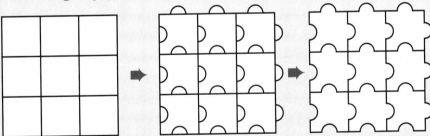

What do you think?

2. Suppose you put a design on the left and right sides that was different from the bump design. Would the design still fit together?

Application

Make an Escher-like drawing for each change shown.

3.

4.

5.

Study Guide and Review

Communicating Mathematics

Match each illustration with the best description. An answer may be used more than once.

1.  2.  3. 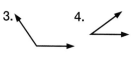 4.   5.

a. congruent
b. similar
c. hexagon
d. compass
e. protractor
f. right angle
g. obtuse angle
h. acute angle

6.   7.   8.   9. ≅ 10. ~

Complete.

11. The ___?___ is the point of an angle where the two sides meet.
12. 360 equal-sized parts called ___?___ are used to measure angles.
13. The instrument used to measure angles is called a ___?___ .
14. To ___?___ something is to separate it into two congruent parts.
15. A ___?___ is a polygon with four sides.
16. A ___?___ is a polygon with eight sides.

17. Tell how the measures of the angles of similar polygons are related.
18. Explain why a two-sided figure cannot be called a polygon.

Skills and Concepts

Objectives and Examples

Upon completing this chapter, you should be able to:

Review Exercises

Use these exercises to review and prepare for the chapter test.

● classify angles *(Lesson 9-1)*

Classify each angle as acute, right, or obtuse.

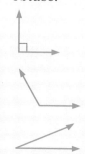

The angle is right since it measures 90°.
The angle is obtuse since it measures between 90° and 180°.
The angle is acute since it measures between 0° and 90°.

Classify each angle as acute, right, or obtuse.

19.   20.

21.   22.

23. 147°   24. 38.95°

Objectives and Examples

- measure angles *(Lesson 9-1)*

  Use a protractor to measure angle *ABC*.

   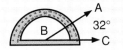

- draw angles and estimate measures of angles *(Lesson 9-2)*

  Tell whether ∠ *MNP* is greater than, less than, or about equal to 100°.

  ∠ *MNP* is less than 100°.

- construct congruent segments and angles *(Lesson 9-3)*

  Construct a line segment congruent to $\overline{AB}$.

  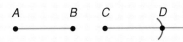

- bisect line segments and angles *(Lesson 9-4)*

  Bisect the angle using a straightedge and compass.

  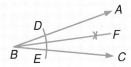

- name two-dimensional figures *(Lesson 9-5)*

  Draw an example of each polygon.

  pentagon   quadrilateral

Review Exercises

Use a protractor to find the measure of each angle.

25.   26.

Tell whether the measure of each angle is greater than, less than, or about equal to the measurement given.

27. 110°   28. 70°

Trace each segment or angle. Then construct a segment or angle congruent to it.

29.   30.

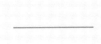

Draw the angle or line segment with the given measurement. Then use a straightedge and compass to bisect each angle or line segment.

31. 5 in.   32. 100°

33. 35°   34. 37 mm

Name each polygon.

35.   36.

Draw an example of each polygon.

37. octagon   38. hexagon

Objectives and Examples	Review Exercises

• describe and define lines of symmetry
(*Lesson 9-7*)

Find the lines of symmetry.

Tell whether the dashed line is a line of symmetry.

39.   40.

41. Tell the number of lines of symmetry for each figure above.

• determine congruence and similarity
(*Lesson 9-8*)

Tell whether each pair of polygons is congruent, similar, or neither.

congruent   similar   neither

Tell if each pair of polygons is congruent, similar, or neither.

42.    43.

Applications and Problem Solving

44. **Sewing** Philana has drawn the figure at the right as a design of a round tablecloth. She wants three more stripes from the center that will bisect the three angles. Copy her design and bisect the angles to show what the new design will look like. (*Lesson 9-4*)

45. There are 32 students in Mrs. Marcum's literature class. Of these, 12 take pre-algebra, 15 take tennis, and 8 take both pre-algebra and tennis. How many students take neither pre-algebra nor tennis? (*Lesson 9-6*)

Curriculum Connection Projects

• **Science** From an aerial photograph of a commercial airport, identify runways that form acute, obtuse, and right angles.
• **Geography** From a world map or globe, find three cities that when connected form an acute triangle, three that form an obtuse triangle, and three that form a right triangle.

Read More About It

Charosh, Mannis. *The Ellipse.*
Jonas, Ann. *Reflections.*
Sleater, William. *The Duplicate.*

9 Test

Classify each angle as acute, right or obtuse.

1.
2.
3.
4. $9.9°$
5. $100°$

Tell whether the measure of each angle is greater than, less than, or about equal to the measurement given.

6. $85°$
7. $48°$
8. $120°$

Trace each segment or angle. Then, using a straightedge and compass, construct a segment or angle congruent to it.

9.
10.
11.

Draw the angle or line segment with the given measurement. Then use a straightedge and compass to bisect each angle or line segment.

12. $4\frac{1}{2}$ cm
13. 3 in.
14. $110°$
15. $50°$

Name each polygon.

16.
17.

18. Draw an example of a quadrilateral.

19. **Sewing** The pattern piece at the right needs to be folded along a line of symmetry in order to cut the piece from material that is folded. Copy the figure and draw the line of symmetry.

Tell the number of lines of symmetry for each figure.

20.
21.

Tell if each pair of polygons is congruent, similar, or neither.

22.
23.
24.

Bonus If a quadrilateral has three acute angles, what must be true about the fourth angle?

Chapter Test

Chapter

9 Academic Skills Test

Directions: Choose the best answer. Write A, B, C, or D.

1. Admission to the water park costs \$10 for adults and \$8 for children. If 12 people paid \$102 for admission, how many were adults and how many were children?

   A. 7 adults, 4 children
   B. 5 adults, 6 children
   C. 4 adults, 8 children
   D. 3 adults, 9 children

2. 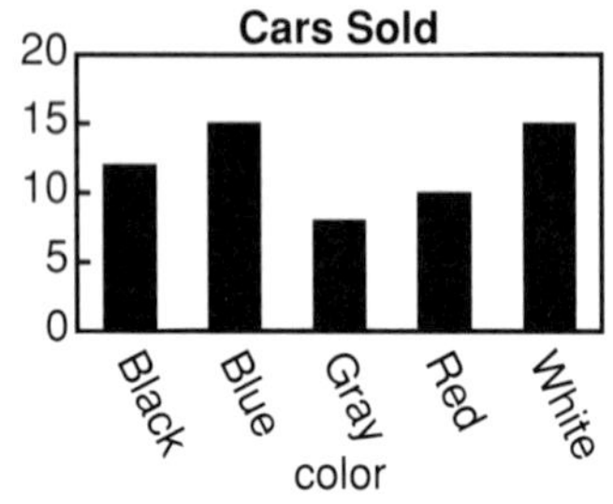

   How many of the cars sold were either white or blue?

   A. 15 B. 20
   C. 30 D. 60

3. $3.18 - 0.213 =$

   A. 1.05
   B. 2.973
   C. 2.967
   D. none of these

4. A football game started at 2:40 P.M. and ended at 5:15 P.M. How long was the game?

   A. 2 h 15 min
   B. 2 h 35 min
   C. 3 h 15 min
   D. 3 h 25 min

5. Gladys cut a ribbon into 5 equal pieces. If each piece was 2.4 meters long, how long was the ribbon before she cut it?

   A. 0.48 m B. 1.2 m
   C. 10.2 m D. 12 m

6. What is the area of the rectangle?

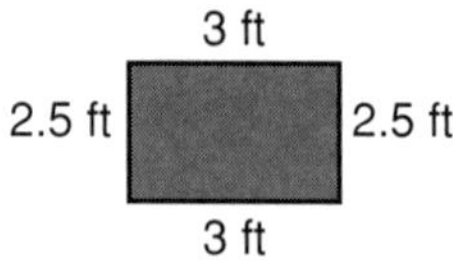

   A. 7.5 square feet
   B. 9 square feet
   C. 11 square feet
   D. 56.25 square feet

7. Allen is 1.6 meters tall. How many centimeters is this?

   A. 0.016 B. 16
   C. 160 D. 1,600

8. $12\frac{2}{3} =$

   A. $1\frac{8}{3}$ B. $\frac{24}{3}$
   C. $\frac{26}{3}$ D. $\frac{38}{3}$

9. If $n = 4.25 \div 1.5$, what is the value of n, to the nearest tenth?

   A. 0.016 B. 16
   C. 160 D. none of these

10. To the nearest whole number, 25.493 rounds to

   A. 20 B. 25
   C. 26 D. 30

11. Which fractions are in order from least to greatest?

   A. $\frac{1}{3}, \frac{3}{8}, \frac{5}{12}$

   B. $\frac{1}{3}, \frac{5}{12}, \frac{3}{8}$

   C. $\frac{3}{8}, \frac{1}{3}, \frac{5}{12}$

   D. $\frac{5}{12}, \frac{3}{8}, \frac{1}{3}$

12. What is $3\frac{2}{5}$ rounded to the nearest half?

   A. $1\frac{1}{5}$ B. 3

   C. $3\frac{1}{2}$ D. 4

13.

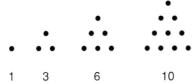

   1   3   6   10

   What is the next number in the pattern?

   A. 14   B. 15
   C. 18   D. 20

14. If $\frac{5}{16} + a = \frac{10}{16}$, what is the value of a?

   A. $\frac{5}{16}$ B. $\frac{15}{16}$

   C. $1\frac{5}{16}$ D. 5

15. $\frac{3}{4} \times \frac{4}{5} =$

   A. $\frac{3}{5}$ B. $\frac{15}{16}$

   C. $1\frac{2}{5}$ D. 12

16. Each serving of a pizza is $\frac{1}{16}$ of the pizza. If $\frac{3}{4}$ of the pizza is left, how many servings are left?

   A. 3   B. 4
   C. 12   D. 16

Test-Taking Tip

You can prepare for taking standardized tests by working through practice tests like this one. The more you work with questions in a format similar to the actual test, the better you become in the art of test taking.

Do not wait until the night before taking a test to review. Allow yourself plenty of time to review the basic skills and formulas that are tested. Find your weaknesses so that you can ask for help.

17. Which angle is acute?

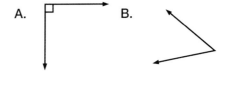

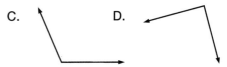

18.

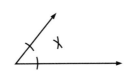

   This drawing shows how to

   A. construct an angle bisector
   B. construct an angle congruent to a given angle
   C. construct a segment bisector
   D. construct a segment congruent to a given segment

19. What is the value of 2^5?

   A. 10   B. 25
   C. 32   D. 64

Ratio, Proportion, and Percent

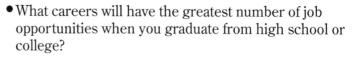

Spotlight on Careers

Have You Ever Wondered...

- What careers will have the greatest number of job opportunities when you graduate from high school or college?

- What are the most popular college majors?

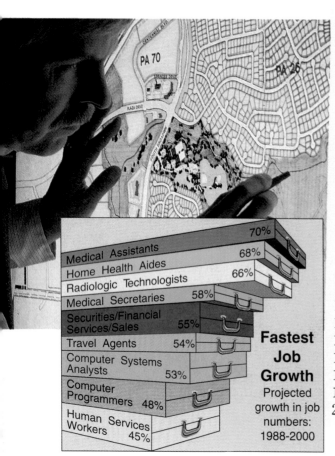

Medical Assistants 70%
Home Health Aides 68%
Radiologic Technologists 66%
Medical Secretaries 58%
Securities/Financial Services/Sales 55%
Travel Agents 54%
Computer Systems Analysts 53%
Computer Programmers 48%
Human Services Workers 45%

Fastest Job Growth
Projected growth in job numbers: 1988-2000

MOST POPULAR MAJORS AMONG INCOMING FRESHMEN, FALL 1988

Major	Percent
1. Business administration	7.3%
2. Accounting	6.1
3. Elementary education	4.9
4. Management	4.9
5. Psychology	4.1
6. Political science	3.2
7. Communications	2.9
8. Electrical engineering	2.9
9. Marketing	2.9
10. Premedicine, predental, preveterinary	2.8
11. Nursing	2.5
12. Arts	2.1
13. Finance	2.0
14. Therapy	2.0
15. General biology	1.8
16. Mechanical engineering	1.8
17. Aeronautical engineering	1.7
18. Computer science	1.7
19. Secondary education	1.7
20. Law enforcement	1.6

Looking Ahead

In this chapter, you will see how mathematics can be used to answer questions about careers of the future.

The major objectives of the chapter are to:

- determine whether two ratios are equal
- solve problems by using cross-products or drawing diagrams
- identify and draw similar figures
- express fractions and decimals as percents and vice versa
- find the percent of a number

Chapter Project

Careers
Work in a group.

1. Interview your friends to find out the careers they are interested in.

2. Make a chart or graph showing your results.

3. Ask several adults about their careers. Compare what each person does with the career he or she wanted at your age. Make a chart showing the results.

JIM DAVIS   10-2   © 1984 United Feature Syndicate, Inc.

10-1A Ratios

A Preview of Lesson 10-1

Objective

Explore ratios.

Materials

2 identical square sheets of paper
scissors

Have you ever put together a puzzle with seven pieces called a tangram? The tangram was first developed in China. It was a very popular puzzle in the 1800s. In this lab, you will construct a tangram and compare the sizes of the puzzle pieces.

Try this!

Work with a partner.

Use one sheet of paper for folding. Put the other sheet aside.

- Fold the top left corner of the paper to the bottom right corner. Crease the paper and cut along the fold. Label one of the triangles A. Place triangle A beside your uncut square.

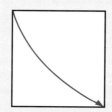 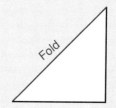 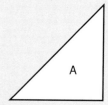

- How does the size of triangle A compare to the size of the uncut square?
- Using the unmarked triangle, place the longest side parallel to the edge of your desk as shown below. Fold the left corner to the right corner. Make a crease. Unfold the triangle.
- Then fold the top corner to the bottom edge along the crease.
- Crease and cut on this second crease line. Label the small triangle B.

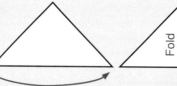 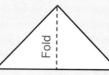 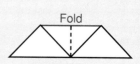

What do you think?

1. How does the size of triangle B compare to the size of the remaining piece?

2. a. How does the size of triangle B compare to the size of the uncut square?
   b. How many times larger is the uncut square?
   c. If the area of triangle B is 1 square unit, what is the area of the uncut square?

- Put triangle B aside.
- Place the remaining piece with the longest side facing you and cut on the crease as shown below. Use the piece on the left, folding the bottom left corner to the bottom right corner. Crease and cut on this fold. Label the triangle C and the square D.

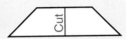

What do you think?

3. Compare the size of triangle C to the size of square D.
4. Compare the size of triangle C to the size of the uncut square.
5. a. Compare the size of square D to the size of the uncut square.
   b. What fraction could be used to show the part of the uncut square that square D is?
   c. If the area of square D is 2 square units, what is the area of the uncut square?

- Place triangle C and square D aside.
- Place the remaining piece with the longest side facing you. Fold the bottom left corner to the top right corner. Crease and cut along the crease. Label the triangle E and the parallelogram F.

What do you think?

6. Compare the size of triangle E to the size of parallelogram F.
7. Compare the size of triangle E to the size of the uncut square.
8. a. Compare the size of parallelogram F to the size of the uncut square.
   b. Compared to parallelogram F, how many times larger is the uncut square?

10-1 **Ratios and Rates**

Objective

Express ratios and rates as fractions.

Words to Learn

ratio
rate

Read My Lips, Kid!™, a charade-like board game, was chosen as the favorite game by 24 out of 36 students in a recent survey. You can compare these two numbers using a ratio. A **ratio** is a comparison of two numbers by division. The ratio that compares 24 to 36 can be written in several ways.

24 to 36   24:36   24 out of 36 $\frac{24}{36}$

A common way to express a ratio is as a fraction in simplest form.

$$\frac{24}{36} = \frac{2}{3} \quad \begin{array}{l} \text{\scriptsize ÷ 12} \\ \text{\scriptsize ÷ 12} \end{array}$$

The GCF of 24 and 36 is 12. Divide the numerator and denominator by the GCF.

So, $\frac{2}{3}$ of the students, or 2 out of 3 students, chose *Read My Lips, Kid!*™ as their favorite game.

Example 1 *Connection*

Geometry The tangram from the Mathematics Lab is shown at the right. If triangle C has an area of 1 square unit, then square D has an area of 2 square units. Write a ratio that compares the area of triangle C to the area of square D.

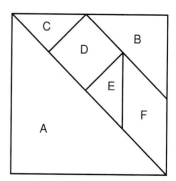

The ratio can be written in several ways.

1 to 2 or 1:2 or $\frac{1}{2}$ ← *area of C*
                                  ← *area of D*

Example 2

Express the ratio *twelve used cars out of 20 cars* as a fraction in simplest form.

$$\frac{12 \text{ used cars}}{20 \text{ cars}} = \frac{3 \text{ used cars}}{5 \text{ cars}} \quad \begin{array}{l} \text{\scriptsize ÷ 4} \\ \text{\scriptsize ÷ 4} \end{array}$$

The GCF of 12 and 20 is 4. Divide the numerator and denominator by 4.

If the quantities that you are comparing have different units of measure, then the ratio is called a **rate.** For example, $\frac{100 \text{ miles}}{2 \text{ hours}}$ compares the number of miles traveled to the number of hours the trip took. Usually rates are expressed in a *per unit* form.

Example 3 *Problem Solving*

Travel   Tina Clark drove 180 kilometers in 3 hours. Use a rate to compare these numbers. Express the rate in kilometers per hour.

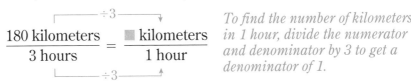

$$\frac{180 \text{ kilometers}}{3 \text{ hours}} = \frac{\blacksquare \text{ kilometers}}{1 \text{ hour}}$$

To find the number of kilometers in 1 hour, divide the numerator and denominator by 3 to get a denominator of 1.

Tina drove 60 kilometers per hour.

Checking for Understanding

Communicating Mathematics

Read and study the lesson to answer each question.

1. **Tell** the difference between ratio and rate.
2. **Tell** how the sentence *15 out of 24 students in Ms. Noel's class are girls.* can be written as a ratio.

Guided Practice

Write each ratio in three ways.

3. 13 out of 16 lockers have combination locks.
4. 25 out of 36 cookies had at least 8 chocolate chips.

Express each ratio as a fraction in simplest form.

5. win 16 games out of 24
6. 12 red ants out of 18 ants
7. 2 double-yolked eggs in a dozen eggs
8. 15 white pairs of socks out of 24 pairs

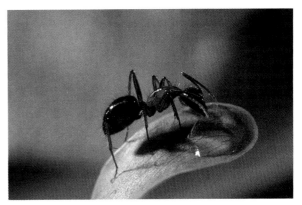

Express each ratio as a rate.

9. 160 miles on 5 gallons of gasoline
10. 6 apples for $2
11. 400 kilometers in 5 hours

Compare the areas of the tangram pieces in Example 1 on page 352 by writing each ratio as a fraction in simplest form.

12. C to A
13. F to A
14. D to A
15. D to C
16. D to F
17. F to B

Exercises

Express each ratio as a fraction in simplest form.

18. 4 blue houses in 14 houses on Ron's block

19. 85 out of 100 questions correct

20. 10 drums in a 75-instrument band

21. 36 honor students out of 96 students

22. $1\frac{1}{4}$ cup raisins in $2\frac{1}{3}$ cups of batter

Using the letters of the phrase "RATIOS AND RATES," write the ratios comparing the numbers of letters as a fraction in simplest form.

23. N to S

24. D to T

25. T to S

26. I to A

27. R to A

28. vowels to consonants

29. **Geometry**   Tell whether the pair of polygons at the right are congruent, similar, or neither. *(Lesson 9-8)*

30. Write the number twenty-seven trillion, twenty-seven thousand, twenty-seven in standard form. *(Lesson 2-1)*

31. Measure the segment ——————————————— to the nearest sixteenth of an inch. *(Lesson 6-7)*

32. **Volunteering**   Angela is planning a party for the residents of a local nursing home. She needs the help of about $\frac{2}{5}$ of her class. If there are 32 students in her class, about how many helpers does she need? *(Lesson 8-1)*

33. **Geography**   6 out of 50 states are in the New England region. What fraction of the United States is in the New England region?

34. **Consumer Math**   *Pop Secret®* popcorn claims that only 250 out of every 500 popcorn kernels meet its high standards. Write a ratio comparing the numbers of good kernels to the number examined in simplest form.

35. **Critical Thinking**   Mr. Kareem stated that 18 out of 24 students in his class received an 85% or higher on the last social studies test. He instructed the class to write a ratio comparing the number of students below 85% with the total number of students. Explain why the following answers are incorrect.

a. 18:24   b. 24:6   c. 24:18

36. **Critical Thinking**   Draw two squares on grid paper. The ratio of the area of one square to the area of the other square should be 1:4.

37. **Journal Entry**   Design an advertisement for a product using a ratio.

10-2 Solving Proportions

Objective

Solve proportions by using cross products.

Words to Learn

proportions
cross products

LOOKBACK

You can review equivalent fractions on page 204.

Have you ever heard an advertisement on TV that sounds like this?

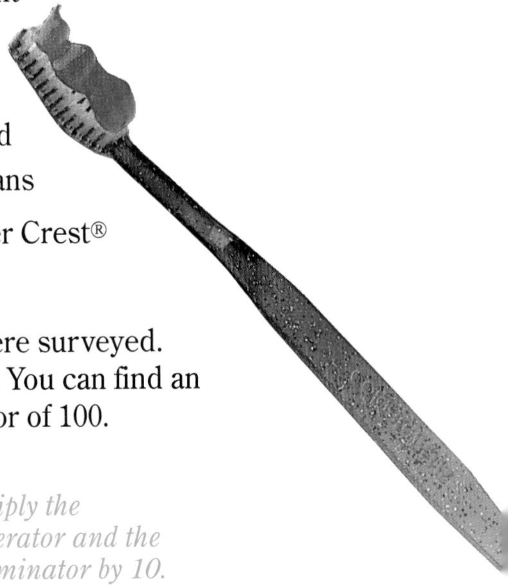

8 out of 10 dentists prefer Crest®.

The ratio *8 out of 10* can be expressed as the fraction $\frac{8}{10}$. The statement means that $\frac{8}{10}$ of the dentists surveyed prefer Crest® to other brands of toothpaste.

Most likely, more than 10 dentists were surveyed. Suppose 100 dentists were surveyed. You can find an equivalent fraction with a denominator of 100.

$$\frac{8}{10} = \frac{80}{100}$$

Multiply the numerator and the denominator by 10.

You can estimate that 80 out of 100, or 80 dentists, prefer Crest®. The two fractions $\frac{8}{10}$ and $\frac{80}{100}$ are equivalent.

An equation stating that two ratios are equivalent is called a **proportion.** One way to determine if two ratios are equivalent is to find their **cross products.** If the cross products of two ratios are equal, then the ratios are equivalent.

$$\frac{8}{10} \diagdown \frac{80}{100} \qquad \begin{array}{l} 8 \cdot 100 = 800 \\ 10 \cdot 80 = 800 \end{array}$$

In the proportion shown above, 8×100 and 80×10 are cross products.

Examples

Use cross products to determine whether each pair of ratios forms a proportion.

1 $\frac{4}{5}, \frac{12}{15}$

$$\frac{4}{5} \diagdown \frac{12}{15}$$

$$4 \times 15 \overset{?}{=} 5 \times 12 \qquad \textit{Write cross products.}$$
$$60 = 60 \qquad \textit{Multiply.}$$
$$\text{So, } \frac{4}{5} = \frac{12}{15}.$$

2 $\frac{3}{7}, \frac{5}{21}$

$$\frac{3}{7} \diagdown \frac{5}{21}$$

$$3 \times 21 \overset{?}{=} 7 \times 5$$
$$63 \neq 35$$
$$\text{So, } \frac{3}{7} \neq \frac{5}{21}.$$

You can use cross products to find a missing value of a proportion.

Mental Math Hint
• • • • • • • • • •

In some cases, you can solve a proportion mentally by using equivalent fractions.
$$\frac{3}{4} = \frac{x}{16}$$
THINK: $4 \times 4 = 16$
$3 \times 4 = 12$
So, $x = 12$.

Examples

Solve each proportion.

3 $\dfrac{6}{7} = \dfrac{n}{42}$

$6 \times 42 = 7 \times n$ *Write cross products.*
$252 = 7n$ *Multiply.*
$\dfrac{252}{7} = \dfrac{7n}{7}$ *Divide.*
$36 = n$

The solution is 36.

4 $\dfrac{1.2}{d} = \dfrac{2.4}{3}$

$1.2 \times 3 = d \times 2.4$
$3.6 = 2.4d$
$\dfrac{3.6}{2.4} = \dfrac{2.4d}{2.4}$
$1.5 = d$

The solution is 1.5.

Example 5 · *Problem Solving*

Entertainment On *Family Feud*, a TV game show, a survey said that 40 out of 100 people chose chocolate as their favorite flavor of ice cream. If there were actually 120 people in the audience when the survey was taken, how many would have chosen chocolate?

First, set up a proportion of two ratios that compare the number of people who chose chocolate to the number of people surveyed. Let c represent the number of people out of 120 who chose chocolate.

no. who chose chocolate → $\dfrac{40}{100} = \dfrac{c}{120}$ ← *no. who chose cholocate*
total →   ← *total*

Because the ratios are equivalent, the cross products must be equal.

$$\frac{40}{100} \times \frac{c}{120}$$

$40 \times 120 = 100 \times c$ *Write the cross products.*
$4,800 = 100c$ *Multiply.*
$\dfrac{4,800}{100} = \dfrac{100c}{100}$ *Divide each side by 100.*
$48 = c$

48 people out of 120 would have chosen chocolate.

Checking for Understanding

Communicating Mathematics

Read and study the lesson to answer each question.

1. **Write,** in your own words, a definition for proportion.
2. **Tell** how to determine if these two ratios form a proportion. $\dfrac{3}{8}, \dfrac{375}{1,000}$

Guided Practice

Use cross products to determine whether each pair of ratios forms a proportion.

3. $\dfrac{2}{13}, \dfrac{4}{26}$ 4. $\dfrac{7}{12}, \dfrac{8}{15}$ 5. $\dfrac{1}{6}, \dfrac{4}{24}$

Solve each proportion.

6. $\dfrac{5}{9} = \dfrac{n}{54}$

7. $\dfrac{4}{5} = \dfrac{8}{d}$

8. $\dfrac{3}{5} = \dfrac{x}{100}$

9. $\dfrac{5}{11} = \dfrac{c}{88}$

10. $\dfrac{3}{10} = \dfrac{18}{a}$

11. $\dfrac{b}{12} = \dfrac{42}{72}$

Exercises

Independent Practice

Use cross products to determine whether each pair of ratios forms a proportion.

12. $\dfrac{8}{25}, \dfrac{24}{75}$

13. $\dfrac{21}{28}, \dfrac{3}{12}$

14. $\dfrac{28}{35}, \dfrac{4}{5}$

Solve each proportion.

15. $\dfrac{6}{10} = \dfrac{n}{100}$

16. $\dfrac{1}{4} = \dfrac{16}{s}$

17. $\dfrac{16}{25} = \dfrac{x}{100}$

18. $\dfrac{23}{20} = \dfrac{c}{100}$

19. $\dfrac{3}{5} = \dfrac{a}{80}$

20. $\dfrac{3}{5} = \dfrac{24}{b}$

21. $\dfrac{n}{100} = \dfrac{12}{60}$

22. $\dfrac{k}{100} = \dfrac{7}{25}$

23. $\dfrac{6}{9} = \dfrac{10}{n}$

24. $\dfrac{8}{12} = \dfrac{a}{21}$

25. $\dfrac{25}{100} = \dfrac{12}{b}$

26. $\dfrac{36}{k} = \dfrac{45}{100}$

State a proportion for each of the following.

27. If you can buy 3 CDs for $41.97, how many can you buy for $69.95?

28. If you watch 2 hours of TV per day, how many hours would you watch in 30 days?

Mixed Review

29. Express the ratio, 2 student council representatives out of 28 students in the class, as a fraction in simplest form. *(Lesson 10-1)*

30. **Geometry** Draw a line segment that is $2\frac{1}{2}$ inches long and then bisect it using a straightedge and compass. *(Lesson 9-4)*

31. Evaluate the expression $12 - 8 \div 2 + 1$. *(Lesson 1-7)*

Problem Solving and Applications

32. **Sports** A player's batting average in baseball is the number of hits he would make in 1,000 times at bat.

   a. If a player makes 2 hits in 5 times at bat, how many hits can he expect to make in 20 times at bat?

   b. What is his batting average?

33. **Education** There are 2 teachers for every 57 students at Elmwood Middle School. If there are 855 students, how many teachers are at the school?

34. **Geometry** Using the tangram puzzle that you made on pages 350 and 351, which ratios of areas form proportions?

   a. C:D, E:B

   b. E:C, B:D

   c. D:A, E:F

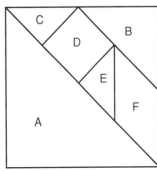

35. **Critical Thinking** If 17 out of 110 people said they liked canoeing, how many people out of 100 would that be?

10-2B Equal Ratios

A Follow-Up of Lesson 10-2

Objective

Build equal ratios.

Materials

pattern blocks

Use the following lab to help you "build" equal ratios.

Try this!

- Use the triangle pattern block as the unit of measure.

- Build another shape that has a ratio equal to each ratio shown below.

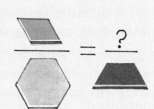

a. $\dfrac{\triangle}{\text{trapezoid}} = \dfrac{\triangle\triangle}{?}$ b. $\dfrac{\text{rhombus}}{\text{hexagon}} = \dfrac{?}{\text{trapezoid}}$

c. $\triangle : \text{rhombus} = \text{rhombus} : ?$

d. $\text{trapezoid} : \text{hexagon} = ? : \text{trapezoid}$

- Build another shape that has a ratio *not* equal to each ratio shown below.

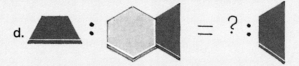

e. $\dfrac{\text{rhombus}}{\text{trapezoid}} = \dfrac{\triangle}{?}$ f. $\dfrac{\triangle}{\text{trapezoid}\,\triangledown} = \dfrac{?}{\text{rhombus}}$

What do you think?

1. How did you find the equal ratios?
2. How did you know when two ratios were *not* equal?

10-3 Draw a Diagram

Objective

Solve problems by drawing a diagram.

Donita and her friends met at the entrance to the Manley Amusement Park. They paid their admission and walked 2 kilometers north to the Pirate ride. After they rode the ride, they walked east 1 kilometer to the Fun House. Next, the girls walked 0.5 kilometers northeast to the refreshment stand to buy drinks. From there they walked 1 kilometer southwest to the roller coaster. Donita decided she wanted to ride the Pirate ride again. Which direction did they need to walk?

Explore

What do you know?
The group met at the entrance.
They walked 2 kilometers north to the Pirate ride.
Then they walked 1 kilometer east to the Fun House.
Next they walked 0.5 kilometers northeast to the refreshment stand.
Finally, they walked 1 kilometer southwest to the roller coaster.

What do you need to find?
You need to know which direction they need to walk to get from the roller coaster to the Pirate ride.

Plan

To help you understand and picture the information, you can draw a diagram.

Solve

Draw the diagram. Label each part of the girls' path and indicate the direction North. The diagram shows that the girls need to walk in a northwest direction to get from the roller coaster to the Pirate ride.

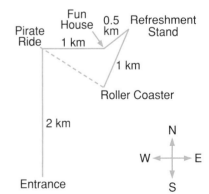

Examine

By looking at the diagram and directional markings, you can see that the girls need to walk in a northwest direction.

Checking for Understanding

Read and study the lesson to answer each question.

1. **Draw** another amusement ride in your diagram of Manley Amusement Park. Give directions to a friend explaining how to get from the entrance to the new ride.

Solve by drawing a diagram.

2. Five students competed in an egg toss competition. Julie came in second, and David finished behind Marco. Keela finished ahead of Marco, and Reynelda won first place. In what order did they finish?

3. Al has a piece of lumber that is 140 inches long. He wants to cut it into 20-inch long pieces. How many cuts does he need to make?

4. The Berstein family lives in Littleton. Kayla works in Parker. There is no direct route from Littleton to Parker, so Kayla goes through either Stony Creek or Castle Rock. There is one road between Stony Creek and Castle Rock. How many different ways can Kayla drive to work?

Problem Solving

5. Dwayne is going to wallpaper one wall in his bedroom. The wall is 12 feet wide and 8 feet tall and has a door that is 7 feet tall and 3 feet wide. If wallpaper comes in rolls 1 yard wide by 12 yards long, how many rolls will he need to wallpaper the wall?

Strategies

● ● ● ● ● ● ●

Look for a pattern.

Solve a simpler problem.

Act it out.

Guess and check.

Draw a diagram.

Make a chart.

Work backward.

6. A number multiplied by itself is 625. What is the number?

7. The Avarillo's backyard is 130 feet by 90 feet. Their manual lawn sprinkler can water an area 30 feet by 30 feet. How many times does Mr. Avarillo need to move the sprinkler to water the entire backyard?

8. What are the next two numbers in the pattern: 56, 43, 48, 35, 40, ... ?

9. Valerie rode her bicycle 3 kilometers from school to her piano lesson. Then she rode 1 kilometer to the library. From the library, she went home. Valerie traveled 8 kilometers in all. How far is it from Valerie's house to the library?

10. Ben uses a tent for summer camping. Each of the four sides of the tent needs 3 stakes to secure it to the ground. How many stakes does he need for the whole tent?

11. List all the lengths and widths of all the rectangles that can be created using seventy-six 1×1 titles.

10-4 Scale Drawings

Objective

Find actual length from a scale drawing.

Words to Learn

scale drawing

You can review similar figures on page 337.

A **scale drawing** shows an object exactly as it looks, but it is generally either smaller or larger. The scale gives the ratio which compares the lengths on the drawing to the actual lengths of the object. Since the scale drawing and the original figure have corresponding sides proportional and corresponding angles congruent, they are similar figures.

The drawing of the bicycle has a scale of 1 inch = 2 feet. The distance between the centers of the wheels or wheelbase measures $1\frac{3}{4}$ inches. What is the actual distance?

$1\frac{3}{4}$ inches

You can write a proportion to find the actual distance, *d.*

$$\begin{array}{cc} \textit{length in drawing} \rightarrow \\ \textit{actual length} \rightarrow \end{array} \quad \frac{1 \text{ inch}}{2 \text{ feet}} = \frac{1\frac{3}{4} \text{ inches}}{d \text{ feet}} \quad \begin{array}{c} \leftarrow \textit{length in drawing} \\ \leftarrow \textit{actual length} \end{array}$$

$$1 \times d = 2 \times 1\frac{3}{4} \qquad \textit{Find the cross products.}$$

$$d = 3\frac{1}{2}$$

The actual wheelbase is $3\frac{1}{2}$ feet or 42 inches.

When am I ever going to use this?

Interior Designer

Creating residential designs often involves a balance of color, texture, lighting, scale, and proportion. Designers use scale drawings to give their clients an idea of what each room will look like after it is decorated. For more information, contact your local chapter of the American Society of Interior Designers.

Example

Find the actual diameter of the wheels of the bicycle above.

On the scale drawing, the diameter *(d)* of the wheels measures $1\frac{1}{8}$ inches. Write a proportion to solve the problem.

$$\frac{1 \text{ inch}}{2 \text{ feet}} = \frac{1\frac{1}{8} \text{ inch}}{d}$$

$$1 \times d = 2 \times 1\frac{1}{8} \qquad \textit{Find the cross products.}$$

$$d = 2\frac{1}{4}$$

The diameter of the wheels is $2\frac{1}{4}$ feet or 27 inches.

Checking for Understanding

Communicating Mathematics

Read and study the lesson to answer each question.

1. **Tell** whether a scale drawing of a shopping mall would be smaller or larger than the actual mall. Explain.

2. **Write** the name of the part of a drawing that must always be given in a scale drawing.

Guided Practice

3. The drawing of the car has a scale of $\frac{1}{4}$ inch = 28 inches. Use a ruler to measure the drawing to the nearest $\frac{1}{8}$ inch and compute the actual measurements of the car.

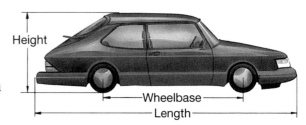

	Drawing	Actual
a. length		
b. height		
c. wheelbase		
d. wheel diameter		

Exercises

Independent Practice

4. The floor plan at the right has a scale of 1 centimeter = 6 feet. Use a centimeter ruler to measure each room to the nearest tenth of a centimeter. Then compute the actual measurements.

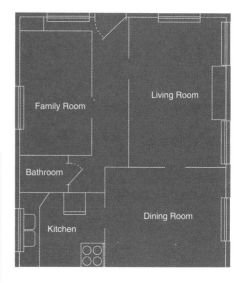

Rooms	Drawing		Actual	
	length	width	length	width
a. living				
b. dining				
c. kitchen				
d. bath				
e. family				

5. The map at the right has a scale of $\frac{1}{4}$ inch = 10 miles. Use a ruler to measure each map distance to the nearest $\frac{1}{4}$ inch. Then compute the actual measurements.

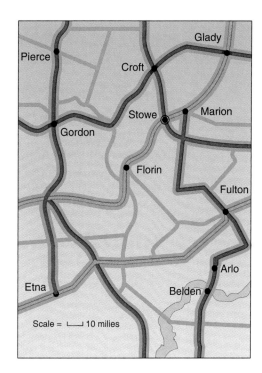

	Map	Actual
a. Arlo to Belden		
b. Glady to Marion		
c. Marion to Fulton		
d. Fulton to Etna		
e. Etna to Gordon		

Scale = ⌊___⌋ 10 miles

Mixed Review

6. Find the product of $\frac{3}{8}$ and $\frac{7}{15}$. *(Lesson 8-2)*

7. Divide 13.56 by 2.4. *(Lesson 5-3)*

Problem Solving and Applications

8. **Computers**   An engineer made an 8-inch by 10-inch scale drawing of a computer chip. If the scale on the drawing was 4 inches to 1 centimeter, find the dimensions of the chip.

9. **Yearbook**   An original 5- by 8-inch photo must be reduced to $1\frac{1}{4}$ by 2 inches to fit in the yearbook. What is the scale of the reduced photo to the original in simplest form?

10. **Critical Thinking**   If you were asked to do a scale drawing of the Statue of Liberty, which scale would you use: 1 cm : 1 m or 1 cm : 1 mm?

10 Mid-Chapter Review

1. Express the ratio *40 out of 60 questions* in simplest form. *(Lesson 10-1)*

2. Solve the proportion $\frac{7}{6} = \frac{h}{12}$. *(Lesson 10-2)*

Solve by drawing a diagram. *(Lesson 10-3)*

3. Audra made a list of her father's parents, grandparents, great grandparents, and great-great grandparents. If Audra did not list any stepgrandparents, how many people are listed?

4. Refer to the map at the top of the page. Measure the distance from Gordon to Pierce. Then compute the actual measurement. *(Lesson 10-4)*

Choosing a Personal Stereo

Situation

Your school had a fund-raising drive selling recycled wrapping paper to the community. As the top salesperson, you will receive a personal stereo which you will select from a catalog. Different stereos, made by different manufacturers, all have different features. Which will you choose?

Hidden Data

Does each personal stereo come with a manufacturer's warranty? Do all stereos come with rechargeable and disposable battery options?

Key	Manuf.	Model	AM/FM Stereo	Clock/ Alarm	Cassette	Bass Boost	Auto Reverse	Graphic EQ	Dolby NR	Ref.	Your Price
13	Nelson	AC2120	Yes	Yes	Yes	Yes	No	No	No	$69.99	$47.99
14	Walkabout	WMFX30	Yes	No	Yes	No	Yes	No	No	$54.95	$49.99
15	Walkabout	WMFX33	Yes	No	Yes	Yes	Yes	No	No	$62.95	$54.99
16	Supersonic	RQV162	Yes	No	Yes	Yes	Yes	3-band	Yes	$69.95	$54.99
17	Walkabout	WMF2031	Yes	Yes	Yes	No	No	No	No	$69.95	$59.99
18	Walkabout	WMF2081	Yes	Yes	Yes	No	Yes	No	No	$79.95	$69.99
19	Walkabout	WMF2041	Yes	No	Yes	No	No	No	No	$89.95	$79.99

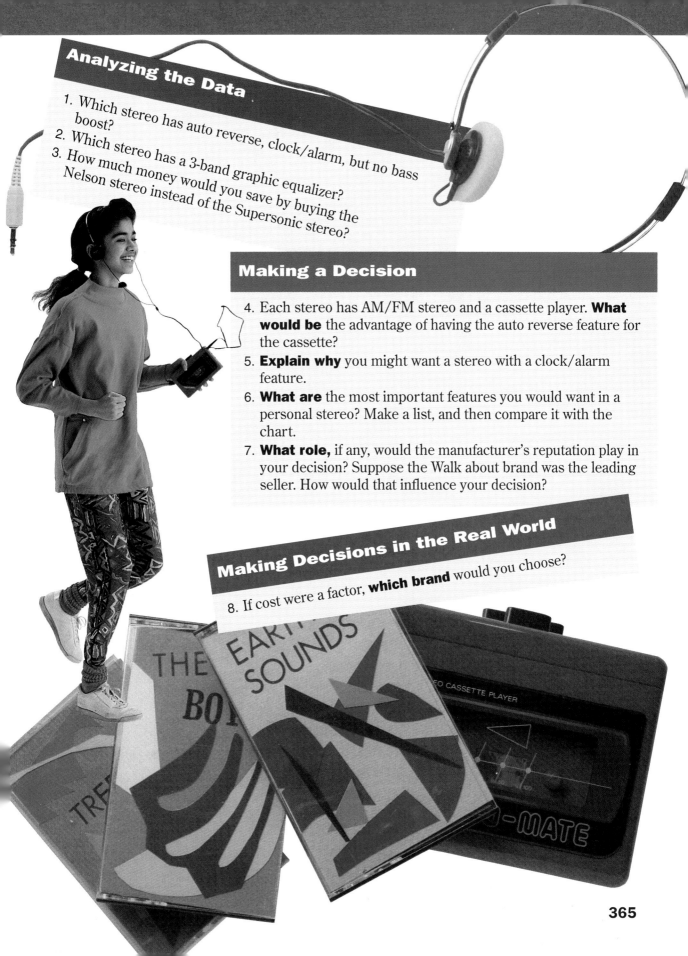

1. Which stereo has auto reverse, clock/alarm, but no bass boost?
2. Which stereo has a 3-band graphic equalizer?
3. How much money would you save by buying the Nelson stereo instead of the Supersonic stereo?

Making a Decision

4. Each stereo has AM/FM stereo and a cassette player. **What would be** the advantage of having the auto reverse feature for the cassette?
5. **Explain why** you might want a stereo with a clock/alarm feature.
6. **What are** the most important features you would want in a personal stereo? Make a list, and then compare it with the chart.
7. **What role,** if any, would the manufacturer's reputation play in your decision? Suppose the Walk about brand was the leading seller. How would that influence your decision?

Making Decisions in the Real World

8. If cost were a factor, **which brand** would you choose?

10-5A Percent

A Preview of Lesson 10-5

Objective

Illustrate the meaning of percent using models.

Words to Learn

percent

Materials

10 x 10 grids
colored pencils

In Chapter 6, you used 10×10 grids to help you write decimals as fractions. In this lab, you will use 10×10 grids to help you write fractions as **percents.** The word *percent* comes from Latin words meaning "parts of a hundred."

Try this!

Work with a partner.

● Have one partner name any fraction that has a denominator of 2, 4, 5, or 10. For example, $\frac{3}{4}$ has a denominator of 4.

● Have the other partner separate a 10×10 grid into the number of equal parts named in the denominator of the fraction. One way to separate the grid into 4 equal parts is shown below.

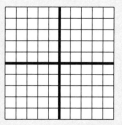

● Have the first partner color in the number of parts named in the numerator of the fraction.

● Have the other partner read the fraction as parts of a hundred. In the example, $\frac{3}{4}$ is the same as $\frac{75}{100}$. Now read this fraction as a percent. $\frac{75}{100}$ is read *75 percent,* or 75%.

● Keep a written record of the original fraction, the hundredths fraction, and the percent.

● Reverse roles and repeat the activity using a different fraction.

What do you think?

1. How did you decide how to separate your grid?

2. How can you use the cross product rule to see if your original fraction and the hundredths fraction are equivalent?

3. Tell whether the grid at the right represents the fraction $\frac{3}{4}$. Explain your reasoning.

Name the percent shaded on each chart.

4.

5.

Extension

6. Shade 36 squares on a 10×10 grid. Write the number of shaded squares as a fraction and a percent.

7. Describe what you think would be an efficient method of changing fractions into percents. Try your method with some "tough" fractions, such as $\frac{3}{8}$ or $\frac{2}{3}$.

8. Would the same method work for changing percents into fractions? Explain.

9. Copy and complete the diagrams so each one shows 60%.

a.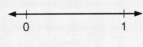

c.

b.

d.

10-5 Percents and Fractions

Objective

Express percents as fractions and vice versa.

Words to Learn

percent

DID YOU KNOW

The movie *Ghost* was the top movie in 1990 with rentals to distributors totaling $94,000,000.

If you think it is expensive to buy a movie ticket now, hold on to your seat! Entertainment executives were asked to predict the average movie ticket price by the year 2000. The predictions are as follows.

The predictions are given as percents. A **percent** is a ratio that compares a number to 100. All of the percents in the circle graph can be expressed as fractions. To do this, express the percent as a fraction with a denominator of 100 and then simplify.

Example 1

58% of the executives said that a movie ticket would cost between $10 and $15 in the year 2000. Express the percent as a fraction and simplify.

58% means "58 out of 100".

$$\frac{58}{100} = \frac{29}{50}$$

Simplify.
Divide the numerator and denominator by 2, the GCF of 58 and 100.

58% is the same as $\frac{29}{50}$.

Mini-Lab

Work in groups of four.

Materials: pencil, paper, and a wastebasket

We can also express fractions as percents.

- All members should stand behind a line set at 15 feet from the wastebasket.
- Have each member throw wadded paper balls into the basket. Each member will have five tries.
- Have one member record a ratio for the number of shots made out of the number of tries for the group.
- Calculate the percent of shots made for the group.

Talk About It

a. How did you find the percent?

b. How does your group's percent compare to other groups?

Mental Math Hint

• • • • • • • • • •

When changing $\frac{1}{4}$ to a percent, think of finding $\frac{1}{4}$ of 100 parts. $\frac{1}{4}$ = 25%

Examples

2 Express $\frac{1}{4}$ as a percent.

$$\frac{1}{4} = \frac{n}{100}$$ *Set up a proportion.*

$1 \times 100 = 4 \times n$ *Find the cross products.*

$100 = 4n$

$100 \div 4 = 4n \div 4$ *Divide.*

$25 = n$

So, $\frac{1}{4}$ is equivalent to 25%.

3 Express $\frac{3}{2}$ as a percent.

$$\frac{3}{2} = \frac{d}{100}$$

$3 \times 100 = 2 \times d$

$300 = 2d$

$300 \div 2 = 2d \div 2$

$150 = d$

So, $\frac{3}{2}$ is equivalent to 150%.

Checking for Understanding

Communicating Mathematics

Read and study the lesson to answer each question.

1. **Show** the fraction $\frac{2}{5}$ on a 10×10 grid like the one used in the Mathematics Lab on page 366.

2. **Tell** what percent represents the fraction $\frac{17}{25}$.

Guided Practice

Write a fraction that represents the number of sections shaded. Then express each fraction as a percent.

3.

4.

5.

Express each percent as a fraction.

6. 1%   7. 6%   8. 34%   9. 20%

Exercises

Independent Practice

Express each fraction as a percent.

10. $\dfrac{63}{100}$ 11. $\dfrac{1}{5}$ 12. $\dfrac{1}{4}$ 13. $\dfrac{1}{10}$

14. $\dfrac{12}{25}$ 15. $\dfrac{4}{5}$ 16. $\dfrac{37}{50}$ 17. $\dfrac{13}{20}$

18. Express seven tenths as a percent.

Express each shaded section as a fraction and as a percent.

19.   20.   21.

Express each percent as a fraction.

22. 66%   23. 12%   24. 45%   25. 80%

26. 85%   27. 16%   28. 4%   29. 110%

30. Express seventy-seven percent as a fraction.

Use a 10 × 10 grid to shade the amount stated in each fraction. Then express each fraction as a percent.

31. $\dfrac{3}{5}$ 32. $\dfrac{3}{10}$ 33. $\dfrac{3}{20}$ 34. $\dfrac{9}{50}$

35. 2 out of 3 students wear tennis shoes. What percent of students wear tennis shoes?

Mixed Review

36. Estimate the product of 52.9 and 99.8. *(Lesson 4-1)*

37. Estimate $2\dfrac{2}{7} - 1\dfrac{1}{4}$. *(Lesson 7-2)*

38. Find the greatest common factor of 45 and 75. *(Lesson 6-4)*

39. Find $3\dfrac{4}{5} \times 7\dfrac{1}{2}$. *(Lesson 8-3)*

Problem Solving and Applications

40. **Environment** The graph at the right shows the types of garbage disposed of (in percents) in an average community. What fraction represents the compostable garbage that can be made into fertilizer?

41. **Critical Thinking** A wealthy man made a will leaving $\dfrac{1}{2}$ of his fortune to his wife, $\dfrac{1}{3}$ to his son, $\dfrac{1}{7}$ to his nephew, and the remainder to his cat. What portion to the nearest whole percent of the man's fortune was left to the cat?

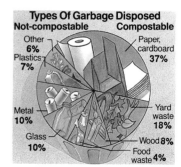

Types Of Garbage Disposed

Not-compostable   Compostable

Other 6%
Plastics 7%
Metal 10%
Glass 10%

Paper, cardboard 37%
Yard waste 18%
Wood 8%
Food waste 4%

10-6 Percents and Decimals

Objective

Express percents as decimals and vice versa.

DID YOU KNOW

The record time for one person making a bed is 28.2 seconds. The record was set following strict rules at the Australian Bedmaking Championship.

What type and size of bed do you sleep on? Did you know that not everyone sleeps on a frame bed and mattress? In Japan and Latin America, many people sleep on the floor on straw mats. In hot climates, some people sleep in hammocks.

When buying a bed, most people purchase one that is twin size. The graph below shows the bed sizes and the percentage bought in each size.

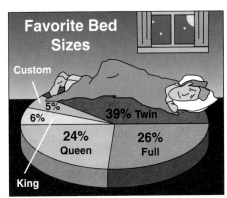

Favorite Bed Sizes

Custom 5%
6%
24% Queen
King
39% Twin
26% Full

Percents can be renamed as decimals. Remember that percent means parts of one hundred or hundredths. To express a percent as a decimal, rewrite the percent as a fraction with a denominator of 100. Then express the fraction as a decimal.

24% who prefer queen-sized beds can be thought of as "24 out of 100," $\frac{24}{100}$, or 0.24.

Examples

Mental Math Hint

• • • • • • • • • •

In Examples 1-3, the decimal point was moved two places to the left as a result of dividing by 100.

1 What part of the people surveyed preferred twin-sized beds? Express your answer as a decimal.

From the graph, 39% prefer twin-sized beds.

$39\% = \frac{39}{100}$ *Write the percent as a fraction with a denominator of 100.*

$= 0.39$ *Write the fraction as a decimal.*

The graph shows that 0.39 of the people surveyed preferred twin-sized beds.

Express each percent as a decimal.

2 16%

$16\% = \frac{16}{100}$

$= 0.16$

3 0.4%

$0.4\% = \frac{0.4}{100}$ $\quad \frac{0.4}{100} \times \frac{10}{10} = \frac{4}{1,000}$

$= \frac{4}{1,000}$

$= 0.004$

To express a decimal as a percent, first express the decimal as a fraction with a denominator of 100. Then express the fraction as a percent.

Mental Math Hint
• • • • • • • • • •
In Examples 4-5, the decimal point was moved two places to the right as a result of multiplying by 100.

Examples

Express each decimal as a percent.

4 0.84

$$0.84 = \frac{84}{100}$$
$$= 84\%$$

5 0.313

$$0.313 = \frac{31.3}{100}$$
$$= 31.3\%$$

Checking for Understanding

Communicating Mathematics

Read and study the lesson to answer each question.

1. **Write** Frances' score as a decimal if Frances received an 84% on her last math test.

2. **Write** the percent of people who prefer king-sized beds as a decimal.

3. **Tell** whether the sentence 8% = 0.8 is correct.

4. **Tell** whether each decimal is more or less than 100%.
   a. 0.45   b. 0.139   c. 2.89   d. 1.25   e. 0.05   f. 0.5

Guided Practice

Express each decimal as a percent.

5. 0.37   6. 0.7   7. 0.125   8. 0.07

Express each percent as a decimal.

9. 33%   10. 40%   11. 37.5%   12. 4.8%

13. Express forty-six percent as a decimal.

Exercises

Independent Practice

Express each decimal as a percent.

14. 0.84   15. 0.17   16. 0.03   17. 0.06

18. 0.006   19. 0.578   20. 0.21   21. 1.25

22. Express six and thirty-two hundredths as a percent.

Express each percent as a decimal.

23. 23%   24. 99%   25. 65%   26. 8%

27. 1%   28. 87.5%   29. 125%   30. 4.2%

31. **Geometry**   Trace ∠*ABC* at the right. Then construct an angle congruent to it. *(Lesson 9-3)*

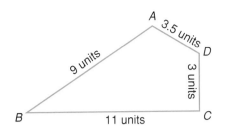

A   3.5 units   D

9 units

3 units

B   11 units   C

32. If the figure in Exercise 31 is a scale drawing with a scale 1 unit = 100 meters, what is the actual length of $\overline{AB}$? *(Lesson 10-4)*

33. **Smart Shopping**   Mrs. Masaru called different stores to get prices for a new telephone. The prices were $19.95, $22.00, and $24.50. What was the mean of these prices? *(Lesson 2-7)*

34. Divide 226.6 by 22. *(Lesson 5-4)*

35. **Taxes**   There is an 8% sales tax in Taylor Township. For each dollar spent, how much money must be paid for the sales tax?

36. **Taxes**   Canadians pay $0.15 sales tax on each dollar.
    a. What percent sales tax do they pay?
    b. Is it greater or less than the sales tax in your community?

37. **Critical Thinking**   Write a word problem that can be solved using percents. Trade with a friend and solve his or her problem. Express your answer as a percent and as a decimal.

38. **Journal Entry**   Explain in your own words how you can tell whether a decimal is greater than 100% or less than 1%.

39. **Mathematics and Laser Technology**   Read the following paragraphs.

> In the 1980s, Philips introduced a color laser videodisc system. Videodiscs are similar to long playing records in size and shape. A videodisc is played on a videodisc player which is hooked up to a television. The videodisc's surface is covered with microscopic pits with codes that carry information about the picture and sound. A laser beam in the player reads the codes. The beam is reflected by the disc to a detector that produces an electrical signal. This signal travels to the television where it is converted into a picture.

> In 1982, Philips and Sony produced the first audiodisc, called a compact disc. The compact disc is based on the same principle as the videodisc.

Vince has a compact disc library that is 35% rap music, 10% rock music, 18% jazz, and 24% rhythm and blues. The remainder of his library is classical music. What percent of his collection is classical music?

10-7 Estimating with Percents

Objective

Estimate a percent of a number.

Morton's Bike Shop is having a sale on mountain bikes. Every mountain bike in the store is on sale for 75% of the regular price. About how much would you pay for a bike that normally sells for $198?

You can estimate the answer because the word *about* tells you that an exact answer is not required.

Express 75% as a fraction. $75\% = \dfrac{75}{100} = \dfrac{3}{4}$ *Simplify.*

Then estimate the product of the fraction and the price.

$$\dfrac{3}{4} \times 200 = \dfrac{3}{4} \times \dfrac{200}{1} \quad \text{\textit{Think: \$198 is close to \$200}}$$

$$= \dfrac{3}{4} \times \dfrac{\overset{50}{200}}{1} \quad \text{\textit{Divide 4 and 200 by their GCF, 4.}}$$

$$= 150 \quad \text{\textit{Multiply.}}$$

You will pay about $150 for the mountain bike.

Sometimes when you estimate, you should round the percent to a fraction that is easy to multiply.

Problem Solving Hint

• • • • • • • • • •

You can also make a model to help you find $\dfrac{3}{4}$ of 200.

0%		50% 75% 100%

0   100 150 200

Examples

Estimate each percent.

1 30% of 152

30% is close to $33\tfrac{1}{3}\%$ or $\dfrac{1}{3}$.

Round 152 to 150.

$$\dfrac{1}{3} \times 150 = \dfrac{1}{3} \times \dfrac{\overset{50}{150}}{1}$$

$$= 50$$

2 25% of 2,113

25% is $\dfrac{1}{4}$.

Round 2,113 to 2,000.

$$\dfrac{1}{4} \times 2,000 = \dfrac{1}{4} \times \dfrac{\overset{500}{2,000}}{1}$$

$$= 500$$

3 Estimate the percent of the figure that is shaded.

14 out of 20 squares are shaded.

$\dfrac{14}{20}$ is about $\dfrac{15}{20}$ or $\dfrac{3}{4}$.

$\dfrac{3}{4} = 75\%$ About 75% of the figure is shaded.

TEEN SCENE

Do you like to ride a bicycle? You'd have to cycle for 24 minutes, at 13 mph, to burn off the calories in a 12-ounce can of regular cola.

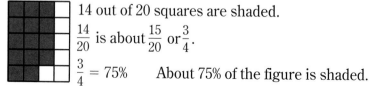

Checking for Understanding

Communicating Mathematics

Read and study the lesson to answer each question.

1. **Tell** how you would estimate 40% of 80.
2. **Tell** how you would estimate a 15% tip on a $17.90 meal.

Guided Practice

Estimate each percent.

3. 7% of 11
4. 5% of 22
5. 44% of 100
6. 34% of 30
7. 17% of 25
8. 48% of 50
9. 31% of 22
10. 27% of 80
11. 19% of 60

Estimate the percent of each figure that is shaded.

12.

13.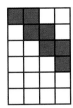

14.

Exercises

Independent Practice

Estimate each percent.

15. 13% of 20
16. 8% of 15
17. 21% of 100
18. 48% of 200
19. 39% of 70
20. 61% of 150
21. 1.5% of 10
22. 2.3% of 22
23. 97% of 310
24. *True* or *false:* 22% of 1,500 is about 300.

Use the chart at the right to estimate the percents.

25. numbers from 1 through 25 that contain *only* even digits
26. numbers from 1 through 25 that contain *only* odd digits
27. numbers from 1 through 25 that contain the digit 3

1	2	3	4	5
6	7	8	9	10
11	12	13	14	15
16	17	18	19	20
21	22	23	24	25

Mixed Review

28. **Health** Three weeks after his accident, the doctor told Jim that he had 45% of the use of his left leg back. What fraction is that? *(Lesson 10-5)*

29. Evaluate $ab - c$, if $a = 3$, $b = 16$ and $c = 5$. *(Lesson 1-6)*

30. Express 1.35 as a percent. *(Lesson 10-6)*

31. Express $\frac{40}{45}$ in simplest form. *(Lesson 6-5)*

32. **Statistics** In 1990, 28.8% of the United States population was between the ages of 25 and 44 years old. If the population was 248,709,873, about how many people were in that age group?

33. **Yearbook** The figure at the right shows the design of a yearbook page that measures $8\frac{1}{2}$" by 11". If $33\frac{1}{3}$% is an acceptable amount of white space, or area without copy and art, does the page at the right appear to have more or less than the acceptable level of white space?

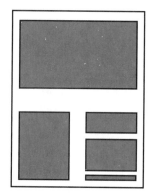

34. **Critical Thinking** What item does not belong? Explain.
    a. 50% of 22
    b. 22% of 50
    c. $\frac{1}{2}$% of 22

CULTURAL KALEIDOSCOPE

Katherine Davalos Ortega

On Septemeber 12, 1983, President Ronald Reagan nominated Katherine Davalos Ortega as the 38th Treasurer of the United States. She was the tenth woman and the second Hispanic person to hold this position.

Ms. Ortega graduated with honors from Eastern New Mexico University and entered the business world. Her career led her to banking, and in 1975 she became the first woman president of the Hispanic-owned Santa Ana State Bank in California. Her business ability and dedication led her to Washington, D.C.

As Treasurer of the United States, Ms. Ortega was responsible for directing and overseeing the branches of our government that directly affect the circulation of the country's money; from printing and circulation to selling bonds. She supervised over 5,000 employees and was responsible for a yearly budget in excess of $340 million.

10-8 Percent of a Number

Objective

Find the pecent of a number.

According to the chart at the right, 25% of the new pet owners in 1990 got their pets from friends. If 2,000 people were surveyed, how many people got their pets from friends?

Pet Source	Percent
Breeders	30%
Pet stores	8%
Animal shelter	14%
Friends	25%
Other	23%

To find 25% of 2,000, you could change the percent to a fraction or to a decimal, and then multiply it by the number. Or you could use a model or a calculator.

Method 1

Change the percent to a fraction.

$25\% = \frac{25}{100}$ or $\frac{1}{4}$

$\frac{1}{4} \times 2,000 = 500$

Method 2

Change the percent to a decimal.

$25\% = \frac{25}{100}$ or 0.25

$0.25 \times 2,000 = 500$

Method 3

Use a model.

0%   25%   50%   100%

0   500   1,000   2,000

Method 4

Use a calculator.

1 $\div$ 4 $\times$ 2000 $=$ 500

500 out of 2,000 people got their pets from friends.

Example 1

How many of the surveyed people got their pets from pet stores? According to the chart, 8% of the 2,000 people surveyed got their pets from pet stores.

$8\% = \frac{8}{100}$ or 0.08 *Change the percent to a decimal.*

$0.08 \times 2,000 = 160$ *Multiply by the number of people surveyed, 2,000.*

160 out of the 2,000 people got their pets from pet stores.

Example 2

How many people got their pets from animal shelters? According to the chart, 14% of the 2,000 people surveyed got their pets from animal shelters.

$14\% = \dfrac{14}{100}$ or $\dfrac{7}{50}$ *Change the percent to a fraction.*

$\dfrac{7}{50} \times 2,000 = \dfrac{7}{50} \times \overset{40}{\dfrac{2,000}{1}}$ *Divide 50 and 2,000 by the GCF, 50.*

$= 280$ *Multiply.*

280 out of the 2,000 people got their pets from animal shelters.

Example 3 *Problem Solving*

Sales   In 1990, the athletic shoe industry sold $11.7 billion worth of shoes. 1.5% of the sales were from shoes that sold for $54.50-$64.49. How much of the sales were from shoes priced between $54.50 and $64.49?

$1.5\% = \dfrac{1.5}{100}$ or $\dfrac{15}{1,000}$

$= 0.015$

0.015×11.7 *Multiply.*

$0.015 \boxed{\times} 11.7 \boxed{=} \boxed{0.1755}$

$0.1755 billion, or $175.5 million, was made from shoes priced between $54.50 and $64.49.

Checking for Understanding

Communicating Mathematics

Read and study the lesson to answer each question.

1. **Tell** in your own words how you change a percent to a fraction.
2. **Tell** in your own words how you change a percent to a decimal.
3. **Tell** what number is shown by the shaded part of the figure.

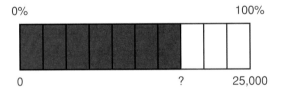

Guided Practice

Find the percent of each number.

4. 20% of 60
5. 30% of 40
6. 40% of 70
7. 11% of 100
8. 9% of 75
9. 16% of 40
10. What is 35% of 80?
11. Find 60% of 365.

Exercises

Independent Practice

Find the percent of each number.

12. 41% of 120   **13.** 13% of 50   **14.** 5% of 125

15. 25% of 90   **16.** 33% of 600   **17.** 66% of 55

18. 0.5% of 20   **19.** 125% of 99   **20.** 98% of 6

21. The Spartans won 68% of their 25 games. How many games did they win?

Mixed Review

22. Estimate 31% of 15. *(Lesson 10-7)*

23. **Attendance** On Thursday, $\frac{2}{7}$ of the student body was absent due to the flu. On Friday, an additional $\frac{3}{7}$ was absent. What portion was absent on Friday? *(Lesson 7-3)*

24. **Algebra** Evaluate x^6 if $x = 3$. *(Lesson 4-2)*

Problem Solving and Applications

25. **Statistics** Students in the United States attend school an average of 180 days a year. Some people in this country want to extend the school year. The graph at the right shows the results from a 1991 poll of 1,500 adults.

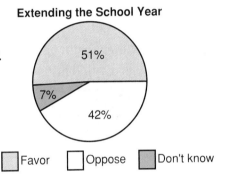

Extending the School Year

51%

7%

42%

☐ Favor   ☐ Oppose   ☐ Don't know

    a. How many people were in favor of extending the school year?

    b. How many were against extending the school year?

26. **Nutrition** A healthy diet contains at most 30% of the calories from fat. A normal active teenage girl requires about 2,000 calories per day. How many calories may be from fat in a healthy diet for a teenage girl?

27. **Number Sense** Which of the following items does not belong? Explain your answer.

    a. 20% of 60   **b.** 40% of 30   **c.** 10% of 120   **d.** 5% of 30

28. **Critical Thinking** Use a protractor and calculator to assign a percent to each section of the circle. Round to the nearest percent.

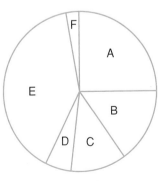

29. **Journal Entry** Write a complete sentence that provides the additional information needed to solve this problem: The school population in my district increased 3%. What is the new school population?

30. **Data Search** Refer to pages 348 and 349. Suppose 1,225 students are admitted to a college. About how many of those students will choose a major in elementary education?

Chapter

10 Study Guide and Review

Study Guide and Review

Communicating Mathematics

Choose the letter that best matches each phrase.

1. a comparison of two numbers by division
2. a comparison of two numbers having different units of measure
3. a ratio that compares a number to 100
4. shows an object exactly as it looks, but is either smaller or larger
5. an equation stating that two ratios are equal
6. in the proportion $\frac{a}{b} = \frac{c}{d}$, the numbers ad and bc
7. Explain how to change a fraction to a percent.
8. If a pie is divided into 8 pieces, tell what percent one piece of the whole pie represents.

a. scale drawing
b. proportion
c. cross products
d. rate
e. percent
f. ratio

Skills and Concepts

Objectives and Examples

Upon completing this chapter, you should be able to:

- express ratios and rates as fractions *(Lesson 10-1)*

  Express the ratio "3 winners out of 12 competing" as a fraction in simplest form.

  $$\frac{3 \text{ people winning}}{12 \text{ people competing}} \overset{\div 3}{\underset{\div 3}{=}} \frac{1}{4}$$

- express ratios as rates *(Lesson 10-1)*

  Express 150 miles in 3 hours as a rate in miles per hour.

  $$\frac{150 \text{ miles}}{3 \text{ hours}} \overset{\div 3}{\underset{\div 3}{=}} \frac{\blacksquare \text{ miles}}{1 \text{ hour}} \qquad \begin{array}{c} 50 \text{ miles} \\ per \text{ hour} \end{array}$$

Review Exercises

Use these exercises to review and prepare for the chapter test.

Express each ratio as a fraction in simplest form.

9. 7 A-students in a class of 28
10. 18 baby girls out of 30 babies
11. 11 rotten potatoes out of 20

Express each ratio as a rate.

12. 75 pounds on 5 square inches
13. Melissa's hair grows 3 inches in 6 months
14. 189 pounds of garbage in 6 weeks

380 **Chapter 10** Study Guide and Review

Objectives and Examples

- solve proportions by using cross products *(Lesson 10-2)*

Solve the proportion $\frac{9}{12} = \frac{g}{8}$.

$$9 \times 8 = 12 \times g$$
$$72 = 12g$$
$$\frac{72}{12} = \frac{12g}{12}$$
$$6 = g$$

- find actual length from a scale drawing *(Lesson 10-4)*

1 unit = 2 feet

11 units

6 units

$$\begin{array}{c} drawing \to \\ actual\ length \to \end{array} \frac{1\ \text{unit}}{2\ \text{feet}} = \frac{11\ \text{units}}{d\ \text{feet}} \begin{array}{c} \leftarrow drawing \\ \leftarrow actual\ length \end{array}$$

$$1 \times d = 2 \times 11 \quad \text{The actual}$$
$$d = 22 \quad \text{length is 22 feet.}$$

- express percents as fractions and vice versa *(Lesson 10-5)*

Express $\frac{3}{10}$ as a percent.

$$\frac{3}{10} = \frac{n}{100}$$
$$3 \times 100 = 10 \times n$$
$$300 = 10n$$
$$300 \div 10 = 10n \div 10$$
$$30 = n$$

So, $\frac{3}{10} = 30\%$.

- express percents as decimals and vice versa *(Lesson 10-6)*

Express 21% as a decimal.
$$21\% = \frac{21}{100}$$
$$= 0.21$$

Review Exercises

Solve each proportion.

15. $\frac{7}{11} = \frac{m}{33}$

16. $\frac{12}{20} = \frac{15}{k}$

17. $\frac{g}{20} = \frac{9}{12}$

18. $\frac{10}{12} = \frac{25}{h}$

19. If the drawing at the left is of Obi's bedroom, find the actual width of the room.

20. If the doorway to Obi's room is actually 3 feet wide, how wide will it be on the drawing?

Express each percent as a fraction.

21. 3%

22. 17%

23. 65%

24. 25.5%

25. 150%

26. 48%

Express each fraction as a percent.

27. $\frac{7}{100}$

28. $\frac{1}{4}$

29. $\frac{3}{5}$

30. $\frac{9}{10}$

31. $\frac{11}{20}$

32. $\frac{5}{25}$

Express each percent as a decimal.

33. 2.2%

34. 38%

35. 150%

36. 66%

Express each decimal as a percent.

37. 0.003

38. 1.3

39. 0.65

40. 0.591

Objectives and Examples	Review Exercises

- estimate a percent of a number
  (Lesson 10-7)

  Estimate 29% of 29.

  29% is close to $33\frac{1}{3}$% or $\frac{1}{3}$.
  Round 29 to 30.

  $$\frac{1}{3} \times 30 = \frac{1}{3} \times \frac{\overset{10}{30}}{\underset{1}{1}}$$
  $$= 10$$

Estimate each percent.

41. 19% of 99

42. 27% of 82

43. 48% of 48

- find the percent of a number
  (Lesson 10-8)

  Find 20% of 50.

  $20\% = \frac{20}{100} = \frac{1}{5}$ $\qquad$ $20\% = \frac{20}{100}$ or 0.20

  $\frac{1}{5} \times 50 = 10$ $\qquad\qquad$ $0.20 \times 50 = 10$

  20% of 50 is 10.

Find the percent of each number.

44. 19% of 99

45. 40% of 150

46. 5% of 40

Applications and Problem Solving

47. **School**   Michael had a graduation party which 75% of his class attended.
    a. What fraction of the class is that? *(Lesson 10-5)*
    b. If there were 32 students in Michael's class, how many attended his party? *(Lesson 10-8)*

48. **Marketing**   A store manager displays 36 cans of cat food in a triangular shape. How many cans are on the bottom row? *(Lesson 10-3)*

49. **Travel**   Alonso drove the first 300 miles of his trip in 6 hours. What is his rate of speed? *(Lesson 10-1)*

Curriculum Connection Projects

- **Automotive**   Change the sizes of a set of automotive wrenches, from $\frac{1}{4}$, $\frac{3}{8}$, ..., $\frac{15}{16}$, to decimals.

- **Sports**   Find the ratio of your height in inches to those of five of your favorite basketball players or your favorite athlete in another sport.

Read More About It

Duck, Mike. *Write Your Own Program, Graphics.*
Roth, Charlene Davis. *The Art of Making Puppets and Marionettes.*
Weiss, Ellen and Friedman, Mel. *The Tiny Parents.*

10 Test

Write each ratio as a fraction in simplest form.

1. two black puppies out of a litter of 6

2. 56 virus-infected people out of a test group of 100

Express each ratio as a rate.

3. 1,000 Italian lira to 90 cents

4. 1,000 million acres of U.S. farmland to 2 million farms

Solve each proportion.

5. $\dfrac{4}{6} = \dfrac{x}{15}$ 6. $\dfrac{m}{12} = \dfrac{12}{16}$ 7. $\dfrac{10}{p} = \dfrac{4}{14}$

8. Refer to the scale drawing at the right of a very small machine part. The scale is 1 unit = 0.01 millimeters. Find the actual length of the side labeled *s*.

12 units

s: 20 units

9. **Maps**   The scale on Olivia's map reads, "1 inch = 50 miles." She measures the distance on the map from Dallas to Houston and finds it is just under 5 inches. About how far apart are those cities?

Express each fraction as a percent.

10. $\dfrac{4}{5}$

11. $\dfrac{9}{100}$

12. $\dfrac{13}{25}$

Express each percent as a fraction and as a decimal.

13. 80%

14. 7%

15. 0.1%

Express each decimal as a percent.

16. 0.012

17. 0.92

18. 3.1

Estimate each percent.

19. 9.5% of 51

20. 49% of 26

21. 300% of 9

Find the percent of each number.

22. 60% of 35

23. 49% of 26

24. 2% of 50

Solve by drawing a diagram.

25. Barbara visits an elderly neighbor every other day. She reads stories to children at the library every third day. How many times in the next three weeks will she do both on the same day if she just did both yesterday?

Bonus   In 1984 in Washington County, 2,000 people voted for Ronald Reagan and the other 500 people voted for Walter Mondale.

    a. What ratio represents the portion of voters that voted for Mondale?

    b. What percent voted for Mondale?   c. for Reagan?

Area and Volume

Spotlight on Food

Have You Ever Wondered...

- How most people like their pizza crust?

- Which fast-foods have the least amount of fat?

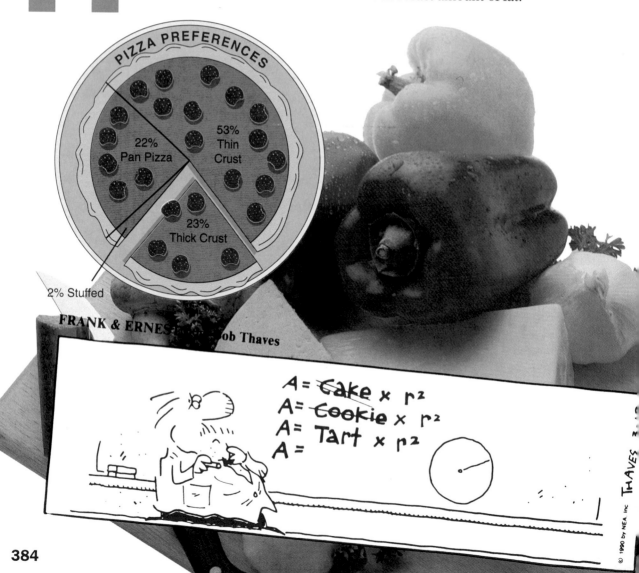

PIZZA PREFERENCES

22% Pan Pizza

53% Thin Crust

23% Thick Crust

2% Stuffed

FRANK & ERNEST ob Thaves

$A =$ ~~Cake~~ $\times r^2$
$A =$ ~~Cookie~~ $\times r^2$
$A =$ Tart $\times r^2$
$A =$

© 1990 by NEA, Inc.   THAVES 3-13

Food

Work in a group.

1. Make a list of foods. Group them into sections. For example, you could use fruits, vegetables, and cereals.

2. Ask your classmates, friends, and family to choose a favorite from each category. Keep a record of your data.

3. When you have completed your poll, make a circle graph for each category.

Looking Ahead

In this chapter, you will see how mathematics can be used to answer questions about foods and circle graphs.

The major objectives of the chapter are to:

- estimate the area of irregular figures

- find the area of parallelograms, trapezoids, triangles, and circles

- construct circle graphs

- identify and draw three-dimensional figures

- find the surface area and volume of rectangular prisms

CUTTING THE FAT FROM THE FAST-FOOD MENU

Place	Best Bets	Calories	Fat (grams)
McDonald's	Mclean Deluxe (no cheese)	320	10
	Chicken salad with 1 package lite dressing	162	4.5
Burger King	BK Broiler (no sauce)	267	8
	Chunky Chicken Salad with 1 package reduced-calorie dressing	172	5
Pizza Hut	Cheese pizza with mushrooms, bell pepper, tomatoes, onions	492	18
Taco Bell	Bean burrito (no cheese)	447	14
	Chicken burrito (no cheese)	334	12

11-1A Area of Irregular Shapes

A Preview of Lesson 11-1

Objective

Find the area of irregular shapes.

Materials

centimeter grid paper
pencil

Many things in life come in irregular shapes. For example, a puddle, a leaf, and your hands are all irregular shapes. In previous chapters, you learned how to find the area of regular shapes. Now, you will find the area of irregular shapes.

Activity One

Work with a partner.

- On a sheet of centimeter grid paper, draw an outline of your shoe.

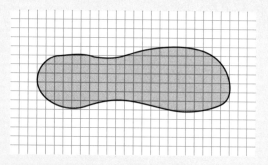

- Count the number of whole squares within the outline of your shoe. Record this number.

- Count the number of squares that touch the outline anywhere. Add this to the number of whole squares within the outline. Record this number.

What do you think?

1. Estimate the area of your shoe by finding the mean of the two numbers.

2. Can you think of another way to find the area of your shoe? Explain your answer.

Activity Two

- On a sheet of centimeter grid paper, draw an outline like the one shown at the right.

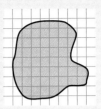

- Draw a rectangle that encloses most of the figure.

- Count squares to find the length of the rectangle.

- Count squares to find the width of the rectangle.

What do you think?

3. Estimate the area of the figure by finding the area of the rectangle.

Extension

4. On a sheet of centimeter grid paper, draw an outline of your hand. Estimate its area. Explain which method you used and why.

Estimate the area of each figure.

5.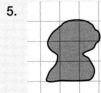

6.

7.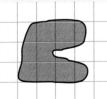

8.

9.

10.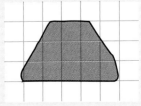

11-1 Area of Parallelograms

Objective

Find the area of parallelograms.

Words to Learn

base
height

DID YOU KNOW

The fastest goal scored in field hockey was scored in 1971 by John French of England, 7 seconds after the game began.

In the United States, field hockey is an exciting sport played mainly by women. It is popular among men in other countries. The object of the game is to hit the ball with a stick into the opponent's goal cage. The field where this game is played measures 100 yards long and 60 yards wide. What is the area of the field?

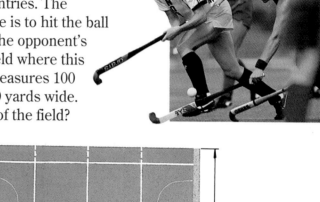

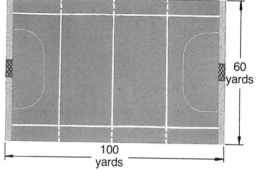

LOOK BACK

You can review the area of rectangles on page 142.

The field is in the shape of a rectangle so you can find its area by multiplying its length by its width.

$A = l \times w$
$A = 100 \times 60$ *Replace l with 100 and w with 60.*
$A = 6,000$

The area of the field is 6,000 square yards.

A parallelogram is a quadrilateral with two pairs of parallel sides. The **base** of a parallelogram is any of its sides. The **height** of the parallelogram is the distance from the base to its opposite side.

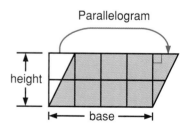

Parallelogram

height

base

A rectangle is a special type of parallelogram. For this reason, the formula for finding the area of a parallelogram is very similar to the formula for finding the area of a rectangle.

Work with a partner.

Materials: grid paper, pencil, and scissors

- On a sheet of grid paper, draw a parallelogram.
- Cut out the parallelogram.
- Draw a line to represent the height of the parallelogram and cut along the height.
- Reassemble the sections to form a rectangle.

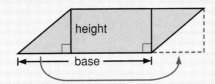

Talk About It

a. Compare the area of the rectangle formed with the area of the original parallelogram.

b. The height of the parallelogram is the same as what part of the rectangle? The base of the parallelogram is the same as what part of the rectangle?

c. In your own words, state a formula for finding the area of a parallelogram.

Area of a Parallelogram	**In words:** The area (A) of a parallelogram equals the product of its base (b) and height (h). **In symbols:** $A = bh$.

Examples

1 Find the area of a parallelogram if the base is 4 units and the height is 7 units.

$A = bh$
$A = 4 \cdot 7$
$A = 28$

The area is 28 square units.

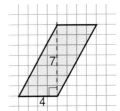

2 Find the area of the parallelogram shown below.

$A = bh$
$A = 6.5 \cdot 4.2$

6.5 ⊠ 4.2 ⊟ 27.3

$A = 27.3$

The area is 27.3 square inches.

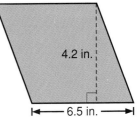

Checking for Understanding

Communicating Mathematics

Read and study the lesson to answer each question.

1. **Tell** why area is expressed in square units.
2. **Draw** a parallelogram that is *not* a rectangle.
3. **Explain** why a square is a parallelogram.

Guided Practice

Find the area of each parallelogram.

4.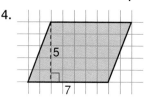

5.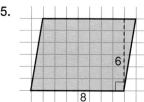

6.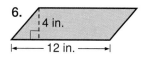

7. Find the area of a parallelogram if the base is 7 meters and the height is 6 meters.

Exercises

Independent Practice

Find the area of each parallelogram.

8.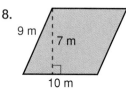

9.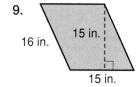

10.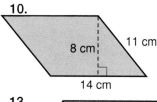

11.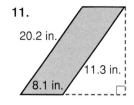

12.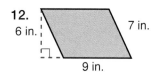

13.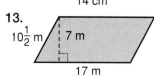

14. Find the area of a parallelogram if the base is 12 centimeters and the height is 20 centimeters.

Mixed Review

15. Which fraction is greater, $\frac{13}{40}$ or $\frac{3}{7}$? *(Lesson 6-9)*

16. Multiply 21.19 and 4.5. *(Lesson 4-5)*

Problem Solving and Applications

17. **Geography** The state of Tennessee resembles a parallelogram. Its northern

   border measures 442 miles and the distance between the northern and southern borders is 115 miles. Estimate its area.

18. **Geometry** Is every parallelogram a rectangle? Is every rectangle a parallelogram? Explain your answer.

19. **Critical Thinking** What happens to the area of a parallelogram if
    a. its height is doubled?
    b. both its base and its height are doubled?

11-2 Area of Triangles

Objective

Find the area of triangles.

Miguel Barcenas is replacing the siding on a triangular portion of his house that was damaged by high winds. What is the area of the section that Mr. Barcenas needs to replace?

You can find the formula for the area of a triangle from the formula of the area of a parallelogram.

Mini-Lab

Work with a partner.

Materials: paper and pencil, scissors

- Make two tracings of the triangle on a sheet of paper.

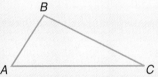

- Cut out both tracings of the triangle.
- Assemble the sections to form a parallelogram.

Talk About It

a. Compare the area of one of the triangles to the area of the parallelogram.

b. The height of the parallelogram is the same as what part of the triangle? The base of the parallelogram is the same as what part of the triangle?

c. In your own words, state a formula for finding the area of a triangle.

DID YOU KNOW

The world's windiest place is The Commonwealth Bay, George V Coast, Antarctica, where winds reach 200 mph.

The base of a triangle can be any one of its sides. The height of a triangle is the distance from the opposite vertex to the base. The area of the parallelogram is $A = bh$. So, the area of a triangle is one-half the area of the parallelogram, or $\frac{1}{2} bh$.

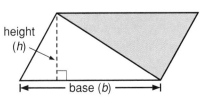

Area of a Triangle	**In words:** The area(A) of a triangle equals half the product of its base (b) and height (h). **In symbols:** $A = \frac{1}{2}bh$

You can use the formula to find the area of the triangular section that Mr. Barcenas needs to replace.

$A = \frac{1}{2}bh$

$A = \frac{1}{2} \cdot 40 \cdot 20$ *Replace b with 40 and h with 20.*

$A = \frac{1}{2} \cdot 800$

$A = 400$

The area that Mr. Barcenas needs to replace is 400 square feet.

Examples

1 Find the area of a triangle whose base is 6 inches and whose height is 11 inches.

$A = \frac{1}{2}bh$

$A = \frac{1}{2} \cdot 6 \cdot 11$ *Replace b with 6 and h with 11.*

$A = 33$

The area is 33 square inches.

11 in.

6 in.

2 Find the area of the triangle shown below.

The base is 22 cm; the height is 15 cm.

$A = \frac{1}{2}bh$

$A = \frac{1}{2} \cdot 22 \cdot 15$

$1 \; \boxed{\div} \; 2 \; \boxed{\times} \; 22 \; \boxed{\times} \; 15 \; \boxed{=} \; 165$

$A = 165$

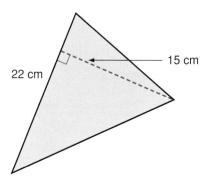

22 cm

15 cm

The area is 165 square centimeters.

Checking for Understanding

Read and study the lesson to answer each question.

1. **Tell** why the area of a triangle can be defined as half the area of a parallelogram.

2. **Tell** how to find the area of a triangle if you know the length of its sides and its height.

Guided Practice   Find the area of each triangle.

3. base, 4 m
   height, 8 m

4. base, 10 in.
   height, 7 in.

5. base, 11 cm
   height, 6 cm

6.

7.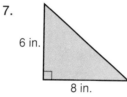

8.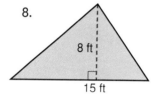

9. Find the area of a triangle if the base is 8 meters and the height is 14 meters.

Exercises

Find the area of each triangle.

10. base, 2 cm
    height, 28 cm

11. base, 8 m
    height, 12 m

12. base, 16 in.
    height, 13 in.

13.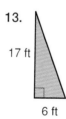

14. 5.4 in. 15.3 in.

15.

16. What is the area of a triangle whose base is 42 inches and whose height is 35 inches?

17. Find the area of a triangle with a base of 32 miles and a height of 20 miles.

Mixed Review   18. Find the next two numbers in the sequence $75, 72\frac{1}{2}, 70, 67\frac{1}{2}, \ldots$
(Lesson 8-5)

19. Order the decimals 2.9, 2.38, 2.474, 2.91, 2.88 from greatest to least.
(Lesson 3-4)

20. **Sports** The figure shown at the right shows the measurements of the rectangular playing field of the Arena Football League. Find the area of triangle A.

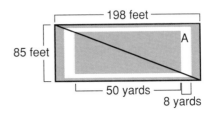

198 feet

85 feet

A

50 yards

8 yards

21. **Critical Thinking** Draw a triangle with an area of 16 square inches.

22. **Geography** The Bermuda Triangle is formed by an imaginary line segment drawn from a point near Melbourne, Florida, to Bermuda, to Puerto Rico and back to Melbourne. The diagram at the right shows the distances between each point. What is the area enclosed by the Bermuda Triangle?

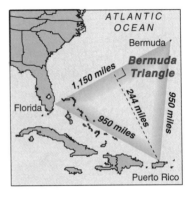

ATLANTIC OCEAN

Bermuda

Bermuda Triangle

1,150 miles

244 miles

950 miles

950 miles

Florida

Puerto Rico

23. **Journal Entry** Find a triangular figure in a newspaper, atlas, book, or magazine. Find the measure of the base and the height of the figure, or make up your own measurements. Then find the area of the triangle.

Save Planet Earth

Energy Efficient Lighting One of the most effective ways to save energy, money, and reduce pollution is to replace incandescent light bulbs with fluorescent bulbs. Substituting a fluorescent light for an incandescent bulb will keep a half-ton of CO_2 (carbon dioxide) out of the atmosphere over the life of the bulb. The initial cost of replacing bulbs is high, $15 to $25, but a fluorescent bulb lasts 10 times longer than an incandescent bulb and uses only one-fourth the energy. This amounts to a savings of about $40 in electricity costs over the bulb's lifetime.

How You Can Help

- Turn off lights when not in use.
- Keep light bulbs clean—dirt absorbs light and uses more energy.
- Write for a free copy of "Energy Efficient Lighting."
  U.S. Dept. of Energy
  P.O. Box 8900
  Silver Springs, MD 20907

11-3 Area of Circles

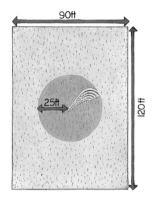

Objective

Find the area of circles.

Angelica and Jason bought a sprinkler to water their backyard. The sprinkler rotates in a circular motion. Their backyard is 90 feet by 120 feet. If they adjusted the sprinkler to cover a radius of 25 feet, what is the area watered by the sprinkler?

To answer this question, you can use the formula for the area of a circle. To find this formula, you can use the formula for the area of a parallelogram.

Mini-Lab

Work with a partner.
Materials: paper plates, scissors, pencil

- Fold a paper plate into eighths.
- Cut out each piece.
- Assemble the pieces to form a "parallelogram."

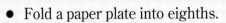

Talk About It
a. What is the height of the parallelogram?
b. What is the base of the parallelogram?
c. How can you find the area of the parallelogram?

The base of the parallelogram is equal in length to one-half of the circumference of the circle, $\frac{1}{2}C$, and the height is equal to the radius (r) of the circle. With this information, and the formula for the area of a parallelogram, you can find the area of a circle.

LOOK BACK

You can review circumference on page 149 and π on page 150.

$A = bh$ *Formula for the area of a parallelogram.*

$A = (\frac{1}{2}C)r$ *Replace b with $\frac{1}{2}C$ and h with r.*

$A = \frac{1}{2}(2\pi r)r$ *Replace C with $2\pi r$.*

$A = \pi \cdot r \cdot r$ *Simplify.* $\frac{1}{2} \cdot 2 = 1$

$A = \pi r^2$ *Simplify.* $r \cdot r = r^2$

Area of a Circle	**In words:** The area *(A)* of a circle equals the product of π and the square of the radius *(r)*.
	In symbols: $A = \pi r^2$

You can apply this formula to find the area that would be watered by the sprinkler.

$A = \pi r^2$
$A = \pi \ (25)^2$ *Replace r with 25.*

$\boxed{\pi}$ $\boxed{\times}$ 25 $\boxed{x^2}$ $\boxed{=}$ 1963.4954

$A \approx 1,963.5$

The area that would be watered is about 1,963.5 square feet.

Example *Problem Solving*

Science The earthquake that shocked the Bay Area of California on October 17, 1989, had its epicenter at Loma Prieta near Santa Cruz, California. The diameter of the earthquake's horizontal waves that shocked the area measured about 120 miles. How large was the area affected by this earthquake?

Estimation Hint

· · · · · · · · · ·

You can estimate the area of a circle by squaring the radius and multiplying by 3. Thus, the area of the circle in the Example is about 3 × 60 × 60 or about 10,800 square miles.

The diameter of the circular area is 120 miles. So the radius is 60 miles.

$A = \pi r^2$
$A = \pi \times (60)^2$ *Replace r with 60.*

$\boxed{\pi}$ $\boxed{\times}$ 60 $\boxed{x^2}$ $\boxed{=}$ 11309.734

The area affected by the earthquake was about 11,310 square miles.

Checking for Understanding

Communicating Mathematics

Read and study the lesson to answer each question.

1. **Tell** why you express the area of a circle in square units.

2. **Draw** a circle with a radius of 2 inches. Using the estimation hint above, find its area.

Guided Practice

Find the area of each circle. Use 3.14 for π. Round to the nearest tenth.

3. radius, 3 inches

4. radius, 6 yards

5. radius, 9 feet

6. diameter, 8 inches

Find the area of each circle. Use 3.14 for π. Round to the nearest tenth.

7. diameter, 14 feet

8. diameter, 16 miles

9.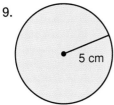
5 cm

10.
7 m

Exercises

Independent Practice

Find the area of each circle. Use 3.14 for π. Round to the nearest tenth.

11.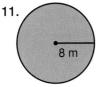
8 m

12.
$18\frac{1}{2}$ ft

13.
2.5 in.

14. radius, 7 inches

15. radius, 11 inches

16. radius, 16 feet

17. radius, 25 centimeters

18. diameter, 6 inches

19. diameter, 18 feet

20. diameter, 42 meters

21. diameter, 32 feet

22. Find the area of a circle if the diameter is 20 meters.

23. What is the area of a circle with a radius of 22 inches?

Mixed Review

24. Round $7\frac{3}{8}$ to the nearest half. *(Lesson 7-1)*

25. Express $5\frac{3}{8}$ as an improper fraction. *(Lesson 6-6)*

26. Use the distributive property to find the product of 5.7 and 60 mentally. *(Lesson 4-4)*

Problem Solving and Applications

27. **Construction** A new theater is under construction. The stage for this theater will be in the shape of a semicircle. If the radius of the semicircle measures 30 feet, what will the area of the stage be?

30 ft

28. **Sports** The shape of an Olympic ice arena is formed by a rectangle and two semicircles. What would be the area of an ice arena with the given dimensions?

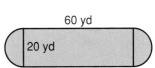
60 yd
20 yd

29. **Critical Thinking** If the radius of a circle is doubled, what happens to its area?

30. **Data Search** Refer to pages 384 and 385. Find the area of the circle in the comic strip if the radius is 2.5 centimeters. Round to the nearest tenth.

DATA SEARCH

Cooperative Learning

11-3B Making Circle Graphs

A Follow-Up of Lesson 11-3

Objective

Construct circle graphs.

Materials

paper
colored pencils
compass
protractor
calculator

Graphs are often used to display information because they are easier to read and understand. Circle graphs are used to compare parts of a whole. Usually the information is expressed in percents.

Try this!

In 1989, foreign visitors spent $43 billion in the United States. The following chart shows the share of international travelers' spending in the United States. Use the information to make a circle graph.

Estimation Hint

●●●●●●●●●●

Round all degree measurements to the nearest tenth before you make the circle graph.

Country	Percentage of Total Spent by International Travelers
Australia	2.6%
Britain	9.6%
Canada	14.0%
France	3.0%
Germany	5.7%
Japan	17.4%
Mexico	9.7%
Other countries	38.0%

- Find the number of degrees for each category in the chart. To do this, multiply the decimal equivalent of each percent by 360°. For example, for Australia, the number of degrees equals 0.026 × 360° or 9.36°.
- Use your compass to draw a circle and a radius.
- Use your protractor to draw each angle. Repeat this step for each category in the chart.
- Color each section of the circle and label it. Give a title to your graph.

TEEN SCENE

Teens spend about $56 million on shoes and apparel each year.

What do you think?

1. Do you think that the information is more clearly understood when displayed in a circle graph? Explain your answer.
2. Why were you able to represent this data in a circle graph?

3. What type of data would you *not* be able to represent in a circle graph? How might you represent such data?

Applications

4. **Statistics**   A recent survey gave the following percents of eighth graders who say they are home alone for various numbers of hours per day. Make a circle graph.

Never home alone	13%
Less than 1 hour	32%
1 to 2 hours	28%
2 to 3 hours	13%
3 or more hours	14%

5. **Geography**   The following list shows the percent of Earth's water that is in each of the oceans. Represent the data in a circle graph.

Pacific Ocean	49%
Atlantic Ocean	26%
Indian Ocean	21%
Arctic Ocean	4%

6. **Science**   Use the graph at the right to answer each question.

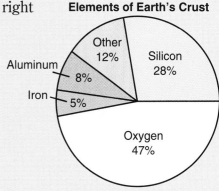

Elements of Earth's Crust

   a. Which element makes up almost $\frac{1}{2}$ of Earth's crust?

   b. Which element is found in five percent of Earth's crust?

   c. How many major elements are found in Earth's crust?

   d. Which element makes up about $\frac{1}{3}$ of Earth's crust?

Extension

7. Make a circle graph to represent your classmates' eye colors.

8. Make a circle graph to represent how you spend your allowance.

11-4 Three-Dimensional Figures

Objective
Identify three-dimensional figures.

Words to Learn
three-dimensional figure
face
edge
vertex
rectangular prism
base
pyramid
cone
cylinder
sphere
center

The salt that we use to flavor foods comes from the mineral *salt*. Table salt is in the form of clear crystals shaped like cubes.

A salt crystal is an example of a three-dimensional figure. A **three-dimensional figure** is a figure that encloses a part of space. The flat surfaces of a three-dimensional figure are called **faces.** The faces intersect to form **edges.** The edges intersect to form **vertices.**

Mini-Lab

Work with a partner.

Materials: straws and gumdrops

- Build a figure like the one shown at the right. Use the straws to make the edges of the faces of the figure. Use the gumdrops to make the vertices.

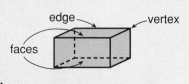

Talk About It

a. How many straws did you use?
b. How many gumdrops did you use?
c. What is true of all of the faces of this figure?

A figure like the one you made in the Mini-Lab is a model of a rectangular prism. A **rectangular prism** is a three-dimensional figure with six faces that are shaped like rectangles. The faces on the top and the bottom are called **bases.** A rectangular prism has a total of 6 faces, 12 edges, and 8 vertices.

rectangular prism

A **pyramid** is different from a prism because it only has one base. The base can have the shape of any polygon, which is used to name the pyramid. For instance, a pyramid with a base in the shape of a square is called a square pyramid. The faces of a pyramid are triangular and all meet at a point.

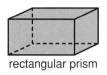

pyramid

Mini-Lab

Work with a partner.

Materials: straws and gumdrops

- Use straws and gumdrops to build a square pyramid.

Talk About It

a. How many straws did you use?

b. How many gumdrops did you use?

c. What is the shape of the base?

d. What is the shape of each face?

cone

Some three-dimensional figures have curved surfaces. As a result, they do not have faces or edges. A **cone** has curved surfaces. Its base is a circle and it has one vertex.

cylinder

A **cylinder** also has curved surfaces, but it has two circular bases and no vertices.

center
sphere

A **sphere** is a three-dimensional figure with no faces, bases, edges, or vertices. All of its points are the same distance from a given point called the **center.**

Checking for Understanding

Communicating Mathematics

Read and study the lesson to answer each question.

1. **Tell** the difference between a prism and a pyramid.

2. **Tell** the difference between a cone and a pyramid.

3. **Tell** how many triangular faces you would have in a pentagonal pyramid.

Guided Practice

Name each figure.

4.

5.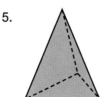

6.

State the number of faces, edges, and vertices in each figure.

7. rectangular prism
8. triangular pyramid
9. cone

Exercises

Independent Practice

Name each figure.

10.

11.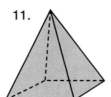

12.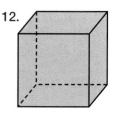

Use straws and gumdrops to model each figure.

13. a cone
14. a triangular pyramid
15. a cylinder

State the number of faces, edges, and vertices for each figure.

16. square pyramid
17. cylinder
18. sphere
19. pentagonal pyramid

Mixed Review

20. **Food** Mrs. Mystal bought one quart of cole slaw. If each serving is $\frac{1}{2}$ cup, how many servings will she have? *(Hint: 1 qt = __?__ c) (Lesson 8-6)*

21. Find the quotient of 4,507.72 and 8.5. *(Lesson 5-1)*

22. Express the ratio *24 pairs of socks for three brothers* as a rate. *(Lesson 10-1)*

Problem Solving and Applications

23. **Critical Thinking** Explain what the term *three-dimensional* means.

24. **Science** The figure at the right is a mineral called *gypsum*.
    a. How many faces do you see in its shape?
    b. What polygons form the faces?

Mid-Chapter Review

1. What is the area of a parallelogram with a base of 50 feet and a height of 77 feet? *(Lesson 11-1)*

2. Find the area of a triangle if the base is 21 inches and the height is 15 inches. *(Lesson 11-2)*

Find the area of each circle. Use 3.14 for π. Round to the nearest tenth. *(Lesson 11-3)*

3. radius, 2 inches
4. diameter, 11 yards
5. diameter, 88 meters

6. *Apatite* is a mineral composed by crystals that have the shape of a prism with six rectangular faces. How many edges and vertices do these crystals have? *(Lesson 11-4)*

11-4B Three-Dimensional Figures

A Follow-Up of Lesson 11-4

Objective

Draw three-dimensional figures.

Materials

dot paper
pencil

When solving a problem that involves a three-dimensional figure, sometimes it is helpful to draw the figure before solving the problem. In this lab, you will use dot paper to help you draw a three-dimensional figure.

Try this!

Work with a partner.

Draw a rectangular prism that is 3 units long, 2 units high, and 4 units deep.

- Use dot paper to draw a rectangle with a base of 3 units and a height of 2 units. This will represent the front side of the rectangular prism.

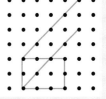

- From each vertex of the rectangle above, draw a line passing through 4 dots.

- Connect all the points to form the rectangular prism.

What do you think?

1. Which of the faces of the figure represent the bases of the rectangular prism?
2. Would the drawing look different if you turned the figure over on its side?
3. How would you draw a prism with a base in the shape of a pentagon?

Extension

Use dot paper to draw each figure.

4. a rectangular prism that is 5 units long, 1 unit high, and 5 units deep
5. a rectangular prism that is 6 units long, 4 units high, and 3 units deep
6. a cylinder that has been cut and unfolded
7. Describe the shapes you drew in Exercises 4-6.

Mathematics Lab 11-4B Three-Dimensional Figures   **403**

11-5 Surface Area of Rectangular Prisms

Objective

Find the surface area of rectangular prisms.

Words to Learn

surface area

Carla and her father want to build a trunk for her magazine collection. She wants the trunk to have these measurements: a height of 20 inches, a length of 30 inches, and a width of 24 inches. Carla draws a diagram of the trunk before they start building it.

❝ When am I ever going to use this? ❞

If you were going to cover the trunk with contact paper, how much would you need? You need to calculate the surface area to find out.

Mini-Lab

Work with a partner.

Materials: centimeter grid paper, pencil, scissors, ruler, and a calculator

- On a sheet of centimeter grid paper, draw a flat pattern like the one shown at the right. Use the measurements shown.
- Fold the pattern on the dashed lines to form a three-dimensional figure. Tape the edges.

Talk About It

a. What figure did you form?

b. What is the area of the faces of the figure you formed?

The **surface area** of a rectangular prism is the sum of the areas of its faces. Each face is in the shape of a rectangle.

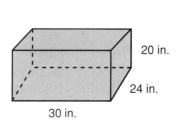

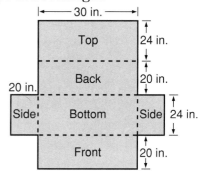

You can find the area of the six rectangular faces and add them together to find the surface area of the trunk.

front $30 \times 20 = 600$ square inches
back $30 \times 20 = 600$ square inches
top $30 \times 24 = 720$ square inches
bottom $30 \times 24 = 720$ square inches
right side $24 \times 20 = 480$ square inches
left side $24 \times 20 = 480$ square inches

Add the areas:
$600 + 600 + 720 + 720 + 480 + 480 = 3{,}600$

The surface area of Carla's trunk is 3,600 square inches.

Example 1   *Problem Solving*

Woodworking   Jeremy is building several enclosed storage containers for his CD collection. Each container will measure $5\frac{1}{2}$ inches by 10 inches by $5\frac{1}{2}$ inches. When each container is completed, he will paint the outside of each. If a can of paint will cover 360 square inches, how many containers can he paint with two cans?

Find the area of each face.

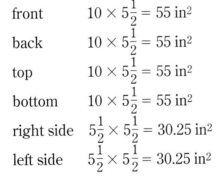
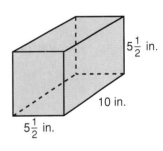

front $10 \times 5\frac{1}{2} = 55$ in²

back $10 \times 5\frac{1}{2} = 55$ in²

top $10 \times 5\frac{1}{2} = 55$ in²

bottom $10 \times 5\frac{1}{2} = 55$ in²

right side $5\frac{1}{2} \times 5\frac{1}{2} = 30.25$ in²

left side $5\frac{1}{2} \times 5\frac{1}{2} = 30.25$ in²

<aside>
Mental Math Hint
• • • • • • • • • •
Remember, in² means square inches.
</aside>

Add the areas.
$55 + 55 + 55 + 55 + 30.25 + 30.25 = 280.50$

The surface area of one container is 280.5 in².

Now find how many containers Jeremy can cover with two cans of paint. Two cans will cover 2×360, or 720 square inches. The surface area of each container is 280.5 square inches, so divide 720 by 280.5.

     720 ⌈÷⌉ 280.5 ⌈=⌉ 2.5668449

Two cans of paint will cover two storage containers and a little over half of a third container.

Example 2

Find the surface area of the rectangular prism shown below.

Find the area of each face.

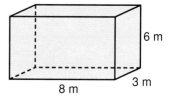

front	$8 \times 6 = 48$ m²
back	$8 \times 6 = 48$ m²
top	$8 \times 3 = 24$ m²
bottom	$8 \times 3 = 24$ m²
right side	$6 \times 3 = 18$ m²
left side	$6 \times 3 = 18$ m²

Add the areas.
$48 + 48 + 24 + 24 + 18 + 18 = 180$

The surface area is 180 m².

Checking for Understanding

*Communicating
Mathematics*

Read and study the lesson to answer each question.

1. **Tell** a real-world situation where you need to know a surface area.

2. **Draw** a rectangular prism and label its three dimensions 2 inches, 7 inches, and 5 inches.

Guided Practice Find the surface area of each rectangular prism.

3.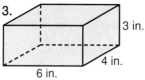

4.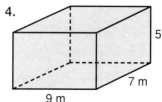

5.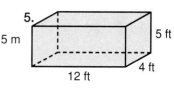

6. length = 2 mm
   width = 5 mm
   height = 4 mm

7. length = 3 cm
   width = 5 cm
   height = 6 cm

8. length = 7 yd
   width = 4 yd
   height = 3 yd

9. What is the surface area of a rectangular prism whose dimensions are 9 feet by 5 feet by 11 feet?

Exercises

Independent Practice

Find the surface area of each rectangular prism.

10.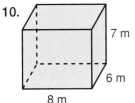
7 m
6 m
8 m

11.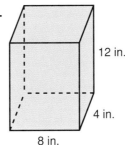
12 in.
4 in.
8 in.

12.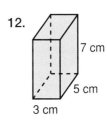
7 cm
5 cm
3 cm

13.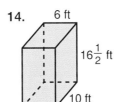
8 in.
8 in.
8 in.

14.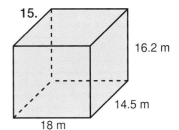
6 ft
$16\frac{1}{2}$ ft
10 ft

15.
16.2 m
14.5 m
18 m

16. length = 5 cm
    width = 7 cm
    height = $12\frac{1}{2}$ cm

17. length = 8 m
    width = 10 m
    height = 15.5 m

18. length = 10 in.
    width = 14 in.
    height = 20 in.

19. Find the surface area of a rectangular prism that has the following dimensions: 21 centimeters by 13 centimeters by 17 centimeters.

Mixed Review

20. Find the area of a circle with a diameter of 6 meters. *(Lesson 11-3)*

21. Estimate the product of $\frac{3}{4}$ and 37. *(Lesson 8-1)*

22. Find the LCM of 15, 20, and 8. *(Lesson 6-8)*

Problem Solving and Applications

23. **Geometry**   Write a formula for finding the surface area of a cube if an edge of the cube is *s* units long.

24. **Critical Thinking**   A cube and a square pyramid have congruent bases and the same heights. Which will have the larger surface area? Explain your answer.

25. **Interior Design**   A rectangular room is 16.5 feet long, 8 feet high, and 12 feet wide. How much wallpaper is needed to cover the walls, not taking into account doorways or windows?

26. **Journal Entry**   Write a real-life problem that can be solved by finding the surface area of a rectangular prism.

11-6 Volume of Rectangular Prisms

Objective

Find the volume of rectangular prisms.

Words to Learn

volume

Carla has decided to store her magazines in boxes inside the trunk. She wants to keep each subscription together. She has collected boxes that will stack neatly, and completely fill her trunk.

The drawing shows the measurements for the trunk using the boxes as units of measure. How many boxes will Carla be able to put in the trunk?

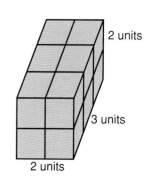

2 units

3 units

2 units

Mini-Lab

Work with a partner.

Materials: a medium-sized box and sugar cubes

- Estimate how many cubes will stack neatly, and completely fill the box.
- Fill the box with cubes.

Talk About It

a. Compare your estimate with the actual number of cubes.

b. Suppose you do not have enough cubes to fill the entire box. However, you have enough to cover the bottom of the box with one layer of cubes, with some cubes left over. Can you determine how many cubes are needed to fill the entire box? Describe your method.

c. Suppose you do not even have enough cubes to cover the bottom of the box. Describe a method to determine how many cubes are needed to fill the box.

From the figure above, you can see that $2 \times 3 \times 2$ or a total of 12 boxes will fit inside Carla's trunk. This is the volume of the trunk. **Volume** is the amount of space that a three-dimensional figure contains.

Volume is normally expressed in cubic units. Since the original dimensions of Carla's trunk given on page 404 were expressed in inches, the volume of Carla's trunk could be expressed in cubic inches, or in^3.

Volume of a Rectangular Prism	**In words:** The volume (V) of a rectangular prism equals the product of its length (l), its width (w), and its height (h).

In symbols: $V = lwh$

Examples

1 Find the volume of Carla's trunk in cubic inches.

length, 30 inches
width, 24 inches
height, 20 inches

$V = lwh$
$V = 30 \times 24 \times 20$
$V = 14,400$

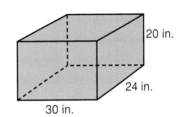

The volume of Carla's trunk is 14,400 in^3.

2 Find the volume of the rectangular prism shown below.

First, identify the length, width, and height.
length, 6 cm; width, 7 cm; and height, 14 cm

$V = lwh$
$V = 6 \times 7 \times 14$
$V = 588$

The volume is 588 cm^3.

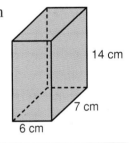

Checking for Understanding

Communicating Mathematics

Read and study the lesson to answer each question.

1. **Tell** the difference between surface area and volume.
2. **Tell** why volume is expressed in cubic units.
3. **Tell** if the order in which you multiply the length, width, and height when finding volume affects your answer. Explain your answer.

Guided Practice

Find the volume of each rectangular prism.

4.
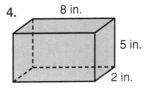

5.
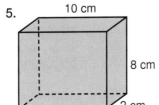

6.
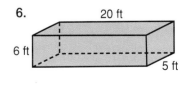

Find the volume of each rectangular prism.

7. length, 4 mm
   width, 3 mm
   height, 6 mm

8. length, 7 in.
   width, 5 in.
   height, 8 in.

9. length, 10 m
   width, 5 m
   height, 12 m

Exercises

Independent Practice

Solve.

10. A box of chocolate chip cookies measures 12 inches by 2 inches by 8 inches. What is the volume of the box?

11. Suppose your locker is 3 feet high, 16 inches wide, and 1 foot deep. How many boxes described in Exercise 10 will fit into your locker? Explain.

Find the volume of each rectangular prism.

12.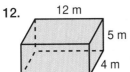
12 m, 5 m, 4 m

13.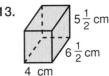
$5\frac{1}{2}$ cm, $6\frac{1}{2}$ cm, 4 cm

14.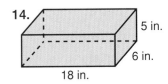
5 in., 6 in., 18 in.

15.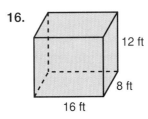
12 in., 8 in., $10\frac{1}{2}$ in.

16.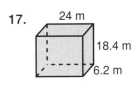
12 ft, 8 ft, 16 ft

17. 24 m, 18.4 m, 6.2 m

18. length, 8 ft
    width, 6 ft
    height, 10 ft

19. length, 5 in.
    width, 11 in.
    height, 13 in.

20. length, 12 mm
    width, 15.5 mm
    height, 16.5 mm

21. Find the volume of a rectangular prism whose length is 15 meters, width is 22 meters, and height is 26 meters.

Mixed Review

22. Find the sum of $\frac{7}{8}$ and $\frac{5}{6}$. Write the answer in simplest form. *(Lesson 7-5)*

23. Find the difference of 52.8 and 6.75. *(Lesson 3-8)*

24. Use a protractor to draw an angle whose measure is 60°. *(Lesson 9-2)*

Problem Solving and Applications

25. **Pets**   A rectangular aquarium measures 28 inches by 16 inches by 12 inches. If the aquarium is filled to a height of 10 inches, what is the volume of water in the aquarium?

26. **Critical Thinking**   A rectangular goldfish container has a volume of 288 cubic inches. What are possible reasonable dimensions of the container?

11-7 Make a Model

Objective

Solve problems by making a model.

The Brooke Middle School Math Club invited parents to visit the class and participate in an activity with their children. Parents and students need to build a prism that has a volume of 48 sugar cubes. The length of the prism is to be three times the height. The width of the prism is to be twice the height.

Explore What do you know?
The volume is 48 sugar cubes.
The length must be three times the height.
The width must be twice the height.

Plan You can solve the problem by making a model.
Start with 48 sugar cubes. Use the clues you have to find the length, height, and width.

Solve

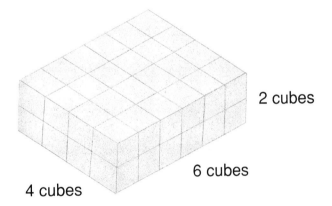

2 cubes

6 cubes

4 cubes

Examine Your goal was to build a prism having a volume of 48 sugar cubes. Its length was to be three times the height, and its width was to be two times the height.

$l = 3h$	$w = 2h$	$V = lwh$
$6 \stackrel{?}{=} 3(2)$	$4 \stackrel{?}{=} 2(2)$	$48 \stackrel{?}{=} 6 \times 4 \times 2$
$6 = 6$	$4 = 4$	$48 = 48$

Since all the conditions are met, your answer checks.

Boxes To Go sells unassembled boxes. Patrick needs to buy a box that will neatly hold a shirt that he wants to mail to a friend. Which pattern for a box should he use?

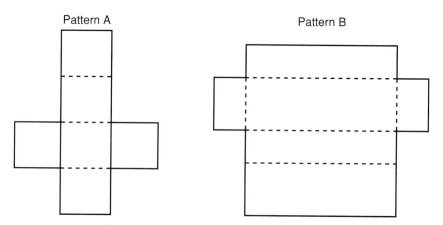

Examine the unfolded patterns. Use paper to make a model of each pattern.

Patrick should choose Pattern B because a shirt will fit neatly inside this box.

Checking for Understanding

Communicating Mathematics

Read and study the lesson to answer each question.

1. **Tell** why a model was helpful in solving the problem in the lesson introduction.
2. **Write** a list of real-life situations where it would be helpful to make a model to solve a problem.

Guided Practice

Solve by making a model.

3. Marietta is making a pattern of a box to mail a rolled-up poster to her sister. When the pattern is put together it will form a triangular prism. Make a model for this pattern.
4. A rectangular prism can be formed using exactly 8 cubes. Find the length, width, and height of the prism.
5. A construction worker wants to arrange 12 cement bricks into the shape of a rectangle with the smallest perimeter possible. How many bricks will be in each row?

Problem Solving

Practice Solve using any strategy.

6. Frankie, Sally, and Bob play football, softball, and basketball. One of the boys is Sally's brother. No person's sport begins with the same letter as their first name. Sally's brother plays football. Which sport does each person play?

Strategies

• • • • • • •

Look for a pattern.
Solve a simpler problem.
Act it out.
Guess and check.
Draw a diagram.
Make a chart.
Work backward.

7. Daniel is making a pyramid-shaped display of laundry detergent at the grocery store where he works. Each box of detergent is in the shape of a rectangle. There are six boxes in the fifth, or bottom, row of the display. If there is one less box in each of the rows above, how many boxes of detergent will Daniel use to make the display?

8. A number is tripled and 14 is added. The result is 23. What is the number?

9. Ten 2-inch cubes fit in a carton. What are the possible dimensions of the carton?

10. When George took some friends to dinner, the bill, without a tip, came to $179.75. The service was so good that he decided to leave a 20% tip. How much was the tip?

11. **Mathematics and Art** Read the following paragraphs.

In sculpture, clay and wax are the most common modeling materials, and the artist's hands are the main tools used to shape the clay. Modeling is an ancient form of sculpture, originating in Egypt and the Middle East. The first sculptures were made 35,000 years ago.

Modeling is said to be an additive process compared to carving, in which portions of the sculpture are cut away. Unlike carvings, corrections and changes can be made during modeling. The finished product, often fired clay or preserved wax, is not as permanent as a stone or wood carving. Modeled work may be reproduced in stone or metal.

Tami is making a sculpture using 8 rectangular prisms. She stacks each one on top of the other to form one long prism that she paints blue. How many faces of the original rectangular prisms are not painted?

11 Study Guide and Review

Study Guide and Review

Communicating Mathematics

Choose the letter of the figure or part of a figure that best matches the description.

1. edge of a three-dimensional figure
2. pyramid
3. base of a three-dimensional figure
4. base of a parallelogram
5. cylinder
6. face
7. sphere
8. cone
9. parallelogram
10. height
11. side
12. vertex

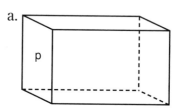
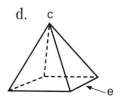

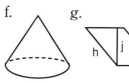

Skills and Concepts

Objectives and Examples

Upon completing this chapter, you should be able to:

- find the area of parallelograms
  (Lesson 11-1)

  Find the area.
  $A = bh$
  $A = 6 \cdot 5$
  $A = 30$

  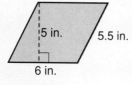

  The area of the parallelogram is 30 in².

- find the area of triangles
  (Lesson 11-2)

  Find the area.
  $A = \frac{1}{2}bh$

  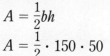

  $A = \frac{1}{2} \cdot 150 \cdot 50$
  $A = 3{,}750$

  The area of the triangle is 3,750 in².

Review Exercises

Use these exercises to review and prepare for the chapter test.

Find the area of each parallelogram.

13.   14.

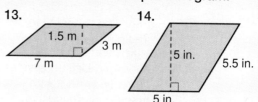

15. base, 12.5 cm; height, 7 cm

Find the area of each triangle.

16.   17.

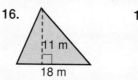
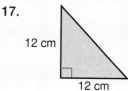

Objectives and Examples

- find the area of circles
  (Lesson 11-3)

  Find the area. Use 3.14 for π.

  $A = \pi r^2$
  $A = \pi \cdot 7^2$
  $A = \pi \cdot 49$
  $A \approx 154$ m^2

Review Exercises

Find the area of each circle. Use 3.14 for π.
Round to the nearest tenth.

18. 11 m

19. radius, 150 km

20. 14 cm

21. diameter, 7 in.

- identify three-dimensional figures
  (Lesson 11-4)

rectangular prism   cylinder   pyramid

cone   sphere

Name each figure.

22.

23.

- find the surface area of rectangular
  prisms (Lesson 11-5)

  front $8 \times 5 = 40$ in^2
  back $8 \times 5 = 40$ in^2
  top $8 \times 4 = 32$ in^2
  bottom $8 \times 4 = 32$ in^2
  right side $5 \times 4 = 20$ in^2
  left side $5 \times 4 = 20$ in^2
  $40 + 40 + 32 + 32 + 20 + 20 = 184$
  The surface area of the rectangular
  prism is 184 in^2.

  5 in.   4 in.   8 in.

Find the surface area of each rectangular
prism.

24. length, 6 in.
    width, 10 in.
    height, 2 in.

25. length, 9 mm
    width, 7 mm
    height, 8 mm

26.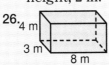
    4 m   3 m   8 m

27.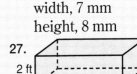
    2 ft   9 ft   4 ft

- find the volume of rectangular prisms
  (Lesson 11-6)

  Find the volume of the rectangular
  prism above.
  $V = lwh$
  $V = 8 \times 4 \times 5$
  $V = 160$
  The volume is 160 in^3.

Find the volume of each rectangular
prism.

28. The prism in Exercise 24.

29. The prism in Exercise 25.

30. The prism in Exercise 26.

31. The prism in Exercise 27.

Study Guide and Review

Applications and Problem Solving

32. **Parades** The Future Leader's Club is making a float for the homecoming parade. The flat surface will be shaped like the drawing below and covered with reflective paper. How many square meters of paper will be needed to cover it?
   (Hint: $\frac{1}{2}$ a circle, a triangle, plus a rectangle)
   (Lessons 11-2, 3 and 4-8)

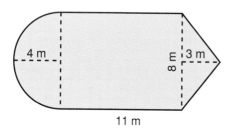

33. **Gifts** Shim has a large package to wrap for his sister's birthday present. The box is 4 feet long, 2.5 feet wide, and 2 feet high. How many square feet of wrapping paper does he need? *(Lesson 11-5)*

34. **Digging** A plumber digs a rectangular hole in the ground to reach the Duggan's water pipes. The hole is dug straight down, so that the hole has parallel sides. If the hole is 3 meters deep, 1.5 meters wide, and 2 meters long, what volume of dirt must be removed? *(Lesson 11-6)*

35. **Geometry** A cube is made of 27 small cubes. The outside of the cube is painted red. If the cube is taken apart, how many small cubes are totally unpainted? *(Lesson 11-7)*

Curriculum Connection Projects

- **Consumer Awareness** Find the dimensions and the area of a wide-screen TV.
- **Drafting** Find the surface area and volume of your school or home.

Read More About It

Roth, Arthur. *The Iceberg Hermit.*
Sharp, Richard. *The Sneaky Square and 113 Other Math Activities for Kids.*
Taylor, Mildred D. *Song of the Trees.*

11 Test

Find the area of each figure.

1.
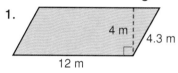
4 m | 4.3 m | 12 m

2.

5 in.

3.
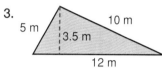
5 m | 10 m | 3.5 m | 12 m

4.
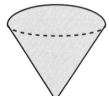
1 mi | 1.7 mi | 4 mi

5.
11 cm

6.
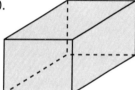
7.3 ft | 2 ft | 7 ft

7. **Gardening** Keisha plans to plant bulbs in a circular flower bed that has a radius of 2 meters. If she is to plant 40 bulbs for every square meter, how many bulbs should she buy?

Name each figure.

8.
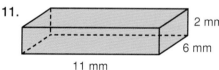

9.

10.

Find the surface area of each rectangular prism.

11.
2 mm | 6 mm | 11 mm

12.

8 cm | 4 cm | 2 cm

13. length, 11 inches; width, 8 inches; height, 5.5 inches

14. length, $4\frac{1}{2}$ meters; width, 2 meters; height, 5 meters;

15. Find the volume of the rectangular prism in Exercise 11.

16. Find the volume of the rectangular prism in Exercise 12.

17. Find the volume of the rectangular prism in Exercise 13.

18. Find the volume of the rectangular prism in Exercise 14.

19. **Pools** A rectangular diving pool is 20 feet by 15 feet and 8 feet deep. How much water is required to fill the pool?

20. A rectangular prism is formed using exactly 10 cubes. How many different prisms can be formed?

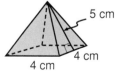
5 cm | 4 cm | 4 cm

Bonus Draw the square pyramid unfolded. Find the surface area.

Chapter Test

Investigations with Integers

Spotlight on Cars

Have You Ever Wondered. . .

- How many cars are produced each year?
- What the percentage of car owners is who own foreign cars?

Passenger Car Production by Make

Companies and models	1989	Companies and models	1989
Chrysler Corporation		General Motors Corporation	
Plymouth	258,847	Chevrolet	1,196,953
Dodge	453,428	Pontiac	732,177
Chrysler	203,624	Oldsmobile	520,981
Total	**915,899**	Buick	470,052
		Cadillac	293,589
		Total	**3,213,752**
Ford Motor Company		Honda	**362,274**
Ford	1,165,886	Mazda	**216,501**
Mercury	297,678	Nissan	**115,584**
Lincoln	213,517	Toyota	**231,279**
Total	**1,677,081**		
		Total	**6,732,370**

1904

1928

1939

1948

1900

1915

1930

1945

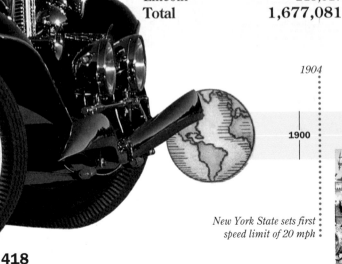

New York State sets first speed limit of 20 mph

Walt Disney creates first Mickey Mouse

Automatic transmission and air conditioning are introduced

Tubeless tires are introduced

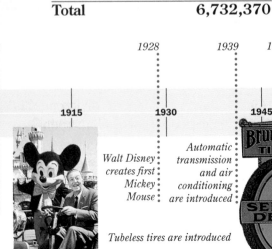

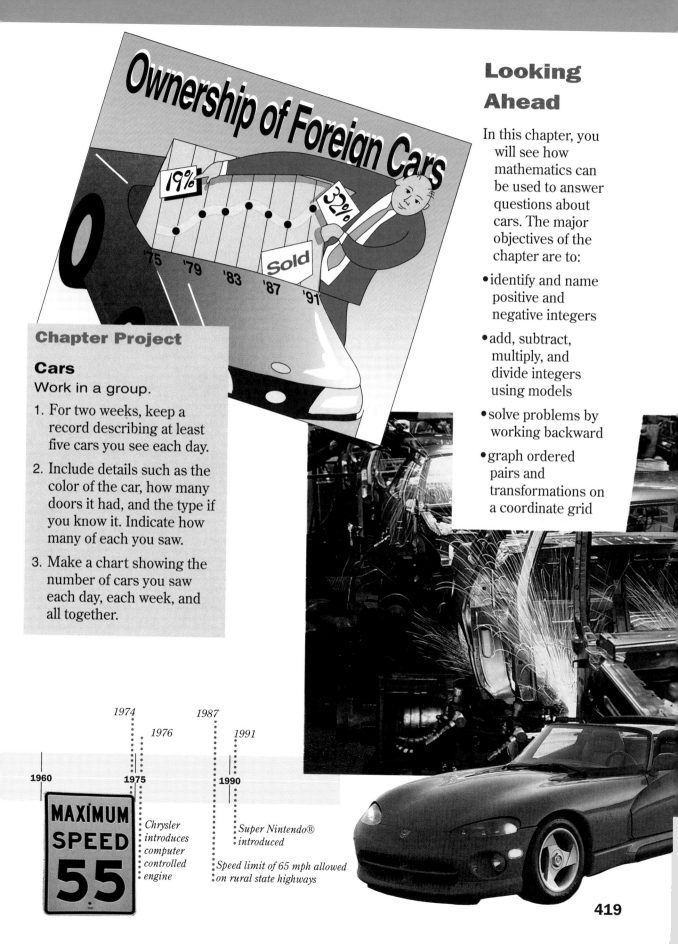

Ownership of Foreian Cars

19%

32%

Sold

'75 '79 '83 '87 '91

Looking Ahead

In this chapter, you will see how mathematics can be used to answer questions about cars. The major objectives of the chapter are to:

- identify and name positive and negative integers

- add, subtract, multiply, and divide integers using models

- solve problems by working backward

- graph ordered pairs and transformations on a coordinate grid

Chapter Project

Cars

Work in a group.

1. For two weeks, keep a record describing at least five cars you see each day.

2. Include details such as the color of the car, how many doors it had, and the type if you know it. Indicate how many of each you saw.

3. Make a chart showing the number of cars you saw each day, each week, and all together.

1974
1976
1987
1991

1960
1975
1990

MAXIMUM SPEED 55

Chrysler introduces computer controlled engine

Super Nintendo® introduced

Speed limit of 65 mph allowed on rural state highways

12-1 **Integers**

Objective

Identify, name, and graph integers.

Words to Learn

integer
positive integer
negative integer
opposite

During the winter, if it is 20° F, it might feel colder depending on how fast the wind is blowing. This effect is called the windchill factor. For a temperature of 20° F, if the wind is blowing at 10 miles per hour, it will feel like it is 2° F. If the wind is blowing at 15 miles per hour, then it will feel like it is −6° F.

The numbers 2 and −6 are integers. An **integer** is any number from the set {..., −3, −2, −1, 0, 1, 2, 3, ...} where ... means continues without end.

Integers that are greater than zero are called **positive integers.** Integers that are less than zero are called **negative integers.** Zero itself is neither positive nor negative. You can show positive and negative numbers on a number line.

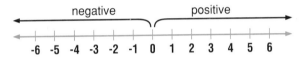

Negative integers are written with a − sign.   *Positive integers can be written with or without a + sign.*

You can graph integers on a number line by drawing a dot.

Examples

1 Graph −4 on the number line.

You read −4 as negative four.

2 Graph +3 on the number line.

You read +3 as positive three or three.

3 Write an integer to describe each situation.

  a. The temperature rose 7 degrees.
     You write +7.
  b. The scuba diver dove 500 feet.
     You write −500.
  c. The running back gained 11 yards on the play.
     You write +11.

Each integer has an opposite. **Opposite** integers are the same distance from zero in opposite directions on the number line. Zero is considered to be the starting point.

Example 4

Write the opposite of each integer.

a. $+2$

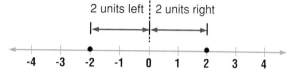

The opposite of $+2$ is -2.

b. -6

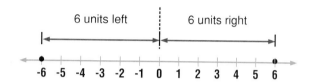

The opposite of -6 is $+6$.

Checking for Understanding

Communicating Mathematics

Read and study the lesson to answer each question.

1. **Tell** the integers graphed on the number line at the right.

2. **Show** the number -10 and its opposite by graphing on a number line.

3. **Write** about a situation at home that you could describe using positive and negative integers.

Guided Practice

Write the integer represented by each letter on the number line.

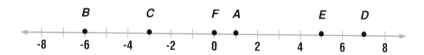

4. D 5. F 6. B
7. A 8. C 9. E

Write an integer to describe each situation.

10. 5 degrees below zero

11. a loss of 7 pounds

Write the opposite of each integer.

12. -8 13. 9 14. 12 15. -15

Exercises

Draw a number line from –10 to 10. Graph each integer on the number line.

16. –1

17. 4

18. –7

19. –9

20. 2

21. 8

Write an integer to describe each situation.

22. a gain of 10 yards

23. 3 feet below sea level

24. positive twelve

25. 2 degrees above zero

Write the opposite of each integer.

26. –3

27. 32

28. 1

29. –21

30. 16

31. 13

32. –87

33. –53

34. **Geometry** Find the volume of a rectangular prism that is 12 meters long, 8.6 meters wide, and 5 meters high. *(Lesson 11-8)*

35. Write *seven hundredths* as a decimal. *(Lesson 3-1)*

36. **Algebra** Solve the equation $\frac{11}{12} - x = \frac{5}{12}$. *(Lesson 7-4)*

37. **Geometry** Name the polygon at the right by the number of sides. *(Lesson 9-5)*

38. **Economics** In 1990, about one million laptop and notebook computers were sold. In 1991, about two million of these computers were sold. Use an integer to show the increase in units sold from 1990 to 1991.

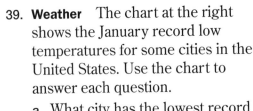

39. **Weather** The chart at the right shows the January record low temperatures for some cities in the United States. Use the chart to answer each question.

   a. What city has the lowest record low temperature?

   b. What city has the highest record low temperature?

Record Low Temperatures		
	Weather	Jan.
AK	Juneau	-22
AR	Little Rock	-4
CA	San Diego	29
CO	Denver	-25
CT	Hartford	-26

40. **Critical Thinking** Compare a number line to a thermometer. How are they alike? How are they different?

41. **Journal Entry** Make a time line of the 20th century. Let 0 represent this year. Mark as negative integers the years that represent events important to you that happened before this year. Mark as positive integers the years for important events you expect to happen in the future. Describe the event represented by each integer.

12-2 Comparing and Ordering Integers

Objective

Compare and order integers.

Ana and Frank are balancing their checkbooks. Ana has a balance, in dollars, of −10, and Frank has a balance of −5. Who has less money in their checkbook?

To answer this question, we need to compare the two numbers. Many real-life situations such as a balance in a checking account can be represented with integers.

You can use a number line to compare the numbers −5 and −10. On a number line, the number to the left is always less than the number to the right.

Since −10 is to the left of −5, −10 < −5.
Ana has less money than Frank.

Mental Math Hint
• • • • • • • • • •
On a number line, any positive integer is to the right of a negative integer. So, when comparing a positive and a negative integer, the positive integer will always be greater.

Examples

Replace each ▪ with <, >, or =.

1 −5 ▪ −7.

Graph −5 and −7 on a number line.

−7 is to the left of −5, so −5 > −7.

2 0 ▪ −8.

Graph 0 and −8 on a number line.

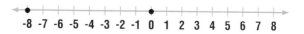

−8 is to the left of 0, so 0 > −8.

Example 3

Order the integers –3, 0, –5, and 1 from least to greatest.

One way to order the integers is to graph each number on a number line first.

Then write the integers as they appear on the number line from left to right.

–5, –3, 0, and 1 are in order from least to greatest.

LOOK BACK

You can review median on page 65.

Example 4   *Connection*

Statistics   Find the median of the following temperatures.

Record Low Temperatures (°F) Pittsburgh, Pennsylvania (by month)											
J	F	M	A	M	J	J	A	S	O	N	D
–18	–12	–1	14	26	34	42	39	31	16	–1	–12

List the temperatures in order from least to greatest.
–18, –12, –12, –1, –1, 14, 16, 26, 31, 34, 39, 42

Since there are two middle numbers, 14 and 16, the median is
$\frac{14 + 16}{2} = \frac{30}{2}$ or 15.

The median for the monthly record low temperatures for Pittsburgh, Pennsylvania, is $15°F$.

Checking for Understanding

Communicating Mathematics

Read and study the lesson to answer each question.

1. **Tell** which integer graphed on the number line is greater.

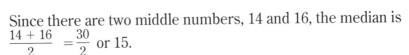

2. **Write** how to determine when an integer is greater than or less than another integer.

3. **Show** that $-2 > -5$ by using a number line.

Guided Practice

Replace each ▓ with <, >, or =.

4. $+2 \ \blacksquare \ +3$

5. $-7 \ \blacksquare \ -9$

6. $-4 \ \blacksquare \ +4$

7. $0 \ \blacksquare \ 1$

8. $-77 \ \blacksquare \ -99$

9. $2 \ \blacksquare \ -22$

10. Which is greater, 7 degrees above zero or 8 degrees below zero?

Order each set of integers from least to greatest.

11. −9, 0, 5, −4

12. −1, 1, 11, −101

13. 3, −3, 4, −5, 6, −7

14. 4, −5, −45, −54

Exercises

Independent Practice

Replace each ▨ with <, >, or =.

15. 6 ▨ −66

16. −7 ▨ −700

17. −45 ▨ −45

18. 44 ▨ −5

19. −578 ▨ −89

20. −91 ▨ −2

21. 76 ▨ −239

22. −999 ▨ −99

23. 0 ▨ −45

24. Which is greater, negative 4 or negative 44?

25. Which is higher, 5 feet below sea level or 4 feet above sea level?

26. Is zero greater than, less than, or equal to negative twelve?

Order each set of integers from least to greatest.

27. 56, −56, −40

28. 0, −12, −3, −5, −64

29. −12, 0, −3, 9, 98, −10, 54

30. −1, 0, −11, −101, −111, 101, 11

31. 0, 54, 2, −21, −12, −18, 64

32. 75, −3, −4, 12, 0, 9, −10

Mixed Review

33. Write an integer to describe a temperature of twelve degrees below zero. *(Lesson 12-1)*

34. Express the number $\frac{5}{8}$ as a percent. *(Lesson 10-5)*

35. **Travel**   The Chestneys' car takes an average of 12.8 gallons of gasoline each fill-up. On their trip from Chicago to Atlanta, they expect to fill up five times. About how much gasoline will they buy? *(Lesson 4-1)*

36. Find the greatest common factor of 40 and 36. *(Lesson 6-4)*

Problem Solving and Applications

37. **Weather**   The coldest temperature ever recorded for Chicago, Illinois, was −27° F. The coldest temperature ever recorded for Great Falls, Montana, was −37° F. Which city had the colder temperature ever recorded?

38. **Critical Thinking**   Why is any negative integer less than any positive integer?

39. **Statistics**   The table below shows the lowest recorded temperatures for each month in Chicago, Illinois, in degrees Fahrenheit. Find the median of the temperatures.

J	F	M	A	M	J	J	A	S	O	N	D
−27	−17	−8	7	24	36	40	41	28	17	1	−25

12-3A Zero Pairs

A Preview of Lesson 12-3

Objective

Model integers.

Words to Learn

zero pair

Materials

counters of two colors
mat

You can use counters to help you understand integers. In this lab, yellow counters will represent positive integers, and red counters will represent negative integers. When you pair one counter of each color, the result is zero. We call this pair of counters a **zero pair.**

Try this!

Work with a partner.

- Place four yellow counters on the mat to represent the integer +4.

- Place four red counters on the same mat to represent the integer −4.

- Pair the positive and negative counters. Remove as many zero pairs as possible.

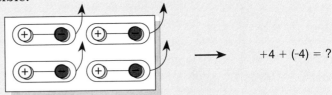

What do you think?

1. How many counters do you have left on the mat?
2. What is the sum of +4 and (−4)?
3. How are +4 and −4 related?
4. What is the sum of any pair of opposites?
5. What is the value of a zero pair?
6. If you remove a zero pair from the mat, what effect does it have on the value of the counters left on the mat?

12-3 Adding Integers

Objective

Add integers using models.

DID YOU KNOW

The tallest apartment building in the United States is the Metropolitan Tower on West 57th Street in New York City. It has 78 stories. The top 48 stories are residential.

Micah lives on the second floor of his apartment building in New York City. He got on the elevator at his floor and rode it up 5 floors to visit his cousin Carlos. He and Carlos then rode the elevator down 3 floors to visit their friend Jeff. What floor are the boys on now?

You have already worked with comparing and ordering integers. To find out what floor Micah, Carlos, and Jeff are on, you will need to add integers. You can add integers using models.

Use yellow counters to represent positive integers and red counters to represent negative integers.

Step 1. Use 2 positive counters to represent +2, the second floor, where Micah got on the elevator.

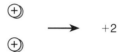

Step 2. Use 5 positive counters to represent +5, the five floors Micah rode up to visit Carlos.

Step 3. Place all of your counters together on a mat.

You have a total of 7 positive counters. The counters represent the integer +7. So, Micah rode to the seventh floor to visit his cousin Carlos.

Step 4. Use 3 negative counters to represent −3, the three floors Micah and Carlos went down to visit Jeff. Place the 3 negative counters on the mat with the 7 positive counters.

$7 + (-3)$

Step 5. Pair the positive and negative counters. Remove as many zero pairs as possible since their value is zero.

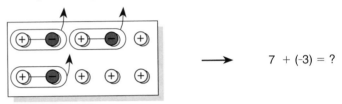

$7 + (-3) = ?$

How many counters are left on your mat? What color are the counters that are left?

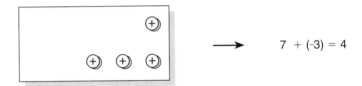

$7 + (-3) = 4$

There are a total of 4 positive counters left on the mat.
$7 + (-3) = 4$
So, the boys are on the fourth floor.

Example 1

Use counters to find −9 + 3.

Place 9 negative counters on the mat to represent −9. Place 3 positive counters on the same mat to represent adding 3. Pair the positive and negative counters. Remove as many zero pairs as possible.

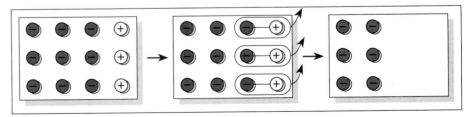

Count the counters left on the mat. You have 6 negative counters left on the mat. This represents −6. So, $-9 + 3 = -6$.

2 Use counters to find $-6 + (-6)$.

Place 6 negative counters on the mat to represent -6. Place 6 more negative counters on the same mat to represent adding -6. Since there are no positive counters, you cannot remove any zero pairs.

Count all the counters on the mat. You have 12 negative counters left on the mat. This represents -12.
So, $-6 + (-6) = -12$.

3 Find $-3 + 3$.

Place 3 negative counters on the mat to represent -3. Place 3 positive counters on the same mat to represent adding 3. Pair the positive and negative counters. Remove as many zero pairs as possible.

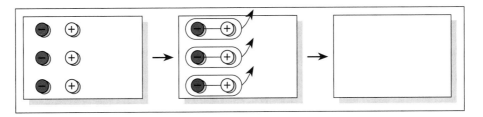

Count all the counters on the mat. You have no counters left on the mat. This represents 0. So, $-3 + 3 = 0$.

Checking for Understanding

Communicating Mathematics

Read and study the lesson to answer each question.

1. **Show** $-2 + 3$ using counters.
2. **Write,** in your own words, how counters can help you tell whether the sum of two integers will be negative or positive.
3. **Tell** the sum represented by the model below.

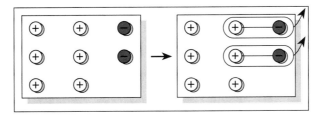

Use counters to find each sum.

4. $-2 + 8$ 5. $4 + (-6)$ 6. $10 + (-1)$

7. $-3 + (-9)$ 8. $-7 + 2$ 9. $-8 + 4$

State whether each sum is positive or negative.

10. $-5 + 7$ 11. $-8 + (-8)$ 12. $9 + (-7)$

Exercises

Independent Practice

State whether each sum is positive or negative.

13. $7 + (-5)$ 14. $-6 + (-4)$ 15. $5 + (-9)$

16. $-3 + (-2)$ 17. $8 + (-7)$ 18. $-12 + 13$

Find each sum. Use or draw counters if necessary.

19. $-4 + (-9)$ 20. $-5 + (-8)$ 21. $-13 + 9$

22. $-9 + 12$ 23. $-4 + 0$ 24. $12 + (-15)$

25. $7 + (-4)$ 26. $-8 + 17$ 27. $-4 + 16$

28. $-9 + 5$ 29. $13 + (-3)$ 30. $-11 + (-18)$

Mixed Review

31. Which is greater: 66 or -75? *(Lesson 12-2)*

32. **Algebra**   Solve mentally: $b \times 6 = 42$. *(Lesson 1-9)*

33. **Algebra**   Evaluate a^3 if $a = 2$. *(Lesson 4-2)*

34. **Tips**   Tips at the big holiday banquet totalled $1,218.75. Thirteen servers split the total evenly. How much did each server receive? *(Lesson 5-2)*

Problem Solving and Applications

35. **Measurement**   A scuba diver dove 12 feet below sea level, then went down another 21 feet.

   a. At what distance from sea level is the diver now? Write an addition sentence to find the answer.

   b. Use an integer to represent the sum.

36. **Critical Thinking**   How can you use a number line to find $-3 + (-9)$?

37. **Golf**   Arnie played in a two-day golf tournament. On day one, Arnie finished 2 over par. On day two, Arnie finished 3 under par. How did Arnie finish for the two days?

38. **Oceanography**   A submarine traveling at 1,200 meters below sea level descends an additional 1,050 meters. How far below sea level is the submarine now?

39. **Critical Thinking**   Explain why the following sentence is true or false. The sum of a negative integer and a positive integer is always negative. Give an example to support your answer.

Objective

Subtract integers using models.

DID YOU KNOW

The woman in the photo is Bonnie Anderson. She is a WTVJ-Channel 4 News Reporter in Miami, Florida.

You may have heard of news reporters reporting the news from high above a city. But what about reports from under water? In Miami, Florida, a news reporter puts on her scuba gear and jumps in while a cameraperson goes along to film her.

Problem Solving Hint

• • • • • • • • • •

You can use the strategy draw a diagram to help you solve this problem.

Suppose the reporter was on a pier that was 5 feet above the surface of the water. After jumping in, she descends to 40 feet below the surface, or −40 feet. What is the difference between 5 and −40?

A number line can help you visualize this situation.

The number line shows that there is a 45 foot difference between the two numbers. That is, $5 - (-40) = 45$.

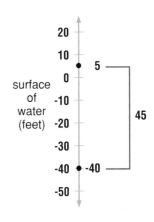

In Lesson 12-3, you used models to find the sums of integers. You can also use models to find the differences of integers.

Example 1

Use counters to find $8 - 2$.

Place 8 positive counters on the mat to represent $+8$. Remove 2 positive counters from the mat to represent subtracting 2.

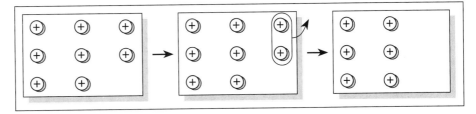

Count all the counters on the mat. You have 6 positive counters left on the mat. This represents $+6$. So, $8 - 2 = 6$.

Example 2

Use counters to find −7 − (−3).

Place 7 negative counters on the mat to represent −7. Remove 3 negative counters to represent subtracting −3.

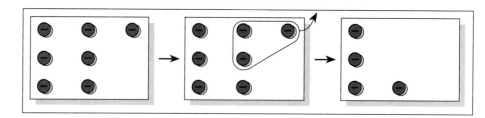

Count all the counters left on the mat. You have 4 negative counters on the mat.
This represents −4. So, −7 − (−3) = −4.

Sometimes, you need to add zero pairs in order to subtract. When you add zero pairs, the value of the integers on the mat does not change.

Example 3

Find 6 − (−2).

Place 6 positive counters on the mat to represent +6. To subtract −2, you must remove 2 negative counters. But there are no negative counters on the mat to remove. Add 2 zero pairs to the mat. The value on the mat does not change. You can now remove 2 negative counters.

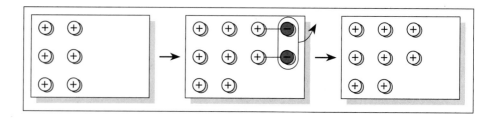

Count all the counters left on the mat. You have 8 positive counters on the mat.
This represents +8. So, 6 − (−2) = 8.

Example 4 *Problem Solving*

Weather What is the change in temperature if it dropped from 3 degrees above zero to 4 degrees below zero?

To solve this problem, you need to find $3 - (-4)$. Place 3 positive counters on the mat to represent $+3$. To subtract -4, you must remove 4 negative counters. Since there are no negative counters to remove, you must add 4 zero pairs to the set. Now remove 4 negative counters.

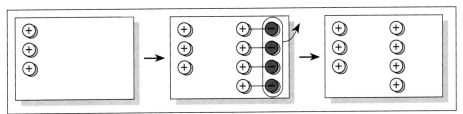

Count all of the counters left on the mat. You have 7 positive counters on the mat. This represents $+7$. So, $3 - (-4) = 7$. The temperature dropped 7 degrees.

Checking for Understanding

Communicating Mathematics

Read and study the lesson to answer each question.

1. **Draw** a model showing how to find $4 - 5$.
2. **Tell** the subtraction sentence represented by the model below.

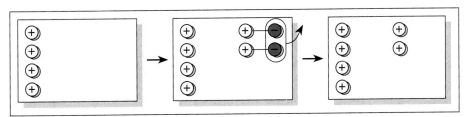

Guided Practice

Use counters to find each difference.

3. $4 - (-5)$ 4. $-8 - (-7)$ 5. $-5 - (-2)$
6. $9 - 10$ 7. $8 - (-4)$ 8. $-3 - (-1)$
9. $-5 - 5$ 10. $3 - 5$ 11. $-7 - 2$

Exercises

Independent Practice

Find each difference. Use or draw counters if necessary.

12. $4 - (-6)$ 13. $-12 - (-10)$ 14. $10 - (-9)$
15. $-12 - 9$ 16. $-5 - (-3)$ 17. $-4 - 2$
18. $-1 - (-1)$ 19. $-9 - (-5)$ 20. $-8 - 12$
21. $-16 - 16$ 22. $-10 - 9$ 23. $5 - (-15)$

24. Find $8 + (-10)$. *(Lesson 12-3)*

25. Write the fraction $\frac{100}{112}$ in simplest form. *(Lesson 6-5)*

26. **Geometry** Trace the line segment at the right. ─────────────
    Then use a compass and straightedge to bisect the segment.
    (Lesson 9-4)

27. **Sports** The Branson Middle
    School football team is 2 yards
    from their opponent's goal line.
    On the next two plays, their
    offense loses 17 yards and then
    gains 11 yards. How many
    yards are they from the goal line
    now?

28. **Geography** Mt. Whitney is
    14,494 feet above sea level. Death
    Valley is 282 feet below sea level.
    What is the difference between
    these two elevations?

29. **Critical Thinking** If you subtract a greater number from a lesser number,
    will the difference be positive or negative? Explain your answer.

30. **Journal Entry** Use a map and/or a geography textbook to find the
    elevation of two famous mountains. Write the difference between the two
    elevations.

31. **Mathematics and Social Studies** Read the following paragraphs.

Elisha Graves Otis (1811-1861) was an American inventor best known
for designing the first elevator. In 1852, his employer sent him to
Yonkers, New York, where he designed and installed what he called
the "safety hoist," the first elevator equipped with an automatic safety
device to prevent it from falling. He then set up a small elevator shop,
selling his first freight elevator on September 20, 1853.

After he demonstrated how the elevator worked at the Crystal Palace
in New York City, many orders began coming in. In 1857, he installed
the first safety elevator in a store in New York. It could carry about six
people.

In 1889, after Otis' death, the Otis Company installed the first electric
elevator. It was lifted by cables driven by an electric motor. The
invention of the elevator encouraged the building of skyscrapers.

Jessica delivers mail to employees in a large office tower. She gets on the
elevator on the 52nd floor. She rides down 16 floors and then rides up 32
floors. What floor is Jessica on now?

12-5 Multiplying Integers

Objective
Multiply integers using models.

Mr. Carr's doctor would like him to lose weight. The doctor has prescribed a diet and exercise plan for Mr. Carr to follow. With this plan, he will lose about 2 pounds per week. If Mr. Carr needs to be on the plan for 6 weeks, how much weight does his doctor want him to lose?

You can represent Mr. Carr's weekly loss as –2. So, to answer the question, you need to multiply 6 and –2.

Remember that multiplication is repeated addition. If Mr. Carr loses 2 pounds per week for 6 weeks, you can write this as follows.

$$-2 + (-2) + (-2) + (-2) + (-2) + (-2) \text{ or}$$
$$6 \times (-2)$$

addition or multiplication

Use counters. $6 \times (-2)$ means to *put in* 6 sets of 2 negative counters. Place them on the mat.

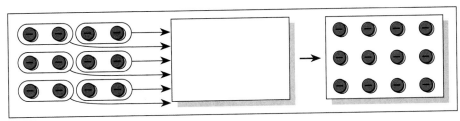

There are 12 negative counters on the mat. This represents –12. So, $6 \times (-2) = -12$. Mr. Carr's doctor wants him to lose 12 pounds.

> **When am I ever going to use this?**

Nutritionists research the effects of various foods and encourage people to eat healthy meals.

A nutritionist must know how to use integers, proportions, and measurements. He/she must have a college degree to work in a medical facility or in a weight-loss center.

For more information, contact:
National Health Council
250 Fifth Avenue
Suite 1118
New York, NY 10118

Example 1

Use counters to find 2×4.

In whole number multiplication, 2×4 means two sets of four. You can write 2×4 as $+2 \times (+4)$. $+2 \times (+4)$ means to *put in* two sets of four positive counters. Place them on the mat.

Count all of the counters on the mat. There are 8 positive counters on the mat. This represents +8. So, $2 \times 4 = 8$.

2 Use counters to find -2×5.

Since -2 is the opposite of 2, -2×5 means to *remove* 2 sets of 5 positive counters. But you cannot remove counters if you don't have any to remove. You must first add 2 sets of 5 zero pairs. Then you can remove 2 sets of 5 positive counters.

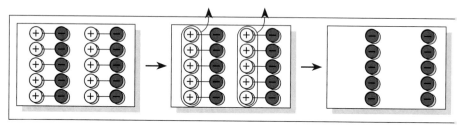

Count all of the counters left on the mat. There are 10 negative counters left on the mat. This represents -10. So, $-2 \times 5 = -10$.

3 Find $-3(-2)$.

Since -3 is the opposite of 3, $-3(-2)$ or -3×-2 means to remove 3 sets of 2 negative counters. You must first add 3 sets of 2 zero pairs. Then you can remove 3 sets of 2 negative counters.

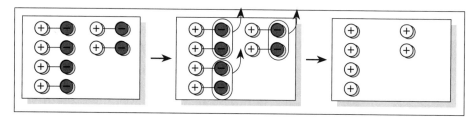

Count all of the counters on the mat. There are 6 positive counters on the mat. This represents $+6$. So, $-3(-2) = 6$.

Checking for Understanding

Communicating Mathematics

Read and study the lesson to answer each question.

1. **Tell** what happened when you multiplied two negative integers.
2. **Draw** a model to show how to find $-3 \times (-3)$.
3. **Tell** the multiplication sentence represented by the model below.

Guided Practice **Use counters to find each product.**

4. $5 \times (-7)$
5. $-7 \times (-7)$
6. $-2 \times (-8)$
7. $-5(-1)$
8. $2(3)$
9. $10(-1)$
10. $-8(-9)$
11. $9(-3)$
12. $-6(-7)$

Exercises

Independent Practice **Find each product. Use or draw counters if necessary.**

13. $5 \times (-9)$
14. $-9 \times (-3)$
15. $-4(20)$
16. $-14(5)$
17. $8(-3)$
18. $4(-12)$
19. $-2(-11)$
20. $9(-5)$
21. $-6(-6)$
22. $5(-8)$
23. $-4(-9)$
24. $-3(-13)$

25. What is the product of -5 and 3?
26. Find the product of -7 and -4.

Mixed Review
27. Use counters to find $-6 - (+5)$. *(Lesson 12-4)*
28. Estimate 31×194 using patterns. *(Lesson 1-4)*
29. Express $\frac{7}{9}$ as a repeating decimal. *(Lesson 6-11)*

Problem Solving and Applications
30. **Measurement**   A submarine is descending at a rate of 8 feet per second. How far below the surface will the submarine be in 3 minutes?
31. **Measurement**   The temperature is falling at a rate of 3 degrees per hour. If it is $48°$ F at 6:00 P.M., what will the temperature be at 2:00 A.M.?

12 Mid-Chapter Review

Write the integer represented by each letter on the number line. *(Lesson 12-1)*

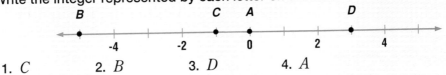

1. C
2. B
3. D
4. A

Replace each ■ with <, >, or =. *(Lesson 12-2)*

5. $+3 \; ■ \; +5$
6. $-8 \; ■ \; -10$
7. $-5 \; ■ \; +5$

Find each sum, difference, or product. Use or draw counters if necessary.
(Lessons 12-3, 12-4, and 12-5)

8. $-5 + 6$
9. $-6(3)$
10. $8 + (-5)$
11. $7 - (-2)$
12. $3(-5)$
13. $-6 - 5$

12-6 Work Backward

Objective
Solve problems by working backward.

Brandy, Vance, Jaquita, and Neal play on the same softball team. During a recent three-game series, Vance had the most hits. Jaquita had 7 fewer hits than Vance. Brandy had 2 more hits than Jaquita. Neal had twice as many hits as Brandy. Neal had 8 hits. How many hits did Vance have?

Explore   What do you know?
Vance had the most hits.
Jaquita had 7 fewer hits than Vance.
Brandy had 2 more hits than Jaquita.
Neal had twice as many hits as Brandy.
Neal had 8 hits.

What do you need to find?
You need to find out how many hits Vance had.

Plan   Since you need to know how many hits Vance had, use the information you do know and work backward to solve this problem.

Solve   Make a flowchart.

$$\boxed{?} \rightarrow \boxed{-7} \rightarrow \boxed{+2} \rightarrow \boxed{\times 2} \rightarrow \boxed{8}$$

Vance   Jaquita   Brandy   Neal   Neal

Use inverse operations to *undo* each step.

$$\boxed{?} \leftarrow \boxed{+7} \leftarrow \boxed{-2} \leftarrow \boxed{\div 2} \leftarrow \boxed{8}$$

Vance   Jaquita   Brandy   Neal   Neal

⌐ Start ⊦

Undo the multiplication. $8 \div 2 = 4$
Undo the addition. $4 - 2 = 2$
Undo the subtraction. $2 + 7 = 9$
Vance had 9 hits.

Examine   Assume that Vance started with 9 hits. Jaquita had $9 - 7$ or 2 hits. Brandy had $2 + 2$, or 4 hits. Neal had 4×2, or 8 hits. Your answer checks.

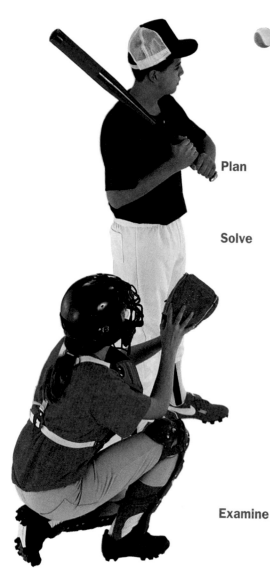

Checking for Understanding

Communicating Mathematics

Read and study the lesson to answer each question.

1. **Tell** how to use the work backward strategy to solve problems.

2. **Write** a paragraph explaining why it is sometimes necessary to use inverse operations when working backward.

Guided Practice

Solve using the work backward strategy.

3. In the Trivia Bowl, each finalist must answer four questions correctly. Each question is worth twice as much as the question before it. The fourth question is worth $6,000. How much is the first question worth?

4. Christa, Parker, Dom, and Luann each handed in term papers on the Industrial Revolution. Parker's paper was 4 pages shorter than Christa's. Dom's paper was 7 pages longer than Parker's. Luann's paper was half as long as Dom's. Luann's paper was 14 pages long. How long was Christa's paper?

5. Melissa has a dental appointment at 7:30 A.M. tomorrow. She wants to get there 10 minutes early. It takes her one hour to get ready in the morning and 40 minutes to drive to the dentist's office. What time should Melissa plan to get up tomorrow?

Problem Solving

Practice

Solve using any strategy.

6. Bargain Books bookstore arranges its best sellers in the front window. How many different ways can they arrange 5 new best sellers?

Strategies

• • • • • • •

Look for a pattern.
Solve a simpler
problem.
Act it out.
Guess and check.
Draw a diagram.
Make a chart.
Work backward.

7. Tina bought some watercolor paints and brushes. She spent $3.75 on brushes and five times that amount on paint. She had $6.89 left. How much money did she have before she went shopping?

8. The sum of fifteen and three times a number is 63. Find the number.

9. The Wilder Middle School sixth grade held a used book sale. Mariah sold 7 more books than Ariel. Brianna sold half as many books as Mariah. Franklin sold 5 more books than Brianna. Franklin sold 9 books. How many books did Ariel sell?

10. Marty bought some grapefruits and some oranges. Grapefruits sell for $0.89 and oranges sell for $0.25. Marty spent $5.56. How many of each did he buy?

12-7 Dividing Integers

Objective

Divide integers using models and patterns.

Ricardo and his friends are playing football. In three consecutive plays, Ricardo's team has advanced the ball a total of 12 yards. If his team gained the same number of yards per play, how many yards were gained each play?

To solve this problem, you need to find $12 \div 3$. You can solve this using positive counters. $12 \div 3$ means to separate 12 into 3 equal-sized groups.

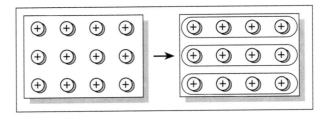

There are 3 groups of 4 positive counters each.
So, $12 \div 3 = 4$. This means that Ricardo's team gained 4 yards per play.

Example 1 *Problem Solving*

Sports   Suppose that Ricardo's team lost a total of 15 yards for 3 penalties. If each penalty was for the same number of yards, how many yards was each penalty?

You can represent losing 15 yards as −15. To answer the problem, you need to find $-15 \div 3$. This means to separate −15 into 3 equal-sized groups.

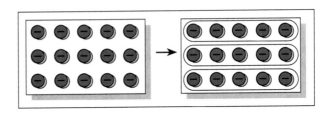

There are 3 groups of 5 negative counters each.
So, $-15 \div 3 = -5$. This means that each penalty was a loss of 5 yards.

2 Use counters to find 8 ÷ 2.

8 ÷ 2 means to separate 8 positive counters into 2 equal-sized groups.

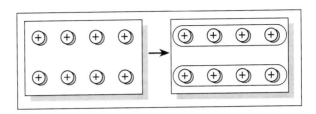

There are 2 groups of 4 positive counters each. So, 8 ÷ 2 = 4.

3 Use counters to find −10 ÷ 5.

−10 ÷ 5 means to separate −10 into 5 equal-sized groups. Use negative counters to show −10.

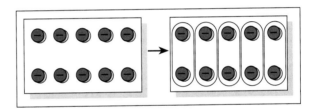

There are 5 groups of 2 negative counters each. So,
−10 ÷ 5 = −2.

LOOKBACK

You can review inverse operations on page 179.

You can also use the relationship of multiplication and division to help find quotients.

8 × 4 = 32, so 32 ÷ 4 = 8.

Look at the relationships below.

a. 6 × (−2) = −12, so −12 ÷ (−2) = 6.

b. −8 × 2 = −16, so −16 ÷ 2 = −8.

Do you notice a pattern? Notice that when you divide a negative number and a positive number, the quotient is negative.

Example 4

Use patterns to find $18 \div (-9)$.

$-2 \times (-9) = 18$, so $18 \div (-9) = -2$.   *The quotient is negative.*

Checking for Understanding

Communicating Mathematics

Read and study the lesson to answer each question.

1. **Tell** the division sentence shown below.

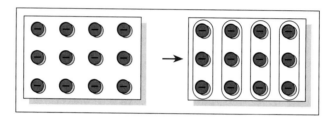

2. **Tell** if the quotient is positive or negative when you divide two negative numbers.

Guided Practice

Use counters or patterns to find each quotient.

3. $14 \div 2$
4. $30 \div (-15)$
5. $36 \div (-2)$
6. $-9 \div 3$
7. $-50 \div 2$
8. $-81 \div 9$
9. $48 \div (-16)$
10. $-7 \div (-1)$
11. $-250 \div (-5)$

Exercises

Independent Practice

Find each quotient. Use counters or patterns if necessary.

12. $45 \div (-3)$
13. $65 \div (-5)$
14. $-28 \div 4$
15. $-63 \div (-9)$
16. $-70 \div (-7)$
17. $-94 \div 2$
18. $108 \div (-9)$
19. $75 \div (-25)$
20. $-21 \div (-3)$
21. $-288 \div 36$
22. $-33 \div 11$
23. $-81 \div (-9)$

Mixed Review

24. Find the product of -3 and -7. *(Lesson 12-5)*

25. On December 19, the sunrise was at 7:19 A.M., and the sunset was at 4:28 P.M. How many hours of daylight were there? *(Lesson 7-8)*

26. **Patterns** Find the next two numbers in the sequence 34, 36.5, 39, 41.5, ... *(Lesson 8-5)*

Problem Solving and Applications

27. **Statistics** The chart at the right shows low temperatures reported for one week. What is the mean temperature for the week?

28. **Critical Thinking** Find values of x and y for which both statements are true.

   ● x and y are negative.   ● $x \div y = 15$

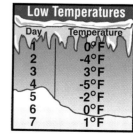

Low Temperatures	
Day	Temperature
1	0°F
2	-4°F
3	3°F
4	-5°F
5	-2°F
6	0°F
7	1°F

442   **Chapter 12** Investigations with Integers

12-8 The Coordinate System

Objective

Graph ordered pairs of numbers on a coordinate grid.

Words to Learn

coordinate system
coordinate grid
x-axis
y-axis
ordered pairs
coordinates

TEEN SCENE

Disney-MGM Studios has an Indiana Jones Epic Stunt Spectacular where scenes from the action-packed Indiana Jones movies are recreated live. If you attend this event, you may be chosen as an "extra" for an exciting chase scene.

Have you ever wanted to be the star of a movie? If you visit Walt Disney-MGM Studios Theme Park, you could become one.

A map of the theme park can be placed on a **coordinate system,** or **coordinate grid.** A horizontal number line and a vertical number line meet at their zero points to form a coordinate system. The horizontal line is called the *x*-**axis,** and the vertical line is called the *y*-**axis.**

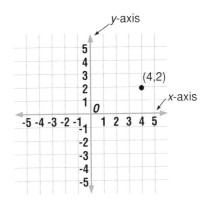

coordinate system

The location of any point in the coordinate system can be named using an ordered pair of numbers. An **ordered pair** gives the **coordinates** of the point. The first number in an ordered pair is the coordinate on the *x*-axis, or horizontal number line. The second number in the ordered pair is the coordinate on the *y*-axis, or vertical number line. An ordered pair is written in this form:

$$(4, 2)$$

first coordinate ⌐ ⌐ *second coordinate*

Example 1

Name the ordered pair for point *A*.

Start at 0. Move right along the *x*-axis until you are under point *A*. Since you moved two units to the right, the first coordinate of the ordered pair is 2.

Now move up along the *y*-axis until you reach point *A*. Since you moved up 4 units, the second coordinate of the ordered pair is 4.

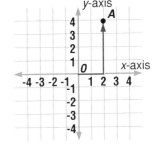

The ordered pair for point *A* is (2, 4).

Example 2

Name the ordered pair for point *R*.

Start at 0. Move left 5 units along the *x*-axis. The first coordinate of the ordered pair is –5. Now move up 3 units along the *y*-axis. The second coordinate of the ordered pair is 3. The ordered pair for point *R* is (–5, 3).

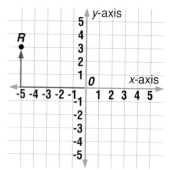

You can also graph a point on a coordinate grid. To graph a point means to place a dot at the point named by an ordered pair.

Examples

Graph each point.

3 *D*(1, 2)

Start at 0. Move 1 unit to the right on the *x*-axis. Then move two units up along the *y*-axis to locate the point. Place a dot and label the dot *D*.

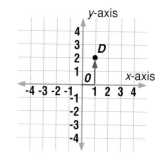

4 *J*(3, –7)

Start at 0. Move 3 units to the right along the *x*-axis. Then move 7 units down along the *y*-axis to locate the point. Place a dot and label the dot *J*.

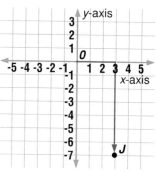

Checking for Understanding

Communicating Mathematics

Read and study the lesson to answer each question.

1. **Tell** how to graph the ordered pair (–7, –9).

2. **Draw** any point on a coordinate grid. Name the ordered pair for that point.

Guided Practice

Name the ordered pair for each point.

3. *P*   4. *T*

5. *R*   6. *M*

7. *Q*   8. *S*

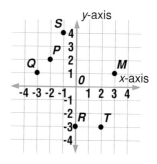

On graph paper, draw a coordinate grid. Then graph and label each point.

9. $A(3, 5)$ **10.** $B(1, -5)$ **11.** $C(-6, 5)$ **12.** $D(-4, -4)$

Exercises

Independent Practice

Name the ordered pair for each point.

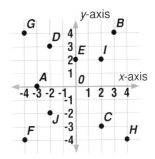

13. A **14.** B

15. C **16.** D

17. E **18.** F

19. G **20.** H

21. I **22.** J

On graph paper, draw a coordinate grid. Then graph and label each point.

23. $E(3, -3)$ **24.** $F(5, -8)$ **25.** $A(-4, 7)$ **26.** $M(-2, 1)$

27. $R(-5, -4)$ **28.** $S(2, 6)$ **29.** $T(4, 0)$ **30.** $V(-3, -2)$

Mixed Review

31. Find the quotient of 32 and −4. *(Lesson 12-7)*

32. **Statistics**   Find the median of the data: 23, 19, 22, 22, 20, 21, 19, 22. *(Lesson 2-7)*

33. **Geometry**   Find the area of the parallelogram at the right. *(Lesson 11-1)*

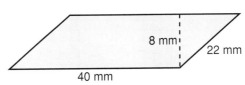

34. **Real Estate**   Mrs. Bates, a real estate agent with East End Realty, earned a commission of 1.5% of the sale price of the last home she sold. If the home sold for $127,500, how much money did she make? *(Lesson 10-8)*

Problem Solving and Applications

35. **Geography**   Longitude and latitude are used to locate places on a map or globe. Find the longitude and latitude of the place where you live.

36. **Art**   Draw a coordinate grid. Number each axis from −10 to 10. Draw a picture on your grid. List the ordered pairs of points where sides of the drawing meet. Have a classmate draw your picture by graphing your ordered pairs.

37. **Critical Thinking**   On graph paper, draw a map of your neighborhood. Give the coordinates for where you live.

38. **Data Search**   Refer to pages 418 and 419. How many Ford, Dodge, and Buick cars were produced in 1989?

Lesson 12-8 The Coordinate System   **445**

12-9A Waxed Paper Transformations

A Preview of Lesson 12-9

Objective

Find transformations using waxed paper.

Words to Learn

transformation
translation
reflection

Materials

waxed paper
scissors
notebook paper
pen

Coordinate grids can be used to graph **transformations,** or movements, of figures. This lab will help you understand transformations.

Activity One

Work with a partner.

- Trace triangle *A* and point *B* onto a sheet of notebook paper. Line up the bottom of triangle *A* with a line on your notebook paper.

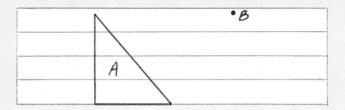

- Cut or tear a rectangular-shaped piece of waxed paper large enough to cover triangle *A*.
- Place the waxed paper piece on top of triangle *A*.
- Trace triangle *A* onto the waxed paper with a pen by pressing down on the waxed paper so an imprint is made.
- Slowly slide the imprint to point *B*. Be sure the bottom of the waxed paper triangle slides along the line on your notebook paper. Stop when the top vertex of triangle *A* is on point *B*.

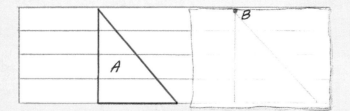

- Trace over the imprint, then remove the waxed paper. (You will need to press down fairly hard with your pen so that you can see the imprint on your notebook paper.)
- Trace over this imprint on your notebook paper with your pen. This movement is called a **translation,** or slide of triangle *A*.

What do you think?

1. Describe what happened when you made the translation.
2. How does triangle *A* compare with the new triangle?

Activity Two

- Trace triangle *C* and line *m* onto a sheet of notebook paper.

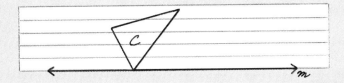

- Cut or tear a piece of waxed paper large enough to cover triangle *C* and line *m*. Place the waxed paper on top of your drawing.

- Trace triangle *C* and line *m* onto the waxed paper with a pen by pressing down on the waxed paper so an imprint is made.

- Lift the waxed paper up. Turn the waxed paper upside down, toward your body. Lay the waxed paper on your notebook paper so that line *m* on your waxed paper is lined up with the line *m* on your notebook paper.

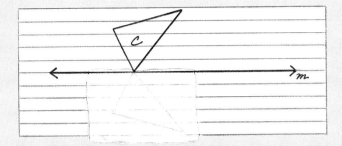

- Trace over the triangle imprint, then remove the waxed paper.

- Trace over this imprint on your notebook paper with your pen. This movement is called a **reflection,** or flip of triangle *C*.

What do you think?

3. Describe what happened when you made the reflection.
4. How does triangle *C* compare with the new triangle?

12-9 **Graphing Transformations**

Objective

Graph transformations on a coordinate grid.

Ray Brown is a "rock n' roll tailor". He has made clothes for more than one hundred rock stars, including Cher and Bon Jovi.

Before Mr. Brown cuts out a pattern, he chooses the fabric. Then, he uses chalk to trace the pattern onto the fabric.

Cher's real name is Cherilyn La Pierre.

The fabric below shows a transformation, or movement of a figure. In Mathematics Lab 12-9A, you learned about two kinds of transformations, a translation and a reflection.

Remember, when you slide a figure from one location to another without changing its size or shape, the new figure is called a *translation image*.

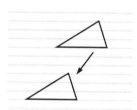

When you flip a figure over a line without changing its size or shape, the new figure is called a *reflection image*.

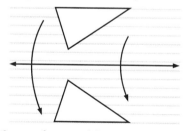

In Lesson 12-8, you learned how to graph ordered pairs on a coordinate grid. You can also graph transformations on a coordinate grid.

Mini-Lab

Work with a partner.

Materials: graph paper, a sheet of paper, scissors, pencil

- On graph paper, draw a coordinate grid.

- Graph points $A(-2, 3)$, $B(-1, 1)$, and $C(-3, 1)$.

- Connect the points to make $\triangle ABC$ as shown at the right.

- On a separate sheet of paper, trace $\triangle ABC$ and cut it out.

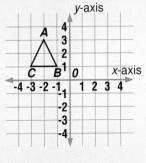

- Place the cut-out triangle on $\triangle ABC$.

- Slide the cut-out triangle 5 units to the right. Label the new vertices A', B', and C'. Then draw $\triangle A'B'C'$.

Talk About It

a. What are the coordinates of A', B', and C'?

b. What can you say about $\triangle ABC$ and $\triangle A'B'C'$?

Example 1

The vertices of rectangle *RSUT* are $R(-4, 3)$, $S(-1, 3)$, $U(-1, 1)$ and $T(-4, 1)$. Find the coordinates of the vertices of the rectangle after it has been translated 3 units to the right and 4 units down.

- On graph paper, draw a coordinate grid.

- Use the coordinates to graph and label the vertices of rectangle *RSUT*.

- Draw rectangle *RSUT*.

- Translate each vertex 3 units to the right and 4 units down. The coordinates of the new vertices are $R'(-1, -1)$, $S'(2, -1)$, $U'(2, -3)$, and $T'(-1, -3)$.

- Label the new vertices R', S', U', and T'.

- Then draw rectangle $R'S'U'T'$.

Lesson 12-9 Geometry Connection: Graphing Transformations **449**

You can also graph reflections on a coordinate grid.

Mini-Lab

Work with a partner.

Materials: graph paper, pencil, and MIRA®

- On graph paper, draw a coordinate grid.
- Graph points $D(-3, 2)$, $E(0, 3)$, and $F(0, 1)$.
- Connect the points to make $\triangle DEF$ as shown at the right.
- Place the MIRA® on the y-axis. Draw the reflection of $\triangle DEF$ you see on the other side of the y-axis.
- Label the new vertices D', E', and F'.

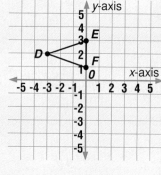

Talk About It

a. What are the coordinates of D', E', and F'?

b. What can you say about $\triangle DEF$ and $\triangle D'E'F'$?

c. How is using a MIRA® like flipping a piece of waxed paper?

Example 2

The vertices of trapezoid *KLNM* are $K(1, 2)$, $L(3, 2)$, $N(4, 1)$, and $M(0, 1)$. Find the coordinates of the vertices of the trapezoid after it has been reflected over the x-axis.

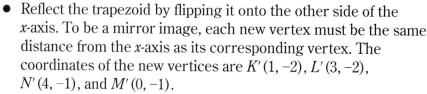

- On graph paper, draw a coordinate grid.
- Use the coordinates to graph and label the vertices of trapezoid *KLNM*.
- Draw trapezoid *KLNM*.
- Reflect the trapezoid by flipping it onto the other side of the x-axis. To be a mirror image, each new vertex must be the same distance from the x-axis as its corresponding vertex. The coordinates of the new vertices are $K'(1, -2)$, $L'(3, -2)$, $N'(4, -1)$, and $M'(0, -1)$.
- Label the new vertices K', L', N', and M'.
- Then draw trapezoid $K'L'N'M'$.

Checking for Understanding

Communicating Mathematics

Read and study the lesson to answer each question.

1. **Tell** how a translation and a reflection are different.

2. **Write** two methods you could use to graph a reflection.

Guided Practice

Tell whether each transformation is a translation or a reflection.

3.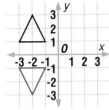

4.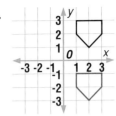

Exercises

Independent Practice

5. Draw △*RST* on a coordinate grid. The vertices are *R*(–4, –1), *S*(–5, –2), and *T*(–4, –4). Then draw its reflection over the *y*-axis.

6. Draw △*TUV* on a coordinate grid. The vertices are *T*(5, 1), *U*(3, 5), and *V*(5, 6). Then draw its reflection over the *y*-axis.

7. Name the coordinates of the vertices for the figure at the right. On graph paper, draw a translation image 5 units to the left. What are the new coordinates of the vertices of the translation image?

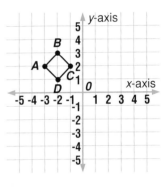

Draw each triangle whose vertices have the coordinates given below. Then draw the reflection over the *x*-axis.

8. *A*(–1, –2), *D*(–1, –4), *C*(–2, –3)

9. *M*(2, 1), *N*(2, 2), *P*(4, 4)

Mixed Review

10. Graph points *G*(1, 5) and *H*(2, –3) on a coordinate grid. *(Lesson 12-8)*

11. Multiply 12.39 by 86.7. *(Lesson 4-5)*

Problem Solving and Applications

12. **Geometry** Draw a quadrilateral with an area of 14 square units on a coordinate grid. Then, draw a translation image 7 units to the left and 6 units down.

    a. What are the coordinates of the translation image?

    b. What is the area of the translation image?

13. **Critical Thinking** If the first coordinate of each vertex of the figure at the right is multiplied by –1 and an image of the figure having these new coordinates is drawn, would the image be a translation or a reflection?

Study Guide and Review

Communicating Mathematics

Tell whether each statement is *true* or *false*. If false, change the statement to make it true.

1. The opposite of the number 7 is $\frac{1}{7}$.

2. The number −44 is farther to the left than −10, so −44 < −10.

3. When adding two negative integers, the answer will always be negative.

4. When subtracting two negative integers, the answer will always be negative.

5. When dividing two negative integers, the answer will sometimes be negative.

6. When multiplying two negative integers, the answer will never be negative.

7. The ordered pair (2, 3) is located at the same point as (3, 2).

8. Each axis of the coordinate grid is a number line.

9. A translation is any movement of a figure.

10. When you flip a figure, the transformation is called a reflection.

11. Explain the difference between a translation and a reflection.

Skills and Concepts

Objectives and Examples

Upon completing this chapter, you should be able to:

- identify, name, and graph integers
  (Lesson 12-1)

  The opposite of 5 is −5.

  To describe a loss of 3, use −3.

  The graph of −4 is

Review Exercises

Use these exercises to review and prepare for the chapter test.

Write the opposite of each integer.

12. +6   13. −11

Write an integer to describe each situation.

14. a temperature drop of 13°

15. a gain of 7 yards on the play

16. Draw a number line. Graph −6 on the number line.

Objectives and Examples

- compare and order integers
  (Lesson 12-2)
  Compare −7 and −9.

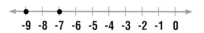

  Since −7 is to the right of −9, −7 > −9.
  Order from least to greatest.
  −5, 0, 7, −1, 2

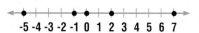

  The order is −5, −1, 0, 2, 7.

- add integers using models
  (Lesson 12-3)
  Find −4 + 6.

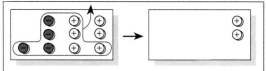

- subtract integers using models
  (Lesson 12-4)
  Find −4 − 6.

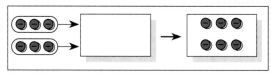

- multiply integers using models
  (Lesson 12-5)
  Find 2(−3).

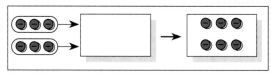

- divide integers using models and
  patterns *(Lesson 12-7)*
  Find 14 ÷ (−2).
  −7 × (−2) = 14, so 14 ÷ (−2) = −7.

Review Exercises

Replace each ▨ with <, >, or =.

17. +4 ▨ +7   18. +1 ▨ −6
19. −9 ▨ +11   20. −7 ▨ −12

Order each set of integers from least to greatest.

21. −15, 0, −1, 6, −7, 10

22. 44, −404, −40, −44, 40

Use counters to find each sum.

23. −2 + (−5)   24. −7 + 1
25. 3 + (−4)   26. −6 + 0
27. −5 + 7   28. 0 + (−2)

Use counters to find each difference.

29. −2 − (−5)   30. −7 − 1
31. 3 − (−4)   32. 6 − 9
33. 8 − (−3)   34. −4 − 7

Use counters to find each product.

35. 6 × (−2)   36. −3 × (−5)
37. −4 × 3   38. 5 × 0
39. −5(2)   40. 7(−1)

Find each quotient.

41. 35 ÷ (−7)   42. −36 ÷ 9
43. −40 ÷ 5   44. 18 ÷ (−3)
45. −30 ÷ (−5)   46. −24 ÷ (−6)

Objectives and Examples

- graph ordered pairs of numbers on a coordinate grid *(Lesson 12-8)*

Graph $A(2, -1)$.
right 2, down 1

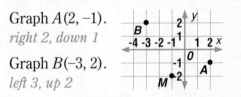

Graph $B(-3, 2)$.
left 3, up 2

Name the ordered pair for point M.
left 1, down 2 $\rightarrow M(-1, -2)$

- graph transformations on a coordinate grid *(Lesson 12-9)*

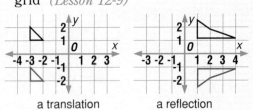

a translation   a reflection

Review Exercises

On graph paper draw a coordinate grid. Then graph and label each point.

47. $P(5, 6)$ 48. $Q(-2, 4)$
49. $R(-2, -5)$ 50. $S(7, -1)$

51. Give coordinates of point G.

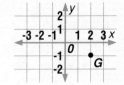

52. What kind of transformation is shown at the right?

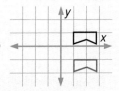

53. Draw triangle ABC with vertices $A(-1, 2)$, $B(0, 5)$, $C(3, 1)$. Reflect $\triangle ABC$ over the x-axis.

Applications and Problem Solving

54. **Weather**   Low temperatures were recorded for one week in January as follows: $-5°$F, $0°$F, $1°$F, $-3°$F, $-3°$F, $-2°$F, $-4°$F. Find the median temperature for the week. *(Lesson 12-2)*

55. **Consumer Math**   Shelby has money in her savings account. She made withdrawals of $100, $43, and $67. Her balance is now $245. How much did she have to begin with? *(Lesson 12-6)*

56. **Crafts**   LuAnne wants to make a two-sided ornament out of fabric. She uses a pattern to cut one side of the ornament. Should she use a reflection or a translation of the pattern for the other side of the ornament? *(Lesson 12-9)*

Curriculum Connection Projects

- **Automotive**   Use the information on a container of antifreeze to make a graph. Let the x-axis represent the number of quarts of antifreeze and the y-axis represent the temperature.

- **Drafting**   Draw a square on a coordinate plane. Label the coordinates of each vertex. Graph a translation and a reflection of your geometric shape.

Read More About It

Chetwin, Grace. *Collidescope.*
Lofting, Hugh. *The Voyages of Dr. Dolittle.*

Chapter 12 Test

1. Write an integer to describe a loss of 12 dollars.
2. Write the integer that is the opposite of -7.
3. Graph -3 on a number line.

Replace each ▦ with <, >, or =.
4. 8 ▦ -12
5. -77 ▦ -777
6. -127 ▦ -9
7. Order $-10, 0, -12, 3, -1,$ and 11 from least to greatest.

Use counters to find each sum or difference.
8. $-5 + (-7)$
9. $6 - (-19)$
10. $-30 + 10$
11. $-4 - (-9)$

12. **Weather** The temperature at 6:00 A.M. was $-5°$ F. What was the temperature at 8:00 A.M. if it had risen 7 degrees?

Find each product or quotient.
13. $21 \div (-7)$
14. $6 \times (-7)$
15. $-60 \div (-6)$

16. **Health** While Jay was recovering from the flu, his temperature dropped $1°$ steadily each hour for 3 hours. Write this as a multiplication of two integers. What was the total drop?
17. Rafael, Vicky, Chet, and Ling went to the drugstore and bought magazines. Vicky spent $5 more than Rafael. Chet spent 3 times as much as Vicky. Ling spent $8 less than Chet. Ling spent $13. How much did Rafael spend?

On graph paper, draw a coordinate grid. Then graph and label each point.
18. $A(5, -1)$
19. $B(-2, -3)$
20. $C(-4, 5)$

Name the ordered pair for each point.
21. K
22. R
23. M

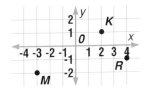

Draw $\triangle ABC$ on a coordinate grid. The vertices are $A(2, 3)$, $B(-4, 1)$, and $C(-1, 0)$. Then complete each transformation.
24. a translation 3 units down
25. a reflection over the x-axis

Bonus If y is an integer, what is the least value that y^4 can have? Explain your answer in words.

Directions: Choose the best answer. Write A, B, C, or D.

1. What is 0.125 written as a fraction in simplest form?

   A. $\frac{1}{8}$

   B. $\frac{1}{4}$

   C. $\frac{25}{100}$

   D. $1\frac{1}{4}$

2. What is the perimeter of a triangle with sides of $5\frac{1}{8}$ in., $3\frac{3}{4}$ in., and $3\frac{3}{4}$ in.?

   A. 13 in.

   B. $12\frac{5}{8}$ in.

   C. $11\frac{1}{2}$ in.

   D. $4\frac{1}{8}$ in.

3. What are the next two numbers in the sequence 1,024, 256, 64, 16, …?

   A. 8, 4

   B. 8, 2

   C. 4, 1

   D. 4, 0

4. Two figures that are the same size and shape are

   A. similar

   B. congruent

   C. neither congruent nor similar

   D. bisectors

5. $\frac{1}{4} =$

   A. 4%

   B. 20%

   C. 25%

   D. 40%

6. How could you estimate a 20% tip on $3.25?

   A. Find $\frac{1}{2}$ of $3.00.

   B. Find $\frac{1}{4}$ of $3.20.

   C. Find $\frac{1}{5}$ of $3.00.

   D. Find $\frac{1}{20}$ of $4.00.

7. Which is the best estimate of the area of $\triangle XYZ$?

   A. 6 square units

   B. 10 square units

   C. 12 square units

   D. 20 square units

8. A round window in Jodie's house is 40 inches in diameter. To the nearest square inch, what is the area of the window?

   A. 126 square inches

   B. 251 square inches

   C. 1,256 square inches

   D. 5,024 square inches

9. What integers are graphed below?

   A. −5, −1, 2   B. −7, −1, 2

   C. 5, 1, 2   D. −7, 1, 2

10. What are the coordinates of point G on the graph?

    A. (2, −3)

    B. (3, −2)

    C. (−3, 2)

    D. (−2, −3)

The questions on this page involve comparing two quantities, one in Column A and one in Column B. In some questions, information related to one or both quantities is centered above them.

Directions: Write A if the quantity in Column A is greater. Write B if the quantity in Column B is greater. Write C if the quantities are equal. Write D if there is not enough information to decide.

Column A	Column B
11. $2b$	$b + b$
12. 3 million	20,000,000
13. the area of a square with a side of 6 inches	the area of a rectangle with a length of 6 inches
14. $0.3 \div 6$	$6 \div 0.3$
15. the greatest prime factor of 24	the greatest prime factor of 36
16. $\frac{2}{3} + \frac{2}{3}$	1
17. $11\frac{2}{5}$	$10\frac{7}{5}$
18. the number of quarts in $2\frac{1}{2}$ gallons	12
19. measure of $\angle x$	measure of $\angle y$

Test-Taking Tip

In test questions like the ones on this page, treat each quantity separately. Do any indicated mathematical operations. Try to put both quantities in the same form. Then see if you can more easily compare the quantities.

Substitute values for any variables. Be sure to use many types and combinations of numbers. Use positive and negative numbers. Do not make assumptions.

If the use of As and Bs in both the columns is confusing, change the column names to another pair of letters or numbers, such as x and y or I and II.

Column A	Column B
20. $\frac{18}{40}$	$\frac{45}{100}$
21. 20% of 50	50% of 20
22. area of triangle A	area of triangle B

Column A	Column B
23. number of faces on a rectangular prism	number of faces on a rectangular pyramid
24. -9	2

Chapter

13

An Introduction to Algebra

Spotlight on Sound

Have You Ever Wondered. . .

- What it means when sounds are described in decibels?
- What substances sound travels through the fastest?

Speed of Sound

Substance	Speed (m/s)
Rubber	60
Air at 0°C	331
Air at 25°C	346
Cork	500
Lead	1,210
Water at 25°C	1,498
Sea water at 25°C	1,531
Silver	2,680
Copper	3,100
Brick	3,650
Wood (Oak)	3,850
Glass	4,540
Aluminum	5,000
Steel	5,200
Stone	5,971

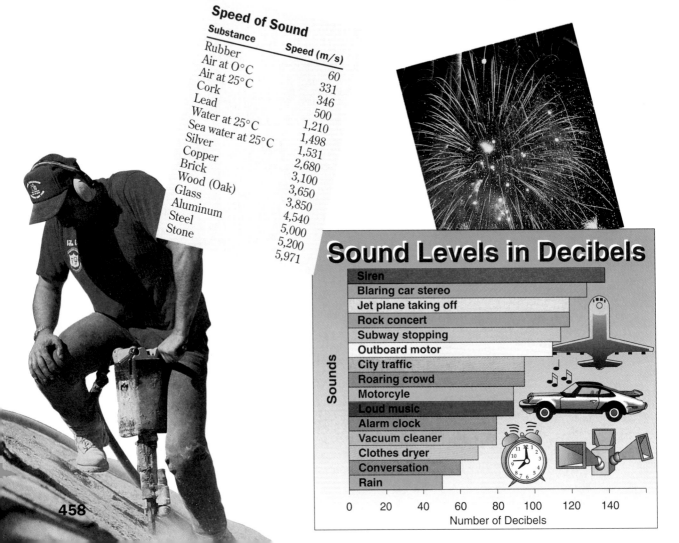

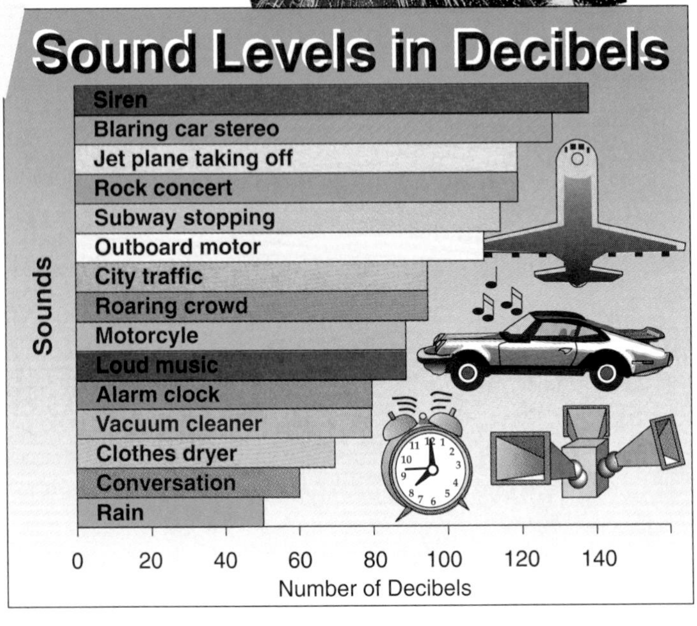
Sound Levels in Decibels
Siren
Blaring car stereo
Jet plane taking off
Rock concert
Subway stopping
Outboard motor
City traffic
Roaring crowd
Motorcyle
Loud music
Alarm clock
Vacuum cleaner
Clothes dryer
Conversation
Rain
Sounds
0
20
40
60
80
100
120
140
Number of Decibels

Looking Ahead

In this chapter, you will see how mathematics can be used to answer questions about sound. The major objectives of the chapter are to

- solve addition, subtraction, multiplication, and division equations

- solve problems using an equation

- complete function tables and graph functions

- identify and solve inequalities

Chapter Project

Sound
Work in a group.

1. You probably hear more sounds than you realize. For one week, keep track of how long you watch television and listen to the radio each day.

2. Calculate the total time you spend listening to either source in one week. Show your calculations.

3. Repeat the process for at least two other sounds.

4. Present your findings to the class.

For Better or For Worse®

by Lynn Johnston

459

13-1A Solving Equations Using Addition and Subtraction

A Preview of Lesson 13-1

Objective

Solve equations involving addition and subtraction using models.

In this lab, you will use cups and counters to help you solve equations involving addition and subtraction.

Materials

cups
counters
equation mat

Activity One

Work with a partner.

Solve $x + (-4) = -7$.

- Use a cup to represent x. Place 4 negative counters beside the cup on the left side of the mat to represent −4. Place 7 negative counters on the right side of the mat to represent −7.

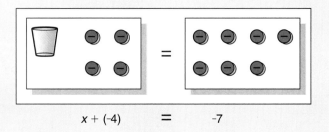

$$x + (-4) \quad = \quad -7$$

- To solve the equation, you need to get the cup by itself on one side of the mat. To do this, remove 4 negative counters from each side.

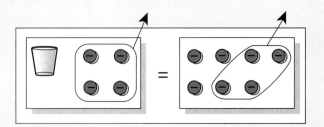

What do you think?

1. How many counters remain on the right side of the mat?
2. What is the value of x, or the solution to the equation?

Activity Two

Solve $x + 2 = -6$.

- Use a cup to represent x. Place 2 positive counters beside the cup on the left side of the mat to represent 2. Place 6 negative counters on the right side of the mat to represent –6.

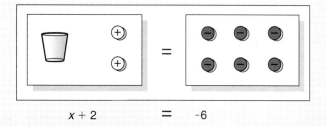

$$x + 2 \qquad = \qquad -6$$

LOOK BACK

You can review zero pairs on page 426.

- To get the cup by itself, you need to remove 2 positive counters from each side. But there are no positive counters on the right side of the mat. Add 2 negative counters to each side to make 2 zero pairs on the left side of the mat.

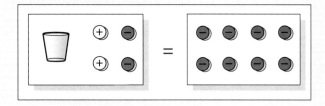

- Now, remove the zero pairs from the left side. Remember that adding or removing zero pairs does not change the value of a side.

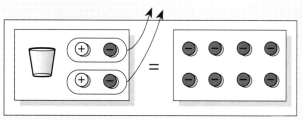

What do you think?

3. How many counters remain on the right side of the mat?
4. What is the value of x, or the solution to the equation?

Extension

Use cups and counters to solve each equation.

5. $x + 4 = 7$ 6. $x + (-5) = -10$ 7. $x + 6 = -18$

Mathematics Lab 13-1A Solving Equations Using Addition and Subtraction   **461**

13-1 Solving Addition and Subtraction Equations

Objective

Solve equations involving addition and subtraction using models.

Jose and Zoraida got together last weekend and played 6 Nintendo® games. They played 4 on Saturday, and the rest on Sunday. How many did they play on Sunday?

You can write an equation that describes the situation. Let *g* represent the number of games played on Sunday.

<u>number of games on Saturday</u>   *plus*   <u>number of games on Sunday</u>   *equals*   <u>total number played during the weekend</u>

$$4 \quad + \quad g \quad = \quad 6$$

LOOK BACK

You can review solving equations on page 32.

To solve an equation means to find a value for the variable that makes the equation true. In this case, you need to find the value of *g*. Notice that 4 has been added to *g*. To solve the equation, you need to get the cup, or variable, by itself.

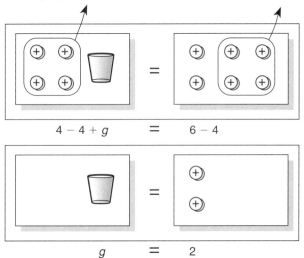

$$4 + g \quad = \quad 6$$

To get the cup by itself, subtract 4 from each side.

$$4 - 4 + g \quad = \quad 6 - 4$$

$$g \quad = \quad 2$$

You should always check your solution by replacing the value of *g* in the original equation.

Check: $4 + g = 6$

$4 + 2 \stackrel{?}{=} 6$ *Replace g with 2.*

$6 = 6$ ✓

The solution is 2. Jose and Zoraida played 2 games on Sunday.

Example 1

Use cups and counters to solve $y - 3 = 4$.

Rewrite as an addition equation. Remember that subtracting an integer is the same as adding its opposite.

$$y - 3 = 4 \rightarrow y + (-3) = 4$$

Use a cup to represent y. Place 3 negative counters beside the cup on the left side of the mat to represent -3. Place 4 positive counters on the right side of the mat to represent 4.

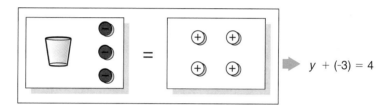

To get the cup by itself, remove 3 negative counters from each side. Since there are no negative counters on the right side, add 3 positive counters to each side to make 3 zero pairs on the left side. Then remove the 3 zero pairs.

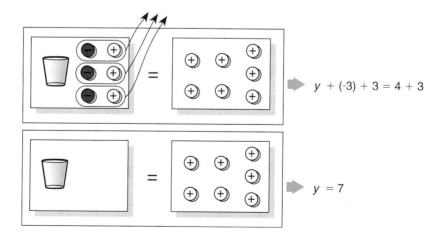

Check: $y - 3 = 4$

$7 - 3 \stackrel{?}{=} 4$ *Replace y with 7.*

$4 = 4$ ✓

The solution is 7.

Example 2 *Problem Solving*

Games In the card game *Clubs,* it is possible to get a negative score. Suppose your friend, Rey, got a score of 2 in the first hand. His total score after two hands was −1. What was his score in the second hand?

You can write an equation to solve this problem.

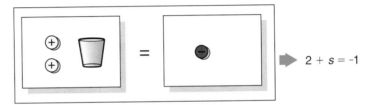

score of first hand	*plus*	*score of second hand*	*equals*	*total score*
2	+	s	=	−1

Use a cup to represent s. Place 2 positive counters beside the cup on the left side of the mat. Place 1 negative counter on the right side of the mat.

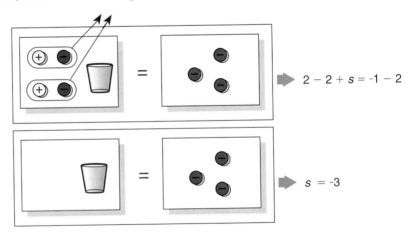

$2 + s = -1$

To get the cup by itself, remove 2 positive counters from each side. Since there are no positive counters on the right side, add 2 negative counters to each side to get 2 zero pairs on the left side. Then, remove the zero pairs.

$2 - 2 + s = -1 - 2$

$s = -3$

Rey's score in the second hand was −3. *Be sure to check your answer.*

Checking for Understanding

Communicating Mathematics

Read and study the lesson to answer each question.

1. **Tell** how you would solve the equation $x + 4 = -2$.
2. **Tell** how you know that 4 is not a solution of $b - 3 = -7$.
3. **Tell** how you know that −9 is a solution of $-9 + z = -18$.

4. **Show** how the model represents $x - 3 = 12$.

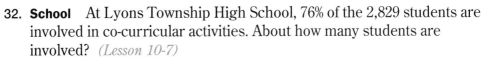

Guided Practice   Use cups and counters to solve each equation.

5. $x - 1 = -3$ 6. $x - 2 = 3$ 7. $x + 3 = -2$
8. $a - 6 = 4$ 9. $c + 2 = -7$ 10. $4 + e = 11$

Exercises

Independent Practice   Solve each equation. Use cups and counters if necessary.

11. $x + 3 = 5$ 12. $x + 3 = -6$ 13. $y + (-3) = -5$
14. $x + (-3) = -1$ 15. $x + 4 = -3$ 16. $y + 6 = 12$
17. $x + 3 = 9$ 18. $z - 7 = 9$ 19. $d + 1 = -2$
20. $m + 6 = -2$ 21. $4 + n = 6$ 22. $1 + p = 4$
23. $f + 6 = -8$ 24. $g - 3 = -2$ 25. $5 + r = -6$
26. $z - 2 = -10$ 27. $x + 7 = 9$ 28. $8 + c = -12$

29. What is the value of t if $t + 3 = -4$?
30. If $x - 8 = 14$, what is the value of x?

Mixed Review   31. Find $-24 \div 8$. *(Lesson 12-7)*

32. **School**   At Lyons Township High School, 76% of the 2,829 students are involved in co-curricular activities. About how many students are involved? *(Lesson 10-7)*

33. Find $46.23 \div 2.3$. *(Lesson 5-4)*
34. Order the integers -6, 6, -606, and 0 from least to greatest. *(Lesson 12-2)*

Problem Solving and Applications   35. **Number Patterns**   Examine the numbers given in the following table. What number should be in the first cell?

	-2	1	4	7	10	13

36. **Critical Thinking**   Replace the boxes with the numbers 2, 3, 7, 8, and 9 to make a true sentence. Use each number exactly once.

$$\blacksquare\,\blacksquare + (-\blacksquare) = \blacksquare\,\blacksquare$$

37. **Sports**   After a gain of 6 yards, Ottis had made 15 yards so far in the game. How many yards had he made before the 6-yard gain?

38. **School**   Janice received 3 extra credit points on her assignment. She now has 9 points. How many points did she have before she was given the extra credit?

39. **Journal Entry**   In this lesson, you solved an equation by adding or subtracting the same quantity to each side. Give an example of an everyday experience where both sides are treated equally.

Lesson 13-1 Solving Addition and Subtraction Equations   **465**

13-2A Solving Equations Using Multiplication and Division

A Preview of Lesson 13-2

Objective

Solve equations involving multiplication and division using models.

Materials

cups
counters
equation mat

You can also use cups and counters to solve equations involving multiplication and division.

Try this!

Work with a partner.

Solve $3x = 12$.

- Use a cup to represent x. Place 3 cups on the left side of the mat to represent $3x$. Place 12 positive counters on the right side of the mat to represent 12.

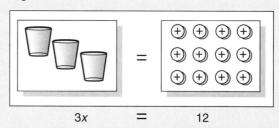

$$3x \quad = \quad 12$$

- To solve the equation, you need to find how many counters are in one cup. To do this, match each cup with an equal number of counters.

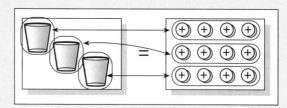

What do you think?

1. How many counters match each cup?
2. What is the value of x, or the solution to the equation?

Extension

Use cups and counters to solve each equation.

3. $\frac{1}{2}x = 3$ 4. $15 = 3x$ 5. $2x = -8$

13-2 Solving Multiplication and Division Equations

Objective

Solve equations involving multiplication and division using models.

TEEN SCENE

In the 1987-88 season, Michael J. Fox of NBC's *Family Ties* won the Emmy for Outstanding Lead Actor in a Comedy Series.

Every year the outstanding achievements in television programming of the preceding year are recognized. The winners of the television awards receive a statue called an Emmy. Each Emmy weighs 9 pounds. In 1990, the awards won by CBS in the Drama-Comedy special category weighed 18 pounds. How many Emmys did CBS win in this category?

You can write an equation that describes the situation above. Let a represent the number of awards won by CBS.

weight of each Emmy	*times*	*number of Emmys won*	*equals*	*total weight*
9	$\times$	a	$=$	18

When the variable is multiplied by a number, divide each side of the equation by that number to get the variable by itself.

$$9a \qquad = \qquad 18$$

Undo the multiplication. Divide each side by 9.

$$\frac{9a}{9} = \frac{18}{9}$$
$$a = 2$$

Check: $9a = 18$

$9(2) \stackrel{?}{=} 18$ *Replace a with 2.*

$18 = 18$ ✓

The solution is 2. CBS won two awards in the Drama-Comedy special category in 1990.

Use cups and counters to solve each equation.

1 $2x = -6$

Use 2 cups to represent $2x$. Place them on the left side of the mat. Place 6 negative counters on the right side of the mat to represent -6. To find the value of one cup, match each cup with an equal number of counters.

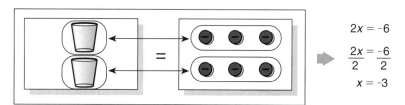

$$2x = -6$$
$$\frac{2x}{2} = \frac{-6}{2}$$
$$x = -3$$

Check: $2x = -6$
$2(-3) \stackrel{?}{=} -6$ *Replace x with –3*
$-6 = -6$ ✓   The solution is -3.

2 $\frac{1}{2}n = -5$

Cut one cup in half to represent $\frac{1}{2}n$. Place it on the left side of the mat. Place 5 negative counters on the right side of the mat to represent -5.

$$\frac{1}{2}n = -5$$

LOOKBACK

You can review reciprocals on page 293.

To get 1 cup on the left side of the mat, you must double the $\frac{1}{2}$ cup. You must also double the counters on the other side of the mat. So, add 5 negative counters to the right side of the mat.

$$2 \times \frac{1}{2}n = 2 \times (-5)$$
$$n = -10$$

Check: $\frac{1}{2}n = -5$
$\frac{1}{2}(-10) \stackrel{?}{=} -5$ *Replace n with –10.*
$-5 = -5$ ✓   The solution is -10.

Example 3 *Problem Solving*

Allowance   Leisa spends one-third of her monthly allowance on snacks. If she spends $4 on snacks every month, how much money does she get for her allowance?

First, write an equation to describe the situation. Let x represent the amount of money that Leisa receives every month.

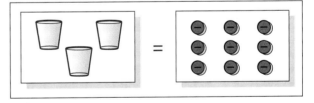

Solve the equation $\frac{1}{3}x = 4$.

$$\frac{1}{3}x = 4$$

$$3\left(\frac{1}{3}x\right) = 3(4) \qquad \textit{Undo the division.}$$
$$\textit{Multiply each side by 3.}$$

$$x = 12$$

Check: $\frac{1}{3}x = 4$

$\frac{1}{3}(12) \stackrel{?}{=} 4$ *Replace x with 12.*

$4 = 4$ ✓

The solution is 12. Leisa receives $12 every month.

Checking for Understanding

Communicating Mathematics

Read and study the lesson to answer these questions.

1. **Tell** what equation is represented by the model at the right. Then solve.

2. **Make a model** to represent the equation $4x = 12$. Then solve the equation.

3. **Tell** what number divided by 3 is equal to 5.

4. **Tell** how you know 100 is not a solution of $2k = 50$.

Guided Practice

Use cups and counters to complete each solution.

5. $2c = 6$
$$\frac{2c}{■} = \frac{6}{■}$$
$$c = ■$$

6. $\frac{1}{3}x = 3$
$$■\left(\frac{1}{3}x\right) = ■(3)$$
$$x = ■$$

7. $\frac{1}{2}k = -2$
$$■\left(\frac{1}{2}k\right) = ■(-2)$$
$$k = ■$$

8. $5e = -15$
$$\frac{5e}{■} = \frac{-15}{■}$$
$$e = ■$$

Use cups and counters to solve each equation.

9. $3g = 18$

10. $2x = -4$

11. $\frac{1}{4}x = -6$

12. $\frac{1}{3}y = -9$

Exercises

*Independent
Practice*

Solve each equation. Use cups and counters if necessary.

13. $4d = 16$ **14.** $5h = 2$ **15.** $2x = -14$ **16.** $\frac{1}{2}y = 8$

17. $\frac{1}{4}m = 3$ **18.** $\frac{1}{5}t = -8$ **19.** $\frac{1}{3}n = -8$ **20.** $6k = -24$

Solve each equation.

21. $8p = 32$ **22.** $7h = -21$ **23.** $\frac{1}{7}t = 8$ **24.** $\frac{1}{3}m = 11$

25. $\frac{1}{4}w = -9$ **26.** $9x = -81$ **27.** $5z = -55$ **28.** $\frac{1}{8}g = -12$

29. Solve for q in the equation $\frac{1}{3}q = -7$.

Mixed Review

30. Solve the equation $y - 11 = -8$. Then check. *(Lesson 13-1)*

31. Multiply mentally using the distributive property: 6.7×30. *(Lesson 4-4)*

32. Science   Helium has a weight of 0.178 kilograms per cubic meter. Express this weight as a fraction in simplest form. *(Lesson 6-10)*

33. Geometry   Find the area of the triangle at the right. *(Lesson 11-2)*

42 m
41 m
18 m

*Problem Solving
and
Applications*

34. Recreation   Kohana rides her bike 4 miles in $\frac{1}{2}$ hour. Use the formula $d = rt$ to find her average speed in miles per hour.

35. Geometry   The area of a rectangle is the product of the length and the width. If a rectangle has a length of 6 inches and an area of 24 square inches, what is the width?

36. Consumer Math   Marta keeps a record of her gasoline usage so she can be reimbursed by her employer. Her car averages 24 miles per gallon. If her odometer shows that she has driven 72 miles on a sales trip, how many gallons of gasoline has her car used?

37. Critical Thinking   If one-fourth of a number is -16, what is the number?

38. Mathematics and Economics   Read the following paragraph.

The labor force is made up of those people who have jobs or who are actively looking for jobs. One of the most important trends in the labor force over the past several decades has been the sharp increase in the number of women who have jobs outside the home. This number has almost doubled since the early 1960s. Many of these women have chosen to pursue careers in areas that have been typically dominated by men.

Margie is a construction worker. Her average weekly earnings in 1992 were 3 times higher than in 1975. She earned $624 per week in 1992. How much did she earn per week in 1975?

13-2B Solving Two-Step Equations Using Models

A Follow-Up of Lesson 13-2

Objective

Solve two-step equations using models.

Materials

cups
counters
equation mat

You can use cups and counters to help you solve equations that involve two operations.

Activity One

Work with a partner. Solve 2x + 3 = 15.

- Place 2 cups on the left side of the mat to represent 2x. Place 3 positive counters beside the cups to represent 3. Place 15 positive counters on the right side of the mat to represent 15.

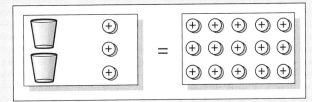

$$2x + 3 \quad = \quad 15$$

- Subtract 3 counters from each side.

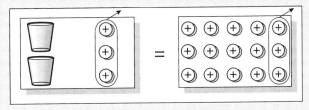

$$2x + 3 - 3 \quad = \quad 15 - 3$$

- Now your equation is 2x = 12. To find how many counters are in one cup, match cups with equal numbers of counters.

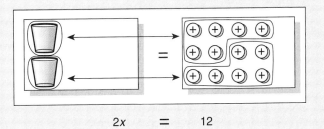

$$2x \quad = \quad 12$$

What do you think?

1. How many counters are matched up with each cup?
2. What is the value of x, or the solution to the equation?

Mathematics Lab 13-2B Solving Two-Step Equations Using Models **471**

Activity Two

Solve 2x + 3 = −9.

- Place 2 cups on the left side of the mat to represent 2x. Place 3 positive counters beside the cups to represent 3. Place 9 negative counters on the right side of the mat to represent −9.

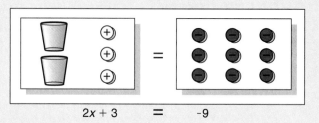

$2x + 3$ = -9

- Add 3 negative counters to each side.

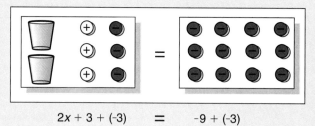

$2x + 3 + (-3)$ = $-9 + (-3)$

- Now, remove all zero pairs.

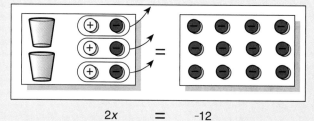

$2x$ = -12

- Now your equation is $2x = -12$. To find how many counters are in one cup, match each cup with an equal number of counters.

What do you think?

3. How many counters are matched up with each cup?
4. What is the value of x, or the solution to the equation?
5. Consider the model below. What counters would you add to both sides to solve the equation? Explain why.

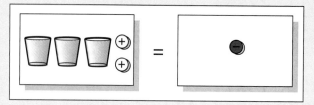

13-3 Use an Equation

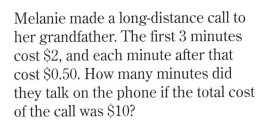

Objective

Solve problems by using an equation.

Melanie made a long-distance call to her grandfather. The first 3 minutes cost $2, and each minute after that cost $0.50. How many minutes did they talk on the phone if the total cost of the call was $10?

Explore   What do you know?
The first 3 minutes cost $2, and each minute after that cost $0.50. The phone call cost a total of $10.

What do you need to find out?
How many minutes they talked on the phone.

Plan   Let m represent the number of minutes they talked at a rate of $0.50 per minute.

Translate the problem into an equation using the variable m.

Solve

The cost of phone call	*is*	*$2 for the first 3 minutes*	*plus*	*$0.50 per minute*	*times*	*the number of minutes.*
10	=	2	+	0.50	×	m

$$10 = 2 + 0.5m$$ *Notice that this is a two-step equation.*

$$10 - 2 = 2 - 2 + 0.5m$$ *Subtract 2 from each side.*

$$8 = 0.5m$$

$$\frac{8}{0.5} = \frac{0.5m}{0.5}$$ *Divide each side by 0.5.*

8 $\boxed{\div}$ 0.5 $\boxed{=}$ ⌐⌐

Melanie and her grandfather talked for 16 minutes at a rate of $0.50 per minute. To find the total length of the call, add the first 3 minutes. The total length of the call was 3 + 16, or 19 minutes.

Examine   Check the solution against the problem. If they talked 19 minutes, the cost would be $2 for the first 3 minutes and $0.50 for each minute after that, or 16 minutes.

$$\$2 + (\$0.50)\,16 = \$10 \checkmark$$

Example

The student council is selling carnations and roses for homecoming. They sell 3 times as many carnations as roses. If they sell 123 carnations, how many roses did they sell?

Let *r* equal the number of roses sold. Write an equation.

Three	*times*	*the number of roses*	*equals*	*the number of carnations.*
3	×	*r*	=	123

$$3r = 123$$
$$\frac{3r}{3} = \frac{123}{3} \qquad \textit{Divide each side by 3.}$$
$$r = 41$$

The student council sold 41 roses.

Checking for Understanding

Communicating Mathematics

Read and study the lesson to answer this question.

1. **Tell** what other problem-solving strategy you could use to solve the Example.

Guided Practice

Solve by using an equation.

2. Lupita went bowling at Great Lanes Sport Center. Shoe rental was $1.50, and games were $2.50 each. If she spent $9.00 on shoe rental and games, how many games did she bowl?

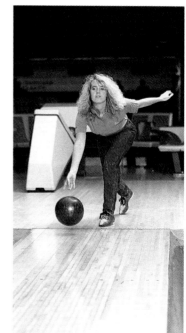

3. The service department at U-Lube-It can work on 14 cars at the same time. On Saturday, 56 cars needed their oil changed. What percent of the customers can be served at the same time?

4. The best-selling item at the class picnic this year was the hot dog. During lunch, 67 were sold. This was 3 more than twice the amount that were sold last year. How many hot dogs were sold at last year's picnic?

Problem Solving

Practice Solve using any strategy.

Strategies

● ● ● ● ● ● ●

Look for a pattern.

Solve a simpler problem.

Act it out.

Guess and check.

Draw a diagram.

Make a chart.

Work backward.

5. A catalog company offers 3 styles of ski sweaters each in 7 different colors. How many combinations of style and color are possible?

6. Jillisa bought a clock radio for $9 less than the regular price. She paid $32 for the item. What was the regular price?

7. How many ounces of fat are in 3 pounds of pastrami that is 15% fat?

8. Science fiction books were the most popular items at the Book Fair. On Monday, 86 science fiction books were sold. This is 8 more than twice the amount that were sold on Wednesday. How many science fiction books were sold on Wednesday?

9. Scott paid $2.10 in sales tax on a Washington Redskins sweatshirt. The total cost was $32.09. What was the price of the sweatshirt before tax?

10. Cassy carves small animals from wood and sells them at local craft shows. At one show she sold 61 animals for $19.90 each. How much money did she earn?

11. A designer wants to arrange 6 glass bricks into a rectangular shape with the least perimeter possible. How many blocks will be in each row?

13 Mid-Chapter Review

Solve each equation. Use cups and counters if necessary.
(Lessons 13-1, 13-2)

1. $c - 8 = 12$ 2. $2y = -8$ 3. $m + 7 = -14$ 4. $\frac{k}{5} = 15$

Solve. *(Lessons 13-1, 13-2, 13-3)*

5. Rodney Peete was sacked for a 12-yard loss. On the next play, he had a gain of 5 yards. What was Mr. Peete's net yardage on the two plays?

6. The temperature at 10:00 P.M. was half what it was at noon. If the temperature at 10:00 P.M. was 25°, what was it at noon?

7. There are 208 students in the sixth grade. Fifty-five of them are members of the marching band. Write an equation to show how many students are not in the marching band. Then solve.

DECISION MAKING

Choosing Software

Situation

The Computer Club at your school has a software budget of $500. As president, you must write a report to the members of the club giving your suggestions for the type of software you will need for the next school year. After your last meeting, the members suggested buying a new word-processing program that includes desktop publishing. Which of the software programs best suits the needs of the club?

Hidden Data

Which software program will work with the hardware you have?
Do you need a mouse to operate the software? Does your computer already have one?
Do you need a special printer to print desktop publishing material?

Analyzing the Data

1. Which programs allow you to use spreadsheets?
2. Which programs can check your spelling for you?
3. Can you get more than one software program and stay within your budget? If so, which ones?

DocumentPerfect 6.0

Spell check, thesaurus,automatic hyphenation, pull down menus and mouse support, view document, columns, spreadsheets/tables, printing/fonts, text and graphics.

No. 423-891
List Price: **$495.00**

Nine-In-One

Word processor, data base, spreadsheet, desktop organizer, speller/thesaurus, graphics, outliner, communications, and hyphenation.

No. 123-876
List: $39.00

The Desktop Publisher

Produce newsletters, ads, reports, catalogs, term papers, brochures, forms, and more. Includes Design ideas, over 85 innovative document layouts and designs for home, school, and business use. Extended print to disk facility enables you to print to disk using your printer configuration.

No. 564-987
List Price: **$249.95**

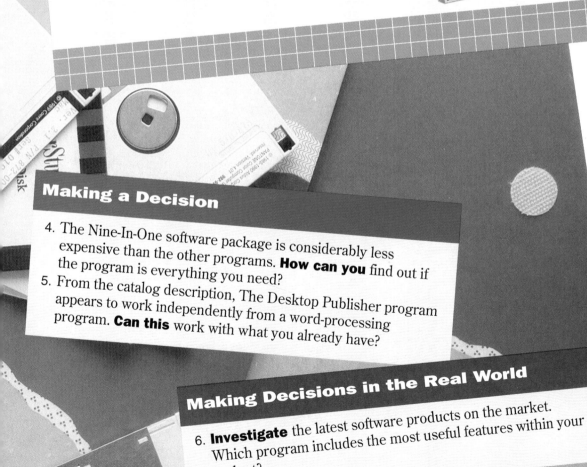

Making a Decision

4. The Nine-In-One software package is considerably less expensive than the other programs. **How can you** find out if the program is everything you need?

5. From the catalog description, The Desktop Publisher program appears to work independently from a word-processing program. **Can this** work with what you already have?

Making Decisions in the Real World

6. **Investigate** the latest software products on the market. Which program includes the most useful features within your budget?

477

13-4A Function Machines

A Preview of Lesson 13-4

Objective

Find the input and output for a given function machine.

Words to Learn

function machine
input
output

Materials

scissors
tape

In this lab, you will use what you have learned about solving equations to help you work with function machines. A **function machine** takes a number called the **input,** performs one or more operations on it, and produces a result called the **output.**

Try this!

Work in groups of three.

Make a function machine for the rule $\boxed{+\ 4}$.

- Take a sheet of paper and cut it in half lengthwise.

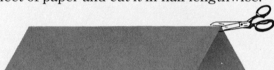

- On one of the halves, cut two slits on each side of the paper. Make each slit at least one inch wide.

- From the other half of the paper, cut two narrow strips lengthwise. The strips should go through the slits of the other half.

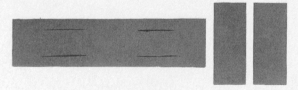

- On one strip, write five consecutive numbers starting with 1. On the other strip, write five consecutive numbers starting with 5. The numbers on both strips should align.

- Place the strips into the slits so that the numbers can be seen. Show 1 and 5. Once they appear, tape the ends of the strips together. When you pull the strips, they should move together.

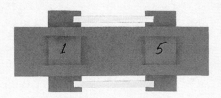

- Mark the left-hand strip *input,* and the right-hand strip *output.* Write the function rule ⏍ + 4 ⏍ between the input and output.

What do you think?

1. What is the output when the input is 3?

2. What is the output when the input is 5?

3. Suppose you added more input numbers to the left strip. What would the output be if the input was 8?

Extension

4. Make a function machine for the rule ⏍ × 3 ⏍ .

5. Make up your own function machine.
   a. You may want to list the input numbers on one strip and leave the other strip blank. Give the rule and then ask the members of the group to fill in the output.
   b. Each member of the group could come up with pairs of inputs and outputs and have the other members determine the rule.

6. Find four sets of inputs and outputs for the function machine at the right.

13-4 Functions

Objective

Complete
function tables.

Words to Learn

function table

DID YOU KNOW

Hundreds of years ago,
sailors pierced their ears
and wore earrings
because they believed the
earrings would keep them
from drowning.

Jillian O'Connor, a 12-year-old
student at Portage Middle School in
Kalamazoo, Michigan, took classes
at the local art center to learn how
to make plastic jewelry. The
earrings that she made became so
popular at her school that she
started a business. The money that
Jillian made depended on how many
pairs of earrings she sold. This
relationship is an example of a
function.

If Jillian sold each pair of earrings for $4, how much did she make
by selling 3 pairs? 4 pairs? 5 pairs?

The answers to these questions can be organized in a **function
table.** To find the amount of sales (output), you need to multiply
each number of pairs (input) by 4.

Input	Function Rule	Output
Number of Pairs Sold (n)	$4n$	Amount of Sales ($)
1	4(1)	4
2	4(2)	8
3	4(3)	12
4	4(4)	16
5	4(5)	20

Jillian made $12 for selling 3 pairs of earrings, $16 for selling
4 pairs, and $20 for selling 5 pairs.

Example 1

Copy and complete the function
table shown at the right. The
rule for this table is to add
3 to each input.

input (n)	output ($n + 3$)
1	■
-2	■
5	■

Replace n in the rule $n + 3$ with 1, −2, and 5.

Replace n with 1.   *Replace n with −2.*   *Replace n with 5.*

$n + 3 = 1 + 3$ $n + 3 = -2 + 3$ $n + 3 = 5 + 3$
$\quad\quad = 4$ $\quad\quad = 1$ $\quad\quad = 8$

The output numbers are 4, 1, and 8.

Example 2

Find the rule for the function table at the right.

input *(n)*	output (▨)
0	0
1	4
2	8
3	12

First, you have to study the relationship between each input and output.

Problem-Solving Hint

• • • • • • • • • • •

Look for a pattern.

input			output
0	$\times 4$	$\rightarrow$	0
1	$\times 4$	$\rightarrow$	4
2	$\times 4$	$\rightarrow$	8
3	$\times 4$	$\rightarrow$	12

The output is 4 times the input.
So, the function rule is $4n$.

Checking for Understanding

Communicating Mathematics

Read and study the lesson to answer each question.

1. **Write** the function rule for the table at the right.

input *(n)*	output (▨)
4	2
2	1
0	0

2. **Tell** the function rule if the output of a function table is 3 less than each input. Use x for the input.

Guided Practice

Copy and complete each function table.

3.
input *(n)*	output *(n + 5)*
4	▨
0	▨
−2	▨

4.
input *(n)*	output $(\frac{n}{2})$
6	▨
8	▨
10	▨

Find the rule for each function table.

5.
n	▨
0	−3
3	0
6	3

6.
n	▨
0	0
3	9
4	12

7.
n	▨
5	0
6	1
7	2

8. If a function rule is $2n + 1$, what is the output for an input of 2?

9. If the input values are 1 and 2 and the respective outputs are 3 and 6, what is the function rule?

Exercises

Copy and complete each function table.

10.

input *(n)*	output *(2n)*
3	■
0	■
−2	■

11.

input *(n)*	output *(n − 6)*
11	■
0	■
−5	■

12.

input *(n)*	output $(\frac{n}{2} + 1)$
−4	■
0	■
4	■

13.

input *(n)*	output *(3n − 1)*
2	■
0	■
−1	■

14. If a function rule is $\frac{3x}{4}$, what is the output for $x = 0.5$?

15. If a function rule is $5n - 3$, what is the output for $n = -4$?

Find the rule for each function table.

16.

n	■
0	6
1	7
2	8

17.

n	■
0	−5
2	−3
4	−1

18.

n	■
3	18
4	24
5	30

19. **School**   Natalie had 3 times as many As on her history quizzes in the second semester as in the first semester. If she had 12 As in the second semester, how many did she have in the first semester? Write an equation and solve. *(Lesson 13-2)*

20. Subtract $3\frac{5}{6}$ from $7\frac{1}{4}$. Write the difference in simplest form. *(Lesson 7-7)*

21. Find $2\frac{3}{5} \div 1\frac{2}{3}$. Write the quotient in simplest form. *(Lesson 8-7)*

22. **Patterns**   Use toothpicks to make the following toothpick triangles.

How many toothpicks will be needed to make 8 triangles? 20 triangles?

23. **Computer Connection**
The spreadsheet at the right shows a function table. Find the outputs to complete the table.

	A	B	C
1	X	X + 1	2X − 5
2	5	6	
3	7		9

24. **Critical Thinking**   Find the rule for the function table at the right.

n	■
4	16
−4	16
0	0
−1	1

13-5 Graphing Functions

Objective

Graph functions from function tables.

Sonar is a device used by oceanographers to locate objects under water. Sonar uses the echo of sound waves to do this. The equation $y = \frac{x}{2}$, where y is the distance and x is the time it takes the echo to return to the ship, will give you the distance below the surface where the object is located. In the equation, $y = \frac{x}{2}$, x is the input and y is the output. The function rule is $\frac{x}{2}$.

LOOK BACK

You can review the coordinate system on page 443.

We can use a coordinate system to graph this equation or function.

Step 1

Record the input and output of the function on a table. We chose 0, 4, 8, and 12 for the input. List the input and output as ordered pairs.

input	function rule	output	ordered pairs
x	$\frac{x}{2}$	y	(x, y)
0	0	0	(0, 0)
4	$\frac{4}{2}$	2	(4, 2)
8	$\frac{8}{2}$	4	(8, 4)
12	$\frac{12}{2}$	6	(12, 6)

Step 2

Graph the ordered pairs from the table in Step 1 on the coordinate grid.

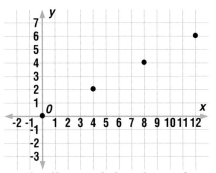

The x-coordinates represent the measured time it takes for the echo to return.

The y-coordinates represent the distance below the surface where the object is located.

"When am I ever going to use this?"

Finding a city on a road map is like graphing an ordered pair on a coordinate grid. To find a city, look at the index on the map. It will give you the coordinates for the city. You can then use these to locate the city on the map. You need to be aware, though, that the coordinates on a map often identify a square area, rather than a single point as on a coordinate grid.

Step 3

Draw the line that contains these points.

The line is the graph of $y = \frac{x}{2}$.

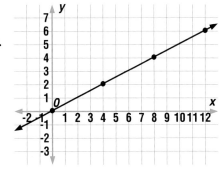

Use the function table at the
right to graph the function.
The function rule is 3x.

input (x)	output (3x)
−2	−6
0	0
2	6
4	12

Step 1 Record the ordered
pairs on a table.

input (x)	output (3x)	(x, 3x)
−2	−6	(−2, −6)
0	0	(0, 0)
2	6	(2, 6)
4	12	(4, 12)

Step 2 Graph the points
that correspond to
the ordered pairs.

Step 3 Draw the line that
contains the points.

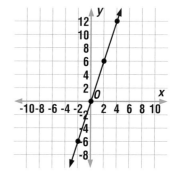

Checking for Understanding

*Communicating
Mathematics*

Read and study the lesson to answer each question.

1. **Write,** in your own words, the three steps you should take to graph a
   function.

2. **Tell** why only those solutions graphed in the first quadrant in the sonar
   problem make sense.

3. **Tell** which axis of a coordinate system represents the output of a function.

Guided Practice

Graph the functions represented by each function table.

4.
input	output
5	3
3	1
1	−1

5.
input	output
−4	1
−2	3
0	5

6.
input	output
2	5
0	−1
−3	−10

Copy and complete each function table. Then graph the function.

7.
input (x)	output (x − 6)
1	▩
4	▩
5	▩

8.
input (x)	output ($\frac{x}{2}$)
8	▩
6	▩
−4	▩

Exercises

Independent
Practice

Copy and complete each function table. Then graph the function.

9.

input *(x)*	output *(3x)*
3	▨
0	▨
2	▨

10.

input *(x)*	output *(x + 2)*
−3	▨
1	▨
4	▨

11.

input *(x)*	output $(\frac{x}{4})$
−4	▨
0	▨
8	▨

12.

input *(x)*	output *(x − 3)*
4	▨
0	▨
−2	▨

13. Make a function table for the rule $x - 2$ using 2, 5, and 8 as the input. Then graph the function.

Make a function table for each graph. Then determine the rule.

14.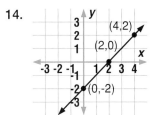

15.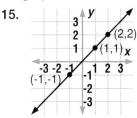

16.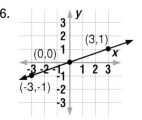

Mixed Review

17. Complete the function table at the right. *(Lesson 13-4)*

input *(n)*	output *(n + 7)*
−5	▨
0	▨
4	▨

18. **Geometry** Draw a rectangular prism. *(Lesson 11-6)*

19. Find $6 - (-5)$. *(Lesson 12-4)*

Problem Solving
and
Applications

20. **Critical Thinking** Which graph most realistically shows the relationship between the time for running a race and the length of the race left to run? Explain your answer.

a.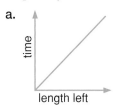

b.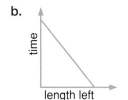

c.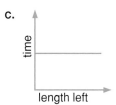

21. **Health** A person's pulse is the number of times the person's heart beats per minute.

a. Take your pulse for 1, 2, and 3 minutes. Record your results in a table.

b. Write the function rule.

c. Make a graph of this function.

Lesson 13-5 Graphing Functions   **485**

22. **Collect Data** Question the traffic division of your police department to find the number of accidents that occur at each hour of the day or night for a certain day.

   a. Graph the results.

   b. Do the number of accidents appear to be a function of the time of the day?

   c. When do most accidents occur?

23. **Critical Thinking** Answer each of the following. Explain your reasoning.

   a. Is batting average a function of the number of hits?

   b. Is the cost of mailing a package a function of the weight of the package?

   c. Is the grade on a math test a function of the color of hair of a person?

   d. Is the cost of heating/air conditioning a function of the temperature?

DATA SEARCH

24. **Data Search** Refer to pages 458 and 459.
   The distance an object travels *(d)* is calculated by multiplying its speed *(r)* by the time it travels *(t)*. Use $d = rt$ to compare the distance sound will travel in 5 seconds in steel with the distance it travels in 5 seconds in air at $0°$C.

25. **Journal Entry** Write a problem that could be solved using the sonar graph on page 483.

CULTURAL KALEIDOSCOPE

I.M. Pei

Ieoh Ming Pei (1917-) immigrated to the United States in 1935 to study architecture at Harvard University. He has become famous for his creative designs which combine the use of light, shadows, and shapes.

One of his most famous designs is the Henry R. Luce Foundation Chapel in Taiwan (1963). It has curved walls that slope upward. The shape of these walls not only give it an unusual appearance, but they also protect the building from typhoons, which are common in this part of the world.

Other notable structures designed by Pei include the John F. Kennedy Library in Massachusetts, the pyramid renovation of the entrance to the Louvre in France, and the National Gallery of Art in Washington, D.C.

13-6 Inequalities

Objective
Identify and solve inequalities.

Words to Learn
inequality

DID YOU KNOW

In the 1880's, women played tennis at Wimbledon in hats and ankle-length dresses.

The American tennis player Jennifer Capriati is the youngest winner ever at Wimbledon. She won her first match at Wimbledon in June of 1990 when she was 14 years old.

If a tennis player who is younger than 14 years old wins at Wimbledon, she or he could be considered the youngest-ever winner. If we let a represent the age of the player, we could write the inequality $a < 14$.

An **inequality** is a mathematical sentence that contains symbols like < or >. The symbols $\leq$ and $\geq$ can also be used in inequalities. They are combinations of an inequality symbol and the equals sign.

Say: x is less than or equal to 3.
Write: $x \leq 3$

Say: y is greater than or equal to 9.
Write: $y \geq 9$

Mini-Lab

Work with a partner.
Materials: ruler

Graph the ages less than 14 on a number line.

- Make a number line like the one shown below.

- Draw an open circle at 14 to show that 14 is not included as a solution. Draw a thick arrow over the numbers to the left to show that any number less than 14 is a solution.

Talk About It
a. Are there ages less than 14 not shown on the number line?
b. What is the oldest someone can be and still be younger than Jennifer Capriati?

The properties used to solve inequalities are very similar to those used to solve equations.

Example 1

Solve $n + 3 > 7$. Check your solution.
Then graph the solution on a number line.

$$n + 3 > 7$$
$$n + 3 - 3 > 7 - 3 \quad \textit{Subtract 3 from each side.}$$
$$n > 4$$

Check: Try 5, a number greater than 4.
$$n + 3 > 7$$
$$5 + 3 \overset{?}{>} 7 \quad \textit{Replace n with 5.}$$
$$8 > 7 \ \checkmark$$

The solution is $n > 4$, all numbers greater than 4.

To show the solution set, draw an open circle at 4. Then draw a thick arrow over the numbers that are solutions.

A circle shows that this point is not included.   *The thick arrow shows that all real numbers to the right of 4 are included.*

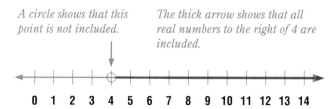

Example 2 *Problem Solving*

Business   Reader's Bookstore makes a profit of \$3 on every dictionary they sell. How many dictionaries would they have to sell to make a profit of at least \$75? Graph the solution on a number line.

Write an inequality.

profit per dictionary	*times*	*number of dictionaries sold*	*at least*	*\$75*
3	×	d	≥	75

$3d \geq 75$

$\dfrac{3d}{3} \geq \dfrac{75}{3}$ *Divide each side by 3.*

$d \ \geq 25$

Check: Try 30, a number greater than 25.
$$3d \geq 75$$
$$3(30) \overset{?}{\geq} 75 \quad \textit{Replace d with 30.}$$
$$90 \geq 75 \ \checkmark$$

Reader's Bookstore must sell at least 25 dictionaries to make a profit of \$75.

Graph the solution on a number line.

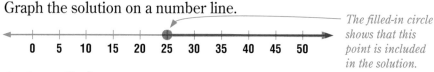

The filled-in circle shows that this point is included in the solution.

Checking for Understanding

Communicating Mathematics

Read and study the lesson to answer each question.

1. **Show** a number line that represents the solution set *all numbers less than 5*.

2. **Write** the inequality graphed on the number line at the right.

-4 -3 -2 -1 0 1 2 3 4

3. **Tell** how you can tell whether 6 is a solution to $9x < 18$.

Guided Practice

Solve each inequality. Graph the solution on a number line.

4. $d + 3 > -4$
5. $f - 6 > 4$
6. $4 + g < 7$
7. $k - 9 < -5$
8. $4y > 24$
9. $6n \leq -30$

Exercises

Independent Practice

Solve each inequality. Graph the solution on a number line.

10. $h - 7 > 2$
11. $5 + b > 2$
12. $d - 11 < 3$
13. $t - 3 > -2$
14. $4n > 8$
15. $7m < 28$
16. $6d \leq -24$
17. $\frac{r}{4} \geq -2$
18. $\frac{e}{3} \leq 12$

Write an inequality for each statement. Then solve the inequality.

19. Three times a number is greater than -18.

20. A number minus five is less than or equal to seven.

Mixed Review

21. Graph the function given by the rule $\frac{x}{2}$. *(Lesson 13-5)*

22. Measure the line segment at the right to the nearest eighth of an inch. *(Lesson 6-7)*

23. **Statistics** Refer to the circle graph on page 103. What part of the trash on U.S. beaches is metal and glass? *(Lesson 3-9)*

Problem Solving and Applications

24. **Jobs** Sonia delivers the weekly neighborhood newspaper. She gets paid 9 cents for each paper she delivers. How many papers would she have to deliver to earn at least $4.50? Graph the solution.

25. **Critical Thinking** The product of an integer and 6 is less than 48. Find the greatest integer that meets this condition.

26. **Smart Shopping** John plans to spend at most $30 for compact discs. He chooses one CD that costs $12. How much can he spend on other CDs? Graph this solution.

27. **Statistics** How many additional visitors could Acadia National Park have had and still have had less than Great Smoky Mountain National Park?

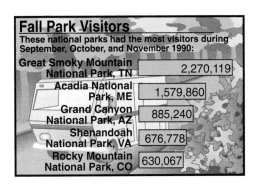

Fall Park Visitors

These national parks had the most visitors during September, October, and November 1990:

Great Smoky Mountain National Park, TN	2,270,119
Acadia National Park, ME	1,579,860
Grand Canyon National Park, AZ	885,240
Shenandoah National Park, VA	676,778
Rocky Mountain National Park, CO	630,067

Study Guide and Review

Communicating Mathematics

Choose the correct term or symbol to complete each sentence.

inequality
multiply
function machine
coordinates
$\leq$
function table
divide
solve
input
$>$

1. To __?__ an equation means to find a value for the variable that makes the equation true.

2. When the variable of an equation is multiplied by a number, __?__ each side by that number to solve the equation.

3. A(n) __?__ is a table of values that uses a mathematical rule to assign an output to a given input.

4. On the graph of a function, every point has __?__ that satisfy the function.

5. A(n) __?__ is a mathematical sentence that contains the symbols < or >.

6. Tell how to solve the equation $\frac{1}{7}m = 3$.

Skills and Concepts

Objectives and Examples

Upon completing this chapter, you should be able to:

- solve equations involving addition and subtraction using models
  (Lesson 13-1)

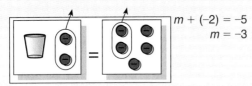

$m + (-2) = -5$
$m = -3$

- solve equations involving multiplication and division using models *(Lesson 13-2)*

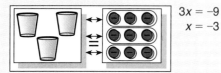

$3x = -9$
$x = -3$

Review Exercises

Use these exercises to review and prepare for the chapter test.

Use cups and counters to solve each equation.

7. $x + 2 = 5$ 8. $x - 1 = 3$

Solve each equation. Use cups and counters if necessary.

9. $p + 7 = 4$ 10. $6 + r = -7$

11. $y - 3 = 11$ 12. $-2 + w = -5$

Solve each equation. Use cups and counters if necessary.

13. $7q = 28$ 14. $\frac{1}{3}d = 9$

15. $\frac{1}{5}q = -11$ 16. $8x = 40$

17. $3m = -15$

Objectives and Examples

Review Exercises

- complete function tables *(Lesson 13-4)*

  a. Complete each function table.

n	n + 3
2	2 + 3 = 5
0	0 + 3 = 3
-4	-4 + 3 = -1

n	$\frac{1}{3}n$
6	$\left(\frac{1}{3}\right)6 = 2$
0	$\left(\frac{1}{3}\right)0 = 0$
-9	$\left(\frac{1}{3}\right)(-9) = -3$

  b. Find the rule for the function table.

  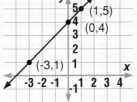

n	▧
-1	3
0	4
3	7

  $$\left.\begin{array}{l} -1 + 4 = 3 \\ 0 + 4 = 4 \\ 3 + 4 = 7 \end{array}\right\} n + 4$$

Copy and complete each function table.

18.

n	n − 2
−1	▨
0	▨
5	▨

19.

n	3n
−1	▨
0	▨
2	▨

20. If a function rule is $2n + 1$, what is the output for $n = -3$? −

Find the rule for each function table.

21.

n	▨
−1	−5
0	−4
3	−1

22.

n	▨
−1	5
0	0
3	−15

- graph functions from function tables
  (Lesson 13-5)

  Complete the function table. Then graph.

x	x + 4
-3	-3 + 4 = 1
0	0 + 4 = 4
1	1 + 4 = 5

Copy and complete each function table. Then graph the function.

23.

x	x + 3
−2	▧
0	▧
2	▧

24.

x	4x
−1	▧
0	▧
2	▧

- identify and solve inequalities
  (Lesson 13-6)

  Solve each inequality. Show the solution on a number line.

  a. $k - 6 < -4$
     $k - 6 + 6 < -4 + 6$
        $k < 2$

  b. $3x \geq 6$
     $\dfrac{3x}{3} \geq \dfrac{6}{3}$
       $x \geq 2$

  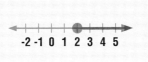

Solve each inequality. Show the solution on a number line.

25. $d - 4 > -1$ 26. $7w \leq -14$

27. $\dfrac{p}{2} \leq 3$ 28. $y + 7 > 9$

29. $2b \geq 5$ 30. $\dfrac{c}{3} < 3$

Study Guide and Review

Applications and Problem Solving

31. **Dieting**   After dieting for 8 weeks and losing 17 pounds, Jeremy weighed 172 pounds. How much did he weigh before the diet? *(Lesson 13-1)*

32. **Age**   Claudia's mother is three times as old as Claudia is this year. If Claudia's mother is 39 years old, how old is Claudia? *(Lesson 13-2)*

33. Keith swam three times as many laps as Dan in the swim meet. Keith swam 24 laps. How many did Dan swim? *(Lesson 13-3)*

34. **Merchandising**   Mr. Toshio got a new shipment of T-shirts to sell at his souvenir shop. The shirts that cost him $4.00, he priced at $7.50; the shirts that cost him $5.50, he priced at $9.00; the shirts that cost him $6.00, he priced at $9.50. Find the function rule he used. *(Lesson 13-4)*

35. **Budgeting**   On her vacation, Mrs. Salgado planned to spend less than $160 each day on her hotel room and meals. She found a hotel room for $75 a night. How much can she spend on meals? *(Lesson 13-6)*

Curriculum Connection Projects

- **Sports**   A bowler receives a handicap (extra points) whenever he or she is competing against better bowlers. Write an equation to find a bowler's final score if the handicap is 12 points.

- **Consumer Awareness**   Rebates are common in the auto industry. Find a rebate being offered on a car in a TV commercial or in a newspaper ad. Write an equation to find the final price of the car after the rebate is subtracted.

- **Consumer Awareness**   Check your family's electric and telephone bills and write an equation for calculating each bill every month.

Read More About It

Holt, Michael. *Maps, Tracks, and the Bridges of Konigsberg: A Book about Networks.*
Kyte, Kathy S. *The Kid's Complete Guide to Money.*
L'Engle, Madeline. *Wrinkle in Time.*

Study Guide and Review

Use cups and counters to solve each equation.

1. $x + (-3) = -1$
2. $r - 2 = 5$
3. $w + 4 = -3$

Solve each equation. Use cups and counters if necessary.

4. $b - 4 = -1$
5. $\frac{1}{10}p = 2$
6. $5m = 30$
7. $\frac{1}{4}w = -8$
8. $x + 10 = -2$
9. $6y = -18$

10. **Smart Shopping** Jeannette bought a sweater. The sale price was $38, which was $15 less than the original price. Find the original price by writing an equation for the problem and solving it.

Complete each function table.

11.

n	$n + 3$
−5	▨
0	▨
1	▨

12.

n	$5n$
−3	▨
−1	▨
1	▨

Find the rule for each function table.

13.

n	▨
−2	−5
0	−3
3	0

14.

n	▨
−2	−8
0	0
3	12

15. **Age** Bret is 12 years old, and his father is 35 years old. When Bret is 20, his father will be 43 years old. Write a function rule for this relationship.

Complete each function table. Then graph the function.

16.

x	$x + 2$
−3	▨
0	▨
3	▨

17.

x	$\frac{1}{2}x$
−2	▨
0	▨
4	▨

Solve each inequality. Show the solution on a number line.

18. $3z < 15$
19. $h - 6 \geq -8$
20. $\frac{f}{4} > -2$

Bonus The function rule $4x$ would have an output of 12 from an input of 3. Find the function rule that is the opposite (input 12, output 3).

Probability

Spotlight on Education

Have You Ever Wondered. . .

- What percent of the population between the ages of 10 and 13 are enrolled in school?

- How many people graduate from high school each year?

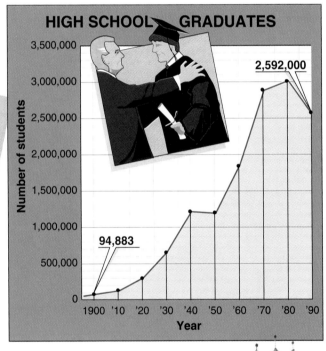

HIGH SCHOOL GRADUATES

2,592,000

94,883

School Enrollment, 1989 (in thousands)		
Age	Enrolled	Percent of Age Group
3 and 4 years	2,898	39.1
5 and 6 years	6,990	95.2
7 to 9 years	10,833	99.2
10 to 13 years	13,598	99.4
14 to 15 years	6,493	98.8
16 and 17 years	6,254	92.7
18 and 19 years	4,125	56.0
20 and 21 years	2,630	38.5
22 to 24 years	2,207	19.9
25 to 29 years	1,960	9.3
30 to 34 years	1,248	5.7
Total	59,236	49.1

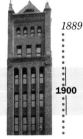

1821   *1847 1851*   *1873*   *1889*

1810   **1840**   **1870**   **1900**

First public high school opened

School separated by grade level for the first time

First public school opened for mentally challenged children

First public kindergarten established

First New York skyscraper completed

Chapter Project

Education
Work in a group.

1. Find out your classmates' favorite and least favorite school subjects. Ask them what they like and dislike about each.

2. Also find out if they like or dislike going to school overall. Find out why or why not.

3. Make a chart presenting your findings to the class.

4. Extend your survey to your friends and family. Add this data to your total. Do the answers of the larger group match those of the original?

Looking Ahead

In this chapter, you will see how mathematics can be used to answer questions about education. The major objectives of the chapter are to:

- find the probability of an event

- solve problems by acting them out

- find and compare experimental and theoretical probabilities

- make predictions

- find the probability of independent events

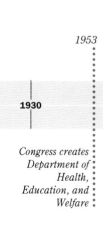

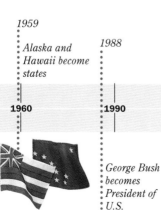

1959

Alaska and Hawaii become states

1953

1988

1930　　　　**1960**　　　　**1990**

Congress creates Department of Health, Education, and Welfare

George Bush becomes President of U.S.

495

14-1A Playing Games

A Preview of Lesson 14-1

Objective
Experiment with fair and unfair games.

Words to Learn
fair game
unfair game

Materials
dice

Have you ever played a game with your friends when someone cried "No fair!"? That usually means that one of the players has tried to take advantage of the rules to have a better chance of winning. Games in which players have an equal chance of winning are called **fair games.** In an **unfair game,** players do not have an equal chance of winning.

In this lab, you will play two different games to learn about fair and unfair games.

Activity One

Work with a partner.

- Rules of the Game:
  Roll two dice.
  Add the two numbers that are face up.
  Player 1 gets one point if the sum is even.
  Player 2 gets one point if the sum is odd.

- The game consists of 40 rolls. Keep track of each sum in a chart like the one shown below.

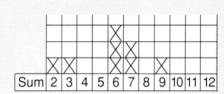

What do you think?

1. How many even sums were there?
2. How many odd sums were there?
3. Do you think the game is fair? Explain why or why not.
4. Your chart should resemble a bar graph. Describe the shape of the graph.
5. Which sum occurred most often? least often?

Extension

6. The addition table at the right can help you analyze the results of your game. Which sum from the addition table occurs most often? least often? Compare these results with Exercise 5.

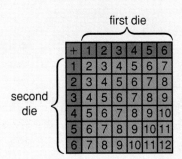

first die

second die

+	1	2	3	4	5	6
1	2	3	4	5	6	7
2	3	4	5	6	7	8
3	4	5	6	7	8	9
4	5	6	7	8	9	10
5	6	7	8	9	10	11
6	7	8	9	10	11	12

Activity Two

Work with a partner.

- Rules of the Game:
  Roll two dice. Multiply the two numbers that are face up. Player 1 gets one point if the product is even; Player 2 gets one point if the product is odd.
- The game consists of 40 rolls. Keep track of each product in a chart like the one shown below.

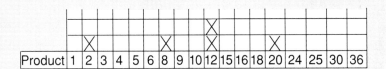

Product	1	2	3	4	5	6	8	9	10	12	15	16	18	20	24	25	30	36

What do you think?

7. How many even products were there?
8. How many odd products were there?
9. Is the game fair? Explain why or why not.
10. Describe the shape of your graph.
11. Which product occurred most often? least often?

Extension

12. Compare this graph with the graph in Activity One. How are they the same? How are they different?
13. If a game is unfair, it can be made fair either by changing the rules or changing the number of points that are awarded to each player. Try to make Activity Two a fair game.

14-1 Probability

Objective

Find the probability of an event.

Words to Learn

probability
outcome
event

In the National Football League, the game begins with the referee tossing a coin at mid-field, while the captains of the two teams look on. The visiting captain calls "heads" or "tails". What is the **probability,** or chance, that the tossed coin will come up "heads"?

Since the coin is fair, there are two equally likely results, or **outcomes.** The coin can either come up "heads" or "tails". Each outcome has the same chance of occurring.

heads

tails

DID YOU KNOW

In tennis, the first server is determined by a racket "toss". The players may use the racket manufacturer's markings on one side of the racket handle as "heads" and the unmarked side as "tails".

An **event** is a specific outcome or type of outcome. In this case, the event is tossing "heads". Out of two possible outcomes, "heads" or "tails", the probability of the coin coming up "heads" is 1 out of 2, or $\frac{1}{2}$. This can be written as $P(\text{heads}) = \frac{1}{2}$.

The probability that an event will happen is somewhere between 0 and 1.

- A probability of 0 means that the event is impossible.

- A probability of 1 means that the event is certain to happen.

- The closer a probability is to 1, the more likely the event is to happen.

You can express the probability of an event as a fraction, decimal, or percent. Since probability is a number, you can picture it on a number line.

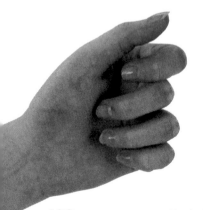

impossible	equally likely	certain
0	$\frac{1}{2}$	1
0%	0.5	100%
	50%	

P(heads)

Mini-Lab

Work with a partner.
Materials: paper, ruler

- Draw a number line like the one shown on page 498.

- Locate each of the following situations on the number line.

1. You will have math homework tonight.
2. A baby born today was a girl.
3. The local meteorologist predicts a 40% chance that it will rain tomorrow.
4. It will snow in your town in August.
5. A student chosen by chance from your class is wearing jeans today.

Talk About It

a. Order the above events from least likely to most likely.
b. Write three events of your own. Add them to your number line.

If the possible outcomes are equally likely to happen, you can use the following expression to find probability.

LOOK BACK

You can review ratio on page 352.

Probability	**In words:** The probability of an event is the ratio of the number of ways an event can occur to the number of possible outcomes.
	In symbols: $P(\text{event}) = \dfrac{\text{number of ways}}{\text{the event can occur}} \Big/ {\text{number of possible outcomes}}$

Example 1

On the spinner at the right there are four equally likely outcomes. Find the probability of spinning yellow.

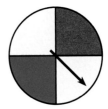

$\dfrac{2}{4}$ ← *number of ways to spin yellow*
← *number of possible outcomes*

Therefore, $P\text{ (yellow)} = \dfrac{2}{4}$ or $\dfrac{1}{2}$.

Example 2

A number cube is marked with 1, 2, 3, 4, 5, and 6 on its faces. Rolling the cube produces six equally likely outcomes.

a. Find $P(1)$.

$\dfrac{1}{6}$ ← *number of ways to roll a 1*
← *number of possible outcomes*

Therefore, $P(1) = \dfrac{1}{6}$.

b. Find $P(even)$.

There are three even numbers: 2, 4, and 6.
$\dfrac{3}{6}$ ← *number of ways to roll an even number*
← *number of possible outcomes*

Therefore, $P(even) = \dfrac{3}{6}$ or $\dfrac{1}{2}$.

Checking for Understanding

Communicating Mathematics

Read and study the lesson to answer each question.

1. **Tell** why an event with a probability of $\dfrac{9}{10}$ is likely to happen.
2. **Write** a sentence that explains what *equally likely* means.
3. **Tell** the probability of an event that is impossible.
4. **Draw** a spinner that shows $P(red) = \dfrac{1}{8}$.

Guided Practice

There are three blue marbles, four red marbles, two green marbles, and three white marbles in a bag. You reach into the bag and, without looking, pull out one marble. Find the probability of each event.

5. $P(blue)$ 6. $P(red)$ 7. $P(green)$
8. $P(yellow)$ 9. $P(blue\ or\ red)$ 10. $P(red\ or\ green)$

Exercises

Independent Practice

A set of 20 cards is numbered 1, 2, 3, ..., 20. Suppose you draw one card without looking. Find the probability of each event.

11. $P(3)$ 12. $P(even)$ 13. $P(2\ digits)$
14. $P(prime)$ 15. $P(whole\ number)$ 16. $P(fraction)$
17. $P(less\ than\ 10)$ 18. $P(greater\ than\ 30)$
19. $P(less\ than\ 1)$ 20. $P(divisible\ by\ 5)$

A set of 26 cards is lettered a, b, c, ..., z. It is equally likely to choose any one card. Find the probability of each event.

21. P(vowel)

22. P(consonant)

23. P(s, t, or u)

24. P(3)

25. Suppose one event has a probability of $\frac{3}{4}$, and another event has a probability of 80%. Which event is more likely to happen?

26. Sue said her chances for an "A" in math are "50-50". What does she mean?

Mixed Review

27. Evaluate $21 - 15 \div 3$. *(Lesson 1-6)*

28. Sally and Carmen put their laundry detergent together to wash their clothes. Sally has $2\frac{2}{3}$ cups of detergent and Carmen has $1\frac{3}{4}$ cups. How much detergent do they have in all? *(Lesson 7-6)*

29. **Geometry** Graph $A(2, -3)$ and $B(-3, 2)$. *(Lesson 12-8)*

30. **Algebra** Solve the inequality $g - 7 < -2$ and graph the solution on a number line. *(Lesson 13-6)*

Problem Solving and Applications

31. **Statistics** The graph below shows the hair colors of 28 students in a certain class. If a student is chosen by chance, find each probability.

   a. P(brown)

   b. P(red)

   c. P(green)

   d. P(black or red)

32. **Advertising** A cereal company is running a contest. Every box of BAMMO cereal contains one of the letters B, A, M, O. In every 20 boxes, there are four Bs, five As, ten Ms, and one O. Janice has A M M O. What is the probability that her next box of cereal will have a B?

33. **Critical Thinking** Explain how you can find the probability of an event *not* occurring if you know that the probability of the event is $\frac{3}{5}$.

34. **Journal Entry** Think about events that happen in your life. Write about an event that is certain, an event that is impossible, an event that is highly likely, and an event that is equally likely to happen or not happen.

14-2 Act It Out

Objective
Solve problems by acting them out.

Words to Learn
experimental probability

Jennifer was opening her birthday presents. She noticed that the cards had fallen off three of the presents. By chance, she matched each present with one of the cards. What is the probability that Jennifer matched at least one present with the correct card?

Sometimes, you can find the probability of an event by collecting data from an experiment. This is called **experimental probability.** When you perform the experiment, you are using an important problem-solving strategy, act it out.

Explore   What do you know?
Three presents did not thave cards attached to them.
What do you need to find out?
Find the probability that Jennifer opens at least one card and present given to her by the same friend.

Plan   Act out the problem by modeling the situation using index cards. Mark the index cards as shown below.

Present A	Present B	Present C	Card A	Card B	Card C

Place the cards in two piles. Turn the cards over and mix them up. Without looking, choose a present and a card. Record the information in a table. Repeat the activity 20 times.

Problem Solving Hint
• • • • • • • • • • •
You can use the make-a-table strategy.

Solve   Some sample outcomes are shown in the table at the right.

Trial	Match ?
1	no
2	yes

If out of 20 trials, Jennifer received a match 5 times, the experimental probability is $\frac{5}{20}$, 0.25, or 25%.

Zillions® magazine is a consumer magazine directed at teenagers. Published by Consumers Union, its purpose is to provide teenagers with information they need to use their money wisely.

Example

The Cougar Pep Club is holding a contest along with their spring magazine sale. For every subscription a student sells, he or she receives a card with a letter on it. The first person to spell the word COUGAR wins a free magazine subscription. There is an equally likely chance of getting a different card each time. Reggie wants to know how many subscriptions he should sell in order to win the contest. How could he act out this problem?

One way to act out this problem is to make a spinner like the one shown at the right. Reggie could spin the spinner and record the data in a table.

A sample is shown below.

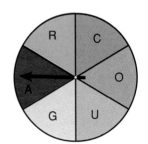

Outcome	Tally	Frequency
C	I	1
O	I	1
U	III	3
G	I	1
A	IIII	4
R	I	1

For this experiment, Reggie concluded that he needs to sell about 11 subscriptions to collect all six cards and win the contest.

Checking for Understanding

Communicating Mathematics

Read and study the lesson to answer each question.

1. **Tell** how the act-it-out strategy is related to experimental probability.

2. **Write** an explanation of another way to act out the problem in the Example using a device other than a spinner.

Guided Practice

Solve using the act-it-out strategy.

3. Toss a coin 10 times. What is the experimental probability of the coin coming up "tails"?

4. How many times do you need to spin this spinner in order to get all eight letters?

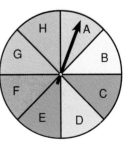

5. Mi-Ling has three different pairs of basketball shoes. She has a white pair with red stripes, a white pair with blue stripes, and a red pair with white stripes. She picks the pairs at random to wear at each game. What is the probability that she wears the white pair with red stripes more than once during a five-day basketball tournament?

Problem Solving

Practice Solve using any strategy.

6. The difference between two whole numbers is 14. Their product is 1,800. Find the two numbers.

Strategies

• • • • • • • •

Look for a pattern.

Solve a simpler problem.

Act it out.

Guess and check.

Draw a diagram.

Make a chart.

Work backward.

7. At the school supply store, 4 pencils and 7 pens cost $1.89. Eleven pencils and 1 pen cost $1. What is the cost of 1 pencil?

8. Jack's birthday this year is Friday, February 13. Gino's birthday is January 18. What day of the week is Gino's birthday this year?

9. At a certain restaurant, prizes are given with children's meals. During a recent promotion, three different prizes are given at random. How many children's meals must be purchased in order to get all three prizes? Use the act-it-out strategy.

10. Here's a famous problem. A snail at the bottom of a 10-foot hole crawls up 3 feet each day, but slips back 2 feet each night. How many days will it take the snail to reach the top of the hole and escape?

COMPUTER

CONNECTION

11. **Computer Connection** Computers have a function, RND, that generates random numbers. With this function, the computer can be programmed to act out tossing a coin or rolling a die many times. The BASIC program below acts out tossing a fair coin.

```
10 INPUT "HOW MANY TRIALS?"; N
20 FOR I = 1 TO N
30 C = INT(RND(I)*2)
40 IF C = 0 THEN PRINT "H"
50 IF C = 1 THEN PRINT "T"
60 NEXT I
```

a. Run the program for 20 trials and find the experimental probability of heads.

b. Run the program for 50 trials and find the experimental probability of heads.

14-2B Experimental Probability

A Follow-Up of Lesson 14-2

Objective

Explore experimental probability.

Materials

paper cup

If you tossed a paper cup in the air 100 times, how many times do you think it would land bottom up? bottom down? on its side? Try the experiment below and find out.

Try this!

Work with a partner.

Assign one student as tosser and one student as recorder.

- Make a recording sheet like the one below.

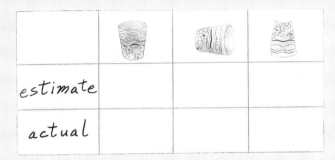

- Before you toss your cup, record your estimate for each way the cup will land if you toss it 100 times.
- Toss the cup into the air 100 times. Make a tally of each toss.

What do you think?

1. Compare your estimates to the actual results. Were you close to the actual results?
2. Explain how you made your estimates.
3. Which way did the cup land most often? Find that probability.
4. Make a new chart to show the results for the entire class. How many throws were there altogether?
5. Which way did the cup land most often for the entire class? Find the probability.
6. Compare your results to the class results. Were the results the same? If not, why might the results be different?

Extension

7. If you threw the cup 10,000 times, how many times would you expect it to land bottom down?

14-3 Making Predictions

Predict actions of a larger group using a sample.

Words to Learn

sample
population
random

The student council at Edgemont Middle School is planning a year-end dance. The entertainment committee will either hire a live band or a DJ to provide music. The committee does not have time to survey every student to find their preference, so they survey a smaller group, or **sample.** They can use the information from the sample to predict how the students in the entire school would feel.

All of the students in the school make up the **population.** How can the entertainment committee survey students so that the sample represents all of the students? Here are some suggestions.

- Survey every tenth student named on the school roster.
- Survey every fifteenth student that enters the building one morning.
- Survey one student from each lunch table.

Each of these methods is a way of making sure that the sample is **random,** or drawn by chance from the population.

When am I ever going to use this?

Market Researcher

A market researcher collects and analyzes data to help businesses make decisions about advertising, pricing, new products, and so on. The sample they choose to survey can make a big difference in the accuracy of the decisions made by the business.

For more information, write to:

Marketing Research Association
111 East Wacker Drive
Chicago, Illinois 60601

Example 1

Talia wanted to find out students' favorite entertainment: concerts, movies, or dances.

a. She surveyed a small group of people standing in line at a movie theater. Is this a random sample?

This is *not* a random sample because those in line at the movie theater may already prefer movies. Also, the people surveyed may not all be students.

b. Talia surveyed every tenth student leaving school at the end of the day. Is this a random sample?

This is a random sample. Talia had no control over the order the students left school. By surveying every tenth student, she broke up any groups that might like the same activity.

Example 2

The Basketball Boosters conducted a random survey of students to find their soft drink preference. Of the 20 students surveyed, 15 preferred cola. The Boosters expect about 150 students to buy a soft drink at the basketball game.

a. What is the probability that the first student who buys a soft drink will buy a cola?

15 out of 20, or $\frac{3}{4}$, prefer cola.

The probability is $\frac{3}{4}$ or 75%.

Estimation Hint

• • • • • • • • • •

THINK: $\frac{3}{4}$ of 150 is a little less than $\frac{3}{4}$ of 160. $\frac{3}{4}$ of 160 is 120.

b. Predict how many colas will be sold at the game.

Since 75% of the sample prefer cola, you can predict that 75% of the population prefer cola.

75% of 150 → 75 [%] [×] 150 [=] 112.5

Of the 150 students that buy a soft drink, about 113 will buy cola.

In the following Mini-Lab, you will do an experiment to predict the number of left-handed students in your school.

Mini-Lab

Work with a partner.
Materials: paper, pencil

DID YOU KNOW

Left-handed people are usually quicker on keyboards and typewriters than right-handed people because they are more equally skilled with both hands. They also have the advantage in some sports like baseball, tennis, and boxing.

- Decide how you will choose a random sample to predict the number of left-handed students in your school.

- With your teacher's permission, conduct your survey.

Talk About It

a. What is the probability that a student selected at random from your sample is left-handed?

b. Predict the number of left-handed students in your school.

c. Compare results with other groups. Explain why different groups may arrive at different conclusions.

Checking for Understanding

Communicating Mathematics

Read and study the lesson to answer each question.

1. **Write** one or two sentences that explain what is meant by *random sample*.

2. **Tell** how you might use a random sample to predict the results of a class election.

3. **Tell** what might happen to your prediction in Exercise 2 if you only surveyed friends of one candidate.

Guided Practice

Tell whether or not each of the following is a random sample. Explain your answer.

Type of Survey	Survey Location
4. favorite sport	baseball game
5. favorite song	every tenth person in phone book
6. favorite performer	people leaving a concert
7. favorite candy bar	people leaving a grocery store

8. In a random survey of 30 students in the school cafeteria, LaVonda found that 12 ordered spaghetti. If there are 450 students that eat the cafeteria lunch, how many will likely order spaghetti?

Exercises

Independent Practice

9. Mickey took a survey of sweatshirt sizes from a random sample of 25 new students at Blendon Junior High School. The results are shown in the chart at the right. Use the results to answer each question.

   a. What is the size of the sample?

   b. What is the probability that a student at Blendon Junior High School wears a large sweatshirt?

   c. Suppose Mickey needs to order 100 sweatshirts for the student store. How many of each size should she order?

Sweatshirt sizes
small   10
medium   8
Large   5
Extra large 2

Mixed Review

10. **Statistics**   What is the mode of the data in the sweatshirt survey above? *(Lesson 2-7)*

11. **Geometry**   What is the surface area of a rectangular prism whose dimensions are 8 feet by 2 feet by 10 feet? *(Lesson 11-5)*

12. Suppose the probability that your math teacher gives you a quiz on Friday is 3 out of 4. The probability that your science teacher gives you a quiz on Friday is 65%. Which event is more likely? *(Lesson 14-1)*

13. **Softball** Karen has made 9 hits in the last 20 times she has batted. If she comes to bat 5 times in today's game, predict how many hits she will make.

14. **Quality Control** A quality control inspector found that 3 out of 50 computer diskettes were defective. Suppose the company manufactured 15,000 diskettes that day. Predict the total number of defective diskettes manufactured that day.

15. **Entertainment** Each year, nearly 3,500,000 people attend the Ohio State Fair, making it the largest fair in the United States. It draws people from nearby states as well as Ohio. The graph below shows United States population distribution by age.

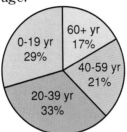

0-19 yr 29%
60+ yr 17%
40-59 yr 21%
20-39 yr 33%

U.S. Population by age

a. The one-millionth visitor to the fair receives a special prize. What is the probability that this visitor is between the ages of 20 and 39?

b. People over age 60 can buy a discount ticket. About how many discount tickets will be sold?

16. **Critical Thinking** A pre-election poll predicted that a certain candidate for the school board would receive 25% of the vote. He actually received 15,248 votes. Estimate how many people voted in the election.

14 Mid-Chapter Review

A box contains 2 red pencils, 4 yellow pencils, and 3 green pencils. One pencil is chosen at random. Find the probability of each event. *(Lesson 14-1)*

1. P(red)
2. P(black)
3. P(red or yellow)

4. Toss a coin 20 times. Find the experimental probability of the coin coming up "heads". *(Lesson 14-2)*

5. **Quality Control** An auto dealer needs to recall a certain model to replace a part that may be defective. Nationally, only 2 autos out of 50 need the part replaced. If the dealer contacts 1,200 owners, about how many replacements will she expect to make? *(Lesson 14-3)*

14-4 **Probability and Area**

Objective

Find probability using area models.

Jill and Servio are using a dart board like the one at the right to play darts. Let's assume that all of the darts hit the board, and that it is equally likely to land on any point on the board. What is the probability that Servio's dart will land on a shaded part of the board?

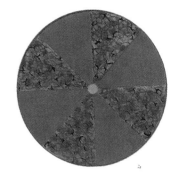

This situation is similar to spinning a spinner. There are eight equally likely outcomes, four of which are shaded.

Therefore, $P(\text{shaded}) = \frac{4}{8}$ or $\frac{1}{2}$.

You can also relate probability to the areas of other geometric shapes.

 Mini-Lab

Work with a partner.

Materials: dot paper, thumbtacks or small beans

- Outline two squares as shown.

- Drop 25 thumbtacks onto the paper.

- Record the number that land within the large square and the number that land within the small square. Do not count those that land outside the squares.

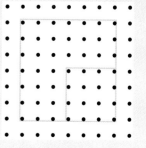

Talk About It

a. Find the ratio $\dfrac{\text{number of thumbtacks in small square}}{\text{number of thumbtacks in large square}}$.

b. Find the ratio $\dfrac{\text{area of small square}}{\text{area of large square}}$.

c. Compare the ratios from parts a and b.

d. Combine your results with another group and compare the ratios.

The results of the Mini-Lab suggest the following conclusion.

$$\frac{\text{number landing in shape}}{\text{number landing in region}} = \frac{\text{area of shape}}{\text{area of region}}$$

Example 1

LOOK BACK

You can review proportions on page 355.

The square at the right represents a dart board. Suppose you throw 500 darts randomly at the board, and that all darts hit the board. How many darts would you expect to land in region B?

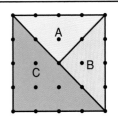

Let n = darts landing in region B.

$$\frac{n}{500} = \frac{4}{16} \quad \begin{array}{l}\leftarrow \textit{area of region B} \\ \leftarrow \textit{area of board}\end{array}$$

$$500 \cdot 4 = 16n$$
$$2,000 = 16n$$
$$125 = n$$

125 darts should land in region B.

Example 2 *Connection*

Calculator Hint

• • • • • • • • • • •

Use the $\boxed{\pi}$ key on your calculator to compute the area of the circle.

Geometry   A sky diver is the feature entertainer for the Monroe High School homecoming football game. If it is equally likely he will land on any part of the field, estimate the probabilty that the diver will land inside the circle.

$$P(\text{circle}) = \frac{\text{area of circle}}{\text{area of field}}$$

$$A = \pi r^2$$
$$\approx 3.14 \times 21 \times 21$$
$$\approx 1,385 \text{ ft}^2$$

$$A = l \times w$$
$$= 160 \times 360$$
$$= 57,600 \text{ ft}^2$$

[Diagram: rectangle 360 ft by 160 ft with a circle of radius 21 ft inside]

$$P(\text{circle}) \approx \frac{1,385}{57,600} \Rightarrow \frac{1,500}{60,000} \quad \textit{1,500 and 60,000 are compatible numbers.}$$
$$\approx \frac{1}{40}$$

The probability that the sky diver will land inside the circle is about $\frac{1}{40}$ or 2.5%.

Checking for Understanding

Communicating Mathematics

Read and study the lesson to answer each question.

1. **Tell** the ratio that compares the area of region A to the area of the large square.

2. **Draw** a square dart board in which the probability of landing on a shaded area is $\frac{2}{3}$.

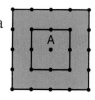

Guided Practice

Each figure represents a dart board. Find the probability of landing in the shaded region.

3.

4.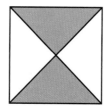

5.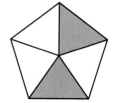

Suppose you throw 100 darts at each dart board below. How many would you expect to land in each shaded region?

6.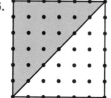

7.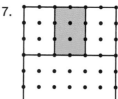

8.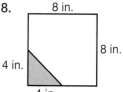

Exercises

Independent Practice

Suppose you throw 100 darts at each dart board below. How many would you expect to land in each shaded region?

9.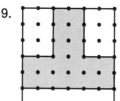

10.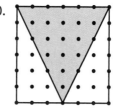

11.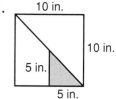

12.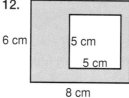

13.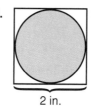

14.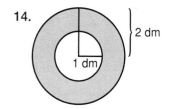

15. Add 15.783 and 390.81. *(Lesson 3-8)*

16. Find the quotient $\frac{5}{6} \div \frac{10}{11}$. *(Lesson 8-6)*

17. In a random survey of 25 students, Juan found that 15 walked to school. If there are 1,250 students in all, predict how many walk to school. *(Lesson 14-3)*

Problem Solving and Applications

18. **Games**   A dart is thrown at the dart board at the right. It is equally likely that the dart lands anywhere on the dart board. What is the probability that it will land in the center region?

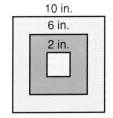

10 in.
6 in.
2 in.

19. **Critical Thinking**   Suppose you made a dart board from the tangram at the right. If you throw 200 darts at the dart board, how many would you expect to land in each region? Assume it is equally likely that the dart lands anywhere in the tangram.

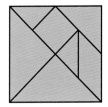

20. **Journal Entry**   Suppose you and your friend are playing darts with the dart board at the right. First, choose a region A or B. You get one point every time you throw a dart and it lands in your region. Which region would you choose? Write a sentence or two to explain your reasoning.

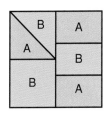

Save Planet Earth

Start Your Own Club   Students are making a significant contribution to saving our planet. Clinton Hill was a sixth-grade student when he organized an environmental activists' club at his school in New Hope, Minnesota. His enthusiasm and concern about the deteriorating environment motivated other students to become involved and help save the planet.

After Clinton's death in 1989, his parents, William and Tessa Hill, established a club called Kids for Saving Earth (KSE). Today there are over 3,600 KSE clubs in schools around the country.

How You Can Help

● Start a KSE Club in your school. For more information and a monthly newsletter, write to KSE Clubs, P.O. Box 47247, Plymouth, Minnesota 55447-0247.

● Organize and implement a recycling program in your school.

14-5 Tree Diagrams

Objective

Find outcomes using tree diagrams.

Words to Learn

tree diagram

Mom's Diner offers these specials for lunch.

José wonders how many ways he can order a burger and shake. He decides to make an organized list.

Beanie Burger - Vanilla
Beanie Burger - Chocolate
Beanie Burger - Orange
Jalapeño Burger - Vanilla
Jalapeño Burger - Chocolate
Jalapeño Burger - Orange

There are six ways to order a burger and shake.

Another way to show all of the possible outcomes is to organize this list into a **tree diagram.**

Problem Solving Hint

• • • • • • • • • •

You can use the draw a diagram strategy.

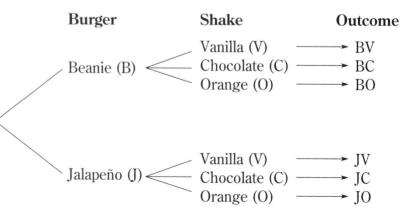

Burger	Shake	Outcome
	Vanilla (V)	BV
Beanie (B)	Chocolate (C)	BC
	Orange (O)	BO
	Vanilla (V)	JV
Jalapeño (J)	Chocolate (C)	JC
	Orange (O)	JO

Example 1

A new car can be ordered in black or red. You may also choose with or without air conditioning, standard or automatic transmission. Draw a tree diagram that shows all of the ways you can order a new car.

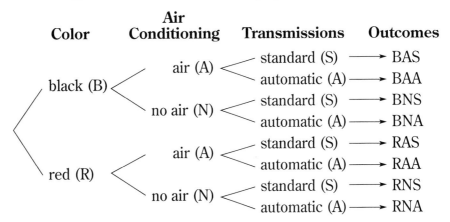

There are eight different combinations.

Example 2 *Problem Solving*

Shopping   Chari wants to buy a navy sweater that is on sale. Before she makes the purchase, she thinks about the blouses and pants that she has at home to wear with the sweater. She has white, red, and fuchsia blouses and gray, khaki, and plaid pants.

a. How many different combinations are there?

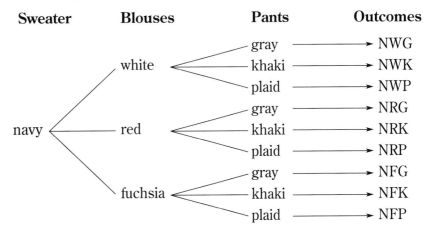

There are nine different combinations.

<aside>
Mental Math Hint
● ● ● ● ● ● ● ● ● ●
You can find the number of outcomes by multiplying the number of choices. In Example 1, there are 2 × 2 × 2 or 8 possible outcomes. In Example 2, there are 3 × 3 or 9 possible outcomes.
</aside>

b. If Chari chooses one blouse and one pair of pants at random one morning, what is the probability that she will wear the red blouse and plaid pants?

One outcome has the red blouse with plaid pants.
There are nine possible outcomes.

Therefore, the probability is $\frac{1}{9}$.

Checking for Understanding

Read and study the lesson to answer each question.

1. **Tell** how the tree diagram at the right can be used to count the different outcomes for tossing a penny and a nickel.

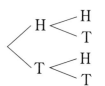

2. **Draw** a tree diagram that shows the different outcomes for tossing a penny, nickel, and quarter.

For each situation, draw a tree diagram to show all the outcomes.

3. a choice of peach or apple pie with a choice of milk, tea, or coffee
4. a choice of five pitchers with a choice of three catchers
5. spinning each spinner once

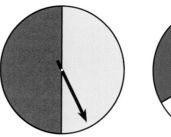
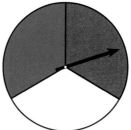

6. tossing a coin and rolling a die

Exercises

For each situation, draw a tree diagram to show all the outcomes.

7. Each spinner is spun once.

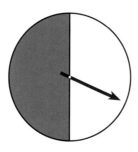
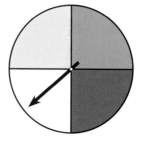
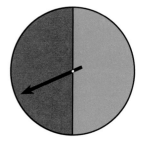

8. Phones come in two styles: wall and desk. They come in four colors: red, white, black, and beige.
9. You can choose either a boy's or girl's bicycle. They have either one, three, or ten speeds.
10. Three students are running for class president. Two students are running for class vice-president.

11. **Geometry**   Find the circumference of a circle with a radius of 5 meters. *(Lesson 4-9)*

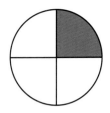

12. Suppose you throw 200 darts at the dart board shown at the right. How many would you expect to land in the shaded region? *(Lesson 14-4)*

Problem Solving and Applications

13. **Food**   At Moretti's Spaghetti House you can order three different pastas with three different sauces. How many possible dinners are there?

14. **Clothing**   Randi has three skirts, four tops, and two sweaters that coordinate. How many days could she come to school in a different outfit?

15. **Travel**   There are three routes between Maria's apartment and her grandmother's apartment. How many ways can Maria go from her apartment to her grandmother's and back again?

16. **Manufacturing**   School notebooks are made in four colors: red, green, blue, and yellow. They come in two styles, striped and plain. Suppose the notebooks are distributed at random.
    a. What is the probability of getting a plain, yellow notebook?
    b. What is the probability of getting a blue notebook?

17. **Critical Thinking**   If a family has four children, are they more likely to have three of one sex or two of each? Show all of your work.

18. **Mathematics and Games**   Read the following paragraph.

    Here's a simple game to play with a friend. Suppose you have three chips marked in the following way:
        Chip 1: A on one side, B on the other
        Chip 2: A on one side, C on the other
        Chip 3: B on one side, C on the other
    You toss all three at the same time. If any two chips match, Player 1 wins; if all three chips are different, Player 2 wins.

    a. Draw a tree diagram to show all of the outcomes.
    b. Would you choose to be Player 1 or Player 2? Explain why.

14-5B Simulations

A Follow-Up of Lesson 14-5

Objective

Experiment with probability using a simulation.

Materials

four pennies
paper cup

A simulation is a way of acting out a problem. In this lab, you will investigate the probability that, in a family with four children, at least two of them are girls.

It is equally likely that a newborn baby will be a girl or a boy. Similarly, it is equally likely that a toss of a coin will result in heads or tails. You will use a coin toss to simulate the children born in a family.

Try this!

Work in groups of four.

- Place four pennies into a cup and toss them onto your desk.
- Count the number of heads. This will represent the number of girls.
- Keep track of each trial in a chart like the one shown below.

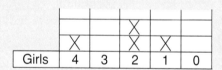

- Repeat until you have 50 trials.

What do you think?

1. Your chart should resemble a bar graph. Describe the shape of the graph.
2. What is the experimental probability that a family of four children will have at least two girls?

Extension

3. Make a tree diagram showing all possible outcomes of the experiment.
4. Using the tree diagram, find the probability that a family of four children will have at least two girls.
5. Compare the experimental probability in Exercise 2 with the probability in Exercise 4.

14-6 Probability of Independent Events

Objective
Find the probability of independent events.

Words to Learn
independent events

Problem Solving Hint
● ● ● ● ● ● ● ● ● ●
You can use the solve a simpler problem strategy.

Heather Pritchard's classmates took a tour of her father's bakery. After the tour, Mr. Pritchard let each student choose one cookie and one brownie from the display case.

The case had cookies and brownies in it as shown in the chart below. If Heather's friend, Sara, chose first, what is the probability that Sara chose a chip cookie and a nut brownie? *You will solve this problem in the Example.*

Baked Good	Nut	Chip	Plain
Cookie	3	5	4
Brownie	8	6	6

The kind of cookie Sara chooses does not affect the kind of brownie chosen. These are called **independent events.**

First, let's consider a simpler problem involving independent events. Suppose two coins are tossed. What is the probability of getting heads on both of them?

Draw a tree diagram showing the probabilities of each outcome. You know the probability of getting heads is $\frac{1}{2}$, and the probability of getting tails is $\frac{1}{2}$.

First Coin   Second Coin

$$
\begin{array}{ccccc}
 & & \frac{1}{2} - H & \longrightarrow & HH \\
\frac{1}{2} \diagup H < & & & & \\
 & & \frac{1}{2} - T & \longrightarrow & HT \\
< & & & & \\
 & & \frac{1}{2} - H & \longrightarrow & TH \\
\frac{1}{2} \diagdown T < & & & & \\
 & & \frac{1}{2} - T & \longrightarrow & TT \\
\end{array}
$$

Multiply to find the probability of getting heads on both coins.

P(heads on both coins) = P(heads on one coin) × P(heads on other coin)

$$
= \frac{1}{2} \times \frac{1}{2}
$$
$$
= \frac{1}{4}
$$

The probability of getting heads on both coins is $\frac{1}{4}$.

<table>
<tr><td>**Probability of Independent Events**</td><td>**In words:** The probability of two independent events is the product of the probability of event A and the probability of event B.
In symbols: $P(A \text{ and } B) = P(A) \cdot P(B)$</td></tr>
</table>

Example

What is the probability that Sara chose a chip cookie and a nut brownie?

$P(\text{chip cookie}) = \dfrac{5}{12}$ ←   *number of ways to choose chip cookie*
  ←   *number of possible outcomes: 3 + 5 + 4 = 12*

$P(\text{nut brownie}) = \dfrac{8}{20}$ ←   *number of ways to choose nut brownie*
  ←   *number of possible outcomes: 8 + 6 + 6 = 20*

Multiply to find the probability.

$$\dfrac{5}{12} \times \dfrac{8}{20} = \dfrac{\overset{1}{5}}{12} \times \dfrac{\overset{2}{8}}{\underset{4}{20}}_{} \quad \begin{array}{l} \textit{The GCF of 5 and 20 is 5.} \\ \textit{The GCF of 8 and 12 is 4.} \end{array}$$

$$= \dfrac{2}{12} \text{ or } \dfrac{1}{6}$$

The probability that Sara chose a chip cookie and a nut brownie is $\dfrac{1}{6}$.

LOOKBACK

You can review multiplying fractions on page 280.

Checking for Understanding

Communicating Mathematics

Read and study the lesson to answer each question.

1. **Tell** how you know when two events are independent.

2. **Tell** what $P(\text{large and green})$ is from the tree diagram at the right.

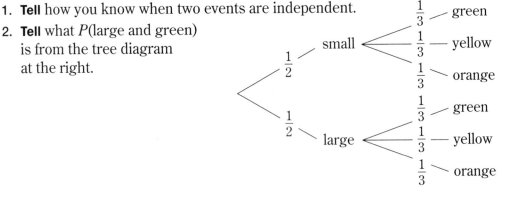

Guided Practice

Solve.

3. Red House Pizza has three different crusts available: thin, thick, and deep dish. There are four toppings from which to choose: pepperoni, sausage, mushroom, and onion. Each choice is equally likely. Jay orders a pizza with one topping.

    a. What is the probability he orders a deep dish pizza?

    b. What is the probability he orders a sausage pizza?

    c. Find $P(\text{deep dish and sausage})$.

4. At Long's Bookstore notebooks come in blue, gold, green, or red. Page dividers come in pink, yellow, tan, or orange.
   a. What is the probability of selecting a blue notebook?
   b. What is the probability of selecting tan dividers?
   c. Find P(blue notebook and tan dividers).

Exercises

Independent Practice Use the two spinners to find the probability of each.

5. P(blue and violet)
6. P(red and white)
7. P(yellow and red)

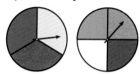

Suppose you flip a coin and roll a die. Find the probability of each of the following.

8. P(heads and 6)   9. P(tails and 1)   10. P(tails and even)
11. P(tails and 0)   12. P(heads and odd) 13. P(heads and prime)

Mixed Review

14. **Automobiles**   Refer to Example 1 on page 515. Suppose Jason likes red cars with standard transmission. What is the probability that the car he chooses at random will be red with a standard transmission? *(Lesson 14-5)*

15. Find $143.52 \div 2.08$. *(Lesson 5-3)*

16. **Geometry**   Trace the angle at the right and use a straightedge and compass to construct another angle congruent to it. *(Lesson 9-3)*

Problem Solving and Applications

17. **Critical Thinking**   Lisa's family cannot decide what to do on Saturday. Lisa flips a coin to determine whether they should stay home or go out. She then rolls a die to determine where to go if they go out. Their choices are the mall, a movie, bowling, a concert, out to eat, or to a baseball game.
   a. What is the probability that Lisa's family will go out?
   b. What is the probability that they go out and go to the mall?
   c. What is the probability that if they go out, they will go to a movie?

18. **Weather**   There is a 60% chance of rain tomorrow. If it does rain, there is an equally likely chance that Michael will read, watch TV, or play table tennis. What is the probability that it will rain and Michael will read?

DATA SEARCH

19. **Data Search**   Refer to pages 494 and 495. About how many more people graduated from high school in 1970 than in 1940?

20. **Critical Thinking**   Five students at Elmwood Middle School have no absences for the first semester. They are able to draw a prize from a bag that contains seven concert tickets and three baseball tickets. Each person keeps what is drawn. Jeff and Sean are the first two to draw. What is the probability that both Jeff and Sean will draw concert tickets?

Communicating Mathematics

Answer *true* or *false*. If false, change the underlined word(s) to make a true statement.

1. Another word for <u>probability</u> is chance.
2. Another word for results is <u>outcomes</u>.
3. When an event is certain, the probability is <u>100</u>.
4. The experimental probability of an event <u>will always be the same</u>, no matter how many times you perform the experiment.
5. When there is not time or capability to survey an entire <u>population</u>, a smaller group can be surveyed. This smaller group is called a <u>prediction</u>. The smaller group must be selected by chance from the population, so that it is a <u>random</u> sample.
6. An <u>area model</u> can be used to find and represent the probability.
7. A good way to find the number of possible outcomes is a <u>tree diagram</u>.
8. When one event's occurring does not affect another event, the events are called <u>dependent events</u>.
9. To find the probability of two independent events, <u>multiply</u> their probabilities.
10. Tell why it is important that a sample be a random one.

Skills and Concepts

Objectives and Examples

Upon completing this chapter, you should be able to:

- find the probability of an event
  (Lesson 14-1)

  On the spinner at the left, there are 6 equally likely outcomes. Find the probability of spinning blue.

  $P(\text{blue}) = \dfrac{1}{6}$ ← *way to spin blue*
  ← *possible outcomes*

Review Exercises

Use these exercises to review and prepare for the chapter test.

Use the spinner at the left to find the probability of each event.

11. $P(\text{anything other than blue})$
12. $P(\text{pink})$ 13. $P(\text{red or white})$

A nickel, dime, and penny are in a pocket. One is drawn out. Find each probability.

14. $P(\text{nickel})$ 15. $P(\text{dime or penny})$

Objectives and Examples

- predict actions of a larger group using a sample *(Lesson 14-3)*

    If a random sample survey found 12 out of 50 surveyed turn down the heat at least 12° at night, predict the number of people out of 10,000 who will do the same thing.

    12 out of 50 → 12 ÷ 50 = 0.24
    or 24%

    24% of 10,000 → 24 % × 10000 = 2400

- find probability using area models *(Lesson 14-4)*

    $$P\text{ (shaded)} = \frac{\text{area of shaded shape}}{\text{area of region}}$$
    $$\text{circle} = \pi r^2$$
    $$= 3.14 \times 3^2$$
    $$\approx 28.26$$

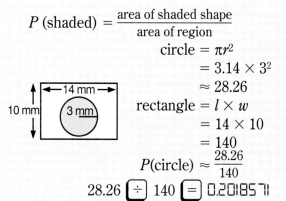

    $$\text{rectangle} = l \times w$$
    $$= 14 \times 10$$
    $$= 140$$
    $$P(\text{circle}) \approx \frac{28.26}{140}$$
    28.26 ÷ 140 = 0.2018571

- find outcomes using tree diagrams *(Lesson 14-5)*

    boy or girl? 6th, 7th, or 8th grade? How many possible outcomes?

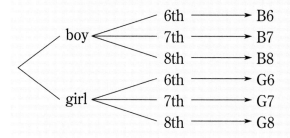

    There are 6 possible outcomes.

Review Exercises

Refer to the survey at the left.

16. 9 out of the 50 surveyed turn the heat down at least 4° at night. Predict the number for the entire population.

17. What could be done to the sample survey to make the results more certain?

How many out of 100 darts would you expect to land in each shaded region?

18.

19.

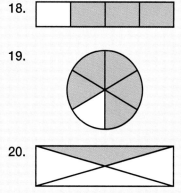

20.

Dinner choices included soup or salad, and beef, chicken, or fish. (Assume equally likely outcomes.)

21. Draw a tree diagram to show the possible combinations of choices.

22. What is the probability of a choice of soup and fish?

23. What is the probability of chicken dinner?

Objectives and Examples	Review Exercises

- find probability of independent events
  (Lesson 14-6)

  Boy or girl? 6th, 7th, or 8th grade?
  Find P(boy and 6th grade)

  $P(\text{boy}) = \frac{1}{2}$, $P(\text{6th grade}) = \frac{1}{3}$

  $P(\text{6th grade boy}) = \frac{1}{2} \times \frac{1}{3} = \frac{1}{6}$

Assume that P(going to a movie) $= \frac{1}{5}$ and P(the sun shining) $= \frac{2}{5}$.

24. Find the probability of going to a movie and the sun shining.

25. Find P(not going to a movie).

26. Find P(sunshine and no movie)

Applications and Problem Solving

27. The cheerleaders at Shaw Middle School have three different skirts that they can wear as part of their uniforms. One skirt is purple, one is white, and one is purple with white trim. The cheerleaders pick the skirts at random to wear. What is the probability that they wear the purple skirt with white trim more than once during a five-day cheerleading workshop? Use the act-it-out strategy. *(Lesson 14-2)*

28. **Planning**   Chicken, hot dogs, and hamburgers will be served at the class picnic. Yongchan, who is in charge of buying the food, conducts a survey of 30 students to determine their preference. Eight preferred chicken, 13 preferred hamburgers, and 9 preferred hot dogs. Out of 750 students in the class, how many will probably prefer each selection? Should he plan on this exactly? *(Lesson 14-3)*

Curriculum Connection Projects

- **Recreation**   Write each class member's favorite summer activity on separate small pieces of paper. Toss into a bag. Calculate the probability of drawing each activity from the bag. Check your probabilities by acting it out.

- **Photography**   A photo developer offers two types of finishes: glossy or matte; and four sizes: 3×5, 4×6, 6×8, or 8×10. Find how many different ways you could have a photo developed.

Read More About It

Rudinstein, Gillian. *Beyond the Labyrinth.*
Weiss, Malcolm E. *Solomon Grundy, Born on Oneday: A Finite Arithmetic Puzzle.*
Hollander, Phyllis and Zander. *Amazing but True Sports Stories.*

A set of 20 cards has two cards numbered 10, three cards numbered 15, three cards numbered 20, and one card each of all the other numbers 1 through 13. Find the probability of each event.

1. $P(8)$

2. $P(15)$

3. $P(3 \text{ or } 10)$

4. $P(\text{a multiple of } 5)$

5. $P(\text{a multiple of } 3)$

6. Six students forgot to put their names on their bus passes before leaving the school office. What is the probability that at least one student receives his or her own pass from the school secretary? Solve by acting it out.

Mr. Levinson needs to reorder posters for his shop. Of his last order, he sold 15 sports posters, 40 rock music posters, 10 modern art, 5 French Impressionist art, and 30 popular actors. He plans to order 200 posters. How many posters should there be of each of the following?

7. rock music

8. sports

9. modern art

Suppose you throw 100 darts at each dart board below. How many would you expect to land in each shaded region?

10.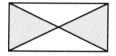

11.

12.

Use the tree diagram to answer the questions.

13. How many possible outcomes are there?

14. What is the probability of a hamburger and fries?

15. What is $P(\text{hot dog})$?

16. Find $P(\text{cole slaw})$.

hot dog — cole slaw ⟶ DC
— fries ⟶ DF
— baked potato ⟶ DB

hamburger — cole slaw ⟶ HC
— fries ⟶ HF
— baked potato ⟶ HB

Suppose you toss a coin and spin the spinner shown at the right. Find the probability of each of the following.

17. $P(\text{tails and blue})$

18. $P(\text{heads and yellow})$

19. $P(\text{no heads and red})$

20. $P(\text{tails and orange})$

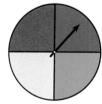

Bonus A bag holds 5 red beads, 8 black beads, and 10 white beads. Akika draws out a black bead and leaves it out of the bag. Then Allen reaches into the bag. What is the probability Allen will draw out a red bead now?

Directions: Choose the best answer. Write A, B, C, or D.

1. Which shows using the distributive property to multiply $3 \cdot 15.4$?

   A. $3(15) + 3(0.4)$

   B. $15.4(3)$

   C. $3(10) \cdot 3(5.4)$

   D. none of these

2. Julio measures 250 grams of salt for a science experiment. What is the weight in kilograms?

   A. 2,500 kg   B. 25 kg

   C. 0.25 kg   D. 0.025 kg

3. Of the students surveyed, $\frac{22}{100}$ are 12 years old and $\frac{18}{100}$ are 13 years old.

   What part of the students surveyed are either 12 or 13 years old?

   A. $\frac{1}{4}$ B. $\frac{1}{5}$

   C. $\frac{2}{5}$ D. $\frac{1}{25}$

4. $\frac{1}{2} + \frac{2}{3} + \frac{1}{4} =$

   A. $1\frac{1}{2}$ B. $\frac{4}{9}$

   C. $1\frac{5}{12}$ D. $\frac{4}{24}$

5. What is the area of a room that is $3\frac{1}{2}$ yards wide and $4\frac{2}{3}$ yards long?

   A. $16\frac{1}{3}$ square yards

   B. 14 square yards

   C. $12\frac{1}{3}$ square yards

   D. none of these

6. How many lines of symmetry does the figure shown at the right have?

   A. 0   B. 1

   C. 2   D. 3

7. Mrs. Porter won 8 of 14 games. Which choice is the ratio of her wins to games played?

   A. $\frac{8}{6}$ B. $\frac{8}{22}$

   C. $\frac{3}{7}$ D. $\frac{4}{7}$

8. A map has a scale of 1 cm = 20 km. Andie measures the map distance from Hertown to Grannytown as 7.2 cm. How far is Hertown from Grannytown?

   A. 720 km   B. 144 km

   C. 27.2 km   D. 20 km

9. What is the area of the parallelogram?

   A. 32 in²

   B. 40 in²

   C. 80 in²

   D. 160 in²

   4 in.   5 in.   8 in.

10. How can you find the surface area of a 3 inch by 4 inch by 5 inch rectangular prism?

    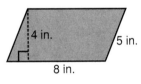

    3 in.   4 in.   5 in.

    A. Multiply $3 \times 4 \times 5$.

    B. Add $2(3 \times 4) + 2(4 \times 5) + 2(3 \times 5)$.

    C. Multiply $2(3 \times 4 \times 5)$.

    D. Add $(3 \times 4) + (3 \times 5) + (4 \times 5)$.

11. If $n = -3$, what is the value of $8 + n$?

    A. -5 B. -3

    C. 5 D. 11

12. $6 \times (-3) =$

    A. -18 B. -9

    C. 9 D. 18

13. What is the solution of the equation $5 + x = 3$?

    A. -8 B. -2

    C. 2 D. 8

14. A rectangular room has an area of 120 square feet. If the room is 8 feet wide, how long is it?

    A. 9.6 ft   B. 15 ft

    C. 96 ft   D. 150 ft

15. What numbers complete the function table?

n	$3n$
10	
0	
-5	

    A. 30, 0, and -30

    B. 30, 20, and -10

    C. 13, 3, and -2

    D. none of these

16. What is the solution of the inequality of $2x > 8$?

    A. all numbers < 8

    B. all numbers > 8

    C. all numbers > 6

    D. all numbers > 4

17. A set of cards is numbered 1 through 50. A card is drawn at random. What is the probability the card drawn is greater than 30?

    A. 30%   B. 40%

    C. 50%   D. 60%

Test-Taking Tip

Many problems can be solved without much calculating if you understand the basic mathematical concepts. Always look carefully at what is asked, and think of possible shortcuts for solving the problem.

You may be able to eliminate all or most of the answer choices by estimating. Also, look to see which answer choices are not reasonable for the information given in the problem.

18. What is the probability represented by the shaded area?

    A. 30%

    B. 50%

    C. 60%

    D. none of these

19. John and Maria play the violin. Kimi, Lisa, and Hector play the piano. How many different piano-violin duet combinations are possible?

    A. 4   B. 5

    C. 6   D. 10

20. In a random survey, it was found that 12 out of 40 students were the oldest or only child in their family. How many oldest or only children would you expect if there are 250 students in the school?

    A. 120

    B. 75

    C. 60

    D. 50

EXTENDED PROJECTS HANDBOOK

To The Student

One of the goals of *Mathematics: Applications and Connections* is to give you the opportunity to work with the mathematics that you will likely encounter outside the classroom. This includes the mathematics demanded by many of the courses you will take in high school and by most jobs as well as the mathematics that will be required of a good citizen of the United States.

Equally important, the authors want you to approach the mathematics you will encounter in your life with curiosity, enjoyment, and confidence.

Hopefully, the **Extended Projects Handbook** reflects these goals.

Three of the most important "big" ideas you are working with throughout *Mathematics: Applications and Connections* are the following. The **Extended Projects** includes these "big" ideas.

1. Proportional Reasoning

You probably have a great deal of experience with proportional reasoning. One example is seen in advertisements like "Three out of five dentists prefer sugarless gum." You can use this information to find out how many dentists out of 100 prefer sugarless gum.

You have made connections among the various applications of proportional reasoning and in this way have seen that the various items listed above are all part of a very big idea — proportions.

2. Multiple Representations

Mathematics provides you with many ways to present information and relationships. These include sketches, perspective drawings, tables, charts, graphs, physical models, verbalizing, and writing. You can use a computer to make graphs, data bases, spreadsheets, and simulations.

You have represented information and relationships in many different ways to completely describe various kinds of situations using mathematics.

3. Patterns and Generalizations

Mathematics has been called the science of patterns. You have experience recognizing and describing simple number and geometric patterns. You will be asked to make, test, and then use generalizations about given information in order to help you solve problems.

You may have used an algebraic expression to generalize a number pattern or the idea of similarity to make a scale drawing.

Looking Ahead

Your **Extended Projects Handbook** contains the following projects.

Project 1
Smoke 'Em Out

Project 2
Polling All Voters

Project 3
Design It Yourself

Project 4
Jump High

Project 1

Smoke 'Em Out

There is a lot of discussion today about the dangers of second-hand smoke. Are you aware of the dangers? Do you think others around you are aware? By designing a questionnaire and using it to survey others, you can find out just how aware people are of the dangers of second-hand smoke.

Getting It Together

Before you begin to put your questionnaire together, decide what information would be important to know about second-hand smoke. You will need to visit your library, local Health Sciences Research Center, or the American Cancer Society to find the latest facts and statistics about the dangers of second-hand smoke.

Creating Your Questionnaire

Work in small groups to put together your questionnaire. When writing your questionnaire, include these three main parts.

1. information about who you are interviewing
   For example:
   - male or female?
   - age?
   - do they smoke?
   - do they live with a smoker?

2. *yes* or *no* questions that relate to knowledge about smoking
   For example:
   - Do you know that smoking is hazardous to your health?
   - Do you know that smoking contributes to heart disease?
   - Do you know that quitting smoking now will add ten years to your life?

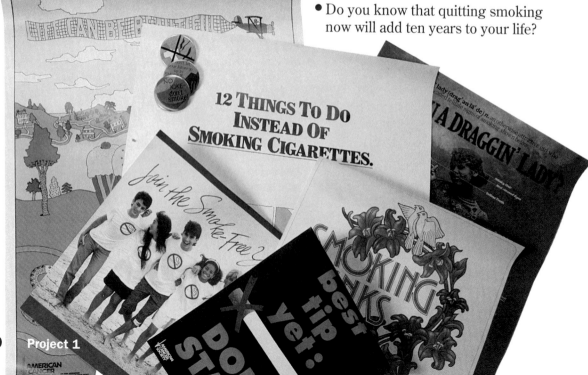

3. *true* or *false* statements about the hazards of smoking

   For example:
   - Smoking may cause low birth weight in newborn babies.
   - Smoking does not cause respiratory problems.
   - Second-hand smoke is hazardous to your health.

In your group, discuss the types of questions described above. Decide how many questions you want to include on your questionnaire. Revise the questions to fit your needs. Ask yourselves:

- Can the questionnaire be improved?
- Do you have enough information for an interesting analysis?
- Will the answers to your questions give you the information you need to determine how aware people are?

Using the Questionnaire

Decide how many people you want to survey. Make copies of the questionnaire. Then decide where you will conduct your survey. Set up a schedule showing when the survey will be conducted and who will conduct the survey.

Be sure to discuss with your group how you will organize the data gathered in each interview.

Ask yourselves these questions:

- What format will you use to summarize the responses to your questions—a chart, a statistical graph, a written report?
- How will you present the results—poster, video, verbal report, written summary?

Analyzing the Data

What does the information you collected tell you? Do your findings show that people are aware of the dangers of second-hand smoke? Is there an age group that is not aware?

EXTENSION: What Lies Ahead?

Kicking the smoking habit not only impacts a smoker's health but also the health of nonsmokers. Use your survey information to design an awareness campaign about the dangers of second-hand smoke.

Polling All Voters

Have you ever thought about how much data is gathered during a political election campaign? Campaign workers, news reporters, and poll takers gather information about issues and people's preferences toward certain candidates to predict winners and trends. In this project, you will interpret data and conduct your own random survey to predict who will win an election.

Interpreting Survey Results

You and a partner have been asked to write and deliver a special 15-minute newscast for your local television station concerning the upcoming election. Your station manager has informed you that a random survey was conducted by the station 7 weeks prior to a presidential election. The responses were tallied and recorded in the charts below.

Question: If you were to vote today, would you re-elect Josie Candidate for president?

BY SEX

	Yes	No	Undecided
Men	58%	24%	18%
Women	32%	60%	8%

BY EMPLOYMENT STATUS

	Working	Not Working
Men	84%	16%
Women	80%	20%

Question: Do you feel that your own lifestyle has improved under the administration of Josie Candidate?

BY SEX

	Yes	No	Undecided
Men	40%	60%	0%
Women	30%	45%	25%

AGE GROUPS OF ALL SURVEY PARTICIPANTS

	Men	Women
18 to 30	44%	56%
31 to 60	52%	48%
61 & over	51%	49%

Analyzing the Data

The data in the survey assumes a margin of error of +/- 3 percentage points.
- What does margin of error mean?
- What is the margin of error for these results?
- Does the margin of error affect Josie Candidate's chances of winning the election? Is it possible that Mr. Candidate, the incumbent, might lose? Explain.
- As a reporter, you must be aware that the survey results could reflect a biased opinion. Focus in on a specific characteristic of the people polled, such as their employment status. What is the ratio of men to women that are employed? How could the rate of employment affect the opinion toward a candidate by an unemployed worker? How could this be reflected in an objective way in your news story?

Getting It Together

Before you begin writing your news story, answer these questions.
1. Who do you think this information would benefit? Who is your audience?
2. How can the information best be used?
3. Is the information specific enough? Is the information important?

4. What does the survey information tell the voter about Josie Candidate?
5. Do you think this sample is a good representation of the entire voting population? Of your listeners? Explain.
6. Would you be influenced to vote for or against a candidate based on the survey information? Why or why not?

Next, write your news story. Decide who will be the anchorperson and who will be the editor. Decide how you will present the information so that it will catch the attention of your audience. Rehearse your story to make it fit within your time limit. Revise your story as needed.

EXTENSION: Class Action

Work with your partner to conduct your own random survey about an upcoming election in your school, community association, local government, or national government. Use your results to predict a winner.

Record the results of your survey in a chart. How many people were in your survey? How can you use these results to project how voters will vote in the actual election? Explain how you predicted a winner using what you know about ratio and proportion.

Project 3
Design It Yourself

Ergonomics, or human-factors engineering, is the science of designing areas in the home or workplace that are comfortable to the people who will spend time in those areas. The layout of the area should make the area efficient and convenient for the person using it.

You and your classmates will act as engineers in developing an ergonomically-designed media room for your school. Once designed, you will write a proposal to your principal outlining the details and costs of your room.

Getting Started

1. Discuss with your group members what electronic equipment will be included in your ideal media room.
2. Investigate the cost of each item in catalogs or newspaper ads. Make a list of your equipment expenses.
3. Find the dimensions of each piece of equipment. Will each item need its own space or will all of the equipment be placed in one area?
4. What type of furniture will be included in the room? Make a list of furniture needed and the cost. Find the dimensions of each piece of furniture.
5. Include costs for building materials and items such as electrical outlets and windows.

Planning Your Design

6. Once you have decided on everything you will need for the room, first estimate, and then decide on the dimensions of the size of room you will need to have space for all of the items. At this stage, pay close attention to detail. You don't want the room to be too large so that everything is too far away, or too small that only one or two people can sit and work comfortably.

7. Make a scale drawing of the media room on grid paper. Be sure to include such things as windows, doors, and electrical outlets. Use the scale drawing as a blueprint to make a model of the room. Draw each item in the room to scale. Cut out each item, place them in your model, and create your ergonomically-designed room.

Analyzing Your Design

8. What adjustments, if any, did you need to make in either the size of the room or in the number or size of the items you want to include in the room?

Finishing Touches

Now that you have your room design, how will you decorate it? Color plays a big part in setting the mood of the room. For example, blue is a cool color and rose is a warm color. Investigate the meaning of other colors and decide on a color scheme.

9. Discuss with your group whether you will need to paint, wallpaper, and/or carpet your room. Be sure to discuss the advantages and disadvantages of each decision.

10. Decide what kind and how much floor covering you will need. Investigate the cost per square yard and calculate the total cost, including tax.

11. Next, decide what kind of paint or wallpaper you will need. Find out how much you will need and calculate the cost, including tax.

12. If there are windows in your media center, decide what , if anything, you will need for them. Look through catalogs to get ideas for curtains, mini-blinds, or shades. Compare the cost of each. Calculate the cost of window coverings and add this to your expenses.

EXTENSION: Writing a Proposal

Once your media room drawing is complete, you are ready to write your proposal to sell your idea to your principal. Your report should include reasons for needing the room, costs of the room, and your layout of the room.

Project

4

Jump Up High

Advertising is the technique used to bring products, services, or opinions to the public's attention for the purpose of persuading the public to respond a certain way to the advertisement. Most advertisements want the public to purchase the product or service. Powerful advertising often involves the use of celebrities or sports figures to influence consumers. Consumers often begin to associate certain people with certain products such as Ray Charles with Pepsi and Mary Lou Retton with Revco. In recent years, advertising for athletic footwear has grown tremendously due to the growing number of NBA players with shoes named after them. The ads often imply that a person can run faster and jump higher with a certain kind of athletic shoe.

In this project, you will explore what is meant by truth in advertising.

Check Out the Ads

Conduct your own experiment to see if what the advertisers are saying is true. Begin by cutting out advertisements for certain athletic shoes, or taking notes on television commercials. Ask yourself,

- What does the ad imply?
- What are the advertisers trying to get you to believe?
- Is the person in the ad someone you recognize?
- Does the advertisement make you want to buy that brand of athletic shoe?

It Must Be the Shoes

Ask volunteers to bring in two or more different kinds of athletic shoes. Set up a 50-meter test course outside. Have volunteers run the course with each kind of shoe, resting between runs.

Record the date run, the weather conditions, the kind of shoe, and the time for each run. Make a chart of the results. Then graph the times of each runner. Analyze the data and discuss how different variables could have affected the results. Ask:

- Did more than one student wear the same kind of shoe?
- Were the running times about the same?
- Do you notice any similarities or differences in your results?
- Can you draw any conclusions?

Next, test the jumping ability of each shoe. You may want to use a basketball setting for this activity. Ask the same volunteers to use the same shoes used in the running experiment. Have each volunteer jump three times. Record the height of each jump for each brand of shoe. Find the average of each set of jumps for each shoe. Ask:

- How did you measure the jump?
- What role does estimation play in this experiment?
- Why is it necessary to find the average of each set of jumps?

Record your results in a graph. Analyze your results.

- Did one shoe perform better than another?
- Were your results different enough to make a generalization?
- Did the volunteer's ability make a difference in your experiment?

After completing both experiments, compare your findings to the advertisements.

- Were the advertisers' claims correct? Was there anything you found that was not true?

- Did you find anything that the advertisers could have mentioned about the shoe to help in its sale?

EXTENSION: Just Do It

Based on your findings, design an advertising campaign for one of the brands you tested. You can create a print advertisement for a newspaper or magazine, or you can develop storyboards for a television commercial. Keep the following objectives in mind when you are designing your advertisement.

- Do not make claims that can be proven false.
- Appeal to the proper audience for your product.
- Use language, color, and facts to appeal to your potential customer.

Extra Practice

Adding Whole Numbers

1. $\begin{array}{r} 40 \\ + \ 8 \\ \hline \end{array}$	2. $\begin{array}{r} 32 \\ + \ 5 \\ \hline \end{array}$	3. $\begin{array}{r} 63 \\ + \ 6 \\ \hline \end{array}$	4. $\begin{array}{r} 41 \\ + \ 8 \\ \hline \end{array}$	5. $\begin{array}{r} 53 \\ + \ 4 \\ \hline \end{array}$
6. $\begin{array}{r} 30 \\ + \ 60 \\ \hline \end{array}$	7. $\begin{array}{r} 20 \\ + \ 50 \\ \hline \end{array}$	8. $\begin{array}{r} 47 \\ + \ 20 \\ \hline \end{array}$	9. $\begin{array}{r} 85 \\ + \ 10 \\ \hline \end{array}$	10. $\begin{array}{r} 56 \\ + \ 33 \\ \hline \end{array}$
11. $\begin{array}{r} 600 \\ + \ 50 \\ \hline \end{array}$	12. $\begin{array}{r} 506 \\ + \ 30 \\ \hline \end{array}$	13. $\begin{array}{r} 225 \\ + \ 40 \\ \hline \end{array}$	14. $\begin{array}{r} 704 \\ + \ 35 \\ \hline \end{array}$	15. $\begin{array}{r} 628 \\ + \ 71 \\ \hline \end{array}$
16. $\begin{array}{r} 500 \\ + \ 200 \\ \hline \end{array}$	17. $\begin{array}{r} 320 \\ + \ 430 \\ \hline \end{array}$	18. $\begin{array}{r} 405 \\ + \ 503 \\ \hline \end{array}$	19. $\begin{array}{r} 342 \\ + \ 127 \\ \hline \end{array}$	20. $\begin{array}{r} 315 \\ + \ 583 \\ \hline \end{array}$
21. $\begin{array}{r} 27 \\ + \ 4 \\ \hline \end{array}$	22. $\begin{array}{r} 76 \\ + \ 9 \\ \hline \end{array}$	23. $\begin{array}{r} 59 \\ + \ 7 \\ \hline \end{array}$	24. $\begin{array}{r} 25 \\ + \ 68 \\ \hline \end{array}$	25. $\begin{array}{r} 24 \\ + \ 48 \\ \hline \end{array}$
26. $\begin{array}{r} 304 \\ + \ 57 \\ \hline \end{array}$	27. $\begin{array}{r} 845 \\ + \ 29 \\ \hline \end{array}$	28. $\begin{array}{r} 637 \\ + \ 36 \\ \hline \end{array}$	29. $\begin{array}{r} 304 \\ + \ 509 \\ \hline \end{array}$	30. $\begin{array}{r} 228 \\ + \ 534 \\ \hline \end{array}$
31. $\begin{array}{r} 83 \\ + \ 56 \\ \hline \end{array}$	32. $\begin{array}{r} 94 \\ + \ 72 \\ \hline \end{array}$	33. $\begin{array}{r} 62 \\ + \ 85 \\ \hline \end{array}$	34. $\begin{array}{r} 380 \\ + \ 270 \\ \hline \end{array}$	35. $\begin{array}{r} 761 \\ + \ 187 \\ \hline \end{array}$
36. $\begin{array}{r} 684 \\ + \ 67 \\ \hline \end{array}$	37. $\begin{array}{r} 495 \\ + \ 48 \\ \hline \end{array}$	38. $\begin{array}{r} 347 \\ + \ 59 \\ \hline \end{array}$	39. $\begin{array}{r} 676 \\ + \ 276 \\ \hline \end{array}$	40. $\begin{array}{r} 733 \\ + \ 197 \\ \hline \end{array}$
41. $\begin{array}{r} 24 \\ 76 \\ + \ 53 \\ \hline \end{array}$	42. $\begin{array}{r} 67 \\ 28 \\ + \ 44 \\ \hline \end{array}$	43. $\begin{array}{r} 55 \\ 89 \\ + \ 23 \\ \hline \end{array}$	44. $\begin{array}{r} 368 \\ 275 \\ + \ 256 \\ \hline \end{array}$	45. $\begin{array}{r} 275 \\ 384 \\ + \ 633 \\ \hline \end{array}$
46. $\begin{array}{r} 4{,}680 \\ +3{,}945 \\ \hline \end{array}$	47. $\begin{array}{r} 5{,}126 \\ +2{,}899 \\ \hline \end{array}$	48. $\begin{array}{r} 2{,}973 \\ +1{,}689 \\ \hline \end{array}$	49. $\begin{array}{r} 52{,}046 \\ +41{,}388 \\ \hline \end{array}$	50. $\begin{array}{r} 96{,}277 \\ +27{,}563 \\ \hline \end{array}$

Extra Practice

Subtracting Whole Numbers

1. $\begin{array}{r} 98 \\ -5 \\ \hline \end{array}$
2. $\begin{array}{r} 87 \\ -4 \\ \hline \end{array}$
3. $\begin{array}{r} 56 \\ -3 \\ \hline \end{array}$
4. $\begin{array}{r} 45 \\ -5 \\ \hline \end{array}$
5. $\begin{array}{r} 29 \\ -7 \\ \hline \end{array}$

6. $\begin{array}{r} 60 \\ -20 \\ \hline \end{array}$
7. $\begin{array}{r} 80 \\ -50 \\ \hline \end{array}$
8. $\begin{array}{r} 56 \\ -40 \\ \hline \end{array}$
9. $\begin{array}{r} 90 \\ -60 \\ \hline \end{array}$
10. $\begin{array}{r} 78 \\ -24 \\ \hline \end{array}$

11. $\begin{array}{r} 798 \\ -45 \\ \hline \end{array}$
12. $\begin{array}{r} 955 \\ -23 \\ \hline \end{array}$
13. $\begin{array}{r} 354 \\ -34 \\ \hline \end{array}$
14. $\begin{array}{r} 865 \\ -52 \\ \hline \end{array}$
15. $\begin{array}{r} 697 \\ -83 \\ \hline \end{array}$

16. $\begin{array}{r} 800 \\ -500 \\ \hline \end{array}$
17. $\begin{array}{r} 650 \\ -300 \\ \hline \end{array}$
18. $\begin{array}{r} 854 \\ -630 \\ \hline \end{array}$
19. $\begin{array}{r} 355 \\ -103 \\ \hline \end{array}$
20. $\begin{array}{r} 695 \\ -132 \\ \hline \end{array}$

21. $\begin{array}{r} 93 \\ -7 \\ \hline \end{array}$
22. $\begin{array}{r} 47 \\ -8 \\ \hline \end{array}$
23. $\begin{array}{r} 54 \\ -5 \\ \hline \end{array}$
24. $\begin{array}{r} 78 \\ -59 \\ \hline \end{array}$
25. $\begin{array}{r} 60 \\ -38 \\ \hline \end{array}$

26. $\begin{array}{r} 760 \\ -36 \\ \hline \end{array}$
27. $\begin{array}{r} 382 \\ -67 \\ \hline \end{array}$
28. $\begin{array}{r} 630 \\ -23 \\ \hline \end{array}$
29. $\begin{array}{r} 460 \\ -248 \\ \hline \end{array}$
30. $\begin{array}{r} 373 \\ -126 \\ \hline \end{array}$

31. $\begin{array}{r} 578 \\ -93 \\ \hline \end{array}$
32. $\begin{array}{r} 247 \\ -83 \\ \hline \end{array}$
33. $\begin{array}{r} 623 \\ -93 \\ \hline \end{array}$
34. $\begin{array}{r} 738 \\ -165 \\ \hline \end{array}$
35. $\begin{array}{r} 954 \\ -372 \\ \hline \end{array}$

36. $\begin{array}{r} 232 \\ -184 \\ \hline \end{array}$
37. $\begin{array}{r} 540 \\ -275 \\ \hline \end{array}$
38. $\begin{array}{r} 727 \\ -538 \\ \hline \end{array}$
39. $\begin{array}{r} 660 \\ -383 \\ \hline \end{array}$
40. $\begin{array}{r} 840 \\ -496 \\ \hline \end{array}$

41. $\begin{array}{r} 315 \\ -227 \\ \hline \end{array}$
42. $\begin{array}{r} 712 \\ -555 \\ \hline \end{array}$
43. $\begin{array}{r} 408 \\ -209 \\ \hline \end{array}$
44. $\begin{array}{r} 705 \\ -509 \\ \hline \end{array}$
45. $\begin{array}{r} 400 \\ -189 \\ \hline \end{array}$

46. $\begin{array}{r} 6{,}791 \\ -899 \\ \hline \end{array}$
47. $\begin{array}{r} 3{,}406 \\ -408 \\ \hline \end{array}$
48. $\begin{array}{r} 5{,}690 \\ -792 \\ \hline \end{array}$
49. $\begin{array}{r} 6{,}243 \\ -4{,}564 \\ \hline \end{array}$
50. $\begin{array}{r} 7{,}092 \\ -6{,}895 \\ \hline \end{array}$

51. $\begin{array}{r} 64{,}700 \\ -3{,}792 \\ \hline \end{array}$
52. $\begin{array}{r} 41{,}905 \\ -4{,}916 \\ \hline \end{array}$
53. $\begin{array}{r} 52{,}009 \\ -7{,}314 \\ \hline \end{array}$
54. $\begin{array}{r} 80{,}490 \\ -60{,}495 \\ \hline \end{array}$
55. $\begin{array}{r} 68{,}418 \\ -39{,}529 \\ \hline \end{array}$

Multiplying Whole Numbers

1. $\begin{array}{r} 40 \\ \times\ 5 \\ \hline \end{array}$
2. $\begin{array}{r} 30 \\ \times\ 6 \\ \hline \end{array}$
3. $\begin{array}{r} 20 \\ \times\ 8 \\ \hline \end{array}$
4. $\begin{array}{r} 60 \\ \times\ 4 \\ \hline \end{array}$
5. $\begin{array}{r} 50 \\ \times\ 7 \\ \hline \end{array}$

6. $\begin{array}{r} 23 \\ \times\ 3 \\ \hline \end{array}$
7. $\begin{array}{r} 44 \\ \times\ 2 \\ \hline \end{array}$
8. $\begin{array}{r} 81 \\ \times\ 6 \\ \hline \end{array}$
9. $\begin{array}{r} 72 \\ \times\ 3 \\ \hline \end{array}$
10. $\begin{array}{r} 61 \\ \times\ 7 \\ \hline \end{array}$

11. $\begin{array}{r} 721 \\ \times\ 4 \\ \hline \end{array}$
12. $\begin{array}{r} 513 \\ \times\ 3 \\ \hline \end{array}$
13. $\begin{array}{r} 234 \\ \times\ 2 \\ \hline \end{array}$
14. $\begin{array}{r} 634 \\ \times\ 2 \\ \hline \end{array}$
15. $\begin{array}{r} 831 \\ \times\ 3 \\ \hline \end{array}$

16. $\begin{array}{r} 46 \\ \times\ 5 \\ \hline \end{array}$
17. $\begin{array}{r} 53 \\ \times\ 7 \\ \hline \end{array}$
18. $\begin{array}{r} 82 \\ \times\ 6 \\ \hline \end{array}$
19. $\begin{array}{r} 27 \\ \times\ 4 \\ \hline \end{array}$
20. $\begin{array}{r} 68 \\ \times\ 8 \\ \hline \end{array}$

21. $\begin{array}{r} 704 \\ \times\ 6 \\ \hline \end{array}$
22. $\begin{array}{r} 409 \\ \times\ 5 \\ \hline \end{array}$
23. $\begin{array}{r} 806 \\ \times\ 8 \\ \hline \end{array}$
24. $\begin{array}{r} 307 \\ \times\ 9 \\ \hline \end{array}$
25. $\begin{array}{r} 208 \\ \times\ 7 \\ \hline \end{array}$

26. $\begin{array}{r} 28 \\ \times\ 10 \\ \hline \end{array}$
27. $\begin{array}{r} 86 \\ \times\ 10 \\ \hline \end{array}$
28. $\begin{array}{r} 51 \\ \times\ 10 \\ \hline \end{array}$
29. $\begin{array}{r} 247 \\ \times\ 10 \\ \hline \end{array}$
30. $\begin{array}{r} 4,328 \\ \times\ 10 \\ \hline \end{array}$

31. $\begin{array}{r} 52 \\ \times\ 20 \\ \hline \end{array}$
32. $\begin{array}{r} 37 \\ \times\ 50 \\ \hline \end{array}$
33. $\begin{array}{r} 26 \\ \times\ 40 \\ \hline \end{array}$
34. $\begin{array}{r} 175 \\ \times\ 30 \\ \hline \end{array}$
35. $\begin{array}{r} 1,469 \\ \times\ 80 \\ \hline \end{array}$

36. $\begin{array}{r} 75 \\ \times\ 19 \\ \hline \end{array}$
37. $\begin{array}{r} 54 \\ \times\ 27 \\ \hline \end{array}$
38. $\begin{array}{r} 45 \\ \times\ 81 \\ \hline \end{array}$
39. $\begin{array}{r} 52 \\ \times\ 64 \\ \hline \end{array}$
40. $\begin{array}{r} 80 \\ \times\ 76 \\ \hline \end{array}$

41. $\begin{array}{r} 89 \\ \times\ 45 \\ \hline \end{array}$
42. $\begin{array}{r} 64 \\ \times\ 37 \\ \hline \end{array}$
43. $\begin{array}{r} 78 \\ \times\ 62 \\ \hline \end{array}$
44. $\begin{array}{r} 56 \\ \times\ 82 \\ \hline \end{array}$
45. $\begin{array}{r} 83 \\ \times\ 59 \\ \hline \end{array}$

46. $\begin{array}{r} 414 \\ \times\ 22 \\ \hline \end{array}$
47. $\begin{array}{r} 321 \\ \times\ 43 \\ \hline \end{array}$
48. $\begin{array}{r} 522 \\ \times\ 34 \\ \hline \end{array}$
49. $\begin{array}{r} 613 \\ \times\ 32 \\ \hline \end{array}$
50. $\begin{array}{r} 202 \\ \times\ 24 \\ \hline \end{array}$

Dividing Whole Numbers

1. $3\overline{)72}$

2. $4\overline{)96}$

3. $2\overline{)78}$

4. $3\overline{)84}$

5. $3\overline{)57}$

6. $6\overline{)918}$

7. $8\overline{)976}$

8. $5\overline{)965}$

9. $7\overline{)903}$

10. $4\overline{)752}$

11. $12\overline{)60}$

12. $17\overline{)51}$

13. $25\overline{)75}$

14. $15\overline{)90}$

15. $24\overline{)72}$

16. $34\overline{)204}$

17. $18\overline{)126}$

18. $27\overline{)135}$

19. $46\overline{)184}$

20. $53\overline{)424}$

21. $24\overline{)240}$

22. $32\overline{)320}$

23. $25\overline{)500}$

24. $17\overline{)510}$

25. $15\overline{)600}$

26. $6\overline{)384}$

27. $23\overline{)483}$

28. $34\overline{)612}$

29. $14\overline{)546}$

30. $48\overline{)720}$

31. $31\overline{)1,953}$

32. $99\overline{)1,881}$

33. $47\overline{)1,927}$

34. $26\overline{)1,742}$

35. $19\overline{)1,045}$

36. $18\overline{)3,672}$

37. $23\overline{)9,223}$

38. $32\overline{)9,824}$

39. $15\overline{)7,545}$

40. $27\overline{)8,154}$

41. $8\overline{)91}$

42. $6\overline{)87}$

43. $5\overline{)99}$

44. $7\overline{)87}$

45. $6\overline{)80}$

46. $8\overline{)685}$

47. $7\overline{)538}$

48. $4\overline{)273}$

49. $6\overline{)580}$

50. $5\overline{)387}$

51. $12\overline{)75}$

52. $23\overline{)97}$

53. $18\overline{)99}$

54. $33\overline{)75}$

55. $27\overline{)56}$

56. $16\overline{)134}$

57. $37\overline{)299}$

58. $53\overline{)483}$

59. $29\overline{)210}$

60. $62\overline{)439}$

61. $49\overline{)29,670}$

62. $84\overline{)25,880}$

63. $32\overline{)38,693}$

64. $26\overline{)80,311}$

65. $46\overline{)92,330}$

66. $100\overline{)706}$

67. $100\overline{)842}$

68. $200\overline{)900}$

69. $500\overline{)705}$

70. $300\overline{)602}$

71. $400\overline{)1,632}$

72. $300\overline{)2,205}$

73. $600\overline{)8,407}$

74. $200\overline{)9,820}$

75. $500\overline{)7,513}$

1 Lesson 1-2

Round to the nearest ten.

1. 46   2. 738   3. 98   4. 163   5. 1,596

6. 19   7. 209   8. 2,960   9. 367   10. 156,999

Round to the nearest hundred.

11. 314   12. 658   13. 1,249   14. 9,860   15. 999

16. 713   17. 2,910   18. 620   19. 12,650   20. 960,715

Round to the nearest thousand.

21. 7,950   22. 9,499   23. 10,653   24. 18,305   25. 105,415

26. 6,710   27. 19,999   28. 6,001   29. 15,610   30. 410

Lesson 1-3

Estimate using front-end estimation.

1. $\begin{array}{r} 36 \\ + 42 \end{array}$ 2. $\begin{array}{r} 56 \\ + 89 \end{array}$ 3. $\begin{array}{r} 62 \\ - 24 \end{array}$ 4. $\begin{array}{r} 91 \\ - 75 \end{array}$ 5. $\begin{array}{r} 300 \\ + 238 \end{array}$

6. $\begin{array}{r} 539 \\ +251 \end{array}$ 7. $\begin{array}{r} 1,896 \\ +2,651 \end{array}$ 8. $\begin{array}{r} 3,145 \\ -2,076 \end{array}$ 9. $\begin{array}{r} 10,915 \\ - 7,423 \end{array}$ 10. $\begin{array}{r} 5,610 \\ +4,916 \end{array}$

11. $54 + 63$ 12. $5,406 + 788$ 13. $4,511 - 653$

14. $23,861 - 7,139$ 15. $25,565 + 19,536$ 16. $9,639 - 183$

17. $1,412 + 36$ 18. $1,563 - 286$ 19. $14 + 657$

Lesson 1-4

Estimate using patterns.

1. 3×60 2. 7×99 3. 5×121 4. 89×6

5. 25×18 6. 93×46 7. 63×5 8. 14×999

9. 376×12 10. 25×12 11. 610×7 12. $25 \times 1,999$

13. $605 \div 6$ 14. $79 \div 10$ 15. $682 \div 65$ 16. $799 \div 42$

17. $858 \div 88$ 18. $2,349 \div 23$ 19. $5,465 \div 6$ 20. $1,880 \div 82$

21. $726 \div 9$ 22. $8,444 \div 101$ 23. $8,133 \div 99$ 24. $419 \div 63$

Lesson 1-5   Estimate using compatible numbers.

1. $34 \div 6$
2. $55 \div 7$
3. $98 \div 11$
4. $26 \div 5$

5. $62 \div 9$
6. $41 \div 6$
7. $49 \div 4$
8. $13 \div 3$

9. $149 \div 5$
10. $157 \div 10$
11. $160 \div 11$
12. $156 \div 17$

13. $5{,}416 \div 7$
14. $7{,}189 \div 8$
15. $6{,}412 \div 8$
16. $39 \div 4$

17. $4{,}789 \div 12$
18. $6{,}113 \div 20$
19. $1{,}796 \div 9$
20. $14 \div 5$

Lesson 1-6   Find the value of each expression.

1. $14 - 5 + 7$
2. $12 + 10 - 5 - 6$
3. $50 - 6 + 12 + 4$

4. $12 - 2 \times 3$
5. $16 + 4 \times 5$
6. $5 + 3 \times 4 - 7$

7. $2 \times 3 + 9 \times 2$
8. $6 \times 8 + 4 \div 2$
9. $7 \times 6 - 14$

10. $8 + 12 \times 4 \div 8$
11. $13 - 6 \times 2 + 1$
12. $80 \div 10 \times 8$

13. $1 + 2 + 3 + 4$
14. $1 \times 2 \times 3 \times 4$
15. $6 + 6 \times 6$

16. $14 - 2 \times 7 + 0$
17. $156 - 6 \times 0$
18. $30 - 14 \times 2 + 8$

Lesson 1-7   Evaluate each expression if $m = 2$ and $n = 4$.

1. $m + m$
2. $n - m$
3. mn
4. $2m$
5. $2n$

6. $2n - 2m$
7. $m \times 0$
8. $64 \div n$
9. $12 - m$
10. $2mn$

Evaluate each expression if $a = 3$, $b = 4$, and $c = 12$.

11. $a + b$
12. $c - a$
13. $a + b + c$
14. $b - a$

15. $c - a \times b$
16. $a + 2 \times b$
17. $b + c \div 2$
18. ab

19. $a + 3b$
20. $a + c \div 6$
21. $25 + c \div b$
22. abc

23. $144 - abc$
24. $c \div a + 10$
25. $2b - a$
26. $2ab$

Lesson 1-9   Name the number that is a solution of the given equation.

1. $q - 7 = 7$; 7, 14, 28

2. $g - 3 = 10$; 7, 10, 13

3. $r - 3 = 4$; 4, 7, 12

4. $t + 3 = 21$; 7, 18, 24

5. $7 + a = 10$; 3, 13, 17

6. $14 + m = 24$; 7, 10, 34

Solve each equation mentally.

7. $b + 7 = 12$

8. $a + 3 = 15$

9. $s + 10 = 23$

10. $9 + n = 13$

11. $20 = 24 - n$

12. $4x = 36$

13. $2y = 10$

14. $15 = 5h$

15. $j \div 3 = 2$

16. $24 \div k = 6$

17. $b - 3 = 12$

18. $42 = 6n$

19. $c \div 10 = 8$

20. $6 = t \div 5$

21. $w \div 2 = 8$

22. $3y = 39$

2 Lesson 2-1   Write the place-value position for each digit in 742,015,908,609,390.

1. 8   2. 5   3. 1   4. 6   5. 2   6. 3   7. 4   8. 7

Write each number in words.

9. 45,680

10. 7,056

11. 480,000,050

12. 542

13. 8,562,002,562

14. 499,001

15. 2,060,500,653,002

16. 1,900

Write each number in standard form.

17. eighty-two thousand, six hundred one

18. five hundred six million, two

19. ten million, four thousand, fifty

20. eight trillion, seventy-two

21. fifty-six thousand, forty-nine

22. seven billion, three hundred

Lesson 2-4   Find the range for each set of data. Then find the best interval and the number to end a scale for a frequency table.

1. 2, 7, 13, 3, 4, 12, 9

2. 11, 15, 13, 18, 19, 20

3. 56, 85, 23, 78, 42, 63

4. 10, 25, 88, 64, 99, 37

5. 165, 167, 169, 164, 170, 166, 167, 165, 169

6. 132, 865, 465, 672, 318, 940, 573, 689

7. 1,450; 7,896; 5,638; 7,142; 4,287; 8,612

Lesson 2-5

Make a bar graph for each set of data.

1. a vertical bar graph

Favorite Subject	
Subject	*Frequency*
Math	4
Science	6
History	2
English	8
Phys. Ed.	12

2. a horizontal bar graph

Final Grades	
Subject	*Score*
Math	88
Science	82
History	92
English	94

Make a line graph for each set of data.

3.

Test	Score
1	62
2	75
3	81
4	83
5	78
6	92

4.

Day	Absences
Mon.	3
Tues	6
Wed.	2
Thur.	1
Fri.	8

Lesson 2-6

Use the following graph to solve each problem.

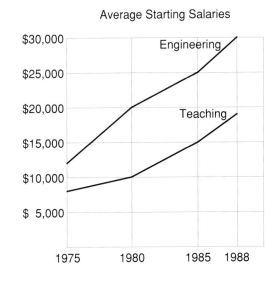

1. Give the expected starting salary in 1988 for:
   a. an engineer
   b. a teacher

2. How much more did an engineer make than a teacher in 1980?

3. Which of these two professions do you think will start with a higher salary in 1995?

4. How much more was the difference in salaries in 1988 than in 1975?

Lesson 2-7

Find the mean, median, and mode for each set of data.

1. 1, 5, 9, 1, 2, 5, 8, 2
3. 1, 2, 1, 2, 2, 1, 2
5. 256, 265, 247, 256
7. 46, 54, 66, 54, 46, 66

2. 2, 5, 8, 9, 7, 6, 3, 5
4. 12, 13, 15, 12, 12, 11
6. 957, 562, 462, 847, 721
8. 81, 82, 83, 84, 85, 86, 87

3 Lesson 3-1

Write each fraction as a decimal.

1. $\frac{4}{10}$
2. $\frac{66}{100}$
3. $\frac{73}{100}$
4. $\frac{5}{100}$
5. $\frac{9}{10}$
6. $\frac{94}{100}$

Write each expression as a decimal.

7. two hundredths

8. sixteen hundredths

9. four tenths

10. two and twenty-seven hundredths

11. nine and twelve hundredths

12. fifty-six and nine tenths

13. two thousand, four hundred seventy-five and six tenths

Lesson 3-2

Write each fraction as a decimal.

1. $\frac{85}{1,000}$
2. $\frac{875}{10,000}$
3. $\frac{1,264}{1,000}$
4. $\frac{527}{10,000}$
5. $\frac{9}{10,000}$
6. $\frac{72}{10,000}$
7. $\frac{24,956}{10,000}$
8. $\frac{999}{1,000}$

Write each expression as a decimal.

9. twenty-seven thousandths

10. one hundred ten-thousandths

11. two ten-thousandths

12. twenty and six hundred thousandths

13. four trillion, six hundred billion, two, and one ten-thousandth

14. twelve thousand, ninety-seven and sixty-two thousandths

Lesson 3-3

Use a centimeter ruler to measure each line segment.

1. _____
2. _____
3. _____
4. _____
5. _____
6. _____
7. ____
8. _____

Lesson 3-4

State the greater number in each group.

1. 0.112 or 0.121

2. 0.9985 or 0.998

3. 0.556 or 0.519

4. 1.19 or 11.9

5. 0.6, 6.0 or 0.06

6. 0.0009 or 0.001

Order each set of decimals from least to greatest.

7.	**8.**	**9.**	**10.**
415.65	0.0256	1.2356	50.12
451.65	0.2056	1.2355	5.012
451.66	0.0255	1.25	50.22
451.56	0.0009	1.2335	5.901
415.56	0.2560	1.2353	50.02

Order each set of decimals from greatest to least.

11.	**12.**	**13.**	**14.**
13.664	26.6987	1.00065	2.014
13.446	26.9687	1.00100	2.010
1.3666	26.9666	1.00165	22.00
1.6333	26.9688	1.00056	22.14

Lesson 3-5

Draw a number line to show how to round each decimal to the nearest tenth.

1. 5.64

2. 0.26

3. 10.39

4. 3.02

Round to the underlined place-value position.

5. 15.$\underline{2}$98

6. 0.002$\underline{6}$325

7. 758.9$\underline{9}$9

8. $\underline{4}$.25

9. 32.65$\underline{8}$3

10. $\underline{0}$.025

11. 1.004$\underline{9}$

12. 9.$\underline{2}$5

13. 67.4$\underline{9}$2

14. 25.1$\underline{9}$

15. 26.$\underline{9}$6

16. 4.00$\underline{0}$98

Lesson 3-6

Estimate using rounding.

1. $\begin{array}{r} 0.245 \\ +0.256 \end{array}$

2. $\begin{array}{r} 2.45698 \\ -1.26589 \end{array}$

3. $\begin{array}{r} 0.5962 \\ +1.2598 \end{array}$

4. $\begin{array}{r} 17.985 \\ -\ 9.001 \end{array}$

5. 0.256 + 0.6589

6. 1.2568 − 0.1569

7. 12.999 + 5.048

Estimate using front-end estimation.

8. $\begin{array}{r} 12.6589 \\ -6.3874 \end{array}$

9. $\begin{array}{r} 0.005698 \\ +0.015963 \end{array}$

10. $\begin{array}{r} 1.26589 \\ +0.76589 \end{array}$

11. $\begin{array}{r} 15.986325 \\ -12.765236 \end{array}$

12. 14.563 + 10.235

13. 0.225 − 0.169

14. 1.222 + 2.599

Lesson 3-8 Add or subtract.

1. $0.46 + 0.72$
2. $13.7 + 2.6$
3. $17.9 + 7.41$
4. $19.2 + 7.36$
5. $0.5113 + 0.62148$
6. $12.56 - 10.21$
7. $0.2154 - 0.1526$
8. $2.3125 + 1.02$
9. $1.025 - 0.58697$
10. $14.526 - 12.654$
11. $2.3568 + 5$
12. $20 - 5.98671$
13. $15.256 + 0.236$
14. $3.7 + 1.5 + 0.2$
15. $0.23 + 1.2 + 0.36$
16. $0.896352 - 0.25639$
17. $25.6 - 2.3$
18. $13.5 - 2.8456$
19. $1.265 + 1.654$
20. $24.56 - 24.32$
21. $0.256 - 0.255$

Lesson 3-9 Solve. Use the circle graph.

1. What subject do the most students prefer?
2. Which is the least popular subject?
3. Which subject received $\frac{1}{4}$ of the votes?
4. Which two subjects together received half of the votes?

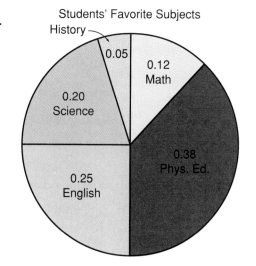

Students' Favorite Subjects

History — 0.05
0.12 Math
0.20 Science
0.38 Phys. Ed.
0.25 English

4 Lesson 4-1 Estimate each product.

1. 19×14
2. 31×72
3. 11×750
4. $14 \times \$1.99$
5. 6×105
6. 39×41
7. 18×72
8. 299×21
9. $9 \times 6{,}846$
10. $\$297 \times 3$
11. 2.4×3.2
12. 7.6×5.1
13. 256×41
14. 11×410
15. 705×816
16. 131×29
17. 96×4
18. $\$26.50 \times 4$
19. $2{,}005 \times 19$
20. 69×74
21. 41×39
22. 21.5×3.9
23. 6.7×4.1
24. 13.2×3.4

Lesson 4-2

Write each product using exponents.

1. $2 \cdot 2 \cdot 2 \cdot 2 \cdot 2$
2. $6 \cdot 6 \cdot 6 \cdot 7 \cdot 7$
3. $9 \cdot 9 \cdot 9 \cdot 9 \cdot 9 \cdot 9 \cdot 10$
4. $k \cdot k \cdot k \cdot l \cdot l \cdot l$
5. $14 \cdot 14 \cdot 6$
6. $3 \cdot 3 \cdot 3 \cdot 3 \cdot y \cdot y$

Write each power as a product.

7. 13^4
8. 9^6
9. $2^3 \cdot 3^2$
10. x^5
11. 169^3
12. $13,410^2$

Evaluate each expression.

13. 5^6
14. 17^3
15. 2^{12}
16. $3^5 \cdot 2^3$
17. $6^4 \cdot 3$
18. $2^2 \cdot 3^2 \cdot 4^2$
19. 176^2
20. $6 \cdot 4^3$

Lesson 4-3   Multiply.

1. $\begin{array}{r} 0.2 \\ \times 65 \\ \hline \end{array}$
2. $\begin{array}{r} 0.73 \\ \times 12 \\ \hline \end{array}$
3. $\begin{array}{r} 0.65 \\ \times 27 \\ \hline \end{array}$
4. $\begin{array}{r} 9.6 \\ \times 13 \\ \hline \end{array}$
5. $\begin{array}{r} 12.15 \\ \times 6 \\ \hline \end{array}$

6. 0.91×16
7. 7×0.265
8. 14×2.612
9. 55×0.003
10. 0.67×21
11. 19×0.111
12. 1.65×72
13. 9.6×101
14. 24×1.201
15. 610×7.5
16. 0.001×6
17. 510×0.0135
18. 9.2×17
19. 14.1235×4
20. 67×2.0356

Lesson 4-4   Find each product mentally. Use the distributive property.

1. 5×18
2. 9×27
3. 8×83
4. 7×21
5. 3×47
6. 2×106
7. 6×34
8. 56×3
9. 27×8
10. 5×3.4
11. 6×40.7
12. 1.5×30
13. 0.9×71
14. 30×2.08
15. 16×7
16. 33×4
17. 0.6×12
18. 80×7.9

Lesson 4-5　　Multiply.

1. 9.6×10.5
2. 3.2×0.1
3. 1.5×9.6

4. 5.42×0.21
5. 7.42×0.2
6. 0.001×0.02

7. 0.6×542
8. 6.7×5.8
9. 3.24×6.7

10. 9.8×4.62
11. 7.32×9.7
12. 0.008×0.007

13. 0.0001×56
14. 4.5×0.2
15. 9.6×2.3

16. 5.63×8.1
17. 10.35×9.1
18. 28.2×3.9

19. 102.13×1.221
20. 2.02×1.25
21. 8.37×89.6

Lesson 4-6　　Find the perimeter of each figure.

1.
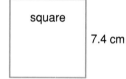

2.
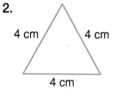

3.
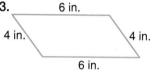

4.
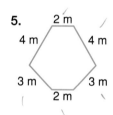

5.

6.
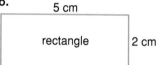

Lesson 4-7　　Find the area of each rectangle or square.

1. $19 \text{ cm} \times 3 \text{ cm}$
2. $5 \text{ in.} \times 3 \text{ in.}$

3. 9 ft square
4. 4.3 m square

5.

6.
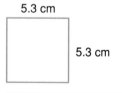

7.

8.
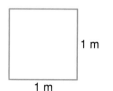

9.

10.

Lesson 4-9

Find the circumference of each circle. Use 3.14 for π. Round answers to the nearest tenth.

1.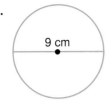
9 cm

2.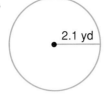
2.1 yd

3.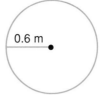
0.6 m

4.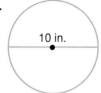
10 in.

5. $d = 5.6$ m

6. $r = 3.21$ yd

7. $r = 0.5$ in.

8. $d = 4$ m

9. $r = 16$ cm

10. $d = 9.1$ m

11. $r = 0.1$ yd

12. $d = 65.7$ m

13. $r = 1$ cm

5 Lesson 5-1

Find each quotient.

1. $6\overline{)1.26}$

2. $8\overline{)23.2}$

3. $6\overline{)89.22}$

4. $15\overline{)54.75}$

5. $13\overline{)128.31}$

6. $9\overline{)2.583}$

7. $47\overline{)11.28}$

8. $26\overline{)32.5}$

9. $8\overline{)7.2}$

10. $9\overline{)3.6}$

11. $7\overline{)1.75}$

12. $25\overline{)167.5}$

13. $37.1 \div 14$

14. $5.88 \div 4$

15. $3.7 \div 5$

16. $41.4 \div 18$

17. $9.87 \div 3$

18. $8.45 \div 25$

19. $7.8 \div 2$

20. $6.3 \div 3$

21. $10.2 \div 3$

Lesson 5-2

Round each quotient to the nearest tenth.

1. $26.5 \div 4$

2. $46.25 \div 8$

3. $19.38 \div 9$

4. $8.5 \div 2$

5. $90.88 \div 14$

6. $23.1 \div 4$

7. $19.5 \div 27$

8. $26.5 \div 19$

9. $46.23 \div 25$

10. $46.25 \div 25$

11. $4.26 \div 9$

12. $18.74 \div 19$

13. $17.9 \div 21$

14. $57.9 \div 14$

15. $21.555 \div 6$

16. $6.435 \div 7$

17. $15.23 \div 8$

18. $1.2356 \div 3$

19. $156.8 \div 25$

20. $19.563 \div 6$

21. $0.125 \div 1$

Lesson 5-3   Estimate first. Then find each quotient.

1. $0.5\overline{)18.45}$ 2. $0.08\overline{)5.2}$ 3. $2.6\overline{)0.65}$ 4. $1.3\overline{)12.831}$

5. $0.87\overline{)5.133}$ 6. $2.54\overline{)24.13}$ 7. $3.7\overline{)35.89}$ 8. $26\overline{)32.5}$

9. $5.88 \div 0.4$ 10. $3.7 \div 0.5$ 11. $6.72 \div 2.4$

12. $9.87 \div 0.3$ 13. $8.45 \div 2.5$ 14. $90.88 \div 14.2$

15. $33.6 \div 8.4$ 16. $25.389 \div 4.03$ 17. $85.92 \div 4.8$

18. $63.18 \div 16.2$ 19. $18.49 \div 4.3$ 20. $9.06 \div 0.003$

21. $1.02 \div 0.3$ 22. $6.4 \div 0.8$ 23. $7.2 \div 0.9$

Lesson 5-4   Find each quotient to the nearest hundredth.

1. $9\overline{)0.36}$ 2. $13\overline{)39.39}$ 3. $45\overline{)0.585}$ 4. $8\overline{)0.24}$

5. $6\overline{)0.312}$ 6. $7\overline{)0.161}$ 7. $7\overline{)7.21}$ 8. $3\overline{)9.18}$

9. $\$0.72 \div 12$ 10. $0.36 \div 9$ 11. $0.56 \div 14$

12. $32.2 \div 8$ 13. $0.3869 \div 5.3$ 14. $0.39 \div 7.8$

15. $0.0426 \div 7.1$ 16. $0.1185 \div 7.9$ 17. $\$0.84 \div 12$

18. $4.544 \div 64$ 19. $0.384 \div 9.6$ 20. $0.2262 \div 8.7$

Lesson 5-5   Write the metric unit that you would use to measure each of the following.

1. a bag of sugar

2. a pitcher of fruit punch

3. the mass of a dime

4. the amount of water in an ice cube

5. a vitamin

6. a pencil

7. the mass of a puppy

8. a bottle of perfume

9. a grain of sand

10. the mass of a car

Lesson 5-6   Complete.

1. 400 mm = _____ cm
2. 4 km = _____ m
3. 660 cm = _____ m
4. 0.3 km = _____ m
5. 30 mm = _____ cm
6. 84.5 m = _____ km
7. _____ m = 54 cm
8. 18 km = _____ cm
9. _____ mm = 45 cm
10. 4 kg = _____ g
11. 632 mg = _____ g
12. 4,497 g = _____ kg
13. _____ mg = 21 g
14. 61.2 mg = _____ g
15. 61 g = _____ mg
16. _____ mg = 0.51 kg
17. 0.63 kg = _____ g
18. _____ kg = 563 g

Lesson 5-7   Solve each equation.

1. $4x = 36$
2. $3y = 39$
3. $4z = 16$
4. $9w = 54$
5. $2m = 18$
6. $42 = 6n$
7. $72 = 8k$
8. $20r = 20$
9. $420 = 5s$
10. $325 = 25t$
11. $14 = 2p$
12. $18q = 36$
13. $40 = 10a$
14. $100 = 20b$
15. $416 = 4c$
16. $45 = 9d$
17. $2g = 0.6$
18. $3h = 0.12$
19. $5k = 0.35$
20. $12x = 144$
21. $9t = 81$
22. $0.16 = 2w$
23. $12r = 36$
24. $3.2 = 8y$

6 Lesson 6-1   State whether each number is divisible by 2, 3, 5, 6, 9, and 10.

1. 89
2. 64
3. 125
4. 156
5. 216
6. 330
7. 225
8. 524
9. 1,986
10. 2,052
11. 110
12. 315
13. 405
14. 918
15. 243
16. 735
17. 1,233
18. 5,103
19. 8,001
20. 9,270

Lesson 6-2   Find the prime factorization of each number.

1. 20
2. 65
3. 52
4. 30
5. 28

6. 155
7. 50
8. 96
9. 201
10. 1,250

11. 72
12. 2,648
13. 32
14. 86
15. 120

16. 576
17. 68
18. 240
19. 24
20. 70

21. 102
22. 121
23. 164
24. 225
25. 54

Lesson 6-4   Find the GCF for each set of numbers.

1. 8, 18
2. 6, 9
3. 4, 12
4. 18, 24

5. 8, 24
6. 17, 51
7. 65, 95
8. 42, 48

9. 64, 32
10. 72, 144
11. 54, 72
12. 60, 75

13. 16, 24
14. 12, 27
15. 25, 30
16. 48, 60

17. 16, 20, 36
18. 12, 18, 42
19. 30, 45, 15

20. 20, 30, 40
21. 81, 27, 108
22. 9, 18, 12

Lesson 6-5   Write each fraction in simplest form.

1. $\dfrac{12}{16}$
2. $\dfrac{28}{32}$
3. $\dfrac{75}{100}$
4. $\dfrac{8}{16}$
5. $\dfrac{6}{18}$

6. $\dfrac{27}{36}$
7. $\dfrac{16}{64}$
8. $\dfrac{9}{18}$
9. $\dfrac{50}{100}$
10. $\dfrac{24}{40}$

11. $\dfrac{32}{80}$
12. $\dfrac{8}{24}$
13. $\dfrac{20}{25}$
14. $\dfrac{4}{10}$
15. $\dfrac{3}{5}$

16. $\dfrac{14}{19}$
17. $\dfrac{9}{12}$
18. $\dfrac{6}{8}$
19. $\dfrac{15}{18}$
20. $\dfrac{9}{20}$

21. $\dfrac{8}{21}$
22. $\dfrac{10}{15}$
23. $\dfrac{9}{24}$
24. $\dfrac{6}{31}$
25. $\dfrac{18}{32}$

Lesson 6-6

Express each mixed number as an improper fraction.

1. $3\frac{1}{16}$ 2. $2\frac{3}{4}$ 3. $1\frac{3}{8}$ 4. $1\frac{5}{12}$ 5. $7\frac{3}{5}$

6. $6\frac{5}{8}$ 7. $3\frac{1}{3}$ 8. $1\frac{7}{9}$ 9. $2\frac{3}{16}$ 10. $1\frac{2}{3}$

11. $3\frac{3}{10}$ 12. $4\frac{3}{25}$ 13. $4\frac{2}{5}$ 14. $6\frac{1}{2}$ 15. $4\frac{5}{6}$

16. $1\frac{1}{100}$ 17. $2\frac{5}{8}$ 18. $3\frac{1}{6}$ 19. $4\frac{3}{5}$ 20. $1\frac{49}{50}$

Lesson 6-7

Draw a line segment for each of the following lengths.

1. $2\frac{1}{4}$ inches   2. $1\frac{3}{8}$ inches   3. $\frac{3}{4}$ inch

4. $1\frac{1}{2}$ inches   5. $3\frac{1}{8}$ inches   6. $2\frac{1}{4}$ inches

Find the length of each line segment to the nearest eighth inch.

7. _____   8. _____

9. _____   10. _____

Lesson 6-8

Find the LCM for each set of numbers.

1. $6, 30$ 2. $14, 42$ 3. $8, 10$ 4. $30, 10$

5. $7, 13$ 6. $28, 42$ 7. $25, 30$ 8. $21, 14$

9. $5, 15$ 10. $13, 39$ 11. $16, 24$ 12. $18, 20$

13. $8, 12$ 14. $12, 15$ 15. $9, 27$ 16. $5, 6$

17. $12, 18, 3$ 18. $12, 35, 10$ 19. $21, 14, 6$

20. $3, 6, 9$ 21. $6, 10, 15$ 22. $15, 75, 25$

Lesson 6-9

Find the LCD for each pair of fractions.

1. $\dfrac{2}{5}$ $\dfrac{4}{15}$

2. $\dfrac{2}{28}$ $\dfrac{15}{42}$

3. $\dfrac{8}{16}$ $\dfrac{14}{24}$

4. $\dfrac{9}{25}$ $\dfrac{21}{30}$

Replace each ● with $<$, $>$, or $=$ to make a true sentence.

5. $\dfrac{1}{2}$ ● $\dfrac{1}{3}$

6. $\dfrac{2}{3}$ ● $\dfrac{3}{4}$

7. $\dfrac{5}{9}$ ● $\dfrac{4}{5}$

8. $\dfrac{3}{6}$ ● $\dfrac{6}{12}$

9. $\dfrac{12}{23}$ ● $\dfrac{15}{19}$

10. $\dfrac{9}{27}$ ● $\dfrac{13}{39}$

11. $\dfrac{7}{8}$ ● $\dfrac{9}{13}$

12. $\dfrac{5}{9}$ ● $\dfrac{7}{8}$

13. $\dfrac{25}{100}$ ● $\dfrac{3}{8}$

14. $\dfrac{6}{7}$ ● $\dfrac{8}{15}$

15. $\dfrac{5}{9}$ ● $\dfrac{19}{23}$

16. $\dfrac{120}{567}$ ● $\dfrac{1}{2}$

17. $\dfrac{5}{7}$ ● $\dfrac{2}{3}$

18. $\dfrac{9}{36}$ ● $\dfrac{7}{28}$

19. $\dfrac{2}{5}$ ● $\dfrac{2}{6}$

20. $\dfrac{5}{9}$ ● $\dfrac{12}{13}$

Lesson 6-10

Express each decimal as a fraction or mixed number in simplest form.

1. 0.5

2. 0.8

3. 0.32

4. 0.875

5. 0.54

6. 0.38

7. 0.744

8. 0.101

9. 0.303

10. 0.486

11. 0.626

12. 0.448

13. 0.074

14. 0.008

15. 9.36

16. 10.18

17. 0.06

18. 0.75

19. 0.48

20. 0.9

21. 0.005

22. 0.4

23. 1.875

24. 5.08

Lesson 6-11

Express each fraction or mixed number as a decimal. Use bar notation to show a repeating decimal.

1. $\dfrac{3}{4}$

2. $\dfrac{2}{5}$

3. $\dfrac{7}{8}$

4. $\dfrac{1}{3}$

5. $\dfrac{4}{9}$

6. $\dfrac{3}{11}$

7. $\dfrac{17}{20}$

8. $\dfrac{5}{6}$

9. $\dfrac{3}{16}$

10. $\dfrac{8}{33}$

11. $\dfrac{7}{12}$

12. $\dfrac{14}{25}$

13. $\dfrac{7}{10}$

14. $\dfrac{5}{8}$

15. $\dfrac{11}{15}$

16. $\dfrac{8}{9}$

17. $\dfrac{15}{16}$

18. $\dfrac{1}{12}$

19. $\dfrac{7}{20}$

20. $\dfrac{5}{18}$

7 Lesson 7-1

Round each number to the nearest half.

1. $\dfrac{11}{12}$ 2. $\dfrac{5}{8}$ 3. $\dfrac{2}{5}$ 4. $\dfrac{1}{10}$ 5. $\dfrac{1}{6}$ 6. $\dfrac{2}{3}$

7. $\dfrac{9}{10}$ 8. $\dfrac{1}{8}$ 9. $\dfrac{4}{9}$ 10. $1\dfrac{1}{8}$ 11. $\dfrac{12}{11}$ 12. $2\dfrac{4}{5}$

13. $\dfrac{7}{9}$ 14. $7\dfrac{1}{10}$ 15. $10\dfrac{2}{3}$ 16. $\dfrac{1}{3}$ 17. $\dfrac{7}{16}$ 18. $\dfrac{5}{7}$

Find the length of each line segment to the nearest one-half inch.

19. _____   20. _____

21. _____   22. _____

Lesson 7-2

Estimate.

1. $14\dfrac{1}{10} - 6\dfrac{4}{5}$ 2. $8\dfrac{1}{3} + 2\dfrac{1}{6}$ 3. $4\dfrac{7}{8} + 7\dfrac{3}{4}$ 4. $11\dfrac{11}{12} - 5\dfrac{1}{4}$

5. $3\dfrac{2}{5} - 1\dfrac{1}{4}$ 6. $4\dfrac{2}{5} + \dfrac{5}{6}$ 7. $4\dfrac{7}{12} - 1\dfrac{3}{4}$ 8. $4\dfrac{2}{3} + 10\dfrac{3}{8}$

9. $7\dfrac{7}{15} - 3\dfrac{1}{12}$ 10. $2\dfrac{1}{20} + 1\dfrac{1}{3}$ 11. $18\dfrac{1}{4} - 12\dfrac{3}{5}$ 12. $12\dfrac{5}{9} + 8\dfrac{5}{8}$

13. $8\dfrac{2}{3} - 5\dfrac{1}{2}$ 14. $3\dfrac{1}{8} - 2\dfrac{3}{5}$ 15. $9\dfrac{2}{7} - \dfrac{1}{3}$ 16. $11\dfrac{7}{8} - \dfrac{5}{6}$

17. $2\dfrac{2}{5} + 2\dfrac{1}{4}$ 18. $8\dfrac{1}{2} - 7\dfrac{4}{5}$ 19. $8\dfrac{3}{4} + 4\dfrac{2}{3}$ 20. $1\dfrac{1}{8} + 7\dfrac{1}{10}$

Lesson 7-3

Add or subtract. Write each answer in simplest form.

1. $\dfrac{2}{5} + \dfrac{2}{5}$ 2. $\dfrac{5}{8} + \dfrac{3}{8}$ 3. $\dfrac{9}{11} - \dfrac{3}{11}$ 4. $\dfrac{3}{14} + \dfrac{5}{14}$

5. $\dfrac{7}{8} - \dfrac{3}{8}$ 6. $\dfrac{3}{4} - \dfrac{1}{4}$ 7. $\dfrac{15}{27} - \dfrac{7}{27}$ 8. $\dfrac{1}{36} + \dfrac{5}{36}$

9. $\dfrac{2}{9} - \dfrac{1}{9}$ 10. $\dfrac{7}{8} + \dfrac{5}{8}$ 11. $\dfrac{9}{16} - \dfrac{5}{16}$ 12. $\dfrac{6}{8} + \dfrac{4}{8}$

13. $\dfrac{1}{2} + \dfrac{1}{2}$ 14. $\dfrac{1}{3} - \dfrac{1}{3}$ 15. $\dfrac{8}{9} + \dfrac{7}{9}$ 16. $\dfrac{5}{6} - \dfrac{3}{6}$

17. $\dfrac{3}{9} + \dfrac{8}{9}$ 18. $\dfrac{8}{40} + \dfrac{12}{40}$ 19. $\dfrac{56}{90} - \dfrac{26}{90}$ 20. $\dfrac{2}{9} + \dfrac{8}{9}$

Lesson 7-4 Solve each equation.

1. $\frac{2}{9} + h = \frac{8}{9}$

2. $\frac{26}{85} + k = \frac{56}{85}$

3. $\frac{12}{16} - w = \frac{8}{16}$

4. $\frac{5}{8} = f + \frac{3}{8}$

5. $x + \frac{5}{7} = \frac{6}{7}$

6. $\frac{2}{6} = p - \frac{5}{6}$

7. $m - \frac{1}{17} = \frac{12}{17}$

8. $\frac{12}{16} = q + \frac{8}{16}$

9. $\frac{9}{57} - \frac{5}{57} = z$

10. $\frac{1}{2} + i = \frac{1}{2}$

11. $\frac{5}{12} = y - \frac{2}{12}$

12. $\frac{2}{7} + d = \frac{5}{7}$

13. $c - \frac{1}{11} = \frac{5}{11}$

14. $\frac{10}{24} = f + \frac{4}{24}$

15. $\frac{5}{7} + \frac{2}{7} = a$

Lesson 7-5 Add or subtract. Write each answer in simplest form.

1. $\frac{1}{3} + \frac{1}{2}$

2. $\frac{2}{9} + \frac{1}{3}$

3. $\frac{1}{2} + \frac{3}{4}$

4. $\frac{1}{4} + \frac{3}{12}$

5. $\frac{5}{9} - \frac{1}{3}$

6. $\frac{5}{8} - \frac{2}{5}$

7. $\frac{3}{4} - \frac{1}{2}$

8. $\frac{7}{8} - \frac{3}{16}$

9. $\frac{9}{16} + \frac{13}{24}$

10. $\frac{8}{15} + \frac{2}{3}$

11. $\frac{5}{14} + \frac{11}{28}$

12. $\frac{11}{12} + \frac{7}{8}$

13. $\frac{2}{3} - \frac{1}{6}$

14. $\frac{9}{16} - \frac{1}{2}$

15. $\frac{5}{8} - \frac{11}{20}$

16. $\frac{14}{15} - \frac{2}{9}$

17. $\frac{9}{20} + \frac{2}{15}$

18. $\frac{5}{6} + \frac{4}{5}$

19. $\frac{23}{25} - \frac{27}{50}$

20. $\frac{19}{25} - \frac{1}{2}$

Lesson 7-6 Add or subtract. Write each answer in simplest form.

1. $5\frac{1}{2} + 3\frac{1}{4}$

2. $2\frac{2}{3} + 4\frac{1}{9}$

3. $7\frac{4}{5} + 9\frac{3}{10}$

4. $9\frac{4}{7} - 3\frac{5}{14}$

5. $13\frac{1}{5} - 10$

6. $3\frac{3}{4} + 5\frac{5}{8}$

7. $3\frac{2}{5} + 7\frac{6}{15}$

8. $10\frac{2}{3} + 5\frac{6}{7}$

9. $15\frac{6}{9} - 13\frac{5}{12}$

10. $13\frac{7}{12} - 9\frac{1}{4}$

11. $5\frac{2}{3} - 3\frac{1}{2}$

12. $17\frac{2}{9} + 12\frac{1}{3}$

13. $6\frac{5}{12} + 12\frac{5}{8}$

14. $8\frac{3}{5} - 2\frac{1}{5}$

15. $23\frac{2}{3} - 4\frac{1}{2}$

Lesson 7-7   Subtract. Write each answer in simplest form.

1. $11\frac{2}{3} - 8\frac{11}{12}$

2. $3\frac{4}{7} - 1\frac{2}{3}$

3. $7\frac{1}{8} - 4\frac{1}{3}$

4. $18\frac{1}{9} - 12\frac{2}{5}$

5. $12\frac{3}{10} - 8\frac{3}{4}$

6. $43 - 5\frac{1}{5}$

7. $8\frac{1}{5} - 4\frac{1}{4}$

8. $14\frac{1}{6} - 3\frac{2}{3}$

9. $25\frac{4}{7} - 21$

10. $17\frac{3}{9} - 4\frac{3}{5}$

11. $18\frac{1}{9} - 1\frac{3}{7}$

12. $16\frac{1}{4} - 7\frac{1}{5}$

13. $18\frac{1}{5} - 6\frac{1}{4}$

14. $4 - 1\frac{2}{3}$

15. $26 - 4\frac{1}{9}$

Lesson 7-8   Complete.

1. 2 h 10 min = 1 h _____ min

2. 3 h 65 min = _____ h 5 min

3. 3 min 14 s = 2 min _____ s

4. 1 h 15 min 10 sec = _____ min 10 sec

Add or subtract.

5. 6 h 14 min
  -2 h   8 min

6. 5 h 35 min 25 sec
  $+$ 45 min 35 sec

7. 5 h   4 min 45 s
  -2 h 40 min   5 s

8. 15 h 16 min
  $-$ 8 h 35 min 16 s

9. 9 h 20 min 10 sec
  $+1$ h 39 min 55 sec

10. 2 h 40 min 20 sec
  $+3$ h   5 min 50 sec

Find the elapsed time.

11. 10:30 A.M. to 6:00 P.M.

12. 8:45 P.M. to 1:30 A.M.

8 Lesson 8-1   Round each fraction to 0, $\frac{1}{2}$, or 1.

1. $\frac{3}{4}$

2. $\frac{5}{8}$

3. $\frac{3}{25}$

4. $\frac{1}{20}$

5. $\frac{5}{11}$

6. $\frac{11}{18}$

7. $\frac{1}{3}$

8. $\frac{4}{9}$

9. $\frac{37}{40}$

10. $\frac{1}{15}$

11. $\frac{7}{20}$

12. $\frac{3}{16}$

13. $\frac{3}{50}$

14. $\frac{41}{45}$

Estimate.

15. $\frac{2}{3} \times \frac{4}{5}$

16. $\frac{1}{6} \times \frac{2}{5}$

17. $\frac{4}{9} \times \frac{3}{7}$

18. $\frac{5}{12} \times \frac{6}{11}$

19. $\frac{3}{8} \times \frac{8}{9}$

20. $\frac{3}{5} \times \frac{5}{12}$

21. $\frac{2}{5} \times \frac{5}{8}$

22. $5\frac{3}{7} \times \frac{4}{5}$

Lesson 8-2   Find each product. Write in simplest form.

1. $\dfrac{5}{6} \times \dfrac{15}{16}$

2. $\dfrac{6}{14} \times \dfrac{12}{18}$

3. $\dfrac{2}{3} \times \dfrac{3}{13}$

4. $\dfrac{4}{9} \times \dfrac{1}{6}$

5. $\dfrac{3}{4} \times \dfrac{5}{6}$

6. $\dfrac{9}{10} \times \dfrac{3}{4}$

7. $\dfrac{8}{9} \times \dfrac{2}{3}$

8. $\dfrac{6}{7} \times \dfrac{4}{5}$

9. $\dfrac{8}{11} \times \dfrac{11}{12}$

10. $\dfrac{5}{6} \times \dfrac{3}{5}$

11. $\dfrac{6}{7} \times \dfrac{7}{21}$

12. $\dfrac{8}{9} \times \dfrac{9}{10}$

13. $\dfrac{7}{11} \times \dfrac{12}{15}$

14. $\dfrac{7}{9} \times \dfrac{5}{7}$

15. $\dfrac{8}{13} \times \dfrac{2}{11}$

16. $\dfrac{4}{7} \times \dfrac{2}{9}$

17. $\dfrac{4}{9} \times \dfrac{24}{25}$

18. $\dfrac{1}{9} \times \dfrac{6}{13}$

19. $\dfrac{4}{7} \times 6$

20. $\dfrac{7}{10} \times 5$

Lesson 8-3   Express each mixed number as an improper fraction.

1. $2\dfrac{1}{3}$

2. $2\dfrac{1}{2}$

3. $2\dfrac{2}{3}$

4. $3\dfrac{1}{2}$

5. $1\dfrac{1}{4}$

6. $4\dfrac{3}{5}$

7. $5\dfrac{5}{6}$

8. $6\dfrac{8}{9}$

9. $2\dfrac{1}{9}$

10. $4\dfrac{7}{15}$

Find each product. Write in simplest form.

11. $3\dfrac{5}{8} \times 4\dfrac{1}{2}$

12. $\dfrac{4}{5} \times 2\dfrac{3}{4}$

13. $6\dfrac{1}{8} \times 5\dfrac{1}{7}$

14. $2\dfrac{2}{3} \times 2\dfrac{1}{4}$

15. $6\dfrac{2}{3} \times 7\dfrac{3}{5}$

16. $7\dfrac{1}{5} \times 2\dfrac{4}{7}$

17. $8\dfrac{3}{4} \times 2\dfrac{2}{5}$

18. $4\dfrac{1}{3} \times 2\dfrac{1}{7}$

Lesson 8-5   Find the next two numbers in each sequence.

1. 14, 21, 28, 35

2. 36, 42, 48, 54

3. 3, 9, 27, 81

4. 2, 6, 10, 14

5. 1,600, 800, 400, 200

6. 5, 10, 20, 40

7. 80, 70, 60, 50

8. 93, 193, 293, 393

9. 2, 8, 32, 128

10. 192, 96, 48, 24

11. 36, 34, 32, 30

12. 3, 7, 11, 15

Lesson 8-6   Name the reciprocal of each number.

1. $\frac{12}{13}$ 2. $\frac{7}{11}$ 3. 5 4. $\frac{1}{4}$ 5. $\frac{7}{9}$ 6. $\frac{9}{2}$ 7. $\frac{1}{5}$

Find each quotient. Write in simplest form.

8. $\frac{2}{3} \div \frac{1}{2}$ 9. $\frac{3}{5} \div \frac{2}{5}$ 10. $\frac{7}{10} \div \frac{3}{8}$ 11. $\frac{5}{9} \div \frac{2}{3}$

12. $4 \div \frac{2}{3}$ 13. $8 \div \frac{4}{5}$ 14. $9 \div \frac{5}{9}$ 15. $\frac{2}{7} \div 7$

16. $\frac{1}{14} \div 7$ 17. $\frac{2}{13} \div \frac{5}{26}$ 18. $\frac{4}{7} \div \frac{6}{7}$ 19. $\frac{7}{8} \div \frac{1}{3}$

20. $15 \div \frac{3}{5}$ 21. $\frac{9}{14} \div \frac{3}{4}$ 22. $\frac{8}{9} \div \frac{5}{6}$ 23. $\frac{4}{9} \div 36$

Lesson 8-7   Find each quotient. Write in simplest form.

1. $\frac{3}{5} \div 1\frac{2}{3}$ 2. $2\frac{1}{2} \div 1\frac{1}{4}$ 3. $4\frac{3}{5} \div 4\frac{1}{5}$ 4. $3\frac{2}{9} \div \frac{3}{4}$

5. $7 \div 4\frac{9}{10}$ 6. $5\frac{1}{9} \div 5$ 7. $1\frac{3}{7} \div 10$ 8. $1\frac{3}{4} \div 2\frac{3}{8}$

9. $3\frac{3}{5} \div \frac{4}{5}$ 10. $8\frac{2}{5} \div 4\frac{1}{2}$ 11. $6\frac{1}{3} \div 2\frac{1}{2}$ 12. $5\frac{1}{4} \div 2\frac{1}{3}$

13. $4\frac{1}{8} \div 3\frac{2}{3}$ 14. $6\frac{1}{4} \div 2\frac{1}{5}$ 15. $2\frac{5}{8} \div \frac{1}{2}$ 16. $4\frac{2}{5} \div 1\frac{1}{9}$

17. $1\frac{5}{6} \div 3\frac{2}{3}$ 18. $21 \div 5\frac{1}{4}$ 19. $18 \div 2\frac{1}{4}$ 20. $12 \div 3\frac{3}{5}$

Lesson 8-8   Complete.

1. 3 gal = _____ pt   2. 24 pt = _____ gal   3. 20 lb = _____ oz

4. 2 gal = _____ fl oz   5. 20 pt = _____ qt   6. 18 qt = _____ pt

7. 2,000 lb = _____ T   8. 3 T = _____ lb   9. 6 lb = _____ oz

10. 9 lb = _____ oz   11. 15 qt = _____ gal   12. 4 pt = _____ c

13. 4 gal = _____ qt   14. 4 qt = _____ fl oz   15. 12 pt = _____ c

16. 10 pt = _____ qt   17. 24 fl oz = _____ c   18. 1.5 pt = _____ c

19. $\frac{1}{4}$ lb = _____ oz   20. 5 T = _____ lb   21. 2 lb = _____ oz

9 **Lesson 9-1**

Use a protractor to find the measure of each angle.

1.
2.
3.
4.

Classify each angle as acute, right, or obtuse.

5.
6.
7.
8.

Lesson 9-2

Use a protractor to draw angles having the following measurements.

1. 165°   2. 20°   3. 90°   4. 41°

Tell whether the measure of each angle is greater than, less than, or about equal to the measurement given.

5. 110°
6. 80°
7. 80°
8. 55°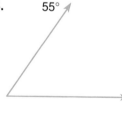

Lesson 9-3

Trace the drawing at the right. Then construct a line segment congruent to each segment or angle named.

1. *BE*   2. ∠*BAE*   3. *BC*

4. ∠*BED*   5. ∠*BCD*   6. *ED*

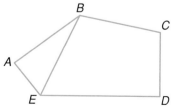

Lesson 9-4 Draw the angle or line segment with the given measurement. Then use a straight edge and compass to bisect each angle or line segment.

1. 3 in.

2. 5 cm

3. 110°

4. 48 mm

5. 70°

6. 33 mm

7. 25°

8. 150°

Lesson 9-5 Name each polygon.

1.

2.

3.

4.

Explain how each pair of figures is alike and how each pair is different.

5.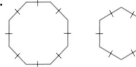

6.

7.

Lesson 9-7 Tell whether the dashed line is a line of symmetry. Write *yes* or *no*.

1.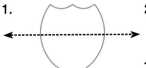

2.

3.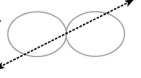

4.

Trace each figure. Draw all lines of symmetry.

5.

6.

7.

8.

Lesson 9-8   Tell whether each pair of polygons is congruent, similar or neither.

1.   **2.**   **3.**

4.   **5.**   **6.**

10 Lesson 10-1   Express each ratio as a fraction in simplest form.

1. 21 sugar cookies out of an assortment of 75 cookies.

2. 10 girls in a class of 25 students.

3. 34 non-smoking tables in a restaurant with 50 tables.

4. 7 striped ties out of 21 ties.

Express each ratio as a rate.

5. $2.00 for 5 cans of tomato soup

6. $200.00 for 40 hours of work

7. 540 parts produced in 18 hours

Lesson 10-2   Use cross products to determine whether each pair of ratios forms a proportion.

1. $\dfrac{3}{10}, \dfrac{7}{25}$ **2.** $\dfrac{5}{12}, \dfrac{3}{8}$ **3.** $\dfrac{12}{16}, \dfrac{9}{12}$ **4.** $\dfrac{5}{4}, \dfrac{125}{100}$ **5.** $\dfrac{4}{5}, \dfrac{80}{100}$

Solve each proportion.

6. $\dfrac{15}{21} = \dfrac{5}{b}$ **7.** $\dfrac{22}{25} = \dfrac{n}{100}$ **8.** $\dfrac{24}{48} = \dfrac{h}{50}$ **9.** $\dfrac{9}{27} = \dfrac{y}{42}$ **10.** $\dfrac{4}{7} = \dfrac{16}{x}$

11. $\dfrac{4}{6} = \dfrac{a}{9}$ **12.** $\dfrac{6}{14} = \dfrac{21}{m}$ **13.** $\dfrac{3}{7} = \dfrac{21}{d}$ **14.** $\dfrac{4}{10} = \dfrac{18}{e}$ **15.** $\dfrac{9}{10} = \dfrac{27}{f}$

Lesson 10-4 The map at the right has a scale of $\frac{1}{4}$ in. = 5 km. Use a ruler to measure each map distance to the nearest $\frac{1}{4}$ inch. Then compute the actual distances.

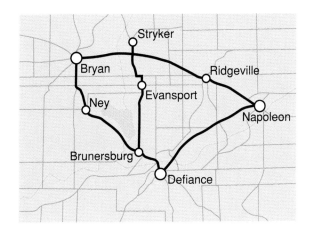

1. Bryan to Napoleon

2. Stryker to Evansport

3. Ney to Bryan

4. Defiance to Napoleon

5. Ridgeville Corners to Bryan

6. Brunersburg to Ney

Lesson 10-5 Express each fraction as a percent.

1. $\frac{77}{100}$ 2. $\frac{3}{4}$ 3. $\frac{17}{20}$ 4. $\frac{3}{25}$ 5. $\frac{3}{10}$

6. $\frac{27}{50}$ 7. $\frac{2}{5}$ 8. $\frac{3}{50}$ 9. $\frac{9}{20}$ 10. $\frac{8}{5}$

11. $\frac{1}{4}$ 12. $\frac{1}{5}$ 13. $\frac{19}{20}$ 14. $\frac{7}{10}$ 15. $\frac{11}{25}$

Express each percent as a fraction in simplest form.

16. 13%   17. 25%   18. 8%   19. 105%   20. 60%

21. 70%   22. 80%   23. 45%   24. 20%   25. 14%

26. 75%   27. 120%   28. 5%   29. 2%   30. 450%

Lesson 10-6 Express each decimal as a percent.

1. 0.02   2. 0.2   3. 0.002   4. 1.02   5. 0.66

6. 0.11   7. 0.354   8. 0.31   9. 0.09   10. 5.2

11. 2.22   12. 0.008   13. 0.275   14. 0.3   15. 6.0

Express each percent as a decimal.

16. 5%   17. 22%   18. 50%   19. 420%   20. 75%

21. 1%   22. 100%   23. 3.7%   24. 0.9%   25. 9%

26. 90%   27. 900%   28. 78%   29. 62.5%   30. 15%

Lesson 10-7

Estimate each percent.

1. 11% of 48

2. 1.9% of 50

3. 29% of 500

4. 41% of 50

5. 32% of 300

6. 411% of 50

7. 149% of 60

8. 4.1% of 50

9. 62% of 200

10. 58% of 100

11. 52% of 400

12. 68% of 30

13. 9% of 25

14. 48% of 1000

15. 98% of 725

16. 1.1% of 700

17. 24% of 80

18. 32% of 92

19. 9% of 32

20. 65% of 89

21. 48% of $23.98

22. 33% of 15

23. 39% of 50

24. 598% of 3

Lesson 10-8

Find the percent of each number.

1. 38% of 150

2. 20% of 75

3. 0.2% of 500

4. 25% of 70

5. 10% of 90

6. 16% of 30

7. 39% of 40

8. 250% of 100

9. 6% of 86

10. 12.5% of 160

11. 9% of 29

12. 3% of 46

13. $66\frac{2}{3}$% of 60

14. 89% of 47

15. 435% of 30

16. 25% of 48

17. 5% of 420

18. 55% of 134

19. 28% of 4

20. 14% of 40

21. 14% of 14

22. 90% of 140

23. 40% of 45

24. 0.5% of 200

11 Lesson 11-1

Find the area of each parallelogram.

1.
3.7 ft
7.5 ft

2.
34 cm
40 cm

3.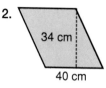
4 m
5 m

4.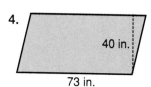
40 in.
73 in.

5.
23 in.
50 in.

6.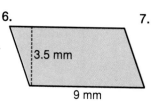
3.5 mm
9 mm

7.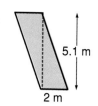
5.1 m
2 m

8.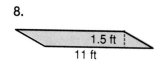
1.5 ft
11 ft

Lesson 11-2   Find the area of each triangle.

1. base, 6 ft
   height, 3 ft

2. base, 4.2 in.
   height, 6.8 in.

3. base, 9.1 m
   height, 7.2 m

4. base, 13.2 cm
   height, 16.2 cm

5.

6.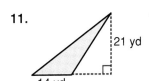

7.

8.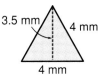

9.

10.

11.

12.

Lesson 11-3   Find the area of each circle. Use 3.14 for π. Round to the nearest tenth.

1. radius, 4 m

2. diameter, 6 in.

3. radius, 16 m

4. diameter, 11 in.

5. radius, 9 cm

6. diameter, 24 mm

7.

8.

9.

10.

11.

12.

13.

14.

Lesson 11-5   Find the surface area of each rectangular prism.

1.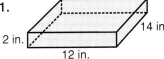

2.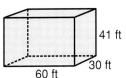

3.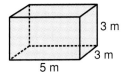

4.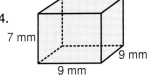

5.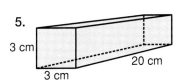

6.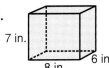

7. length = 4 mm
   width = 12 mm
   height = 1.5 mm

8. length = 16 cm
   width = 20 cm
   height = 20.4 cm

9. length = 8.5 m
   width = 2.1 m
   height = 7.6 m

Lesson 11-6 Find the volume of each rectangular prism.

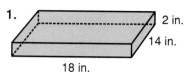

1. 2 in. 14 in. 18 in.

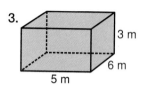

2. 41 ft 38 ft 96 ft

3. 3 m 6 m 5 m

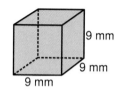

4. 9 mm 9 mm 9 mm

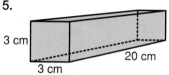

5. 3 cm 3 cm 20 cm

6. 7 in. 9 in. 4 in.

7. length = 8 in.
   width = 5 in.
   height = 2 in.

8. length = 10 cm
   width = 2 cm
   height = 8 cm

9. length = 20 ft
   width = 5 ft
   height = 6 ft

12 Lesson 12-1 Write the integer represented by each letter on the number line.

1. Q 2. T 3. N 4. M 5. V 6. S 7. W

Write an integer to describe each situation.

8. a loss of 15 dollars

9. 9 degrees below zero

Write the opposite of each integer.

10. 7   11. –3   12. 11   13. –9   14. –13   15. 101   16. 0

Lesson 12-2 Replace each ● with <, >, or = .

1. –5 ● –55
5. –898 ● –99
9. –6 ● –7

2. 4 ● –66
6. 0 ● 44
10. 90 ● 101

3. –777 ● –77
7. 56 ● –1
11. 4 ● –2,000

4. –75 ● –75
8. –82 ● –9
12. –3 ● 0

Order each set of integers from least to greatest.

13. 8, 0, –808, –8, –88, 88, –888
14. 0, 3, –21, 9, –89, 8, –65, –56
15. 70, –9, 67, –78, 0, 45, –36, –19
16. 0, –90, –56, –29, –92, –87, –35
17. –239, –999, 458, –29, –77, 200, –818

Lesson 12-3

Find each sum. Use or draw counters if necessary.

1. $-4 + (-7)$
2. $-81 + 0$
3. $17 + (-23)$
4. $-90 + 72$
5. $4 + (-6)$
6. $-12 + 9$
7. $-12 + (-10)$
8. $5 + (-15)$
9. $17 + 9$
10. $18 + (-18)$
11. $-4 + (-4)$
12. $0 + (-99)$
13. $-12 + (-9)$
14. $-8 + 7$
15. $3 + (-66)$
16. $-9 + 16$
17. $-55 + (-33)$
18. $-56 + 56$
19. $-34 + (-34)$
20. $-11 + 6$
21. $-30 + 6$
22. $-5 + (-9)$
23. $18 + (-20)$
24. $-40 + (-40)$
25. $24 + (-4)$
26. $-3 + (-11)$
27. $17 + 9$
28. $-11 + 60$
29. $-6 + (-12)$
30. $80 + 80$
31. $-29 + 20$
32. $4 + (-5)$

Lesson 12-4

Find each difference. Use or draw counters if necessary.

1. $7 - (-4)$
2. $-4 - (-9)$
3. $13 - (-3)$
4. $12 - (-15)$
5. $-9 - 5$
6. $-11 - (-18)$
7. $-4 - (-7)$
8. $-6 - (-6)$
9. $-6 - 6$
10. $17 - 9$
11. $-12 - (-9)$
12. $0 - (-4)$
13. $-78 - 0$
14. $-12 - (-10)$
15. $-32 - (-11)$
16. $33 - (-5)$
17. $15 - (-11)$
18. $-50 - (-60)$
19. $90 - (-10)$
20. $10 - 90$
21. $-58 - 10$
22. $-1 - 40$
23. $0 - (-72)$
24. $8 - 53$
25. $-34 - (-56)$
26. $29 - 29$
27. $-77 - (-77)$
28. $67 - 95$
29. $86 - (-50)$
30. $5 - 8$
31. $11 - 60$
32. $-88 - (-8)$

Lesson 12-5

Find each product. Use or draw counters if necessary.

1. $3 \times (-5)$
2. -5×10
3. $-8 \times (-4)$
4. $6 \times (-3)$
5. -3×27
6. $-1 \times (-14)$
7. $18 \times (-20)$
8. $-5 \times (-7)$
9. $9 \times (-9)$
10. -9×4
11. $-4 \times (-5)$
12. $5 \times (-20)$
13. $-8(18)$
14. $-9(-16)$
15. $27(-3)$
16. $42(3)$
17. $-67(0)$
18. $-56(-1)$
19. $15(-5)$
20. $-20(-30)$
21. $8(-40)$
22. $-12(4)$
23. $-40(-40)$
24. $11(9)$
25. $-6(-12)$
26. $7 \times (-4)$
27. -5×-9
28. -60×11
29. $14(-2)$
30. $45(-4)$
31. $-3(-11)$
32. $-13(3)$

Lesson 12-7

Find each quotient. Use counters or patterns if necessary.

1. $12 \div (-6)$
2. $-77 \div (-11)$
3. $-40 \div 4$
4. $67 \div (-67)$
5. $0 \div (-45)$
6. $45 \div (-9)$
7. $45 \div (-15)$
8. $-60 \div 20$
9. $-63 \div (-7)$
10. $70 \div (-7)$
11. $-40 \div (-8)$
12. $72 \div (-9)$
13. $-18 \div 6$
14. $29 \div (-1)$
15. $-30 \div 6$
16. $-54 \div (-9)$
17. $28 \div (-7)$
18. $-24 \div 8$
19. $24 \div (-4)$
20. $-42 \div 7$
21. $39 \div 13$
22. $-48 \div (-16)$
23. $-19 \div (-1)$
24. $81 \div (-9)$
25. $-125 \div (-5)$
26. $525 \div (-35)$
27. $-36 \div 9$
28. $-42 \div 21$
29. $-900 \div 18$
30. $-32 \div 4$
31. $-27 \div (-9)$
32. $-18 \div 18$

Lesson 12-8

Name the ordered pair for each point.

1. M
2. A
3. D
4. E
5. P
6. Q
7. B
8. C
9. F
10. G
11. N
12. R
13. K
14. H

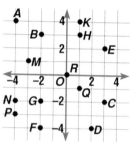

On graph paper, draw a coordinate grid. Then graph and label each point.

15. $S(4, -1)$
16. $T(-3, -2)$
17. $W(2, 1)$
18. $Y(-5, 3)$
19. $Z(-1, -3)$
20. $U(3, -3)$
21. $V(1, 2)$
22. $X(-1, 4)$

13 Lesson 13-1

Solve each equation. Use cups and counters if necessary.

1. $x + 4 = 14$
2. $y - (-7) = 2$
3. $b + (-10) = 0$
4. $a - 10 = -22$
5. $-2 + w = -5$
6. $g - (-1) = 9$
7. $k + (-3) = -5$
8. $c - 8 = 5$
9. $-4 + h = 6$
10. $-7 + d = -3$
11. $z - (-2) = 7$
12. $m + 11 = 9$
13. $n - 1 = -87$
14. $f + (-9) = -19$
15. $p + 66 = 22$
16. $-34 + t = 41$
17. $e + 56 = -24$
18. $j - 15 = -22$
19. $x - 12 = 45$
20. $-29 + a = -54$
21. $17 + m = -33$
22. $b + (-44) = -34$
23. $y - 65 = -79$
24. $w + (-39) = 55$

Lesson 13-2

Solve each equation. Use cups and counters if necessary.

1. $5x = 30$
2. $18w = 2$
3. $\frac{1}{2}a = 7$
4. $2d = -28$
5. $\frac{1}{4}c = -3$
6. $11n = 77$
7. $\frac{1}{3}z = 15$
8. $9y = -63$
9. $6m = -54$
10. $5f = -75$
11. $20p = 5$
12. $\frac{1}{4}x = 16$
13. $4t = -24$
14. $7b = 21$
15. $19h = 0$
16. $22d = -66$
17. $\frac{1}{3}y = 11$
18. $3m = -78$
19. $8x = -2$
20. $9c = -72$
21. $\frac{1}{2}p = 35$
22. $\frac{1}{5}k = 20$
23. $33y = 99$
24. $6z = -5$

Lesson 13-4

Copy and complete each function table.

1.

input (n)	output ($n - 4$)
5	
2	
−1	

2.

input (n)	output ($3n$)
1	
0	
−2	

Find the rule for each function table.

3.

n	
−1	4
0	5
3	8

4.

n	
−6	3
0	0
8	−4

Lesson 13-5

Copy and complete each function table. Then graph the function.

1.

input (x)	output ($x + 1$)
2	
0	
−3	

2.

input (x)	output ($2x$)
2	
0	
−3	

3.

input (x)	output ($x - 3$)
4	
0	
−1	

4.

input (x)	output ($\frac{x}{5}$)
10	
0	
−5	

5.

input (x)	output ($-3x$)
2	
−1	
−2	

6.

input (x)	output ($2x - 3$)
2	
0	
−1	

Lesson 13-6   Solve each inequality. Graph the solution on a number line.

1. $x - 8 > -5$

2. $y + 13 \leq 11$

3. $\frac{a}{3} < -12$

4. $g - 13 \geq 25$

5. $\frac{k}{4} < -7$

6. $6t \geq -54$

7. $w + 121 < 125$

8. $c + 43 \geq 33$

9. $e + 56 > -24$

10. $3t > 18$

11. $-7 + d < -3$

12. $z + 4 < -14$

13. $x - 12 \geq 45$

14. $7b < -21$

15. $5x > 30$

16. $k + (-3) \leq -5$

17. $c - 8 > -5$

18. $4h \leq -24$

19. $9c > -72$

20. $k + 4 \leq -7$

21. $a - 10 > -12$

22. $5 + y > 2$

23. $\frac{z}{2} > -3$

24. $11x < 33$

14 Lesson 14-1   A set of 30 tickets is placed in a grab bag. There are 6 baseball tickets, 4 hockey tickets, 4 basketball tickets, 2 football tickets, 3 symphony tickets, 2 opera tickets, 4 ballet tickets, and 5 theater tickets. One ticket is to be drawn. Find the probability of each event.

1. P (basketball)

2. P (sports event)

3. P (opera or ballet)

4. P (soccer)

5. P (not symphony)

6. P (theater)

7. P (basketball or hockey)

8. P (not a sports event)

9. P (not opera)

10. P (baseball)

11. P (football)

12. P (not soccer)

13. P (opera)

14. P (not theater)

15. P (symphony)

16. P (soccer or football)

17. P (opera or theater)

18. P (hockey)

Lesson 14-3   A random sample of customers at the cookie store showed the results listed below at the left. Use the results to answer each question.

Drink Preferences	
Coffee	5
Tea	1
Milk	3
Cola	9
Diet Cola	7
Lemon-lime	3
Orange	2
No Drink	10

1. What is the size of the sample?

2. What is the probability that a customer will come to the store and order a cola?

3. What is the probability that a customer will come to the store and order coffee or tea?

4. In a day when 1,200 customers come to the store, how many of each drink will probably be sold?

   How many customers will buy no drink?

5. Which is the most popular drink?

Lesson 14-5

For each situation, draw a tree diagram to show all the outcomes.

1. Jane can choose from 3 sweaters. All 3 sweaters can be worn with 4 skirts.

2. Plane, car, and train can be used to get from Columbus to New York. From New York to London, plane or ship can be used.

3. The local sandwich shop serves chicken salad, tuna salad, turkey and cheese, or ham and cheese. With the sandwiches, coffee, milk, juice, or canned pop are served.

4. Five banquet halls and 3 disc jockeys are available for the party.

5. Two interesting TV shows are scheduled from 8:00 until 9:00; four are scheduled from 9:00 until 10:00.

Lesson 14-6

Use the situation above in Exercise 3, Lesson 14-5, to find each probability. Each choice is equally likely.

1. What is the probability of a customer ordering chicken salad and coffee?

2. What is the probability of a customer ordering turkey and cheese?

3. Find P (ham and cheese, not milk).

4. Find P (not tuna salad, juice).

A bag contains 3 quarters and 5 dimes. Another bag contains twelve pennies and 8 nickels. One coin is drawn from each bag. Find each probability.

5. P (dime, nickel)   6. P (quarter)   7. P (quarter, penny)

8. P (not a nickel)   9. P (dime, penny)   10. P (not a dime, nickel)

Glossary

acute angle (311)   Any angle that measures between 0° and 90°.

algebra (27)   A mathematical language that uses letters along with numbers. The letters stand for numbers that are unknown. $10x - 3 = 17$ is an example of an algebra problem.

algebraic expression (27)   A combination of variables, numbers, and at least one operation.

angle (310)   Two rays with a common endpoint form an angle.

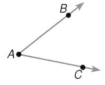

area (142)   The number of square units needed to cover a surface.

average (64)   The sum of two or more quantities divided by the number of quantities; the mean.

bar graph (47)   A graph that is used to compare quantities. The height or length of each bar represents a designated number.

base (119)   The number used as a factor. In 10^3, 10 is the base.

base (388)   Any side of a parallelogram.

base (400)   The faces on the top and the bottom of a three-dimensional figure.

bisect (321)   To separate something into two congruent parts.

cell (138)   A spreadsheet is made up of cells. A cell can contain data, labels, or formulas.

center (149, 401)   The middle point of a circle or sphere. The center is the same distance from all points on the circle or sphere.

centimeter (85)   A metric unit of length. One centimeter equals one-hundredth of a meter.

circle (149)   The set of all points in a plane that are the same distance from a given point called the center.

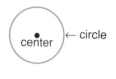

circle graph (103)   A graph used to compare parts of a whole. The circle represents the whole and is separated into parts of the whole.

circumference (149)   The distance around a circle.

clustering (95)   A method used to estimate decimal sums and differences by rounding a group of closely related numbers to the same whole number.

common multiple (213)   A number that is a factor of two or more numbers.

compatible numbers (18)   Two numbers that are easy to divide mentally. They are often members of fact families.

composite number (194)   A number that has more than two factors.

cone (401)   A three-dimensional figure with curved surfaces, a circular base and one vertex.

congruent angles (319)   Angles that have the same angle measure.

congruent figures (337)   Figures that are the same size and shape are congruent. The symbol ≅ means *is congruent to.*

congruent segments (318)   Segments having the same length are congruent.

coordinate grid (443)   Another name for a coordinate system.

coordinates (443)   The numbers associated with a point in the coordinate system. The x-coordinate names the number on the horizontal number line and the y-coordinate names the number on the vertical number line.

574   Glossary

coordinate system (443) Two perpendicular number lines that intersect at their zero points form a coordinate system.

cross products (355) A way to determine if two ratios are equivalent. If the cross products are equal, then the ratios are equivalent and form a proportion. In the proportion $\frac{3}{6} = \frac{4}{8}$, the cross products are 3×8 and 6×4.

cube (119) The product of a number that is multiplied by itself three times. The cube of 2 is 8 because $2 \times 2 \times 2 = 8$.

cup (300) A customary unit of capacity equal to 8 fluid ounces.

cylinder (401) A three-dimensional figure with all curved surfaces, two circular bases and no vertices.

D **data base** (52) A collection of data that is organized and stored on a computer for rapid sorting, categorization, and retrieval.

decagon (326) A polygon having ten sides.

degree (310) The most common unit of measure for angles.

diameter (149) The distance across a circle through its center.

distributive property (129) For any numbers *a, b,* and *c, a(b + c) = ab + ac* and *(b + c)a = ba + ca.*

E **edge** (400) The intersection of faces of a three-dimensional figure.

equation (32) A mathematical sentence that contains the equal sign, =.

equilateral (326) A figure having all sides equal. An equilateral triangle has three congruent sides.

equivalent fractions (204) Fractions that name the same number. $\frac{3}{4}$ and $\frac{6}{8}$ are equivalent fractions.

evaluate (27) To find the value of an expression by replacing variables with numerals.

event (498) A specific outcome or type of outcome.

experimental probability (502) An estimated probability based on the relative frequency of positive outcomes occurring during an experiment.

exponent (119) The number of times the base is used as a factor. In 10^3, the exponent is 3.

F **face** (400) The flat surfaces of a three-dimensional figure.

factor (119) When two or more numbers are multiplied, each number is a factor of the product.

factor tree (195) A diagram that shows the way to find the factors of a number.

fair game (496) Games in which players have an equal chance of winning.

field (52) Each column heading in a data base.

fluid ounce (300) A customary unit of capacity.

foot (210) A customary unit of length equal to 12 inches.

frequency table (50) A table for organizing a set of data that shows the number of times each item or number appears.

front-end estimation (12) A method used to estimate decimal sums and differences by adding or subtracting the front-end digits, then adjusting by estimating the sum or difference of the remaining digits, then adding the two values.

function machine (478) A machine that uses a number called the input, performs one or more operations on it, and produces a result called the output.

function table (480) A table used for organizing input and output data.

G **gallon** (300)   A customary unit of capacity equal to 4 quarts.

gram (174)   The basic unit of mass in the metric system.

greatest common factor (GCF) (199)   The greatest of the common factors of two or more numbers. The greatest common factor of 24 and 30 is 6.

H **height** (388)   The distance from the base of a parallelogram to its other side.

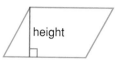

hexagon (327)   A polygon having six sides.

I **improper fraction** (208)   A fraction that has a numerator that is greater than or equal to the denominator.

inch (210)   A customary unit of length. Twelve inches equal one foot.

independent events (519)   Two or more events in which the outcome of one event does not affect the outcome of the other event or events.

inequality (487)   Any mathematical sentence that contains the following symbols : $<$, $>$, $\neq$, $\leq$, $\geq$.

input (478)   Information or data given to a function machine to produce output or results.

integer (420)   The whole numbers and their opposites. . . . , -3, -2, -1, 0, 1, 2, 3, . . .

interval (53)   The difference between successive values on a scale.

K **kilogram** (174)   A metric unit of mass. One kilogram equals one thousand grams.

kilometer (85)   A metric unit of length. One kilometer equals one thousand meters.

L **least common denominator (LCD)** (216) The least common multiple of the denominators of two or more fractions.

least common multiple (LCM) (213)   The least of the common multiples of two or more numbers, other than zero. The least common multiple of 2 and 3 is 6.

length (139)   The longest way an object can be measured; how long an object is.

line graph (57)   A graph used to show change and direction of change over a period of time.

line of symmetry (334)   Any line that divides a shape or object into two matching halves.

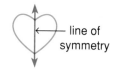

line segment (318)   Two endpoints and the straight path between them. A representation of line segment ST ($\overline{ST}$) is shown below.

liter (175)   The basic unit of capacity in the metric system. A liter is a little more than a quart.

M **mean** (64)   The sum of the data divided by the number of addends; the average.

median (65) The middle number when a set of data is arranged in numerical order. If the data has an even number, the median is the mean of the two middle numbers.

meter (85)   The basic unit of length in the metric system.

metric system (85)   A base-ten system of weights and measures. The meter is the basic unit of length, the gram is the basic unit of weight, and the liter is the basic unit of capacity.

mile (210)   A customary unit of length equal to 5,280 feet, or 1,760 yards.

milligram (174)   A metric unit of mass. One milligram equals one-thousandth of a gram.

milliliter (175)   A metric unit of capacity. One milliliter equals one-thousandth of a liter.

millimeter (85)   A metric unit of length. One millimeter equals one-thousandth of a meter.

mixed number (207)   A number that shows the sum of a whole number and a fraction. $1\frac{1}{2}$, $2\frac{3}{4}$, and $4\frac{1}{8}$ are mixed numbers.

mode (64)   The number or item that appears most often in a set of data.

N **negative integer** (420)   Integers that are less than zero.

O **obtuse angle** (310)   Any angle that measures between 90° and 180°.

octagon (327)   A polygon having eight sides.

opposite (420)   Two integers are opposites if they are represented on the number line by points that are the same distance from zero, but in opposite directions from zero. The sum of opposites is zero.

ordered pair (443)   A pair of numbers where order is important. An ordered pair that is graphed on a coordinate plane is written in this form: (*x*-coordinate, *y*-coordinate).

order of operations (22)   The rules to follow when more than one operation is used. 1. Multiply and divide in order from left to right. 2. Add and subtract in order from left to right.

outcome (498)   One possible result of a probability event. A 4 is an outcome when a die is rolled.

output (478)   The result of input that has had one or more operations performed on it in a function machine.

P **parallel** (326)   Going in the same direction and always being the same distance apart. If lines are parallel, they never meet or cross each other.

parallelogram (326)
A quadrilateral that has both pairs of opposite sides equal and parallel.

pentagon (327)   A polygon with five sides.

percent (366, 368)   A percent is a ratio that compares a number to $\frac{7}{100}$ is read *7 percent* or *7%*.

perimeter (139)   The perimeter of any closed figure is the distance around the figure.

pictograph (48)   A graph used to compare data by using pictures to represent designated quantities.

pint (300)   A customary unit of capacity equal to 2 cups.

place value (44)   A system for writing numbers in which the position of the digit determines its value.

polygon (324)   A simple closed figure in a plane formed by three or more line segments.

population (506)   The entire group of items or individuals from which the samples under consideration are taken.

positive integers (420)   Integers that are greater than zero.

pound (300)   A customary unit of weight equal to 16 ounces.

power (119)   A number expressed using an exponent. The power 3^2 is read *three to the second power*, or *three squared*.

prime factorization (195)   A composite number that is expressed as the product of factors that are all prime numbers. The prime factorization of 12 is $2 \times 2 \times 3$.

prime number (194)   A number that has exactly two factors, 1 and the number itself.

probability (498)   The ratio of the number of ways an event can occur to the number of possible outcomes; how likely it is that an event will occur.

proportion (355)   An equation that states that two ratios are equivalent.

protractor (310)   An instrument used to measure angles.

pyramid (400)   A three-dimensional figure with three or more triangular faces and a base in the shape of a polygon. A pyramid is named for the shape of its base. A pyramid with a base in the shape of a square is called a square pyramid.

rectangular prism (400)   A three-dimensional figure with six rectangular shaped faces. A rectangular prism has a total of six faces, twelve edges, and eight vertices.

repeating decimal (222)   When dividing the numerator by the denominator of a fraction, if a remainder of zero cannot be obtained, the digits in the quotient repeat. 0.33333. . . . is a repeating decimal. It can also be written $0.\overline{3}$. The bar above a digit indicates that the digit repeats.

right angle (310)   An angle that measures 90°.

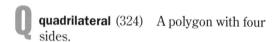

quadrilateral (324)   A polygon with four sides.

quart (300)   A customary unit of capacity equal to 2 pints.

sample (506)   A randomly selected group chosen for the purpose of collecting data.

scale (53)   The set of all possible values of a given measurement, including the least and greatest numbers in the set, separated by the intervals used.

radius (149)   The distance from the center of a circle to any point on the circle.

scale drawing (361)   A drawing that shows an object exactly as it looks but that is generally smaller or larger than actual size. The scale gives the ratio that compares the lengths on the drawing to the actual lengths on the object.

random (506)   Outcomes occur at random if each outcome is equally likely to occur.

range (53)   The difference between the greatest number and the least number in a set of data.

sequence (289)   A list of numbers in a specific order.

rate (353)   A ratio of two measurements having different units.

side (139, 310, 324)   A line segment or surface that encloses something. A triangle has three sides.

ratio (352)   A comparison of two numbers by division. The ratio comparing 2 to 3 can be stated as 2 out of 3, 2 to 3, 2:3, or $\frac{2}{3}$.

similar figures (337)   Figures that have the same shape but different sizes. The symbol ~ means *is similar to*.

ray (319)   A path that extends endlessly from one point in a certain direction. The arrow at the end of the ray indicates that the ray is endless. A representation of ray DE $(\overrightarrow{DE})$ is shown below.

simplest form (205)   The form of a fraction when the GCF of the numerator and denominator is 1. The fraction $\frac{1}{3}$ is in simplest form because the GCF of 1 and 3 is 1.

reciprocal (293)   Any two numbers whose product is 1. Since $\frac{5}{6} \times \frac{6}{5} = 1$, the reciprocal of $\frac{5}{6}$ is $\frac{6}{5}$.

solution (32)   Any number that makes an equation true. The solution for $x - 5 = 8$ is 13.

record (52)   Each row of information in a data base.

solve (32)   To replace a variable with a number that makes an equation true.

Index

E

Edges, 400–402

Equations, 32–34
addition, 460–462, 464–465
division, 468–470
multiplication, 466–470, 490
solving, 32–34, 39, 179–181,
185, 246–248, 269,
460–475, 490
subtraction, 463–465, 490
two-step, 471–473
using, 473–475

Equilateral triangles, 327–329

Equivalent fractions, 204, 355

Equivalent ratios, 355–358, 361

Escher, Maurits Cornelis,
340–341

Estimation, 4–5, 8–19, 229, 527
of area, 386–387
clustering, 95, 109
compatible numbers, 18–19,
39
of differences, 94–97, 100,
109, 236–239, 254,
257–258, 268
front-end, 12–14, 39, 95–97,
109
hints, 33, 122, 158, 165, 241,
254, 315, 396, 398, 507
using patterns, 15–17, 39
with percents, 374–376, 381
predictions, 61–63, 73, 507
of products, 116–118, 122,
126–127, 130, 135–136,
152, 265–266, 274–277,
285, 304, 507
of quotients, 158, 160, 165,
168–169, 171, 297–298
rounding, 5, 9–11, 15–19, 38,
50, 91–96, 109, 116–118,
126, 232–239, 254, 268,
275–276, 304
of sums, 94–101, 109,
236–239, 241, 254, 258,
268

Evaluating expressions, 27–29,
39, 120–121

Events, 498–501
independent, 519–521, 524

Experimental probability,
502–505, 518

Exponents, 119–121, 152

Expressions, 27–29, 39
evaluating, 27–29, 39,
120–121

Extended Projects
Design It Yourself, 534–535
Jump Up High, 536–537
Polling All Voters, 532–533
Smoke 'Em Out, 530–531

F

Faces, 400–402

Factors, 119, 190, 194–196,
199–201, 205, 225
common, 199–200
greatest common, 199–201,
205–206, 219, 225, 281,
285, 352, 368, 378

Factor trees, 195, 225

Fair games, 496–497

Feet, 84, 210

Fields, 52

Fluid ounces, 300

Formulas
for area of circles, 511, 523
for area of parallelograms,
389, 391, 395, 414
for area of rectangles, 107,
142–143, 146, 153, 172,
388, 511, 523
for area of squares, 143, 153
for area of triangles, 392, 414
for circumferences of circles,
150, 155, 229
for distance traveled, 470,
486
for perimeters, 106–107
for perimeters of rectangles,
139–141, 153, 229
for perimeters of squares,
140, 229
using, 106–107
for volume of prisms, 409,
411, 415

Four-step plan, 4–7, 38

Fraction calculators, 205, 251,
281

Fractions, 202–209, 230–260,
274–299
adding, 236–243, 250–256,
258, 268–269
comparing, 216–218, 225
to decimals, 79–83, 221–223,
226, 297, 371
decimals to, 219–220, 226,
372
dividing, 292–299, 305
equivalent, 204, 355
improper, 208–209, 225,
284–285, 297–298
like, 240, 243, 269

multiplying, 274–286,
294–295, 297–298,
304–305
ordering, 218
to percents, 369–370, 372, 381
percents to, 368, 371,
377–378, 381
rounding, 232–239, 254, 268,
275–276, 304
simplest form of, 205–206,
219–220, 225, 251–252,
352–354, 380
subtracting, 236–239,
241–243, 250–260,
268–269

Frequency tables, 50–51,
53–59, 69, 73, 503

Front-end estimation, 12–14,
39, 95–97, 109

Functions, 478–486, 491
graphing, 483–485, 491
machines, 478–479
tables, 480–485, 491

G

Gallons, 300–302

Games, 496–497, 517

Geoboards, 278–279, 292,
324–325

Geometry, 308–330, 334–345
angles, 310–317, 319–320,
322–324, 334, 342–343
area, 107, 142–148, 153,
386–397, 404–407,
414–415, 510–513, 523
circles, 149–151, 395–399,
415, 511, 523
congruent figures, 337–339,
344
perimeters, 106–107,
139–141, 145, 153, 229
polygons, 324–329
quadrilaterals, 324–329
rays, 319
reflections, 447–448,
450–451, 454
scale drawings, 361–363, 381
segments, 318, 320–324, 343
similar figures, 337–339, 344
surface area, 404–407, 415
symmetry, 334–336, 343
three-dimensional figures,
400–413, 415
translations, 340–341, 446,
448–451, 454
triangles, 324–325, 327–329,
337–339, 391–394, 414
volume, 408–411, 415

Trillions, 44–46, 72
Twin primes, 197–198

U

Unfair games, 496–497

V

Variables, 27–29, 39, 457
Venn diagrams, 331–333
Vertices
    of angles, 310, 319
    of cones, 401
    of polygons, 324
    of three-dimensional figures, 400–402
Volume, 408–411, 415

W

Washington, Booker T., 105
Weight, 300–303, 305
    ounces, 300–302, 305
    pounds, 300–303, 305
    tons, 300–302
When Am I Ever Going To Use This?, 10, 68, 85, 135, 158, 171, 221, 232, 289, 318, 361, 404, 435, 483, 506
Whole numbers
    adding, 100
    composite, 194–196, 225
    dividing, 158–159
    factors of, 190, 194–196, 199–201, 205, 224–225
    multiples of, 213–217, 225, 274–275, 304
    multiplying, 116–123
    place value, 44–46
    prime, 194–197, 225
    rounding, 5, 9–11, 15–19, 38, 50, 116–118
Width, 139–140, 142–143

X

x-axes, 443–444, 450

Y

Yards, 210
y-axes, 443–444, 450

Z

Zero, 420–421
    annexing, 88–89, 100–101, 126, 165
    pairs, 426, 428–429, 432–433, 436, 461, 463–464, 471–472
    probability of, 498
    in quotients, 171–173, 185

Photo Credits

Cover: (tl) Comstock, Inc./Stuart Cohen, (tr) Comstock, Inc./Michael Stuckey, (b) Michael Smith Studios

iii, Robert Mullenix; **viii,** (t) Latent Image, (b) ©Richard Stocton/FPG International, Inc.; **ix,** ©Ken Reid/ FPG International, Inc.; **x,** (t) Myron J. Dorf/ The Stock Market, (bl) Comstock, Inc., (br) J. H. Robinson/Photo Researchers, Inc., (bkgd) Mark Gibson; **xi,** Duomo/David Madison; **xii,** (t) Steve Lissau, (b) Grant V. Fantt/The Image Bank; **xiii,** (t) Aaron Haupt, (b) Robert Mullenix; **xiv,** (t) Howard Sochurek/The Stock Market, (m) Robert Mullenix, (b) Jon Love/The Image Bank, (bkgd) file photo; **xv,** Steve Lissau; **xvi,** (t) Michael A. Keller/The Stock Market, (b) Mary Lou Uttermohlen; **xvii,** (t) Brent Turner, (bl) Ken Frick, (br) Robert Mullenix, (bkgd) Glencoe photo; **2,** (t) Duomo/Dan Helms, (bl) Robert Mullenix, (br) The Bettmann Archive; **3,** (m) Tony Craddock/Science Photo Library/Photo Researchers Inc., (bl) Robert Mullenix, (br) courtesy of the Polaroid Corporation; **4,** Skip Comer; **5,** (t) Latent Image, (m) Brent Turner, (b) Studiohio; **7,** Skip Comer; **8,** (l) Latent Image, (r) Doug Martin; **10,** Latent Image; **12,** (t) Skip Comer, (b) Duomo/Al Tielemans; **13,** Sheila Goode; **14,** (t) Robert Mullenix, (b) Lightscapes/The Stock Market; **15-16,** Doug Martin; **17,** David Stoecklein/The Stock Market; **18,** Comstock, Inc.; **21,** (tl) Myron J. Dorf/The Stock Market, (tr) Robin Kennedy/LGI, (b) Mark Tomalty/Masterfile; **22,** Skip Comer; **23,** Ken Frick; **27,** Comstock, Inc; **28,** Gordon Garradd/Science Photo Library/Photo Researchers, Inc.; **30,** Howard Sochurek/The Stock Market; **32,** Isoline Rand/Photo Researchers, Inc. ; **33,** Doug Martin; **35,** (l) Peter Beck/The Stock Market, (r) Donna Ferrato/Black Star; **36,** Skip Comer; **37,** Francois Gohier/Photo Researcher, Inc.; **40,** Skip Comer; **42,** (t) Roy Morsh/The Stock Market, (b) Allen Russell/Profiles West; **43,** (t) Paul Spinelli/Profiles West, (m) Robert Mullenix, (bl) Historical Pictures Service, (br) Tim Courlas; **47-48,** Doug Martin; **49,** ©Richard Stocton/FPG International, Inc.; **50,** Ross Hickson & Associates; **51,** Duomo/David Madison; **53,** Doug Martin; **54,** Ross Hickson & Associates; **61-62,** Doug Martin; **63,** (t) Lee Kuhn/FPG International, (b) Anup & Manoj Shah/Animals Animals; **67,** Skip Comer; **69,** Michael Salaz/The Image Bank; **71,** Superstock; **76,** (l) Universal/Shooting Star, (m) Movie Still Archives, (r) NFL Photos; **77,** (t) Shooting Star International Photo Agency,Inc., (m) Movie Stills Archives, (bl) NASA; **79,** J. Barry O'Rourke/The Stock Market; **81,** Duomo/Paul J. Sutton; **82,** ©Robert A. Tyrrell; **83,** (t) Duomo, (b) Duomo/Richard Dole; **84,** Joe Standart/The Stock Market; **87,** (l) Comstock, Inc., (r) Comstock, Inc.; **89,** Daniel A. Erickson; **91,** Skip Comer; **92,** Ross M. Horowitz/The Image Bank; **93,** Ken Frick; **95,** Skip Comer; **96,** (l) Latent Image, (r) First Image; **97,** Sheila Goode; **98,** (l) Skip Comer, (r) Paul Steel/The Stock Market; **99,** Duomo; **100,** (t) Anne-Marie Weber/The Stock Market, (b) John Kelly/The Image Bank; **101,** Frank P. Rossotto/The Stock Market; **102,** Pictures Unlimited; **103,** Latent Image; **105,** Library of Congress; **106,** ©Peter Gridley/FPG International, Inc.; **107,** ©Hanson Carroll/FPG International, Inc.; **114,** (t) David Barnes/The Stock Market, (b) Andy Sacks/Tony Stone Worldwide; **115,** (t) The Far Side Cartoon by Gary Larson is reprinted by permission of Chronicle Features, San Francisco, CA, (b) Doug Wilson/Westlight; **116,** (t) R. Compillo/The Stock Market, (b) Skip Comer; **118,** Smithsonian Institution; **119,** Edward Lines, Jr./The Shedd Aquarium; **123,** Bob D'Amico/ABC Television; **126,** Duomo/David Madison; **127,** Duomo/Michael Layton; **129-130,** Doug Martin; **132-133,** KS Studios; **135,** Skip Comer; **142,** MAK-I; **145,** ©Ken Reid/FPG International, Inc.; **146,** Doug Martin; **147,** Gerald Zanetti/The Stock Market; **148,** (l) Skip Comer, (r) Comstock, Inc; **149,** NASA; **154,** David Stocklein/The Stock Market; **156,** (t,l) KS Studios, (m) National Baseball Hall of Fame and Museum, (br) Steve Powell/Allsport; **157,** (t,br) Duomo/Paul J. Sutton, (bl) NFL Photo; **159,** Doug Martin; **161,** KS Studios; **162,** (t) Guido Alberto Rossi/The Image Bank, (b) Skillman Photography; **165,** KS Studios; **166,** Benn Mitchell/The Image Bank; **168,** Doug Martin; **171,** Fotex/Shooting Star; **172,** Comstock, Inc./Michael S. Thompson; **174,** (t) Duomo/David Madison, (b) Duomo/Al Tielemans; **175,** (l) Aaron Haupt, (r) Skip Comer; **176,** Historical Pictures Service; **177,** Skip Comer; **179,** (t) KS Studios, (b) Duomo/David Madison; **180,** Latent Image; **182,** Doug Martin; **186,** Focus on Sports; **188,** Tony Freeman/Photo Edit; **189,** (t) GARFIELD reprinted by permission of United Features Syndicate, Inc., (m) KS Studios, (b) Jeff Gnass Photography/The Stock Market; **190,** Co Rentmesster/The Image Bank; **191,** First Image; **194,** Doug Martin; **196,** Robert Mullenix; **197,** Skip Comer; **198,** Brett Froomer/The Image Bank: **199,** George E. Jones III/Photo Researchers, Inc.; **200,** (t) Doug Martin, (b) Studiohio; **204,** Latent Image; **207,** Duomo; **212,** Miami Herald; **213,** Duomo/David Madison; **215,** Latent Image; **216,** KS Studios; **222,** Doug Martin; **230,** (t) ©Fred McKinney/FPG International, Inc., (b) Eric Grave/Science Source/Photo Researchers Inc.; **231,** (t) Richard Price/Westlight, (b) GARFIELD reprinted by permission of United Features Syndicate, Inc.; **232,** Doug Martin; **234,** (t) Skip Comer, (b) Latent Image; **236,** MAK-I; **237,** Latent Image; **238-239,** James Westwater; **239,** Roy Morsch/The Stock Market; **240,** Skip Comer; **241,** KS Studio; **242,** Aaron Haupt; **243,** Michael Melford/The Image Bank; **244,** KS Studios; **245,** Robert

Mullenix; **246,** Wilderness Inquiry; **248,** (t) KS Studios, (b) The Bettmann Archive; **249,** (t) Skip Comer , (b) Latent Image; **250,** Anne Heimann/The Stock Market; **253,** (t) Steve Lissau, (b) Doug Martin; **254,** Grant V. Faith/The Image Bank; **255,** file photo; **256,** Matt Meadows; **258,** Doug Martin; **261,** Duomo; **263,** Ken Frick; **264,** MAK-I; **265,** Skip Comer; **266,** D.W. Productions/The Image Bank; **267,** Duomo/Steven E. Sutton; **270,** Tim Courlas; **272,** (l) Tom Braise/The Stock Market, (r) Doug Martin; **273,** (t) Larry Lee/Westlight, (ml) Robert Mullenix, (mr) Randy Brandon/Peter Arnold Inc., (b) Superstock; **274,** Robert Holland/The Image Bank; **275,** Comstock, Inc.; **276,** KS Studios; **280,** MAK-I; **283,** Phillip Kretchmar/The Image Bank; **284,** Duomo/Mitchell Layton; **287,** Skip Comer; **289,** photo courtesy of the Chicago Symphony Orchestra/Jim Steere; **293,** Robert Mullenix; **294,** Karen Leeds/The Stock Market; **295,** Latent Image; **296,** Skip Comer; **297,** (t) Comstock,Inc., (b) Lloyd Lemmerman; **300,** Calvin Larson/Photo Researchers, Inc.; **301,** Latent Image; **306,** Robert Mullenix; **308,** (t) Bob Leroy/Profiles West, (mr) The Bettmann Archive, (bl) Louis Goldman/Photo Researchers Inc., (br) Michael Holford; **309,** (t) Don Mason/The Stock Market, (bl)Art Resource, (br) Burton McNeely/The Image Bank; **314,** Tim Courlas; **316,** Werner Bokelerg/The Image Bank; **321,** Aaron Haupt; **323,** Pete Saloutos/The Stock Market; **326,** Doug Martin; **330,** Skip Comer; **331,** (t,m) Skip Comer, (b) Tim Courlas; **332,** KS Studios; **333,** Comstock, Inc.; **334,** J.H. Robinson/Photo Researchers, Inc.; **336,** David Cavagnaro; **337,** KS Studios; **348,** (l) Brownie Harris/The Stock Market, (r) R. Ian Lloyd/Westlight; **349,** (t) Chuck O'Rear/Westlight, (bl) GARFIELD reprinted by permission of United Features Syndicate, Inc., (br) Ken Davies/Masterfile; **350,352,** Robert Mullinex; **353,** Raymon A. Mendez/Animals Animals; **355,** Skip Comer; **356,** ©Simon Metz/FPG International, Inc.; **357,** Melanie Carr/Zephyr Pictures/Stock Imagery; **359,** Tim Courlas; **360,** KS Studios; **364,** Robert Mullenix; **365,** (t,b) Robert Mullenix, (m) KS Studios; **368,** Stephen Harvey/LGI; **370,** Studiohio; **371,** KS Studios; **373,** Robert Mullenix; **374,** Michael Kevin Daly/The Stock Market; **376,** Johnson/Gamma Liaison; **377,** (tl) Geiersperger/Stock Imagery, (bl,r) Stock Imagery; **378,** Robert Mullenix; **384,** (t,m) Robert Mullenix, (b) FRANK & ERNEST reprinted by permission of NEA, Inc.; **385,** (t) file photo, (b) Robert Mullenix; **386,** Comstock, Inc.; **388,** Duomo/Al Tielemans; **394,** Antonio Luiz Hamdan/The Image Bank; **396,** David Weintraub/Photo Researchers, Inc.; **399,** Steve Lissau; **400,** William E. Ferguson; **401,** Latent Image; **402,408,410,** Doug Martin; **413,** Harvey Lloyd/The Stock Market; **416,** Doug Martin; **418,** (t) Robert Mullenix, (bl) Ron Kimball, (bm) Disney/Shooting Star, (br) Cindy Lewis; **419,** (bl) Tony Freeman/Photo Edit, (m) Bob Anderson/Masterfile, (br) Cindy Lewis; **420,** Stephen Frink/The Stock Market; **422,** John P. Kelly/The Image Bank; **423,** (t) KS Studios, (b) Skip Comer; **424,** James Blank/The Stock Market; **424,** Skip Comer; **427,** Berenholtz/The Stock Market; **429,** Skip Comer; **431,** George Butch, reprinted with permission from the National Broadcasting Company; **434,** Historical Pictures Service; **434,** Jeffry W. Myers/The Stock Market; **435,** Latent Image; **437,** Scott Stallard/The Image Bank; **438,** KS Studios; **439,** Aaron Haupt; **440,** Duomo/Al Tielemans; **443,** Doug Martin; **445,** Berenholtz/The Stock Market; **448,** Michael Mertz; **458,** (l) Disario/The Stock Market, (r) J. A. Kraulis/Masterfile; **459,** (t) KS Studios, (b) FOR BETTER OR WORSE copyright 1992 Lynn Johnston Production,Inc. Reprinted with permission of Universal Press Syndicate. All rights reserved. **465,** Laima Druskis/Photo Researchers, Inc.; **467,** ©1988 Eugene Adepoari/LGI; **469,** (l) Shiki/The Stock Market, (r) Latent Image; **470,** Richard Dunhoff/The Stock Market; **473,** KS Studios; **474,** (l) Michael Furman/The Stock Market, (r) Richard Gross/The Stock Market, (b) Tim Courlas; **476,** (bl) Jon Love/The Image Bank; **476-477,** Robert Mullenix; **480,** T. J. Hamilton; **486,** Bob Abraham/The Stock Market; **486,** J. Barry O'Rourke/The Stock Market; **487,** Duomo/Paul J. Sutton; **492,** John P. Kelly/The Image Bank; **494,** (t) Robert Mullenix, (bl) Matt Meadows, (bm) North Wind Picture Archives, (br) courtesy of The New York Historical Society; **495,** (t) KS Studios, (bl) Robert Mullenix, (br) Markowitz/Sygma; **496,** Mary Lou Uttermohlen; **498,** (t) Aaron Haupt, (b) KS Studios; **499,** (l) Bud Fowle, (r) Studiohio; **500,** (t) Doug Martin, (b) Skip Comer; **501** Ben Simmons/The Stock Market; **502,** (l) Latent Shay, (rt) Skip Comer, (rb) Skip Comer; **504,** Mary M. Thacher/Photo Researchers, Inc.; **506,** KS Studios; **507,** John Gilmore/The Stock Market; **508,** Nino Mascardi/The Image Bank; **509,** Doug Martin; **511,** Heinz Fischer/The Image Bank; **514,** (t) Stock Imagery, (b)Chris Collins/The Stock Market; **517,** Steve Niedorf/The Image Bank; **519-520,** Skip Comer; **524,** Michael A. Keller/The Stock Market; **528,** KS Studios; **529,** Ken Frick; **530,** Robert Mullenix; **531,** (t) Brent Peterson/The Stock Market, (b) Ken Frick; **532,** MAK-I; **533,** Robert Mullenix; **534,** Matt Meadows; **535,** Robert Mullenix; **536,** Brent Turner; **537,** Mike Powell/Allsport.